# PEDOLOGY

## P. Duchaufour

Translated by T. R. Paton

# Pedology

# PEDOLOGY

Pedogenesis and classification

# P. Duchaufour

## Translated by T.R. Paton

ASSOCIATE PROFESSOR OF EARTH SCIENCES,
MACQUARIE UNIVERSITY,
SYDNEY, AUSTRALIA

London
GEORGE ALLEN & UNWIN
Boston          Sydney

© Masson, Paris, 1977
English edition © George Allen & Unwin, 1982

**Pedology**
under the direction of
Philippe Duchaufour
Honorary Director of the CNRS Pedology Centre
and
Bernard Souchier
Director of the CNRS Pedology Centre

This book is the English translation of *Pédogenèse et classification*, the
first volume of *Pédologie*, published by Masson in 1977. The English
translation of the second volume, subtitled *Constituants et propriétés*, is
published by Academic Press.

**George Allen & Unwin (Publishers) Ltd,**
**40 Museum Street, London WC1A 1LU, UK**

George Allen & Unwin (Publishers) Ltd,
Park Lane, Hemel Hempstead, Herts HP2 4TE, UK

Allen & Unwin Inc.,
9 Winchester Terrace, Winchester, Mass 01890, USA

George Allen & Unwin Australia Pty Ltd,
8 Napier Street, North Sydney, NSW 2060, Australia

First published by George Allen & Unwin in 1982

**British Library Cataloguing in Publication Data**

Duchaufour, Philippe
    Pedology: pedogenesis and classification
1. Soil science
I. Title      II. Pédologie. Pédogenèse et
classification. *English*
631.4        S593
ISBN 0-04-631015-0
ISBN 0-04-631016-9 Pbk

**Library of Congress Cataloging in Publication Data**

Duchaufour, Philippe, 1912–
    Pedology: pedogenesis and classification.

Translation of: Pédogenèse et classification, the
first volume of Pédologie, published by Masson in 1977.
Includes bibliographical references and index.
1. Soil science. 2. Soil formation. 3. Soils–
Classification. I. Title.
S591.D7913 1982       631.4       82–11517
ISBN 0–04–631015–0
ISBN 0–04–631016–9 (pbk.)

Set in 10 on 12 point Times by Preface Ltd, Salisbury, Wilts.
and printed in Great Britain
by Butler & Tanner Ltd, Frome and London

# Preface

This book is the first of two volumes intended to replace the old and now out of print *Précis de pédologie*, the previous three editions of which were produced by the same publisher in 1960, 1965 and 1970.

It was apparent that the term 'précis', which means that the text was necessarily condensed and summarised, no longer corresponded with the present day situation, for pedology has developed considerably in the past 10 years and it now makes use of the most modern and varied research techniques. It has become an entirely separate discipline and has assumed, at least in certain countries, considerable importance. In addition, different schools of thought have developed and their sometimes contradictory viewpoints are presented at many international conferences, which, if valid conclusions are to be reached from them, required considerable space for discussion. Thus, even by being very concise it was no longer possible to deal with the whole of soil science within the space of one volume, so that a two volume format became a necessity.

As soil science is known to have two fundamentally distinct aspects, it has been easy to determine the contents of each volume and also to give each an identity and unity, as well as enabling a different kind of presentation to be made in each case.

The purely pedological aspect of soil science, that is to say the one that deals with soil dynamics as a function of the environment, cannot be divided, for it involves an *environmental synthesis* which of necessity must be the work of a single author who is capable of presenting a general environmental classification of soils, otherwise the necessary unity would be lacking. This volume attempts to do precisely this: the study of pedogenesis and soil classification.

The other aspect concerns all of the physical, chemical and biological properties of soil, which makes it more analytical, less theoretical and more applied, so that its importance to foresters and agronomists is readily apparent. Thus, the second volume deals with *Constituents and properties of soils*, in which, even though reference is made to the connections which exist between each chapter, overall synthesis is more limited than in Volume 1. Of necessity, as only experts in particular fields can write about them in a clear and competent manner, several authors have contributed one or more chapters covering their particular specialist areas.

Thus the two volumes of this pedological series are part of a whole, but at the same time each of them is complete in itself and thus independent of the other. In the English edition, the French text has been brought up to date and

the bibliography completed. For his translation Professor Paton is to be particularly thanked.

It must be particularly mentioned that the completion of this first volume has been possible only by the help of research groups in the Biological Centre of Pedology at Nancy. For the preparation of the manuscript and illustrations Miss E. Jeanroy and M. P. Sueur, whose support has never faltered, are to be greatly thanked.

Ph. Duchaufour

# General key

 slightly decomposed organic layer

 slightly active humic horizon

 active humic horizon with crumb structure

 calcium carbonate

 2 : 1 clays (illite, vermiculite, montmorillonite — with absorbed iron oxides)

 1 : 1 clays (kaolinite)

 ashy or bleached horizon

 accumulation of ferric hydrate (bright ochreous or rusty)

 accumulation of dehydrated ferric iron (red)

 localised precipitation of ferric iron

 iron—manganese concretions

 gley : ferrous iron dominant (greenish grey)

 free alumina

 weathering parent rock

 unweathered siliceous parent rock

 unweathered calcareous parent rock

*Note:*
the abundance of these different materials is indicated by the spacing of the lines or the density of the symbols utilised

# Contents

# List of tables

# List of plates

*Between pages 242–3*

*Part I*

# *THE PHYSICOCHEMICAL PROCESSES OF PEDOGENESIS*

# Introduction and definitions

**Pedology** is a relatively new idea in soil science which was initiated by the Russian school at the end of the last century. Its basis is that soil is no longer considered as an inert material that reflects only the composition of the underlying rock, but that it is formed and developed as a result of the effect of the active environmental factors of climate and vegetation on the mineral material. As a consequence, soil passes through successive phases of youth and then of maturity, leading to a state of stable equilibrium with the natural vegetation. The **organic matter** which this vegetation contributes to the soil has properties that reflect the combined effect of all the environmental factors (climate, vegetation and parent materials) and, depending upon the strength of the bonding established with the mineral material, it determines the kind of soil formation. *During this development, often called* **pedogenesis**, *the initially thin surface soil gradually increases in thickness and successive layers, or* **horizons**, *become differentiated in terms of colour, texture and structure, to form a* **profile**.

A detailed study of the profile allows account to be taken of the effect of various environmental factors in determining the history of the soil, for horizon characteristics are the result of certain biochemical or physicochemical processes which are determined by the environmental factors. This relationship can be expressed as follows:

environment $\longrightarrow$ pedogenic processes $\longrightarrow$ profile characteristics

It is evident that modern, so-called genetic classifications are generally based on this dynamic idea of soil science. This is the reason why it is essential first of all to study the fundamental processes of pedogenesis, by relating them whenever possible to the environment, before tackling the problem of classification, which will be done in the second part of the book.

In the second part, the definition of horizons will be studied in greater detail, but in this short introduction, consideration will be restricted to defining the main horizons in terms of the relationships existing between their formation and the basic process (or processes) involved.

When a soil is colonised by vegetation, the processes that occur most rapidly are connected with the incorporation of organic matter (**A horizon**) into the little-altered mineral material (**C horizon**). Such slightly developed profiles are thus of the AC type.

Subsequently, intermediate mineral **B horizons** are formed at a variable rate and two main types are differentiated – **weathered (B)**, resulting from the weathering of primary to secondary minerals, and **illuvial B**, resulting from the movement of material from the A to the B horizon by infiltrating water. Such completely developed profiles are thus of the A(B)C or ABC type.

These considerations introduce the idea of the successive investigation of the three fundamental processes of pedogenesis – incorporation of organic matter, rock weathering, and movement of materials within soils – which are dealt with in the first three chapters. However, rock weathering is dealt with first because, in certain climates, organic matter is not involved.

It is not until the fourth chapter that a general synthesis is attempted, by considering the *time* factor, for it is only by taking into account the very variable speed at which these processes operate, depending on the soil involved, that a full understanding can be obtained of the relationships that exist between the environment and these processes, which is the basic objective of pedology.

# Chapter 1

# Weathering and clay formation

## I   General introduction

The mineral fraction of soils is produced by the transformation of parent rocks which are subject to a twofold process: (i) physical and mechanical disaggregation without chemical modification of the minerals; (ii) chemical weathering causing a transformation of **primary minerals**, with the formation of **secondary minerals** (particularly **clays**) which make up the **weathering complex**.

Owing to the effects of temperature variations, freezing, erosion agents etc., the first process is particularly characteristic of *cold* or *desert* climates. The second can occur only in the presence of water, which carries active agents such as oxygen, organic acids and carbon dioxide at a sufficiently high temperature. It reaches its maximum intensity in humid equatorial climates and gradually ceases in cold, boreal or alpine climates.

This chemical weathering gives rise to: (i) **soluble materials** that are generally carbonates or bicarbonates, the cations of which either become exchangeable or are leached; also **silica** with a maximum solubility of 100 ppm; (ii) **colloidal gels** by hydration and polymerisation of free heavy cations of aluminium and iron however, the insolubilisation of these heavy cations, which is rapid in most soils, decreases in acid conditions rich in soluble organic matter, which favours complex formation); (iii) microcrystalline entities with **sheet** structure (clays) which fix to their surface iron and aluminium hydroxides.

**The index of weathering** of a soil requires an evaluation of the amount of secondary minerals (the weathering complex) in a soil compared to its total (primary and secondary) mineral content; generally, the more developed the soil, the higher the index.

In the determination of the index it is usual to use a particular element, such as iron or aluminium, as a standard: **the gross index of weathering** of an horizon is then expressed by the ratio:

$$\frac{\text{Al of weathering complex}}{\text{total \% Al}} \quad \text{or} \quad \frac{\text{Fe of weathering complex}}{\text{total \% Fe}}$$

For soils which have been subject to relatively gentle weathering, such as those of temperate climates, the second index is higher than the first, for the ferromagnesian minerals release their iron more quickly, as they are weathered before some of the more resistant aluminous minerals, such as orthoclase. These differences decrease for the soils of hot climates.

In fact, the gross index of weathering, for a particular horizon, often gives an incorrect idea of the degree of weathering, for it does not take account of losses of soluble salts from the profile, nor of the redistribution of certain parts of the weathering complex from one horizon to another. For this reason, Souchier (1971) developed the concept of a 'corrected index of weathering' which specifically takes account of these disturbing factors; as will be seen later, the loss of soluble salts is determined by comparing a particular horizon with that of the original material (C horizon) by means of 'geochemical balances', this loss is then added to the weathering complex of that particular horizon. It is easy to see why, for the corrected index of weathering for certain horizons such as the A2 of a podzol is much higher than the gross index of weathering, for most of the weathering complex has been removed from the horizon.

## 1   Methods of studying weathering
Very varied methods have been used in the study of weathering, such as the comparative analysis of different soil horizons to determine 'balances', or laboratory experiments or thermodynamic calculations.

**Geochemical and mineralogical balances** (Lelong 1967, Souchier 1971, Lelong & Souchier 1972, 1979, Bornand 1978). These balances compare the composition of each horizon of the profile with the composition of the original material (C horizon), the material being supposed to be homogeneous, which it is necessary to verify by the use of parameters such as the grain size of the coarser quartz. Because of variations of density and volume that occur as a result of the weathering, it is necessary to establish these balances as a function of an invariant entity (in fact, one that varies as little as possible, the variation being evaluated and taken into account). This invariant entity is generally quartz for rocks which contain sufficient of it (e.g. granites (Souchier & Lelong 1972)), but aluminium can also be used (Hetier 1975). Lelong considered that the loss of quartz in tropical soils was of the order of 15% to 20% between the top and the bottom of the profile; it is not necessary to disguise the fact that this correction can only be approximate, which gives this method mainly a comparative value. These balances are of two types – geochemical or mineralogical – as will be seen in Fig. 1.2.

This method is best applied to those kinds of weathering that involve **subtraction** or **redistribution** of materials, but is not suited to those situations where **absolute accumulation** has occurred, for example by lateral additions at the base of slopes; in such cases, the index of weathering can be above 100 (which is the proof that it is an absolute accumulation); it is then necessary to consider the balance over the whole of the slope so as to determine to what extent the gains observed down slope balance the calculated losses from up slope – this has been done by Paquet (1969) and Bocquier (1971).

**Laboratory experiments.** The best experimental method of studying weathering is to reproduce in the laboratory the pedoclimatic and drainage conditions of the natural environment by the percolation of solutions through columns filled with parent material, but this method is obviously very slow. Pedro (1964) developed a more rapid method using a system of Soxlet **perfusion** of water at 70°C, which allows the separation of the insoluble **residue** from the **hydrolysate** which then develops fairly rapidly by insolubilisation. The experimental conditions could be altered to simulate various drainage conditions.

This method has been applied by Pedro (1964) and Robert (1970) to crystalline or eruptive rocks, and to the minerals contained in these rocks; by Trichet (1969) to the study of volcanic glasses; and finally by Durand and Dutil (1971) to the study of the weathering of calcareous rocks.

**Thermodynamic calculations** (Tardy 1969, Tardy *et al*. 1973). Tardy and his co-workers have attempted to determine the thermodynamic equilibrium curves existing between ions in solution and crystalline minerals, and have compared them with equilibria of the same kind existing in the natural environment.

The authors showed that such equilibria sometimes correspond to closed systems – that is to say, those that do not allow the removal of saline solutions – and at other times to open systems, in which these solutions are eliminated in the drainage water. In fact, it must be realised that this distinction is difficult and that this type of calculation is better applied to closed than to open systems. In addition, under natural conditions these equilibria are often disturbed by the presence of organic matter in solution, and also by the fact that apart from the two materials discussed – soluble salts and crystalline minerals – there is a third type of material – **amorphous gels** – consideration of which is also important (Trichet 1969). Nevertheless, the equilibrium calculations enable account to be taken of certain incompatibilities existing under natural conditions between certain minerals (for example gibbsite and montmorillonite) so that, when these two minerals co-exist in a soil, which does occur, it will be seen that this 'anomaly' provides pertinent information about an unusual type of pedogenesis.

## 2   General processes of rock weathering

**Hydrolysis**, or the effect of water containing such active entities as hydrogen ions ($H^+$), is the most important process of weathering out of the many which occur and, therefore, the way it operates as a function of the environment will be studied in detail. Other secondary processes, which nevertheless can be important in particular cases, have been studied by many authors: they range from the simple dissolution of saline rocks to **hydration**, which results in water molecules combining with certain slightly hydrated rock minerals, such as ferric oxides, thus aiding in the process of rock decomposition and dis-aggregation, and finally to **oxidation**, which causes the release of ferrous ions ($Fe^{2+}$) contained in certain primary minerals, so disrupting their crystal lattices. The processes of **reduction** occur more rarely but, under hydromorphic and badly aerated conditions, they are responsible for the solution of some ferruginous sandstone cements and hence for their breakdown.

**Application to the main types of rock.** Hydrolysis, as indicated above, is the major weathering process affecting **crystalline rocks**; it leads, by various means that will be studied, to the formation of clays, which represent a generally stable part of the weathering complex. In the case of sedimentary rocks, hydrolysis is also involved in so far as they contain inherited primary materials and to a certain extent even some of the inherited clays, both of which can undergo a variable degree of **transformation**. However, for sedimentary rocks the other processes of dissolution and hydration can be of considerable importance, especially when they contain calcareous materials. It is known that calcium carbonate is dissolved fairly rapidly, depending upon the conditions, by water containing $CO_2$: this is the process of **decarbonation**, which will be studied in greater detail in Chapter 3 and also in Chapter 7.

The mechanism of weathering of calcareous rocks has been studied by several authors: indurated limestones by Ciric (1967), Lamouroux (1971), and Pochon (1978) and chalk by Durand and Dutil (1971). This mechanism differs profoundly according to the nature of the calcareous rock.

**Indurated limestones** are subjected to **surface corrosion**: in each humid period a surface skin is detached, the carbonates are dissolved, silicate impurities (clays) alone persist which are either deposited *in situ* or else moved by running water down slope.

**Unindurated marly limestones** are subject to swelling by hydration of the clay, which results in a physical breakdown of the rock. Calcium carbonate does not disappear; it is freed as a very fine 'active' form. The process of decarbonation then occurs at very variable rate depending upon the nature and amount of organic matter, as well as the pedoclimatic conditions.

**Chalk** is also subject to physical breakdown, but here it is freezing, affecting the absorbed capillary water of the rock, which is the principal motive force. The processes of decarbonation are also involved, but because of the enormous reserves of carbonate that the rock contains (compared to the very small amount of silicate impurities), the soil is not generally subject to a complete decarbonation.

**Resistance of primary minerals to weathering.** Primary minerals of crystalline rocks have a very variable resistance to hydrolysis. **Quartz** is attacked only to a slight extent; however, in hot and humid climates it is corroded. Fine-grained quartz of 'silt' size can be almost entirely dissolved in certain ferralli-tic soils. **Feldspars** weather more quickly the poorer they are in silica: the alkaline feldspars (orthoclase and microcline) weather very slowly (in general, incompletely in a temperate climate), while plagioclases, especially the more calcic, weather fairly rapidly. **Biotite** weathers rapidly, often to vermiculite (Robert 1970), by loss of iron and interlayer potassium ions ($K^+$). On the other hand, **muscovite** is very resistant and it forms sericite by physical subdivision. The various **ferromagnesian minerals** are subject to a simultaneous hydrolysis and oxidation which removes iron in the ferric state ($Fe^{3+}$). The clays formed are almost always magnesium rich, such as primary chlorites, serpentine, and trioctahedral vermiculite.

It is easy from this quick summary to conclude something about the resistance to weathering of actual rocks. Acid rocks, rich in silica (free or combined), are more resistant to weathering than are basic rocks, poor in silica but relatively rich in iron and magnesium which makes them weak from the point of view of chemical weathering.

## 3   Clay formation

Although hydrolysis is the fundamental mechanism of weathering of primary minerals, its action varies considerably depending on the environmental conditions, and particularly on the climate. *Temperature* is an essential factor of hydrolysis: any rise in soil water temperature increases the speed of chemical reactions, which enables two types of weathering to be distinguished; **geochemical weathering**, *resulting in the complete liberation of the mineral constituents – silica, aluminium, bases, etc. – characteristic of tropical climates*; **biochemical weathering**, *gradual and gentle, often incomplete, characteristic of temperate climates; this second type generally preserves the initial crystalline structure, the insoluble residue always being particularly important.*

In each case, the origin of the clay is different: in a hot climate, the clays are most often of **neoformation**, i.e. formed at the expense of the parent materials, from entities freed by complete weathering; in temperate climates, the clays are most often the result of a gradual **transformation** of the primary minerals (particularly of the phyllitic minerals).

The transformation of a mineral to a clay varies in its intensity and amount as a function of the environment. If it is very slight, the term **inheritance** is often used; if the change is considerable, it is either a case of **degradation**, where there is a gradual decrease in the degree of crystallinity and a loss of constituents, or **aggradation**, if there is an addition of elements to a badly crystallised lattice which causes an improvement in the crystal structure; all these terms are derived from Millot (1964). More recently (Tardy 1969, Souchier & Lelong 1972), a new type of transformation has been described which involves ion substitution in a slightly acid, somewhat impeded weathering environment; certain micaceous minerals are thus changed to montmorillonite by the addition of silica.

These general considerations thus lead into a detailed study of the two fundamental types of weathering – geochemical (which is general in hot climates) and biochemical (more characteristic of temperate climates) – in which account will be taken of the great amount of recent work on these topics.

## II   Geochemical weathering: total hydrolysis and neoformation of clay

### 1   Physicochemical mechanism

This type of hydrolysis is specific to tropical soils; it has been reproduced experimentally by Pedro (1964) and Robert (1970); it occurs under *neutral* conditions, in the presence of circulating water (well drained environment) and *in the absence of organic acid anions*. In these conditions, the primary minerals are totally destroyed and their constituents (particularly silica and alumina) are freed, which is the case for both phyllosilicates (micas) and tectosilicates (feldspars). It differs from temperate weathering in that there is a *convergence of the process of weathering, no matter what the type of primary mineral involved*. This total hydrolysis favours the elimination not only of

**basic cations** ($Ca^{2+}$, $Mg^{2+}$, $K^+$, $Na^+$) but also of *silica*, which is almost as mobile (Tardy 1969). *On the other hand, the oxides of iron and aluminium are only slightly mobile and accumulate* in situ. Pedro has noted in his experiments that 100% of the initial iron and 70% of the aluminium is found in the residue (this is as close as possible to a reproduction of the process of **ferrallitisation**).

However, when the elimination of silica slows down owing to poor drainage (impeded or semi-impeded environment), neoformation of clay becomes possible by a recombination of silica and alumina. How does this recombination come about? Is the change from the soluble to the crystalline form direct, or indirect, through the intermediary of amorphous gels? This question has been and still is debatable; if Tardy (1969) and Trichet (1969) are followed, the transition through the 'gel' stage should be, in fact, general for most soils; in this case it is thus possible to write:

$$\text{ions (in solution)} \longrightarrow \text{mixed gels} \longrightarrow \text{crystalline form}$$

Tardy shows, indeed, that during weathering, aluminium changes from 4- to 6-coordination. Now it is known that in this form aluminium does not remain in solution for long and it passes very quickly, by ageing, to the complex ion stage, then to insoluble amorphous alumina (Schwertmann 1969). Trichet has been able to follow experimentally the formation of neoformed phyllites around spherical gel particles when studying the weathering of volcanic glass – the alumina forms the central octahedral layer first; the silica then becomes organised into the marginal tetrahedral layers.

## 2 Influence of the environment on the neoformation of clay

The clays of neoformation resulting from this type of weathering are in fact very variable, both qualitatively and quantitatively. The variation is controlled by two fundamental factors: (i) natural drainage, which removes the freed constituents at a very variable rate; (ii) the differential solubility of these constituents in the environment, which is dependent to a large extent on the pH. In fact, although the pH is evidently related to the richness in basic cations of the weathered materials, it also depends considerably on the drainage conditions; for these cations are the most mobile, and are always the first to be removed in the soil solution, particularly the alkaline earths. Thus, it is normal that a well drained environment (sometimes called **leaching**) acidifies very quickly, which is the opposite of what happens in a semi-impeded or to an even greater extent in an impeded environment; and this occurs, whatever the initial reserve of bases in the original material.

It is important, in these conditions, to know the differential solubility of the two main components of clay (silica and alumina) as a function of pH. In very acid conditions (pH less than 5) the solubility of aluminium is greater than that of silica (Pedro 1964, Brethes 1973), but this is so only in temperate or cold climate, not tropical soils; this point will be returned to later.

Acquaye and Tinsley (1965) showed that, when the pH rises, the solubility of aluminium decreases very sharply around pH 5, which results in an adsorption of silica by the precipitated alumina: the neoformation of clay becomes possible; but at these acid pHs, the quantity of silica adsorbed by the alumina is small. *In horizons of tropical soils where the pH is around 5, and where most of the base and silica have been removed previously by drainage to depth, only the clays poorest in silica, of the kaolinite type, are able to form.*

When the environment is richer in bases and in silica, which is the case in less well drained situations or where the original material is richer in alkaline earths, the phenomena observed are totally different: the adsorption of silica by alumina is much greater, which leads to the neoformation of clay types rich in silica, such as the smectites (montmorillonite); this is the case for certain eutrophic brown soils and vertisols (Leneuf 1959).

Mehlich (1967) has emphasised the role of the bivalent cations $Ca^{2+}$ and $Mg^{2+}$ in the formation of the aluminosilicates richest in silica: the formation of an intermediate amorphous gel stage has again been demonstrated, in particular by Pedro (1964) who, in the experiments referred to previously, at first obtained a gel by the precipitation of the entities dissolved during weathering; the addition to this material of a magnesium salt led to the formation of a clay rich in silica (smectite, of the montmorillonite clay family).

These theoretical or experimental results are in complete agreement with observations and analytical studies carried out in the natural environment by different authors, such as Ségalen (1973), Tardy *et al.* (1973), Boulet (1974), Kantor and Schwertmann (1974), Blot *et al.* (1976).

The three essential minerals found in tropical soils are *gibbsite* (formed in the complete absence of silica), *kaolinite* and *montmorillonite*: *this succession, which reflects an increasing richness in silica of the neoformed clay, is to be seen on the one hand in going from more humid to drier climates and on the other (and more particularly) when going from more 'leaching' environments (the most acid) towards the more impeded (that is to say the less acid).*

Fritz and Tardy (1976) showed that the formation of *gibbsite* (in fact the absence of clay neoformation) occurs in only a special kind of environment: for instance where the parent material is very poor in silica and is very well drained. *Kaolinite* is the dominant clay of the ferrallitic soils, formed under reasonable to semi-impeded drainage conditions, most of the silica produced by weathering having been previously removed. In such an environment, gibbsite resulting from the weathering of primary minerals (**primary gibbsite**) is present always in small amounts and is limited to steep sites with strong lateral drainage (Ségalen 1973). This kaolinite is formed directly and immediately on acid parent materials such as sandstone and acid granites (Blot *et al.* 1976). On basic materials such as calc-alkaline granites, dolerites and diorites for example, the processes are rather different. In these conditions, under slow drainage, the initial formation of montmorillonite appears to precede that of the kaolinite; montmorillonite is in fact present in the deep regolith still relatively rich in alkaline earths and soluble silica; this montmorillonite quickly changes into kaolinite by loss of silica. Kaolinite is dominant in the most acid zone situated immediately above the horizon of weathering.

All of this discussion can be summarised as shown in Table 1.1.

**Table 1.1** Rock weathering.

| Environment | | Impeded, little acidity | Leached, acid |
|---|---|---|---|
| basic rocks: | primary ⎱ minerals ⎰ ⟶ | montmorillonite $\xrightarrow{\text{(loss of silica)}}$ | kaolinite |
| acid rocks: | primary ⎱ minerals ⎰ ⟶ | kaolinite (in all conditions) | |

## III   Biochemical weathering: dominated by gradual hydrolysis and transformations

In temperate climates, hydrolysis is more gradual than in tropical climates, as is demonstrated by the abundance of intermediate forms (Souchier 1971). For the most part, the clays are derived from phyllitic minerals by transformations which may be very limited (inheritance) or more considerable (degradation); an important part is played by the environmental conditions, particularly the *acidity and the presence of organic anions*. Neoformation is generally a minor process in temperate climates; however, in certain cases it can be of some importance. Conversely, in tropical climates, certain slow transformations can occur, such as the very gradual loss of silica from some very resistant 2 : 1 clays: this will be referred to again in Chapter 12.

### 1   Types of hydrolysis in temperate climates (Robert 1970, Robert et al. 1979, Razzaghe-Karimi 1974, Pedro 1976a,b)

Robert, in dealing with the micaceous minerals, has defined very different types of hydrolysis, resulting from transformations of increasing intensity as the environment is increasingly active. *Differing from tropical weathering, it is the kind and quantity of the soluble organic anions which control the amount and the kind of changes observed: thus a part of the aluminium can be dissolved and lost in the drainage water.*

**Neutral hydrolysis (saline solutions present).** Neutral hydrolysis occurs in a non-acid, base-rich environment, which is relatively rare in a temperate climate (calcareous materials). The salts of $Ca^{2+}$ or $Mg^{2+}$ which are present take part in an exchange reaction that leads to a very gradual and incomplete removal of the interlayer $K^+$; there is also a partial oxidation of the $Fe^{2+}$ of the octahedral layers and only a very limited degradation of the sheet structure; this moderate degree of change is very much like that of *inheritance* (simple **microdivision** of the mica particles, see Nguyen Kha 1973).

It is possible to compare this type of hydrolysis with the change from *illite to montmorillonite* by ion substitution which was investigated by Paquet (1969) and Tardy (1969) for fersiallitic soils and by Souchier and Lelong (1972) for certain temperate eutrophic brown soils on diorites. This type of change, which can be taken as being almost an *aggradation*, occurs in the presence of concentrated solutions of silica and alkaline earths. First of all the sheet structure expands as a result of the removal of the interlayer $K^+$, the 4-coordinated aluminium leaves the tetrahedral layer and takes up a 6-coordinate position; the sites thus left vacant in the tetrahedral layer are then occupied by silica.

As will be seen in Chapters 8 and 12, this kind of transformation by the addition of silica to the micaceous minerals is common at impeded sites that dry out seasonally (Seddoh & Pedro 1974, fersiallitic soils and vertisols).

**Acid hydrolysis: acidolysis** (Bruckert 1970, Razzaghe-Karimi 1974, Fig. 1.1). This is the commonest kind of weathering that occurs, for example, in acid brown soils (acid **mull**). In these conditions, with active humus, the soluble organic matter produced by the litter is sometimes abundant, but a study of its

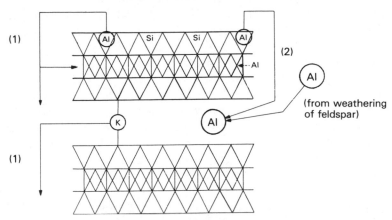

**Figure 1.1**  Transformation processes of clays under acid conditions. (1) Degradation in a 'leaching' environment. (2) 'Aggradation' in an impeded or semi-impeded environment.

development shows that *most of the anions that are able to form complexes, such as citric and oxalic acids and polyphenols, are immobilised or biodegraded within the humus-rich horizons* (Bruckert 1970); only a small amount of organic acids (which have little or no ability to form complexes) and dissolved $CO_2$ reach the (B) horizon. In these conditions, Razzaghe-Karimi (1974) showed that the micaceous minerals undergo a more important transformation: complete removal of the interlayer $K^+$, opening up of the sheet structure, so that illites are changed into **open vermiculites**. In addition, 4-coordinated aluminium is gradually removed from the tetrahedral layer; the Al ions thus formed, together with those coming from the weathering of the feldspars, are in part lost in the drainage; however, certain of them take up an interlayer position as islands of complex aluminium ions $(Al(OH)^n)$, this time the Al being 6-coordinated – this is an **aluminous vermiculite** (Villiers & Jackson 1967).

*Formation of aluminous chlorites by aggradation*
In impeded (or semi-impeded) environments where the Al ions freed by the weathering of minerals other than micas, in particular feldspars, cannot be removed by leaching, the initial degradation can be followed by an undoubted *aggradation*. The alumina islands, previously formed in the interlayer position, join up with one another to form a continuous gibbsitic sheet; the clay assumes a chlorite structure (2 : 1 : 1), but in this case it is an aluminium-rich secondary chlorite very different from a magnesium-rich primary chlorite which occurs in slightly developed, slightly acid soils, rich in ferromagnesium minerals. In certain cases, these secondary chlorites can be well crystallised and relatively stable (Gac 1968, Fig. 1.1).

**Complex formation: complexolysis** (Bruckert 1970, Razzaghe-Karimi 1974, Robert & Vicente 1977, Schnitzer 1981). This particular kind of weathering – which cannot accurately be called hydrolysis – involves certain soluble organic compounds, such as oxalic and citric acids and some phenolic compounds, forming complexes with iron and aluminium (this will be referred to again when the processes of migration are discussed in Ch. 3). It is not only the $H^+$ ions of these compounds that are involved, but also their ability to form complexes which makes them particularly active towards primary minerals as well as to certain clays: *they are capable, in fact, of extracting and mobilising by complex formation the atoms of aluminium and iron from within the crystalline sheet structures.*

Bruckert (1970) showed the great importance of the formation of mobile organometal complexes in the process of podzolisation: the agents of complex formation are produced in great amounts by the slightly active humus of the mor type (or moder); in addition, they are not insolubilised in the surface humus-rich horizons, but are carried into the mineral horizons where they increase the speed of weathering.

Robert (1970), Razzaghe-Karimi (1974) and Boyle *et al.* (1974) have shown experimentally the drastic effect of these complexing agents on micaceous clays; they extract not only interlayer Al, and 4-coordinated Al, but even a considerable part of the aluminium of the octahedral layer, so that the sheet structure is completely destroyed as the constituents become soluble or amorphous. At pH 3, the destruction of biotites and illites is almost total (Robert & Vicente 1977). *This complex formation, or* **complexolysis**, *is the characteristic mode of weathering of the podzols.*

## 2   Resistance of clays to complexolysis

Clays have a variable resistance to the degrading action of complexing agents. Climatic factors are also involved, for it is evident that these agents are more effective in hot than in temperate or cold climates.

While montmorillonites are the least resistant of clays, the kaolinites, on the contrary, appear stable in a temperate climate. Guillet *et al.* (1975) showed, by the method of balances, that kaolins are resistant to the process of podzolisation on Triassic sandstones of the Vosges. It is different in hot, humid climates where the destruction of kaolinite in certain ferrallitic soils has been demonstrated by several authors (Pedro & Berrier 1966, Lelong 1967, Eswaran & Coninck 1971); it is in these ferrallitic soils that crystalline *secondary gibbsite* is formed at the base of the humus-rich horizons (see Ch. 12).

The micaceous clays (illites, vermiculites) are, from this point of view, intermediate in resistance between the kaolinites and montmorillonites of neoformation. In very 'podzolising' environments, most of these clays are destroyed, as Guillet *et al.* showed in their investigation of the podzols of the Vosges. Gogolev and Anastasyeva (1964) and Zaidelman (1974) consider that an important degradation of clays with destruction of the sheet structure occurs in the podzolic soils of the USSR. Coninck *et al.* (1968) have the same opinion as far as the podzols of the Campine are concerned. Sawney and Voigt (1969) have been able to reproduce this degradation experimentally: in the presence of citric acid, the octahedral layer disappears entirely; there remains only a siliceous wafer which is very unstable and cellular and which is destroyed by **amorphisation**, or even complete solution.

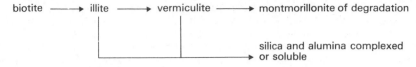

**Figure 1.2**  Degradation of micaceous clays.

However, it appears that certain micaceous clays are particularly resistant to complexolysis, for in these cases intense degradation does not give rise to a totally amorphous product: tetrahedral Al is certainly removed. On the other hand, the octahedral layer appears to be resistant; the gaps formed in the tetrahedral layer cause the crystallinity to decrease at the same time as the negative charge is lowered, which results in an increase of the swelling properties which become similar to those of a montmorillonite. This is a **montmorillonite of degradation**, that Souchier (1971) and Coen and Arnold (1972) have identified in the Λ2 horizon of podzols, which differs profoundly in its origin and properties from montmorillonites of neoformation.

These processes of degradation of micaceous clays by complexolysis can be represented as shown in Figure 1.2.

## 3  Neoformation of clay in temperate climates

Even though the transformation of the phyllitic minerals constitutes the dominant process of clay formation in temperate climates, clay neoformation cannot be ignored, for although always less important than in tropical climates it can in certain cases be far from negligible. This is a difficult problem which has been investigated by many people. The results are equivocal so that as yet it cannot be considered as being definitely solved.

If it is granted, from the preceding discussion, that phyllitic minerals (the micas) weather in a temperate climate by gentle and gradual transformation, they can play only a minimal part in this neoformation; thus the neoformed clays must come from other sources: the tectosilicates (feldspars), volcanic glasses and, to a certain extent, additions from the biogeochemical cycle. Litter supplies to the soil a considerable amount of certain elements, for example silica (Erhart 1973, see also Ch. 3). As far as the feldspars are concerned, Souchier (1971) has shown that, in a temperate humid climate, they are altered and they gradually free their constituents in a soluble or amorphous form; those such as silica and alumina will be able in certain circumstances to form a clay mineral which will be kaolinite, for this is the only one compatible with the acidity of the environment. Confirmation of this theory is provided by the work of Icole (1973) on the neoformation of kaolinite on the terrace soils of the Pyrennean foothills.

However, this simple theory has been attacked by some recent investigations which show the extreme complexity of the processes involved. For example, in thick regoliths the feldspars undergo, as it were prematurely, a complete reorganisation of their structure to form phyllosilicates of a sericitic or montmorillonitic type (Tardy 1969, Hetier 1975, Seddoh & Pedro 1975). It is a case of neoformation occurring outside of the profile and indeed before its formation ('pre-pedogenesis' according to Hetier); the soils thus 'inherit' these prematurely neoformed clays.

Conversely, kaolinites considered as typical clays of neoformation can, in certain very acid environments, be produced by a slow transformation of micaceous clays; they develop either by the slow loss of silica (in a tropical climate) or, on the contrary, by a gain of aluminium; thus 2 : 1 : 1 Al-chlorites are formed which are then split into two 1 : 1 kaolinite sheets (Tardy 1969, Ségalen 1973). Certain kaolinites can be formed in this way in a temperate climate.

In fact, the slowness of primary mineral hydrolysis and crystallisation from gels are not the only factors limiting temperate climate clay neoformation: *the presence of acid organic material in almost all of the horizons in most temperate soils is another much more effective limiting factor.*

From the initiation of 'complexolysis', particularly in well drained environments, there is a massive removal from the horizon of all the constituents from which clay is formed (see Ch. 3). *Under these conditions it is evident that all neoformation is prevented* (Pedro 1976b). But now it has also been shown that the insoluble humic compounds form, with the Al and Fe cations, mixed gels of various types (insoluble complexes, absorption complexes) which to a considerable extent prevent the crystallisation of the mineral gels: this also explains the persistence in temperate soils of amorphous or only slightly crystalline free hydroxides.

In 1964, Beckwith and Reeve showed that organic anions slowed down the adsorption of silica by sesquioxides and stopped all neoformation. Schwertmann (1969) for his part showed that organic material prevented the crystallisation of iron hydrates. More recently, Hetier (1975) has shown that organic material prevents all clay neoformation in the andosols; neoformation can take place in these soils only in two phases of their development: (i) a phase prior to the incorporation of organic material and thus of pedogenesis (prepedogenesis); (ii) a terminal phase, on the other hand, after the mineralisation of part of the organic material, which thus frees the mineral gels.

From this discussion, it is apparent that clay neoformation as a significant characteristic of temperate soils can occur only in certain environments – horizons of slight acidity, lacking in organic matter not too well drained so that the constituents of clay (alumina and especially silica which is more soluble in such an environment than aluminium, Tardy 1969) can be concentrated sufficiently – these conditions are found within certain regoliths of quartz-poor rocks rich in weatherable materials, as shown by Dejou (1967a,b) and Dejou *et al.* (1977), or on material containing volcanic glass (Hetier 1975).

Even in this environment, however, a considerable loss of silica and bases by drainage appears inevitable: *hence the neoformed minerals are poor in silica (halloysite, kaolinite)*; in certain cases, the deficit in silica is such that the alumina crystallises as gibbsite (Green & Eden 1971).

In a temperate climate, neoformed 2 : 1 clays are certainly more rare than 1 : 1 clays. They result, as has been shown, from a prior neoformation occurring in a 'prepedologic' state (within deep regoliths), by reorganisation of structures within the minerals themselves in an environment still sufficiently rich in silica and bases. This very particular type of neoformation is still not well known compared to clay neoformation of the classic type which uses materials previously in solution.

## IV    Influence of environmental factors on weathering

Detailed studies of weathering processes enable the fundamental contrast between the effects of *general climate* and of *local factors* (site) to be specified.

The general climate plays an essential part: it is involved on the one hand by the factor of *water* and on the other by the factor of *temperature*: weathering in a hot and humid climate differs from that which characterises temperate climates, not only in its speed but also by the kind of fundamental physicochemical processes. *The type of climatically controlled weathering is one of the essential elements of climatic soil zonation when combined with latitude.*

But *local* factors – those which characterise the *site* (parent rock, topography, vegetation) – also play an important part, since in particular cases they are able to modify profoundly the climatically controlled processes as a whole. As a general rule, in a temperate climate the abundance of soluble organic products and the strong acidity speed up the weathering, but they slow down the neoformation of clays or even cause the degradation of pre-existing clays. Conversely, a high amount of alkaline-earth cations and strong biological activity slow down weathering, but at the same time favour the neoformation or the conservation of clays that are richer in silica. *In addition, no matter what the climate, neoformation is favoured on basic volcanic rocks compared to acid crystalline rocks.*

### 1   Influence of general climate

On a world scale, it is possible to differentiate three major zones of climatically controlled weathering, of which the first two are directly controlled by the organic matter, while the third is controlled more particularly by the temperature. These three fundamental processes of weathering – **podzolisation, brunification** and **ferrallitisation** – will be more fully dealt with elsewhere in this book. At this point, podzolisation, characterised by strong complexolysis which excludes all neoformation, will not be considered further, but the other two processes will be compared in a more detailed fashion, and from this a third intermediate type will be derived which is characteristic of areas with a strongly contrasted seasonal climate (**fersiallitisation**).

**Temperate climate: brunification.** The slow changes of phyllitic minerals, giving rise to clays of illitic or vermiculite type, constitute the dominant process; therefore, it is possible to refer to it as **bisiallitisation** (Pedro 1964, Novikoff *et al*. 1972). Acid hydrolysis is general and it affects the profile to only a shallow depth, of about 1 m; it is accompanied by a certain loss of elements (alkaline earths, Al, Fe and silica); clay degradation starts when complexolysis occurs, that is to say in a biologically less active and very well drained environment (moder or mor humus type). On the other hand, in an acid but impeded environment, the high concentration of aluminium in solution favours *aggradation* to aluminous chlorites. Neoformation of clay is of little importance and is limited to certain semi-impeded, moderately acid environments, poor in organic matter but rich in rapidly weatherable minerals; it gives rise to halloysites or kaolinites and occurs particularly at depth in regoliths and characterises the earlier phases of 'prepedogenesis'. *In the brown soils, iron hydroxides remain associated with the clays in an amorphous and cryptocrystalline state.*

**Hot and humid climate (equatorial): ferrallitisation.** Neutral hydrolysis (sometimes even weakly alkaline) is general and it affects the parent rocks to a depth of several metres; in these conditions organic matter is involved only slightly. Weathering is total, resulting in a freeing of all constituents – silica, iron, alumina – silica and the bases being preferentially removed (if the environment is sufficiently well drained), while iron and aluminium hydroxides (well crystallised) accumulate in the profile. For the same parent rock the soils of hot climates are richer in iron and aluminium than are temperate soils. Furthermore, *neoformation is almost exclusively the way in which clays are formed, the clays being poor in silica of the kaolinite type.*

In fact, as Pedro (1964) demonstrated, ferrallitic weathering appears to consist of two phases, one rapid, the other one slow. The **rapid phase** involves the weathering of the most susceptible of the primary minerals (**unstable** minerals according to Pedro), accompanied to a greater or lesser extent by the formation of clays, some of which are of the 2 : 1 type. These clays constitute a **metastable** state of the weathering complex and can be subject to a gradual desilication, particularly the 2 : 1 clays. This gradual loss of silica perhaps can be considered as characterising the **slow phase** of ferrallitic weathering. The very slow kaolinisation of 2 : 1 clays in a hot climate has been confirmed by some recent work (Spaargaren 1979).

**Seasonally contrasted subtropical climate: fersiallitisation.** Paquet (1969) and Lamouroux (1971, 1972) defined the characteristics of this weathering typical of mediterranean climates, which appears as an intermediate type between the two preceding ones, but with peculiarities owing to the occurrence of a hot, dry season. During this season, soil solutions rich in silica and alkaline earths rise towards the surface. In addition, because of its rapid mineralisation, organic matter is of little importance in this climate. These conditions are favourable to the neoformation of 2 : 1 montmorillonite type clays. But clays inherited or transformed by aggradation and addition of silica are equally abundant (Novikoff *et al.* 1972); in these conditions no process of degradation is possible. In summary, *inherited or neoformed 2 : 1 clays are dominant in profiles with fersiallitic weathering; in addition, the loss of iron oxides, immobilised* in situ *at these high pHs, is reduced to a minimum*: these iron oxides are generally subject to the process of rubification which will be defined in Chapter 12. It is in this kind of weathering that the free Fe : clay ratio reaches a maximum (Bornand 1978).

Finally, when climatic conditions approach those of the Tropics, with a marked dry season, weathering takes on a character intermediate between that of the ferrallitic and fersiallitic types: the neoformation of kaolinite gradually becomes more and more important as the mean temperature increases (ferruginous tropical soils).

Lelong and Souchier (1972) were able to show, by the method of geochemical and mineralogical balances, the contrast that exists between ferrallitic weathering and temperate weathering (acid brown soil) on the same kind of granite. Figure 1.3 shows the results obtained in these two cases, by comparison with the original material (base 100). In a temperate climate it is evident that weathering is slow and gradual; many of the weatherable (or partially weathered) minerals

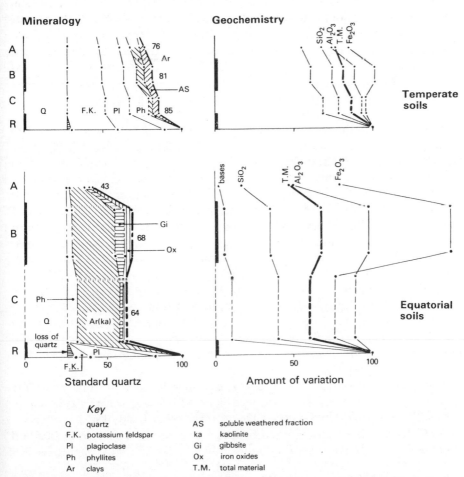

**Figure 1.3** Mineralogical and geochemical balances in temperate and tropical weathering. The amount of variation is the percentage of each element compared to its initial content in granite.

persist; there is a slight amount of clay mainly formed by transformations; three-quarters of the weathering complex is made up of soluble materials which are removed from the profile by drainage.

Under equatorial conditions (lower part of diagram), weathering is complete; the removal of bases and silica by drainage is considerable; neoformed kaolinite is very abundant; apart from quartz, almost no other primary minerals remain. The decrease in clay in the upper part of the diagrams reflects the process of degradation of clay which occurs under well drained, very humid and acid conditions (see following section).

Note that in *arid climates*, weathering becomes very slow or ceases; the environment does not become acid: clays of the 2 : 1 type are mostly *inherited*.

## 2   Influence of local conditions: site and vegetation

In temperate climates, biological activity combined with the speed and the kind of humification exerts a decisive influence on the mode of weathering; this influence decreases considerably in a hot climate. However, a certain slowing down of humification is observed in certain very acid environments with a very humid climate (altitude); it has been shown that under these conditions it is possible for kaolinite to weather in the surface horizons and give rise to secondary gibbsite (Lelong 1967). But this is only one particular case: in hot climates, it is the local drainage and the topography which play the essential role in controlling the kind of weathering.

**Influence of the humus type in temperate climates** (Berthelin *et al*. 1979). **The 'humus' factor** is decisive in the local control of temperate weathering. Carbonate mulls can undergo an accelerated decarbonation owing to the action of organic material (see Ch. 3), but so long as active calcium carbonate persists in the profile, the montmorillonite clays inherited from the parent material are preserved even though they are susceptible to degradation. Non-calcareous forest mull and acid mull favours a fairly moderate type of transformation of the micaceous clays (opening up of the sheets – **vermiculisation**). *Degradation by complexolysis for the most part only occurs in a temperate climate in the presence of an inactive moder or mor humus* (there are certain exceptions in cold and mountainous climates which will be dealt with later). This type of degradation is general in soils with a podzolic tendency; it causes the complete destruction and solution of some micaceous clays and in others the very gradual weathering of the sheets, accompanied by the development of the capacity to swell (montmorillonites of degradation: Guillet *et al*. 1975).

This 'podzolic' degradation of clays is modified in hydromorphic soils: *it is considerably increased in horizons with freely circulating water* (Brinkman *et al*. 1973, Zaidelman 1974, Gury 1976). On the other hand, it is slowed down in horizons that have an impeded drainage as a result of fine grain size and slight permeability, when an aluminous chlorite is produced by 'aggradation'.

**Influence of topography and local drainage in hot climates.** Weathering in a hot climate can be profoundly modified by local factors: *topography*, comparing well drained slopes with lower areas where an impeded environment occurs; *formation of a perched water table* in a very badly drained environment causes a particular kind of silicate hydrolysis (**ferrolysis**).

*Effect of topography and parent material.* The effect of these two factors has been emphasised by many authors who have described soil catenas formed around inselbergs in tropical regions (Paquet 1969, Bocquier 1971, see Ch. 4). The soluble elements (silica, calcium and magnesium salts) removed from the upslope profiles cause an increase in the concentration of solutions down slope, which accentuates the contrast between the two.

These contrasts reach their maximum development on basic rocks under a tropical climate with a dry season: well drained (or leaching) environments of the higher slopes are characterised by a deficit of silica; *kaolinite* is dominant sometimes with some gibbsite (in regions with a humid climate). In contrast, impeded environments of the lower slopes are often characterised by an abundant production of montmorillonite by neoformation and aggradation (vertisols): as development continues, this neoformation occurs further and further up slope – the upslope encroachment of neoformed montmorillonites of Paquet (1969).

*The iron oxides* follow a parallel development to that of aluminium; they occur as free entities to a decreasing extent in going down slope; on well drained slopes, individual crystals are formed (or in combination with gibbsite); in midslope positions in a semi-impeded environment, they form coatings around the kaolinite; finally, at the bottom of the slope, in an impeded environment, they occur within the octahedral layers (ferriferous montmorillonites, Paquet 1969).

On slopes developed on acid rocks, the contrast between upper and lower sites decreases considerably; the conditions are altogether too acid for the neoformation of montmorillonite and only kaolinite is formed. The loss of silica is, however, greater from up slope (a certain amount of gibbsite coexists with the kaolinite) than from down slope where kaolinite alone is formed. In addition, iron oxides in these acid conditions remain as free entities and are much more mobile than in basic conditions; they cannot be integrated within clays of neoformation: they accumulate on lower slopes, where they often lead to the formation of **iron cuirasses** (see Ch. 12).

*Weathering within surface water tables: ferrolysis.* In regions with a hot climate and strongly contrasted seasons, very temporary perched water tables frequently form on platforms or in badly drained depressions. Within these bodies of water there is a massive reduction of iron oxides accompanied by a very characteristic degradation of 2 : 1 clays; this causes the formation of a surface **albic** horizon which is bleached and markedly impoverished in clay, while the clays that persist are transformed to Al-chlorites. In these conditions there is an abundance of complex aluminous ions (**planosols**, see Ch. 11).

Generally, the soils involved in this process are only slightly acid and poor in organic matter, and the little organic matter that does occur is biologically active and gives rise to only few active soluble compounds. Therefore complexolysis cannot be responsible for the process of degradation, which thus must be ascribed to another process – that of **ferrolysis** (Brinkman 1970). The large-scale variation of Eh, which accompanies both the formation and the disappearance of the perched water table, is invoked by Dudal (1973) to explain the formation of the albic horizon; the iron oxides are reduced in the hydromorphic period; the $Fe^{2+}$ ions when they are reoxidised in periods of high Eh, free $H^+$ ions capable of momentarily acidifying the environment and of causing degradation of the octahedral layer of 2 : 1 clays, particularly of the montmorillonites, which are very sensitive to these great variations of Eh and pH.


# V   Conclusion

The study of weathering and clay mineral formation can no more be detached from its environmental context than the study of humification and the development of organic matter: clays are 'dynamic' in the same way as organic matter and soils as a whole. To forget this fundamental point is a grave error: this is the case in the static classifications of clays based on structural similarities which have caused certain authors to make ill founded comparisons, particularly in the choice of names. Certain clays have the same name, yet they have been produced by entirely different genetic mechanisms and they occur in totally different environments. In fact, these clays that have the same name are very different in certain properties that are considered as secondary by the classifiers but have an importance which has been underestimated; two examples will be given – the chlorites and montmorillonites (or smectites).

   *Chlorites* have in common a particular sheet structure – 2 : 1 : 1. But it has been mentioned that the magnesium-rich *primary chlorites* which are inherited from certain parent materials are very unstable and are not resistant to the process of acidification. In contrast, the aluminium-rich *secondary chlorites* are the result of a prolonged process of degradation, followed by one of aggradation in an impeded acid environment; the identity of names should not obscure the genetic contrast that exists between these two types.

   *The example of the smectites and montmorillonites* is even more to the point: montmorillonites (so named because of their 2 : 1 structures and strong swelling capacity) are found in environments where the conditions are as fundamentally contrasted as the A2 horizons of podzols (very acid and very leaching conditions with intense complexolysis) and vertisols (neutral or alkaline conditions, rich in alkaline earths and with impeded drainage). Given such different environmental conditions, it is hardly to be expected that the montmorillonites formed would be identical, although this has been maintained by many authors. However, it has been shown that these two types of montmorillonite, called respectively montmorillonites of degradation (podzols) and true montmorillonites (vertisols), have a very different genetic relationship, the first coming from an extremely powerful degradation of the micas, while the second result from neoformation, or on occasions of an aggradation in a neutral environment rich in silica and cations. Schwertmann (1962) and Ross and Mortland (1966) have shown that despite the similarities, there are in fact important differences between these two types of clay in the nature of their charges, their ability to fix $K^+$ ions, and their degree of crystallinity; thus their distinction in a pedological sense is fundamental.

The study of weathering, and, as will be seen, that of humification and of movement of materials within the soil, must be considered in terms of the soil environment. Among the environmental controls, that of general climate and of local conditions or 'site' will be distinguished in most chapters of this book. After what has been said, the pre-eminence of the part played by general climate is obvious, if the overall distribution of weathering types is considered; this leads to the concept of the **climatic zonation of soils**, which will be referred to frequently.

The succession of three fundamentally different major types of weathering, arranged in sequence from boreal to equatorial climates, is in this respect very significant: active 'complexolysis', excluding all clay neoformation in cold climates; gentle and gradual hydrolysis resulting in moderate transformations, characteristic of temperate or dry climates; total hydrolysis and the predominance of massive neoformation in tropical climates (Fig. 1.4, after Pedro 1976b).

In other words, if the organic matter plays an important active part in the zones with a cold and humid climate, this role is negligible in the zones with a hot or dry climate.

Table 1.2 summarises the major kinds of climatic weathering. To these three fundamental types of weathering, there correspond three different types of B horizon (it should be noted, however, that in the American system, these three types of B horizon are defined in a non-climatic manner): (i) spodic B horizon, where amorphous products accumulate after movement (podzols); (ii) cambic (B) horizon, characteristic of temperate weathering (brown soils); (iii) oxic (B) horizon, typical of ferrallitic weathering. These will be discussed in greater detail in the second part of this book.

Although the local factors of site only seem to have a secondary influence on a world scale, they are nevertheless, in some particular cases, of great importance, as will be discussed in Chapter 4. Here discussion will be restricted to emphasising two examples of particular importance. (i) In *temperate climates* brunification under the influence of mull-type humus can be considered as being the general climatically controlled process: however, under certain local influences, which will be dealt with later, a mor can form and locally induce a podzolisation process comparable to that which is general in a boreal climate. (ii) In *tropical climates*, if neoformation of kaolinite (with or without gibbsite) is the generally climatically controlled process, certain local conditions of site can give rise to entirely different processes – destruction of clays by ferrolysis in very hydromorphic conditions (planosol formation); massive neoformation of montmorillonite in non-acid conditions with a

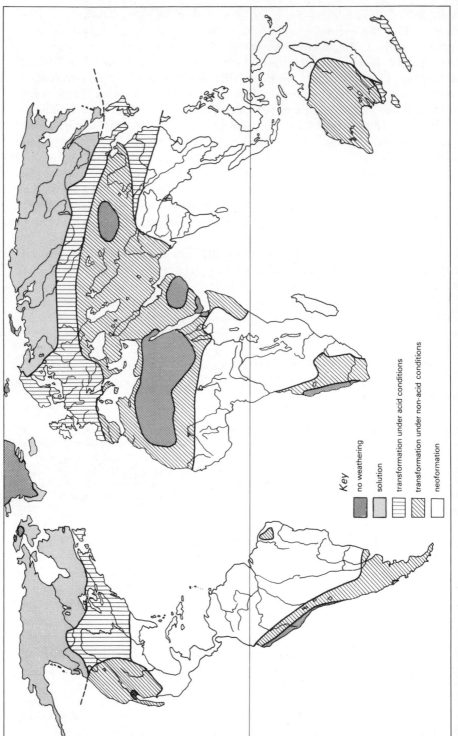

**Figure 1.4**  Distribution and extent of main types of weathering of silicate rocks (after Pedro 1976b).

Key

no weathering

solution

transformation under acid conditions

transformation under non-acid conditions

neoformation

**Table 1.2** Types of climatic weathering.

| Climate | Soil | Weathering type | Type of clay formation | |
|---|---|---|---|---|
| | | | Transformations | Neoformation |
| boreal | podzolic | complexolysis | degradation and solution | nil |
| temperate | podzolic (mor) | complexolysis | degradation and solution | nil |
| | brown (mull) | acid hydrolysis (gradual) | transformation (and moderate degradation) | weak (kaolinite) |
| subtropical | fersiallitic | neutral hydrolysis | (inheritance, moderate transformation) | moderate (montmorillonite) |
| tropical (with a dry season) | well drained tropical ferriginous | total hydrolysis (neutral) | limited inheritance and transformation | moderate kaolinite |
| | poorly drained vertisols | total hydrolysis (neutral) | limited inheritance and transformation | strong montmorillonite |
| humid equatorial | ferrallitic | total hydrolysis (neutral) | no inheritance and transformation | strong kaolinite (gibbsite) |

pedoclimate of seasonally strong contrasts (vertisol formation); and even, as will be seen in Chapter 12, surface podzolisation in very special environmental conditions.

# References

Acquaye, D. K. and J. Tinsley 1965. In *Experimental pedology*, E. G. Hallsworth and D. V. Crawford (eds), 126–39. London: Butterworth.

Beckwith, R. S. and R. Reeve 1964. *Aust. J. Soil Res.* **2**, 33–44.

Berthelin, J., B. Souchier and F. Toutain 1979. In *Alteration des roches cristallines, INRA seminar, Bull. AFES* **2** and **3**, 175–86.

Blot, A., J. C. Leprun and J. C. Pion 1976. *Bull. Soc. Géol. France* (7) **XVIII**, 51–4.

Bocquier, G. 1971. *Genèse et évolution de deux toposéquences de sols tropicaux du Tchad. Interprétation biogéodynamique*. State doct. thesis. Univ. Strasbourg.

Bornand, M. 1978. *Altération des materiaux fluvio-glaciaires. Genèse et évolution des sols sur terrasses quaternaires dans la moyenne vallee du Rhône*. State doct. thesis. Univ. Montpellier, ENSA Montpellier.

Boulet, R. 1974. *Toposéquences de sols tropicaux en Haute-Volta: équilibres dynamiques et bioclimats*. State doct. thesis. Univ. Strasbourg.

Boyle, J. R., G. K. Voigt and B. L. Sawhney 1974. *Soil Sci.* **117** (1), 42–5.

Brethes, A. 1973. *Mode d'altération et différenciation pédogénétique sur leucogranite du massif du Morvan. Comparaison avec le massif vosgien*. Spec. doct. thesis. Univ. Nancy I.

Brinkman, R. 1970. *Geoderma* **3**, 199–206.

Brinkman, R., A. G. Jongmans, R. Miedema and P. Maaskant 1973. *Geoderma* **10** (4), 259–70.

Bruckert, S. 1970. *Influence des composés organiques solubles sur la pédogénèse en milieu acide*. State doct. thesis, Univ. Nancy I; *Ann. Agron.* **21** (4), 421 and **21** (6), 725.

Ciric, M. 1967. *Soviet Soil Sci.* **1**, 57.

Coen, G. M. and R. W. Arnold 1972. *Soil Sci. Soc. Am. Proc.* **36** (2), 342–50.

Coninck, F. de, A. J. Herbillon, R. Tavernier and J. J. Fripiat 1968. *9th Congr. ISSS* Adelaide **IV**, 353–65.

Dejou, J. 1967a. *C.R. Acad. Sci. Paris* **264D**, 37.

Dejou, J. 1967b. *Ann. Agron.* **18** (2), 145.

Dejou, J., J. Guyot and M. Robert 1977. *Evolution superficielle des roches cristallines et cristallophylliennes dans les regions tempérées*. Paris: INRA.

Dudal, R. 1973. *Pseudogley and gley. Trans Comms V and VI ISSS*, E. Schlichting and U. Schwertmann (eds), 275–85.

Durand, R. and R. Dutil 1971. *Ann. Agron.* **22** (4), 397–424.

Erhart, H. 1973. *Itinéraires géochimiques et cycle géologique du silicium*. Paris: Doin.

Eswaran, H. and F. de Coninck 1971. *Pédologie*, Ghent **XXI** (2), 181–201.

Fritz, B. and Y. Tardy 1976. *Bull. Soc. Géol. France* (7) **XVIII**, 7–12.

Gac, J. Y. 1968. *Les altérations de quelques roches cristallines des Vosges. Etude minéralogique et géochimique*. Spec. thesis. Univ. Strasbourg.

Gogolev, I. N. and O. M. Anastasyeva 1964. *Soviet Soil Sci.* **11**, 1144.

Green, C. and M. Eden 1971. *Geoderma* **6**, 315–17.

Guillet, B., J. Rouiller and B. Souchier 1975. *Geoderma* **14** (3), 223–45.

Gury, M. 1976. *Evolution des sols en milieu acide et hydromorphe sur terrasses alluviales de la Meurthe*. Spec. thesis. Univ. Nancy I.

Hetier, J. M. 1975. *Formation et évolution des andosols en climat tempéré*. State doct. thesis. Univ. Nancy I.

Icole, M. 1973. *Géochimie des altérations dans les nappes d'alluvions du Piemont occidental nord-pyrénéen*. State doct. thesis. Univ. Paris VI.

Kantor, W. and U. Schwertmann 1974. *J. Soil Sci.* **25** (1), 67–78.

Lamouroux, M. 1971. *Etude des sols formés sur roches carbonatées. Pédogénèse fersiallitique au Liban*. State doct. thesis. Univ. Strasbourg.

Lamouroux, M. 1972. *Cah. ORSTOM, sér. Pédologie* **X** (3), 243.

Lelong, F. 1967. *Nature et genèse des produits d'altération de roches cristallines sous climat tropical humide (Guyane française)*. State doct. thesis. Fac. Sci. Nancy.

Lelong, F. and B. Souchier 1972. *C.R. Acad. Sci. Paris* **274D**, 1896.

Lelong, F. and B. Souchier 1979. In *Alteration des roches cristallines, INRA seminar, Bull. AFES* **2** and **3**, 267–79.

Leneuf, N. 1959. *L'altèration des granites calco-alcalins et des granodiorites en Côte-d'Ivoire forestière et les sols qui en sont dérivés*. State doct. thesis. Univ. Paris.

Mehlich, A. 1967. *Z. Pflanzener. Bodenk.* **117** (3), 193–204.

Millot, G. 1964. *Géologie des argiles*. Paris: Masson.

Nguyen Kha 1973. *Recherches sur l'évolution des sols à texture argileuse en conditions témperées et tropicales*. State doct. thesis. Univ. Nancy.

Novikoff, A., G. Tsawlassou, J. Gac, B. Bourgeat and Y. Tardy 1972. *Bull. Sci. Géol.* **25** (4), 287–305.

Paquet, H. 1969. *Evolution géochimique des minéraux argileux dans les altérations et les sols des climats méditerranéens et tropicaux*. State doct. thesis. Univ. Strasbourg.

Pedro, G. 1964. *Contribution à l'étude expérimentale de l'altération geochimique des roches cristallines*. Fac. Sci. thesis. Univ. Paris.

Pedro, G. 1976a. *Bull. Soc. Géol. France* (7) **XVIII**, 27–32.

Pedro, G. 1976b. *Science du Sol* **2**, 69–84.

Pedro, G. and J. Berrier 1966. *C.R. Acad. Sci. Paris* **262D**, 729.

Pochon, M. 1978. *Origine et évolution des sols du Haut-Jura suisse*. *Soc. Helvet. Sci. Nat.* Zurich: Fretz.

Razzaghe-Karimi, M. 1974. *Evolution géochimique et minéralogique des micas et phyllosilicates en presence d'acides organiques*. Spec. thesis. Univ. Paris VI.

Robert, M. 1970. *Etude expérimentale de la désagrégation du granite et de l'évolution des micas*. State doct. thesis. Fac. Sci. Paris.

Robert, M. and M. A. Vicente 1977. *C.R. Acad. Sci. Paris* **284D**, 511–14.

Robert, M., M. Razzaghe-Karimi, M. A. Vicente and G. Veneau 1979. In *Alteration des roches cristallines, INRA seminar, Bull. AFES* **2** and **3**, 153–72.

Ross, G. and M. Mortland 1966. *Soil Sci. Soc. Am. Proc.* **30** (3), 337–42.

Sawney, B. L. and G. K. Voigt 1969. *Soil Sci. Soc. Am. Proc.* **33** (4), 625–9.

Schnitzer, M. 1981. In *Migrations organo-minerales dans les sols tempéres*, Coll. Intern., CNRS, Nancy, 1979, 229–34.

Schwertmann, U. 1962. *Beitr. Mineral. Petrogr.* **8**, 199–209.

Schwertmann, U. 1969. *Proc. Intern. Clay Conf.* **1**, 683.

Seddoh, K. and G. Pedro 1974. *Bull. Groupe Fr. Argiles* **XXVI** (1), 107–25.

Seddoh, K. and G. Pedro 1975. *Cah. ORSTOM, Sér. Pédologie* **XII** (1), 7–25.

Ségalen, P. 1973. *L'aluminium dans les sols. Initiations documentations, techniques* **22**. ORSTOM.

Souchier, B. 1971. *Evolution des sols sur roches cristallines à l'étage montagnard. Vosges*. State doct. thesis. Univ. Nancy.

Souchier, B. and F. Lelong 1972. *Sciences de la Terre* **XVII** (4), 353.

Spaargaren, O. C. 1979. *Weathering and soil formation in a limestone area near Pastena (Italy)*. Doct. thesis. Univ. Amsterdam.

Tardy, Y. 1969. *Géochimie des altérations. Etude des arènes et des eaux de quelques massifs cristallins d'Europe et d'Afrique*. State doct. thesis. Univ. Strasbourg.

Tardy, Y., G. Bocquier, H. Paquet and G. Millot 1973. *Geoderma* **10** (4), 271–84.

Trichet, J. 1969. *Contribution a l'étude de l'altération expérimentale des verres volcaniques*. State doct. thesis. Fac. Sci. Paris.

Villiers, J. M. de and M. L. Jackson 1967. *Soil Sci. Soc. Am. Proc.* **31** (4), 473–9.

Zaidelman, F. R. 1974. *Podzolisation and gleyification*. Moscow: Nauka.

# Chapter 2

# The dynamics of organic matter

## I   Introduction

The prime source of soil organic matter is plant debris of all kinds, such as dead leaves and branches, that fall onto the soil and are then biologically decomposed at a variable rate. The slightly altered plant material that covers the mineral soil is the **litter** which, during subsequent decomposition, forms either soluble or gaseous compounds, such as $NH_3$, $CO_2$, nitrates, sulphates etc. by **mineralisation**, or amorphous compounds that bond with minerals, particularly clays, to form the **clay–humus complex**. This considerably altered amorphous organic matter, which is more stable and resistant to biodegradation than fresh organic matter and is more slowly mineralised, is **humus** (*sensu stricto*) and the term **humification** can be used for all the processes involved in changing fresh organic matter into humus.

It should be pointed out that the term humus is often used in a broader and more ecological sense to refer to all organic horizons of the soil, including both the litter and humus horizons (L and A0) that overlie the mineral soil, as well as the mixed horizons (A1 or Ah) which are particularly rich in altered amorphous organic matter. It is in this situation that **raw humus** is used as a more precise term for slightly altered organic matter with little amorphous material.

   **Litter decomposition** is the initial phase of humification, in which mineralisation is dominant while the organic molecules that are not mineralised are **simplified**, often becoming soluble. In contrast, the subsequent phase of *humification* is dominated by the microbial or physicochemical build up of new molecules. In practice, it is almost impossible to distinguish these two phases. (A detailed study of the morphological classification of humus and the microbiological processes is given in Berthelin & Toutain 1982.)

### 1   Aspects of litter decomposition and humification in soils
Soils developing under natural vegetation show considerable differences, in both the manner and the speed at which these two processes operate, that are clearly reflected in the morphology of the humus-rich horizons.

**Litter decomposition.** Litter decomposition occurs at a very variable rate (depending on the ecological nature of the humus, particularly that of forests) that is to be seen in the variable time over which fresh litter persists. In environments of low biological activity, this decomposition is slow, and often a *mor* A0 horizon, overlying the mineral soil, is developed which is characteristically a very acid, brown or black, fibrous organic material. In the case of *moder* (Kubiena 1953), as the biological activity is a little greater, the litter and the A0 horizon are only 2–3 cm thick and are less clearly demarcated from the underlying mineral soil. Finally, in the case of *mull* the humus is biologically active and decomposition is rapid. There is practically no A0 horizon, almost all the humus is incorporated in the mineral soil and it forms an A1 or Ah horizon characterised by *clay–humus* aggregates forming a stable crumb structure.

Several authors have compared the speed of decomposition of forest litter in ecologically different humus types: in very active mull (leaves of alder and ash) the litter decomposes in less than a year. According to Toutain (1974), beech litter giving an acid mull decomposes in nearly two years, but if it forms a moder the decomposition takes seven years. Finally, pine litter forming a thick mor (5–10 cm) takes several decades to decompose and it forms an A0 horizon rich in intermediate compounds with a dark colour and fibrous structure (Guittet 1967).

**Humification.** The humification of the A1 or Ah mixed horizons shows even greater differences, for these are related to the type of neoformed compounds present and their variable resistance to biodegradation which is reflected in differences in horizon thickness, colour, structure etc. Thus, the A1 horizons of moders and mulls have a different structure, the clay–humus aggregates being better developed and more stable in the second case. Certain atlantic-type mulls have little colour and thickness and are relatively poor in organic matter (5%), which is a reflection of a weak resistance to biodegradation, for in this case the mineralisation of the neoformed amorphous compounds is almost as rapid as that which occurs in the litter. In contrast, a chernozem mull is characterised by a black A1 or Ah horizon, 60–80 cm thick, with a very strong crumb structure, which can be taken as indicating that the humic compounds have a much greater resistance to biodegradation in this soil than in atlantic-type mulls.

## 2   The different humus fractions

It is not possible to consider the dynamics of soil organic matter without referring briefly to the characteristic fractions of humus and how they are obtained. As far as the techniques of fractionation are concerned, the reader is referred to Henin and Turc (1950), Tiurin (1951), Kononova (1961), Manil (1961), Manil *et al*. (1963), and the various publications of the Centre of Biological Pedology (CNRS Nancy).

First of all it must be emphasised that often the methods of fractionation give results that are approximations and are difficult to interpret. For example, in the mixed horizons, the separation of fresh and humified organic matter by heavy liquids (bromoform-benzene, density 1.8) gives results that vary with the intensity and duration of the shaking; it is the same when **ultrasonics** are used to separate **inherited humin**. As a final example, the extraction of humic (AH) and fulvic (AF) acids by alkaline reagents requires great care to avoid the oxidation of fresh organic material which increases the quantity of compounds extracted by this method very considerably; therefore it is necessary to adhere strictly to a carefully worked out method.

**Fulvic acids and humic acids.** Fulvic acids (AF) and humic acids (AH) are generally extracted by a sodium pyrophosphate solution or sometimes by a dilute solution of sodium hydroxide. For some soils, the quantity (and also the type) of AF and AH obtained varies considerably according to the reagent employed and also its pH – it is greater at pH 10 than at pH 7; for the humus of andosols, it increases greatly between pH 10 and pH 12. If the results of AF and AH extractions on the whole soil and on fresh and humified organic matter fractions (separated by differential density) are compared, they are found to be different.

Humic acids are precipitated by $H_2SO_4$ at pH 1, which does not affect the fulvic acids. Two major groups of humic acids can be distinguished:

(a) *Brown humic acids*, which have a light colour, are only slightly condensed and are relatively *unstable*. They are weakly bonded to clays and are flocculated slowly by highly concentrated solutions of calcium ions; when subjected to paper electrophoresis, they migrate towards the anode (Jacquin 1963).

(b) *Grey humic acids*, which have a darker colour and are more condensed, are strongly bonded to clays; flocculation by calcium ions is very rapid even at low concentrations; when subject to paper electrophoresis, movement is very weak or completely absent. They are very *stable* and resistant to biodegradation.

**Humin.** The non-extractable fraction (humin), which is residual after the extraction of humic and fulvic acids, has considerable importance, for it often amounts to some 50% or even 70% of the total organic matter. As yet it has been studied to only a slight extent, doubtless because of the difficulties encountered in its extraction and separation from the mineral material. It is known to be very heterogeneous and there are several kinds of humin which have very different behaviour and origins (Duchaufour 1973).

(a) *Inherited or residual humin* ($H_3$ in Figs 2.1, 4 & 8, and Table 2.2) is closely related to fresh organic matter; is mainly composed of lignin affected by demethoxylation and oxidation with an increase of the COOH groups; it forms bonds of little stability with clay, which can be separated by the combined use of ultrasonics and density.

(b) *Insolubilised humin* is formed by the precipitation and irreversible insolubilisation of soluble precursors (phenolic compounds often bound to peptides); it is often separated into two parts, one ($H_1$) extractable with sodium hydroxide after treatment with dithionite and HCl–HF, the other part ($H_2$) remaining non-extractable (Védy 1973, Toutain 1974).

(c) *Microbial humin* (Guckert 1973) is a fraction composed of polysaccharides, polyuronides and amino sugars produced by microbial neo-formation in a very active environment.

(d) *Developed humin* (by maturation): humin with aromatic rings, very strongly polycondensed, poor in functional groups and resistant to normal extracting agents.

It is a remarkable fact that, as shown by Kononova (1961), most of these humic compounds have structures in which the same elementary units are involved, no matter what the size of the molecule (molecular weight varies from 1000 for some AFs to more than 100 000 for some grey AHs) and they consist of a roughly spherical **aromatic nucleus**, derived from phenolic or quinone compounds by polycondensation, surrounded by variably branching chains, with a peptide or saccharide base. The relative importance of the chains and nucleus varies. For example, in some fulvic acids the nucleus is of little importance while the chains are dominant. Conversely, the nucleus is more important in humic acids, where its increase in size can be correlated with the change from the more mobile and less condensed types, brown AH, to the more condensed, grey AH (Scheffer & Schluter 1959, Kononova 1961).

Many authors have attempted to characterise humic compounds in terms of their molecular sizes, by comparing their molecular volume with that of type compounds of known molecular weights. In fact, this has caused confusion between molecular volume and molecular weight as the correlation between the two is not very great; for example brown AHs with a great number of more or less folded chains often have molecular volumes greater than those of grey AHs, even though their aromatic nuclei are, in fact, less condensed (Orlov *et al.* 1973).

The use of the term *humic compounds* by Russian authors (Kononova 1961) is restricted to those which have an aromatic nucleus with a variable degree of condensation. In these terms, not all of the humin types discussed above would be humic compounds, for part of the inherited humin and all of the microbial humin would be excluded. However, a broader definition is to be used in this book in which humin is defined as being all of the non-extractable compounds that have been subject to biochemical change and are bonded to minerals.

## 3   The phases of humification

Litter decomposition is essentially a rapidly operating biological process, humic compounds being formed almost contemporaneously. Because, as stated previously, these two processes cannot be separated either in space or time, they will be looked at together under the heading of **biological humification**. But frequently a much slower second process occurs that particularly affects humic compounds and which is dependent not only on the mineral make up of the soil and the biological conditions, but also, and more particularly, on the *climatic conditions*. This process of **humus maturation** will be investigated in Section IV of this chapter. As will be seen, maturation is also one of the essential causes of soil zonality (which will be studied later in the book).

## II    Biological humification: biochemical processes

Litter is formed from all the dead plant debris which accumulates on the surface of the soil and is subject to a variable rate of decomposition. It is made up of two kinds of material: (i) soluble (or at least can become so very rapidly), such as carbohydrates, tannins, peptides or amino acids resulting from the rapid hydrolysis of protoplasmic proteins; or (ii) insoluble cell membranes composed of lignin, cellulose and hemicellulose that are decomposed very gradually into neoformed soluble compounds but, in many cases, a part remains insoluble and is subject to yet further slow changes. This insoluble fraction remains on the surface or, in certain cases, is incorporated into the soil by mechanical means such as by worm action. However, the soluble compounds – both those *inherited* (existing previously in the cell) as well as those that are *neoformed* – are generally separated fairly quickly from the insoluble residue and they blend with the mineral fraction: *this separation in space of residual insoluble materials from soluble materials, either inherited or neoformed, is an essential factor in humification*.

### 1    Decomposition of the insoluble membranes

These membranes are composed of *celluloses* (and hemicelluloses) and also of *lignin*. Some are impregnated by *bitumens*, in the form of waxes and resins, which inhibit their decomposition (Jambu 1971).

**The hydrolysis of cellulose** or **cellulolysis.** Cellulose and hemicelluloses are subject to a more or less powerful hydrolysis which results in a simplification of the molecules, even to the extent of producing completely soluble sugars:

cellulose $\longrightarrow$ oligosaccharides $\longrightarrow$ soluble sugars

Most of these compounds form the energy source of soil micro-organisms and as a result are completely mineralised, particularly under aerobic conditions (high $CO_2$ production). Others are used in the formation of microbial polysaccharides which, while present to only a slight extent in humic acid chains and a little more in fulvic acids, are the main constituents (80–90%), together with amino sugars, of *microbial humin* (Guckert 1973) which can amount to 10% of the total humin fraction of soils that are both well aerated and biologically very active. In addition, in biologically very active, slightly acid conditions which are rich in easily assimilable peptides and amino acids, certain of these carbohydrates are changed by fungi (such as *Epiococcum nigrum*, *Hendersonula toruloidea*, etc.) into coloured compounds with an aromatic ring structure which, as a result of oxidation by polyphenoloxydases, develop rapidly into dark-coloured humic compounds (microbial melanines: Haider & Martin 1967, Martin *et al.* 1967, 1972).

In contrast, in badly aerated anaerobic conditions the cellulose and saccharides undergo a totally different development in which they are decomposed into aliphatic acids of various kinds and then to methane ($CH_4$) and $H_2$, resulting in a marked lowering of the redox potential. Thus, in these very reducing conditions, cellulose-type compounds disappear without leaving any trace.

**The hydrolysis of lignin** or **ligninolysis**. Lignin is generally degraded more slowly than cellulose, except in certain aerated, nitrogen-rich, moderately acid environments (acid mull) which are very favourable to basidiomycetes of the **white rot** type (Mangenot 1974, Toutain 1974).

White rot is responsible for the rapid disappearance of certain litters in non-calcareous conditions. The ligninolysis which it causes produces small, often soluble phenolic molecules (monomers or polymers) which, as will be seen, are the basic units in the development of insoluble humic compounds.

**Soft** or **brown rot** (Duncan 1959), which seems to be characteristic of neutral or calcareous conditions, changes lignin molecules more gradually and then not into a soluble form but only into more or less dark-coloured materials, by the breaking of side chains and some of the phenolic rings (Mangenot 1974).

The acidity and biological activity of the environment as a whole also play an important part in the process of adsorption of peptide chains by lignin while it is being altered. In the most biologically active environment, this adsorption is important, and undoubted **ligno-protein compounds** are formed (Grabbe & Haider 1971, Sinha 1972, Dupuis & Cheverry 1973). In contrast, this process is reduced in environments with low biological activity such as moder or mor.

The question can be asked, what is the relative speed of lignin and cellulose biodegradation in different humus types? Toutain (1974) and Selmi (1975) tried to answer this question by experiments on beech litter in various environments.

(a) In a calcareous environment, cellulose is biodegraded much more rapidly than lignin, which is, in contrast, preserved within the inherited humins; the water-soluble materials formed in this process are richer in saccharides than in phenolic compounds.
(b) In an acid mull, or moder, the lignin or cellulose of the litter is degraded at about the same rate. However, this is not the case in the A1 horizons where ligninolysis is clearly more important than cellulosis. In addition, in an acid mull, ligninolysis occurs at about five times the rate that it does in a moder, which is reflected in the fact that, despite the production of approximately equal amounts of water-soluble compounds in both cases, those of the acid mull are twice as rich in phenolic compounds as those of the moder (in $g/m^2$).

## 2   Development of soluble or slightly condensed compounds

Some of these compounds are directly inherited from plant cells; others, as stated above, are gradually neoformed. In both cases they infiltrate more or less rapidly into the mineral material. According to Toutain (1974), they can be classified into two groups which have a very different fate in the soil: one group (the saccharides, peptides, etc.) is an energy source which is rapidly biodegraded in being mineralised or incorporated into microbial tissue. The other group consists mostly of variably condensed phenolic compounds that are more resistant to microbial biodegradation, which some of them avoid by polycondensation of the ring structure and then function as basic units for the building of insoluble humic compounds.

These two types of compounds frequently react with one another, as was first shown by Handley (1954) when he demonstrated the formation of polyphenol–protein complexes in soil. In this reaction, certain phenolic compounds combine with proteins or peptides which they protect against biodegradation. They act as a kind of tannin, or indeed as a poison, which prevents, or at least retards, nitrogen mineralisation. Those phenolic compounds which have a tendency to polymerise by oxidation rapidly lose their tanning (or toxic) properties in forming coloured humic compounds (Coulson et al. 1960).

As the properties of these water-soluble phenolic compounds are mainly dependent on their origin, this factor will form the basis of their detailed study. There are, in fact, three possible origins:

(a) inheritance, where the compounds already existed in the protoplasm;
(b) biodegradation of lignin (white rot);
(c) microbial neoformation.

**Phenolic compounds inherited from cells.** From the time of leaf fall in forests, in the autumn, these compounds are carried *en masse* down into the soil by infiltrating water (Toutain 1974), i.e. they play a fundamental role in the initial phase of litter decomposition and humus formation. They can act either as a stimulant or as a curb on biological humification.

The nature and composition of these water-soluble compounds are evidently closely related to those of the litter and hence of the vegetation, while in terms of its mineral composition the soil is only slightly involved. Bruckert et al. (1971) showed that, for Norwegian pine and ash, the nature of the phenolic polymers contained in rain leachates and their tendency to polymerise depend to a much greater extent on the plant species than on the mineral environment, no matter whether it is acid or calcareous. This important problem will be discussed again in Section III of this chapter; at this juncture it will suffice to contrast the tanning action of the **diphenols** of *Calluna* leaves, which prevent all nitrogen mineralisation by forming very resistant **diphenol–protein complexes** (Schvartz 1975) with the melanising phenolic compounds of walnut (Andreux et al. 1971a,b, Mangenot et al. 1966) which have no tanning action and polymerise very quickly by auto-oxidation.

**Phenolic compounds resulting from the degradation of lignin.** By their gradual formation and their movement into the soil, these compounds take over from the inherited phenolic compounds, being more abundant in summer during the phase of strong microbiological activity. Their production reaches its maximum in very aerated, nitrogen-rich, moderately acid mulls, as a result of the action of white rot. They are slightly less abundant in moder and still less so in calcareous mull (Toutain 1974, Selmi 1975).

This case differs from the previous one in that the nature and composition of these neoformed compounds are not related to those of the litter, in fact they vary little from one plant species to another, *but they are closely related in their development to the general environment and in particular to the composition of the mineral material*; for example their rate of insolubilisation in the soil is dependent on the pH, then aeration and, in acid conditions, on the presence of fine clays and free iron. The tanning action of these compounds is more limited than that of some of the inherited compounds and this is also directly related to the insolubilising action of the environment. These factors will be discussed again in Section III of this chapter.

**Phenolic compounds of microbial synthesis.** These are also neoformed compounds but in this case they are produced by microbial synthesis – being the *microbial melanines* of Martin and Haider (1971) which have already been mentioned. They have also been observed by Kilbertus (1970) and appear under the stereoscan as black granules within litter which has been subject to a very active biodegradation; they are formed, in fact, only in a neutral environment rich in assimilable nitrogen and with strong biological activity. According to Filip *et al*. (1974), their composition is very different from those phenolic compounds derived from lignin: *their tanning action preventing nitrogen mineralisation, and thus humification, is very weak or non-existent, while on the other hand, their tendency to polymerise and to form humic compounds is very high*.

In conclusion, it is evident that the production and development of water-soluble phenolic compounds, either from litter or by neoformation, are of considerable importance for they have a decisive influence on pedogenic development. In certain instances, *they prevent all nitrogen mineralisation and slow down humification; in others, on the contrary, they are rapidly insolubilised forming humic cements necessary for structural development and which, at the same time, do not in any way prevent nitrogen mineralisation*.

## 3   The three avenues of biological humification

This concluding part of this section will consider biological humification in terms of three fundamental processes (Duchaufour 1973), in order of decreasing importance.

**Insolubilisation of phenolic precursors (indirect humification).** This has been discussed at length and it occurs at a very variable rate depending on the soil conditions and generally involves condensation of phenolic ring structures. Several processes appear to be involved: (i) increase in the charge of the cations involved leading to a precipitation of the complexes; (ii) adsorption by colloidal or crystalline mineral compounds such as allophane, clay and active iron; (iii) loss by biodegradation of side chains and functional groups (Bruckert 1970) leading to a decrease in charge; and (iv) polycondensation of ring structures (plant melanines, Carballas *et al*. 1972).

*No matter what, the role of mineral compounds such as active calcium carbonate, allophane, clay, and heavy cations, is fundamental to these processes and will be examined in detail later.*

This insolubilisation is particularly characteristic of mixed A1 (or Ah) horizons in biologically active mull soils, producing humic cements that aid clay–humus complex formation, and thus of what are called **constructive** fabrics. Earthworms play an essential part in this process. However, a minor fraction of these phenolic precursors is absorbed by more or less decomposed litter, as shown experimentally by Selmi (1975), which accounts for the small quantities of extractable compounds present in the L and A0 horizons. In environments that biologically are only slightly active and are poor in clays or in cations, this insolubilisation occurs slowly, after more or less considerable movement within the profile has taken place, which is the cause of podzolisation (see Ch. 3, Sec. II).

The humic compounds formed by the insolubilisation and condensation of the phenolic compounds are for the most part extractable by normal reagents – *these are the fulvic and humic acids*. But, depending on the nature and the quantity of the agents of insolubilisation, the ease of extraction of the humic compounds varies and is never complete: raising the pH of the reagent often increases the amount extracted. If iron and fine clays are responsible for the insolubilisation, the amount extracted remains low; pretreatment with dithionite and HCl–HF increases the amount extractable by alkaline reagents, but in all cases there remains a high proportion of non-extractable humin (insolubilised humin). It is possible to represent the effect of the mineral factors in the following way:

$$\text{phenolic compounds} \left\{ \begin{array}{l} \rightarrow\ \text{AF}\ \rightarrow\ \text{AH}\ \rightarrow\ \text{humin} \\ \rule{6cm}{0.4pt}\!\!\!\uparrow \end{array} \right.$$

**Humification by inheritance** (direct humification). This concerns pre-existing insoluble compounds of cell membranes (particularly lignin) which are concentrated by the selective removal of cellulose and transformed by the acquisition of $-COOH$ groups at the expense of $-OCH_3$ groups, which gradually disappear. *Such humic compounds are not extractable and they form inherited (or residual) humin* ($H_3$ in Figs 2.1, 4 & 8 and Table 2.2). In this case, the humification is incomplete; for example, the inherited humin of a calcareous environment (or one saturated in calcium) is simply the result of an incomplete degradation of membranes by brown rot, while that of a very acid environment (moder or mor) forms the major part of the **coprogenic** aggregates of enchytraeid worms and collembola, which are finer and less coherent than those formed by earthworms in mull (Brun 1978).

This inherited humin is not easy to study, for only with difficulty can it be separated from fresh organic matter, on the one hand, and from humin bound to clays, on the other. Within the A1 (or Ah) horizon it is possible to separate the inherited humin from clays by ultrasonic dispersion (using the residual fraction after heavy liquid separation); this is because the bonding between inherited humin and clay is less stable than that between insolubilised humin and clay (Satoh & Yamane 1972, Védy 1973).

Within the litter of purely organic horizons (A0 of mor and peats), it is not possible to separate inherited humin from fresh organic matter by chemical means. Nevertheless, it is possible to follow the formation of inherited humin indirectly by measuring the increase in the exchange capacity of the litters (or peats), as a reflection of their degree of decomposition, for as this increases so do the —COOH groups. This is the method used by Védy (1973) in investigating moder and mull and by Menut (1974) for peats. For example, Védy showed experimentally that the exchange capacity of an acid mull litter was double that of a moder after they had developed under similar conditions. Menut showed that there is a rapid increase in the exchange capacity of peat when it dries out as a result of the lowering of the water table.

Inherited humin, as will be seen, is particularly important in both moder (or mor) and carbonate mull. However, the biological activities of these two environments are very different and it is probable that the nature of the inherited humin differs in these two cases. This is suggested by the fact that there is much more peptide nitrogen fixed by lignin in the calcareous mull than in the moder, or mor (Grabbe & Haider 1971, Toutain 1974).

**Microbial neoformation** (Guckert 1973). This involves microbial polysaccharides synthesised by micro-organisms in a very active environment. Only a few of these are incorporated in the side chains of extractable compounds (AF); most of them (80–90%) are not extractable and they form microbial humin. As stated previously, some authors have objected to this being called humin as it is not of a phenolic type. However, these compounds play a far from negligible role in soils, particularly in those that are cultivated, where these long chains of saccharides and uronides form a network which bonds with clay (Monnier 1965, Guckert 1973) and is involved in the formation of a very porous crumb structure. While these compounds are physically resistant, they are rapidly biodegraded (in comparison to the phenolic types of humin), to be reformed again in the next period of intense biological activity.

Note that the **microbial biomass**, which is difficult to separate, is often equated with microbial humin even though it is of a very different nature.

## III    The effect of environmental factors on biological humification

Environmental factors that influence biological humification can be classified into three groups: (i) those of soil climate, (ii) those of the mineral material, (iii) those of litter composition and thus of vegetation.

The soil climate, which integrates those factors controlled by the general climate (temperature) and those internal to the soil (aeration), has a *dominant influence* which is relatively independent of the other environmental factors. This is not the case for the mineral composition of the soil and that of litter which are more closely interrelated; for instance, the types of humus of the temperate zone are determined by their interaction and will be discussed in the final part of this section.

## 1   Effect of the soil climate

Soil climate results from the interaction of general climatic factors, such as temperature, and local conditions involved in aeration.

**General climate.** It is the *mean temperature* which is the most important factor for, as a rule, *the speed of litter decomposition is roughly proportional to the mean temperature*, provided there is sufficient moisture. Thus in boreal climates, humus with slow decomposition (such as mor) is dominant, while in the humid Tropics only a very active mull is to be found.

In regions of uniform temperature, alternating seasons of very different moisture status increase the speed of decomposition of slightly altered organic matter (a factor that will be discussed again in the section on maturation). This is the reason why mediterranean climates are very favourable to rapid litter decomposition so that mor is practically never found, even in poor acid environments.

At high elevations (in mountains) the low mean temperature at cold sites favours the formation of the boreal type of mor (climatic alpine mor, often with a low C : N ratio). In contrast, at warm sites the strong insolation and radiation intensity lead to marked pedoclimatic alternations in temperature and humidity which accelerate litter decomposition, and a moder or a mull–moder is formed with little accumulation of organic matter.

The influence of rapid variations of microclimate with time is also important. For example, forest clearing (or fire) suddenly exposes a soil which has been previously shaded, causing an acceleration of humus decomposition, particularly of certain kinds of mor. These alternations of phases of shade and light also cause an alternation of the processes of accumulation and mineralisation. Thus raw humus is able to release its immobilised nutritive elements all at once and the normally slow cycle can be rapidly accelerated in certain periods.

**Local conditions of soil climate: aeration and water balance.** A certain amount of moisture is obviously necessary for good litter decomposition. In a very dry environment, moisture is lacking and humification is slow; thus in very dry mediterranean climates or on indurated limestones denuded by erosion and exposed to the Sun (temperate zone mountains) a **xeromoder** is formed.

Conversely, an excess of moisture completely saturates the soil pores with water, preventing oxygen circulation and lowering the Eh, which leads to a marked slowing down of the rate of litter decomposition. In these conditions a hydromorphic humus is formed, with an accumulation of a very slightly humified A0 organic horizon (peats) or, where seasonal phases of better aeration allow greater humification, a **saprist** (see Ch. 11).

The formation of acid humus such as **hydromoder** or **hydromor**, where hydromorphic conditions are not permanent, is characterised by a large-scale production of soluble organic compounds of all kinds which accumulate in the profile during the anaerobic phase (Bloomfield 1975). They migrate down-wards when the profile dries out and then behave as active agents of podzol-isation and complex formation. But they are frequently biodegraded during the summer when aeration is best (Gury 1976).

In an environment in which periodic water table variations are of such long duration that the anaerobic phases are insufficient to prevent a more or less strong humification and even maturation of the organic matter (to be dis-cussed later), an **anmoor** is formed (Kubiena 1953) in which the very abun-dant organic matter is mixed with the mineral material.

## 2 Effect of mineral material (Figs 2.1 and 8)

The percentage base saturation ($S/T$) and the pH of the soil environment on which a litter occurs obviously have an important influence on the general microbial activity and thus on the rapidity of biological humification. In gen-eral, in neutral or slightly acid environments it is more active than in very acid conditions. However, this fundamental relationship does not depend exclu-sively on the mineral materials but on their interaction with the biogeochemi-cal cycle of bases and that of humification itself. It is in fact a very complex chain reaction which will be studied in detail when the movement of material within soils and biogeochemical cycles are considered (see Ch. 3).

It has been noted already that a high level of base saturation is not abso-lutely necessary for rapid litter decomposition and mull formation, in temper-ate climates. If other environmental conditions are favourable, such as the aeration, litter composition and particularly the presence of active iron and fine clay, the action of white rot allows ligninolysis to occur and an *acid mull* develops, which is a particular type of forest humus characteristic of a very desaturated environment. In these acid forest mulls, where free iron and clay play a dominant role, humification is active but humic compounds do not accumulate as they are biodegraded rapidly. It is very different when other minerals are involved, such as active calcium carbonate or amorphous aluminium, for not only is a different direction given to biochemical and humification processes but also humic compounds accumulate in the profile as they are protected against subsequent biodegradation by the *stabilising* action (Laatsch 1963) of these two materials, which iron hydroxides do not have. Thus it is necessary to consider the role of mineral materials from two points of view: *direct action* on humification and *indirect action* by their pro-tective or stabilising effect.

**Active iron hydroxides.** These involve iron hydroxides bound to fine clays, around which they form a thin skin. As shown by Guillet *et al.* (1975), this is the only form of iron which has a positive influence on the process of humifi-cation.

*Direct effects*. Direct effects of iron on humification have been demonstrated both analytically (Souchier 1971, Toutain 1974) and experimentally by laboratory and field studies (Védy 1973, Toutain 1974). Védy's experiments

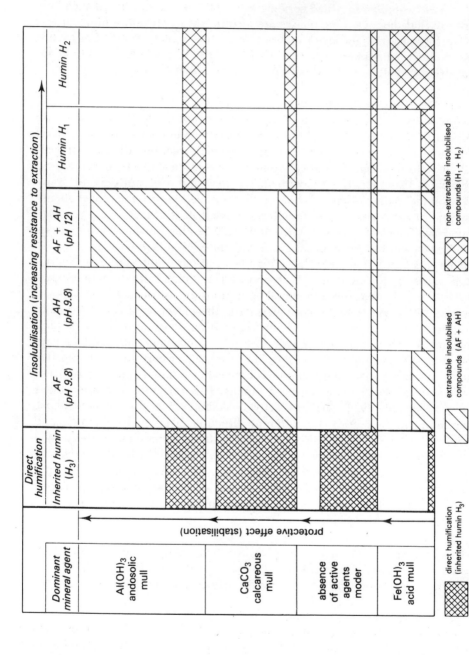

**Figure 2.1** Effect of mineral agents on humification (A1 horizons). Note that area is roughly proportional to average amounts.

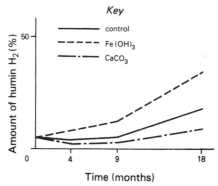

**Figure 2.2** Influence of the addition of ferric hydroxide or calcium carbonate on the formation of insolubilised humin ($H_2$) in an acid moder (Toutain 1974).

involved modified lysimetry on soil columns, which will be discussed later (see Fig. 2.4); Toutain added a precipitate of ferric hydrate to a beech moder placed in cylinders in the laboratory and in the field; by this means it was shown that, under the influence of iron, a moder can gradually take on the character of an acid mull. All these experiments gave absolutely consistent results: *the direct effect of active iron is to be seen in the great increase in the quantity of humic compounds produced by the process of insolubilisation; this includes both extractable compounds (AF and AH) and, more particularly, insolubilised humin*. It is this last fraction which is the most characteristic of the humus formed under the control of iron in an acid environment. As a result of an investigation of about 40 soils in Lorraine, Toutain showed that there was a perfect correlation between the amounts of free iron, of humus, and the content of insolubilised humin; in addition, experimental work enabled him to show *the extreme rapidity of formation of this non-extractable humin – two years being sufficient* (Fig. 2.2).

Thus, iron insolubilises precursor molecules in giving rise to two types of compound: one, which is in minor amounts, is extractable (AF and AH); the other, which is dominant, is non-extractable (humin). In both cases the iron is in a complexed (or chelated) state, but two different types of complex are involved (Theng & Scharpenseel 1975, Huang *et al*. 1977, Jambu *et al*. 1981, Schnitzer 1981).

(a) *Extractable compounds* (AF and AH) are present as 'salt complexes' (see Ch. 3) in which iron hydroxides are associated with complexed aluminous ions, the latter always being present in smaller amounts: these two cations are able to bind to the clays through —OH groups. A general formula can be suggested, as shown in Figure 2.3. *Thus iron and aluminium act as linking cations between the clay and humic compounds*.

(b) *Non-extractable compounds (insolubilised humin)*. These totally insoluble materials are as yet not well known; they are **adsorption complexes**, with a very high cation : anion ratio, in which iron hydroxides alone are

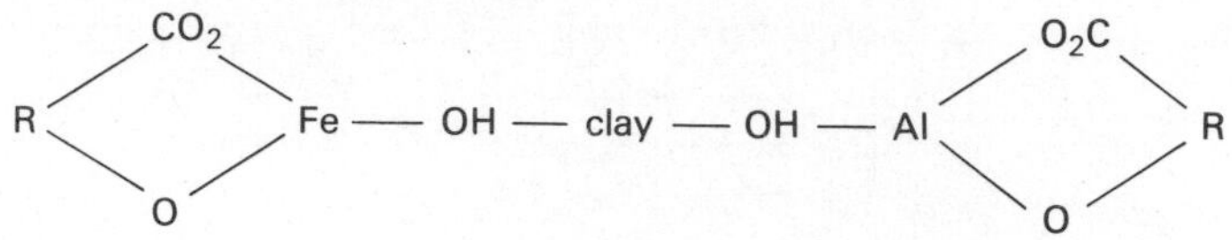

**Figure 2.3**

involved to the exclusion of alumina. The strong polycondensation of phenolic ring structures as well as the loss of reactive —COOH groups are catalysed by $Fe(OH)_3$ (Mayaudon *et al*. 1973, Huang *et al*. 1977).

STABILISING EFFECT. Although the direct effect of iron on humification is considerable (particularly in an acid environment), its indirect stabilising effect is, on the contrary, of little significance. All the humic compounds insolubilised by iron remain susceptible to biodegradation (even humin); their biological turnover is rapid and they do not accumulate in acid forest mulls. Toutain (1974) verified, by incubation experiments, the high rate of mineralisation of C and N in this type of humus. Finally, Andreux (1978) showed experimentally that ferric hydrates catalyse the opening of phenolic rings that accelerate the biodegradation of this humus.

Note that the clays of the atlantic brown forest soils, which are dominantly illite-vermiculites, react with humus mainly through the intermediary of the active iron which is bound to them so that when they occur in only moderate quantities in the profile, their stabilising action is very limited. It is different when semi-swelling (interstratified) clays are involved and particularly so in the case of swelling clays, which are abundant in certain profiles (pelosols, vertisols). Not only is their stabilising action more marked, but also, in certain climates, they catalyse the process of maturation (which will be discussed in Part IV).

**Active calcium carbonate** (Chouliaras *et al*. 1975). This involves finely divided calcium carbonate which is capable of being chemically reactive, particularly with the $CO_2$ and organic acids of the soil solution: the calcium thus dissolved acts either directly by insolubilising the precursors (Lineres 1977) or by reprecipitation of calcium carbonate as a protective skin.

*Direct effects*. Active calcium carbonate favours biological activity in the soil and allows a rapid initiation of litter decomposition but, contrary to what has been accepted for a long time, *biological humification is blocked at an early stage*. This is to be seen in the abundance of slightly developed humic forms in the analysis of such organic matter which consists of (i) calcium-saturated fulvic acid, (ii) fresh organic matter (which is often in excess of 50% in a rendzina A1 horizon), and (iii) inherited humin ($H_3$).

The effect of active calcium carbonate has been investigated in greater detail by experiment (Toutain 1974), as well as by using the stereoscan microscope to compare humic fractions extracted before and after decarbonation (Chouliaras *et al*. 1975). As stated above, the effect is twofold: (i) Calcium carbonate rapidly insolubilises the phenolic precursors while at the same time

being partially dissolved, which in effect accelerates the process of decarbonation (Le Tacon 1976; this will be referred to again in Ch. 3). These insolubilised compounds are only slightly transformed and for the most part they occur
as fulvic acids. Little insolubilised humin is present in calcareous soils, for
(Toutain 1974) the addition of active calcium carbonate to a moder causes
this type of humin to decrease (Fig. 2.2). Because of these circumstances,
almost all of the insolubilised compounds are extractable by complexing
agents such as pyrophosphates, which is not the case when iron is responsible
for the insolubilisation. Where carbonates are very abundant, a preliminary
decarbonation with gentle reagents is necessary to achieve complete extraction, which at the same time causes some of the insolubilised AF to be freed.
(ii) On the other hand, active calcium carbonate is conducive to direct humification, for it forms a protective skin around fresh or slightly altered organic
matter which almost completely prevents subsequent development and gives
rise to inherited humin by undoubted **sequestration**. It should be remembered
that lignin is the essential constituent of inherited humin for it is almost
integrally preserved, only being subject to oxidation (Selmi 1975) while the
celluloses are being hydrolysed.

*Stabilising effect.* In contrast to the effect of iron, active calcium carbonate
and to a lesser degree *exchangeable calcium* in a saturated environment
(Lineres 1977) are very effective stabilisers, protecting organic matter against
microbial biodegradation. Thus in the case of the calcareous mull of a
rendzina, the accumulation of organic matter per unit of surface area is ten
times greater than for an acid mull (this is not so if based on simple percen-
tages as in Figs 2.1 & 8), which is reflected in the rate of nitrogen and carbon
mineralisation in calcareous or eutrophic mulls being about half what it is in
acid mulls (Herbauts 1974, Le Tacon 1976). As far as nitrogen is concerned,
the peptide compounds are insolubilised in two ways – either by blockage
within the AF precipitated by the calcium carbonate for the most soluble
compounds, or by sequestration within carbonate coverings for those which
form bonds with the insoluble lignin.

**Amorphous alumina and complex aluminium ions** (Hetier 1975)

*Direct effects.* According to Hetier, amorphous alumina is the active part of
the very abundant **allophanes** of andosols. This alumina bonds strongly with
organic matter, particularly the precursor compounds, which results in an
almost *immediate insolubilisation* and a *stabilisation* against microbial biodegradation. Both of these effects were demonstrated by Hetier in incubation
experiments with allophane and previously labelled water-soluble compounds
from maize litter.

 There are two kinds of insolubilised materials: fulvic acids associated with
small amounts of mineral material, which is extractable at a pH of 9.8, and
humic acids associated with a considerable amount of more condensed
allophanic material and extractable at a pH greater than 11. Extraction is
almost complete owing to the solubility of aluminium in alkaline conditions.

However, *alumina does not cause, as does iron, the formation of non-extractable insolubilised humin*; the minor amount which is contained in andosols is in fact bound by iron. In andosols, the presence of a minor but by no means negligible amount of *inherited humin* should be noted, which appears to have been sequestered within aggregates of the more condensed form of allophane (see Ch. 6).

*Stabilising effect.* The main fact concerning amorphous alumina is its great protective effect against microbial biodegradation: *the stabilising effect is at a maximum* and organic material accumulates in a spectacular fashion in andosols. Hetier (1975) showed that a considerable proportion of the organic matter accumulating in the central part of allophane–humus aggregates is of the same age as the soil (up to 4000 years).

**Note: the way in which complex aluminous ions act**
One should not confuse the action of the amorphous alumina of andosols, where its presence as a gel is responsible for the adsorption of humic compounds, with that of the aluminous ions ($Al^{3+}$, $Al(OH)^{2+}$, $Al(OH)_2^+$) of the absorbent complex of acid soils, which react mainly through their positive charges (particularly $Al^{3+}$, which is relatively abundant in very acid conditions) and maintain the clay–humus complex in a flocculated state. In addition, when the complexes age, a part of the $Al^{3+}$ becomes the complex ion $Al(OH)_n$ which acts as a 'linking cation' between clay and humus (Schwertmann 1969) and thus reinforces the action of iron. It should not be forgotten, however, that generally these aluminous ions are present only in small amounts compared to the iron oxides which thus play the main part in acid conditions.

## 3   Effect of litter composition
The properties of litter evidently reflect those of vegetation, which thus has an important effect on pedogenesis.

Among the properties of fresh organic matter, the closely interconnected factors of nitrogen content and the kind of water-soluble compounds appear to play a dominant role. Thus, slightly lignified young tissues are rich in nitrogen and water-soluble compounds, which are favourable to microbiological activity and rapid decomposition, while old, hard, lignified tissues are the opposite and decompose slowly, forming mor or at least moder. The presence of bituminous material (waves and resins) in some coniferous mors is an additional cause of slow decomposition (Jambu & Righi 1973, Fustec-Mathon *et al.* 1981).

**The role of nitrogen.** The role of nitrogen as an accelerator of humification is now well known (Jacquin 1963, Dell'Agnola & Ferrari 1979).

(a) Plant tissues with low C : N ratios (less than 25) decompose quickly, humify well and at the same time free a large proportion of mineral nitrogen (**ammonification**); this is the situation for ameliorating litters such as those of alder, false acacia, ash and grasses and legumes among substorey plants (Wittich 1961, Lemee 1973, 1975).
(b) Litter of forest species with an intermediate C : N ratio (30–45), such as oak and beech, decompose more slowly, liberate little mineral nitrogen

and give rise to very different kinds of humus depending on the mineral composition of the soil; either mull or moder (sometimes, in the case of beech, to a mor).

(c) Species forming a litter with a high C : N ratio, such as most of the conifers (pine with a C : N ratio greater than 60), heather and ericaceous plants, or **acidifiers**, in general decompose very slowly and usually form a raw humus or mor in which nitrogen is not mineralised.

**Water-soluble compounds.** The properties of these compounds are closely related to the nitrogen content of the litter: the ameliorating species are rich in water-soluble compounds which are generally capable of stimulating bacterial activity – saccharides, amino acids etc.; in addition *the soluble phenolic compounds produced by these species have a strong tendency to polymerise (or biodegrade); they do not have a tanning reaction with regard to the proteins.* Some **melanising** nitrophilic species (nettle, walnut, melandryum, studied by Mangenot *et al.* 1966 and Carballas *et al.* 1972) produce, as a result of rapid auto-oxidation, dark-coloured humic acids (plant melanines).

Acidifying plants, such as the ericaceous species (*Calluna* for example), are not only poor in nitrogen but they contain phenolic compounds (particularly diphenols) that form *extremely resistant phenol–protein complexes that totally block nitrogen mineralisation* (Handley 1961, Schvartz 1975).

The experiments of Handley (1961) are proof of this: polyphenol–protein complexes were prepared by adding casein to leaf extracts of acidifying mor species (*Calluna vulgaris*) or ameliorating mull species (*Circaea lutetiana*). These complexes were used as a nitrogen source for birch seedlings, the subsequent analysis of which showed the importance of the nitrogen source, for the *Calluna vulgaris* complexes mineralised their nitrogen much more slowly than the *Circaea lutetiana* complexes. The author attributed these differences to the nature of the tannins – condensed in the first case, hydrolysible in the second.

Schvartz completed these experiments by showing that the soluble extracts of *Calluna vulgaris* were indeed rich in diphenols (catechol, Brachet 1975); the addition of a labelled diphenol (hydroquinone) to soluble ash extracts followed by incubation with a mineral soil horizon (acid brown soil) resulted in effective inhibition of nitrogen and carbon mineralisation.

It is a result of its action on *nitrogen nutrition of soil micro-organisms* that litter plays a determinative role in biological humification. It should be remembered that good nitrogen nutrition is necessary for white rot to be active and even more so for fungi which synthesise humic compounds of the microbial melanine type, in non-acid conditions. Thus, contrary to long-held opinion, nitrogen is a more effective stimulant of biological humification than the basic cations that determine the neutrality of the soil, which is demonstrated by the occurrence of acid mulls that are biologically very active despite being very undersaturated with respect to calcium.

## 4   Conclusion: Interaction between the vegetation and the mineral constituents of the soil

Excluding certain soils rich in active calcium carbonate (rendzinas) or in amorphous alumina (andosols) in which the mineral constituents play the dominant role, it can be said that generally it is the vegetation that is the determinative factor in humification. This is well seen in certain long-established facts: (i) convergent evolution of humus formed under the same vegetation on different parent rocks; (ii) conversely, the degrading influence on humus when secondary vegetation replaces the primary vegetation (climax), resulting in acidification and unfavourable humus changes (mor of *Calluna* heaths replacing the mull of primary oak forests). However, apart from this particular example, caused by man's disturbance of the natural equilibrium, the vegetation is not independent of the mineral composition of the soil, for, in general, favourable mineral conditions correspond with an ameliorating vegetation, and unfavourable mineral conditions with an acidifying vegetation in which both factors have an effect on humification. For example, on a rich, well aerated mineral soil material with calcium reserves, natural vegetation is generally ameliorating and the humus is an eutrophic mull.

In contrast, on acid parent rocks with very poor calcium reserves, the plant–mineral material interaction becomes much more complex and difficult to interpret. Thus, in this situation, if sufficient free iron bound to fine clay is present, the rate of humification is increased so as to be comparable to the effect caused by the presence of calcium (or magnesium) ions. In addition, by the indirect process of humification which it favours, free iron is responsible for a concentration of calcium which raises the pH and thus accelerates microbial activity, even on material very poor in this element: this important process will be discussed later (see Ch. 3, Sec. III).

The experiments of Védy (1973) and Toutain (1974) showed that *the effect of free iron bound to clay is a limiting factor in acid conditions. If there is insufficient of it in the soil, no effect is to be seen, even when the vegetation is ameliorating*, and the humic precursors remain soluble and move to lower horizons (**podzolisation**). However, given sufficient free iron bound to clay, the litter is again determinative, with a mull forming under an ameliorating litter, while a moder forms under an acidifying one (Fig. 2.4).

Modified lysimetry field experiments (Védy 1973) allowed the decomposition of two types of litter to be followed over a period of five years – one ameliorating (spruce, beech and fescue grass), the other acidifying (pine and *Calluna*) – on two different materials – one granitic regolith very poor in clay and active iron, the other the middle sandstone of the Vosgesian Trias, richer in these elements (7% clay).

Figure 2.4 shows the results obtained: the humus formed is either moder (under both litters on granite regolith and under acidifying litter on sandstone) or mull (under ameliorating litter on sandstone). When subject to density separation, the light fraction is important in the moder while in the mull it is not (note that this fraction does not appear on the diagram); the inherited or residual humin ($H_3$) is relatively more important in moder than in mull; in contrast, the insolubilised humic compounds ($H_1$ and $H_2$) are twice as abundant, and extractable AF and AH three and a half times as abundant in mull compared to moder. It can be concluded that the active iron bound to clay is indeed a limiting factor, vegetation type being of no importance on granitic regolith poor in these materials. However, on sandstone which contains sufficient of them, the vegetation is the determinative factor, forming either a mull (ameliorating litter) or a moder (acidifying litter).

Toutain (1974) studied experimentally the decomposition of a single forest species – the beech – which, depending on the mineral material, gives rise to very different humus types (acid mull on

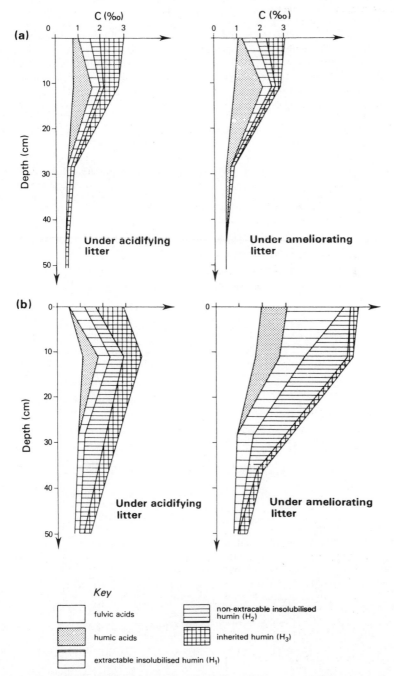

**Figure 2.4** Modified lysimetry experiments (Védy 1973); results after five years: (a) profiles on granite; (b) profiles on sandstone.

materials sufficiently rich in free iron bound to clay, moder on material poor in iron; this occurs on two different types of Rhaetic sandstone of the Lorraine Lias). The litter composition is absolutely comparable at the two sites; however, the humus is of a different composition in each case, similar to that obtained from the two types of litter in the preceding experiment. Here, *the properties of the humus are caused by the mineral constituents, not by those of the litter*. But the mull (on the iron-rich sandstone) differs from the moder (on the iron-poor sandstone) by the same biochemical characteristics as those found in the previous experiment: the mull contains four times more AF and AH than the moder and four and a half times more insolubilised humin $(H_1 + H_2)$, while the moder contains three and a half times more inherited humin than the mull (Toutain & Védy 1975). This confirms the role of active iron which previously had been determined experimentally by the addition of artificial iron hydrates to moder (Toutain 1974).

A more detailed study of the way in which iron reacts with litter-derived, water-soluble compounds enabled the author to show that only the most polymerised coloured compounds were insolubilised by iron in the mull; they are also the poorest in nitrogen and the richest in phenolic compounds. The compounds of small molecular size, uncoloured, poor in phenols but relatively rich in nitrogen and saccharides (low C : N), are not immobilised by iron and are very rapidly biodegraded by micro-organisms. In acid mulls they appear to play an important role as an energy source for the white rot which degrades lignin.

Figures 2.5 and 2.6, taken from Toutain (1974), allow a comparison to be made of the biochemical processes of humification of beech litter in acid conditions, for both mull and moder formation. In the case of the acid mull, Figure 2.5 shows that the insolubilisation of phenolic precursors occurs by the

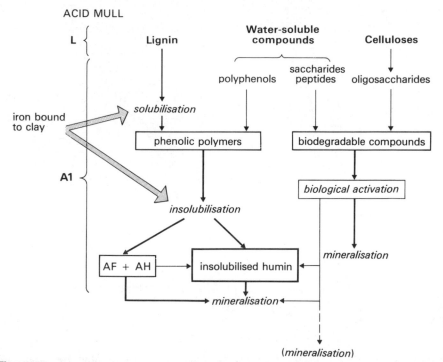

**Figure 2.5** Humification in an acid mull (Toutain 1974). The carbon cycle is entirely restricted to the A1 horizon.

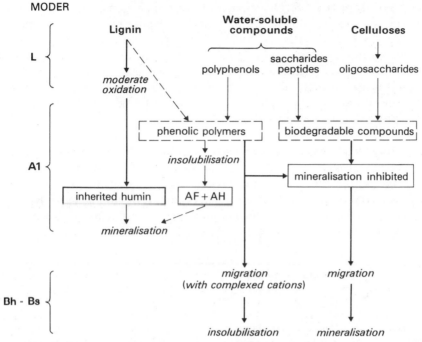

**Figure 2.6** Humification with a moder humus (Toutain 1974). Some compounds not insolubilised in the A1 horizon move into the B horizon.

intervention of active iron and clay, in the A1 horizon; the only things that persist are the water-soluble, biodegradable compounds which are very quickly decomposed: *almost none of the soluble or dispersed compounds is carried into the lower mineral horizons.* The inherited humin is almost non-existent because of the strong degradation of lignin by white rot.

In the case of moder it is different, as the low amount of nitrogen mineralisation does not allow white rot to develop. Lignin is only slightly modified and it persists for the most part as inherited or residual humin. The phenolic precursors are not immobilised by iron, but form slowly biodegradable complexes with proteins capable of migrating in the profile and causing podzolisation.

## IV   The slow development of humic materials: maturation

This slow development (**maturation**) is essentially dependent on the pedo-climate (and thus indirectly on the general climate), the mineral constituents of the soil only playing a secondary role.

### 1   Influence of climate on maturation

The effect of seasonal climatic alternations, in particular that of wetting and drying, has been emphasised for a long time by numerous people. Birch

(1958) and Bernier (1960) have particularly emphasised *the resulting increase in the speed of degradation of the most labile fraction of the organic matter*. An opposing reaction also occurs, which is a polycondensation of ring structures of some humic compounds which leads, in contrast, to a stabilisation *vis-à-vis* biodegradation (for convenience, the term **polymerisation** will be used even though it is not correct). This process is characteristic of continental climates and is very limited in atlantic climates (where only a minor part of the humus is stable; Paul & van Veen 1978).

**Stimulating effect of seasonal alternations.** Contrary to what was first thought, this process, which accelerates biodegradation and mineralisation, only affects the most labile part of the organic matter. According to Bernier and Birch (referred to above) it causes small, readily fermentable molecules to be detached from large humic molecules and to be very quickly mineralised. It is generally the peptides of aliphatic side chains that are involved.

A recent experiment has produced evidence (for a soil of the Vosges) of an extraordinary activation of nitrogen mineralisation, caused by an abrupt rewetting of humus after an exceptionally marked dry season: the amount of nitrogen mineralised is twice that which has been measured in a more normal springtime (Duchaufour *et al.* 1971).

Nguyen Kha (1973) has been able to give a more detailed account of the seasonal development of the labile parts of the non-extractable fraction (humin) of two clay soils – one a tropical vertisol, the other a temperate pelosol. After extraction of AF and AH, both soils were incubated and subjected to seasonal alternations, either of a slight extent in terms of humidity (temperate type) or to a more marked extent (tropical type); only with the strong seasonal contrast of the tropical type was biodegradation accelerated. Further, the *old humin of the vertisols is not really affected by this activation as compared to the young humin of the pelosols*.

**The seasonal processes of polymerisation–depolymerisation.** Other authors (Jagnow 1973, Turenne 1975) have emphasised the seasonal processes of polymerisation–depolymerisation and in particular the seasonal equilibrium variations between AF (with small molecules) and AH (with large molecules). Turenne studied this process in a tropical mull from the coastal savannas of Guyana; it was shown that periods of desiccation are polymerising (AF changing to AH), while those in which moisture increases are depolymerising, with the appearance of more or less soluble polymers of molecular weight about 3000; in very sandy conditions, these precursors of small molecular size can migrate and cause podzolisation.

*However, many observations have tended to show that if the processes of desiccation are very strong, new types of humic compounds are formed that escape the depolymerisation (which is general in the more humid period) and in fact acquire a very great stability.*

Analytical studies by many authors provide adequate proof of this phenomenon. For example, Duchaufour and Dommergues (1963) and Perraud (1971) compared, by paper electrophoresis, the nature of the AH in tropical climates of decreasing humidity and showed that the increase in the AH polymerisation (measured by the ratio of grey AH : brown AH) could be correlated with the increasing importance of the dry season.

The stability (*vis-à-vis* biodegradation) of these grey AHs with strongly condensed ring structures has been confirmed by many authors using [14]C dating (Scharpenseel 1972, Gerassimov 1973). In chernozems, the mean age of the AH increases regularly from the surface downwards, from about 1000 years at the surface to 4000–5000 years at a depth of 80 cm. This is explicable in terms of the existence of two fractions in the humic compounds – one with a rapid turnover which is relatively abundant in the surface, while the other, in contrast, is very stable and occurs almost exclusively at depth (see Ch. 8).

Nguyen Kha (1973) confirmed the stabilising and polymerising effect of dry seasons by laboratory incubation techniques that simulated pedoclimatic variations. These experiments were similar to those described previously when humins were determined and the same soils, a tropical vertisol and a temperate pelosol, were used, but in this case the whole soil was periodically extracted for AF and AH determinations (Fig. 2.7).

To simulate the seasonal variations of climate, a three-month alternation period was used. In the case of the tropical climate, temperature was maintained constantly high and a seasonal alternation of waterlogging and desiccation established, while in the case of the temperate atlantic climate, a constant but moderate humidity was established with a moderate seasonal temperature variation. For the vertisol, the results show a remarkable stability of the AH in all conditions, with AF only decreasing gradually. In the case of the temperate pelosols, however, development is different depending on the type of climate: in a temperate climate the AF and AH appear to be unstable and are subject to phases of polymerisation–depolymerisation, depending on the season; in a tropical climate stabilised AH is formed which increases gradually independently of the seasons.

This development is a result of a polymerisation in the sense of a polycondensation of ring

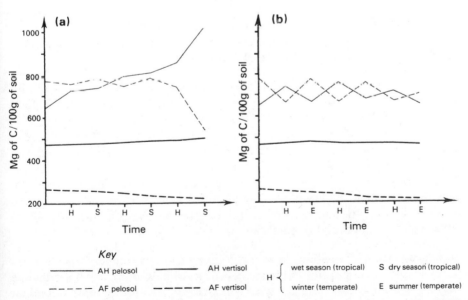

**Figure 2.7** Experimental study of the climatic maturation of humus (Nguyen Kha 1973): (a) tropical conditions; (b) temperate conditions.

structures. Nguyen Kha confirmed by paper electrophoresis that the increase in the ratio of grey AH : brown AH can be correlated with the stabilisation of temperate AH. Chakrabarty *et al.* (1974), Orlov *et al.* (1973) and Kononova and Alexandrova (1973) noted that the maturation of humus is accompanied by a partial disappearance of aliphatic chains (which, as mentioned previously, is reflected in an active C and N mineralisation) and, in contrast, also by a condensation of aromatic rings.

Three conclusions can be drawn from these experiments:

(a) There are two types of AF and AH – those of vertisols formed in a climate of seasonal contrasts which are inert and resistant (except for a small fraction of the AF) to pedoclimatic variations and, in contrast, those of humid temperate soils that can develop either by polymerisation or depolymerisation depending on the conditions.

(b) A pedoclimate of strong contrasts favours the preservation and polymerisation of humic compounds.

(c) *It is the dry season that plays the main role in this process*; the humid summer of the temperate type has the opposite effect, causing a depolymerisation, as Goh and Stevenson (1971) have also stated. This can be reconciled with Turenne's observations on some tropical forms of humus which are not subject to very strong wetting and drying cycles, for their humic compounds remain labile (the converse of the situation in the vertisols) and they are subject to the processes of depolymerisation (described previously) under the influence of humid heat. In a constantly humid equatorial climate, the amount of humus in A1 horizons remains low, despite the abundance of litter, for the humification–mineralisation turnover rate is exceptionally high.

## 2  The influence of the mineral constituents of the soil on maturation

Even though the influence of the mineral constituents of the soil is secondary to that of the climatic factors, it is by no means negligible.

**Bivalent cations.** Maturation and polymerisation are generally associated with a high saturation of the absorbing complex in bivalent cations, particularly $Ca^{2+}$ and also $Mg^{2+}$. Thus it seems reasonable to attribute to them some role in this process, in the same way that it appears almost certain that they play a part in the neoformation of swelling clays, which are generally present in environments favourable to the maturation of humus. The trilogy *stabilised humic acids – $Ca^{2+}$ and $Mg^{2+}$ – swelling clay* effectively characterises the environment most favourable for maturation. However, the action of these bivalent cations must be somewhat indirect for they have a *protective* effect with regard to microbial degradation, as shown by Lineres (1977).

**Stabilising effect of swelling clays.** Illitic clays of brown temperate soils, even though they are important in the process of insolubilisation, are not effective stabilisers. The semi-swelling clays of the pelosols favour the more abundant formation of young inherited humin; thus their stabilising effect is somewhat greater.

*The swelling clays* of soils with strong pedoclimatic contrasts (chernozems, vertisols), in contrast, have a very marked *stabilising action*, which has been demonstrated in much recent work (Scharpenseel 1967, Martinez & Perez-Rodriguez 1969, Anderson 1979, Young & Spycher 1979, Cloos 1981). This reaction appears to be not with the condensed phenolic compounds but more particularly with nitrogen-rich aliphatic compounds, capable of taking up *interlayer positions*, which explains why this type of dark-coloured complex is effectively protected against biodegradation, despite the labile character of this type of organic compound, and has a strong resistance to extraction by normal reagents.

*Thus, within soils with strong seasonal contrasts there are two processes of humification: (i) by polycondensation of ring structures, and (ii) blockage by swelling clays of nitrogen-rich compounds.*

## 3　Conclusion: biological humification and maturation

In terms of this study, it is possible to confirm the existence in soil of two types of humic compounds with very different behaviour and properties. First, there are labile humic compounds, with a rapid turnover marked by a regular alternation of polymerisation–depolymerisation. Their mineralisation is generally rapid, except in environments that are only slightly biologically active and in the presence of stabilisers such as allophane and active calcium carbonate. This type of compound is practically the only one characteristic of humid climates without a pronounced dry season. Secondly, in climates with a pronounced dry season other types of humic compounds appear in considerable quantities, with very condensed ring structures, and are thus naturally stable, or well protected against biodegradation by swelling clays, of which the neoformation (or preservation) is favoured by the same environmental conditions. From the point of view of nitrogen plant nutrition, the first type forms a very labile and thus easily utilised reserve while, in contrast, the second group of compounds only has a very limited function from this point of view. However, to balance this they do have an indirectly favourable effect on the physicochemical properties of the soil in terms of structure, aeration and the absorbent complex (high exchange capacity and a marked affinity for bivalent cations).

## V　Classification of humus

As will be demonstrated generally throughout this book, the role of soil organic matter in the formation and development of profiles is considerable, for not only does organic matter act as the best integrator of environmental factors, but it profoundly influences pedogenic processes by the nature and properties of the organomineral complexes that it forms in the soil. Therefore, *a good classification of humus is an essential preliminary to the development of a really good genetic classification*.

As in the classification of soils, the genetic classification of humus should be based on all of the *biochemical processes* of humus formation which reflect the effects of environmental factors. Until the last few years, however, these processes were not well known and the classification of humus was based essentially on *morphological* criteria and certain simple parameters that took account, in a very approximate way, of general biological activity. Undoubtedly, these criteria can still be used, but they are insufficient and they need to be supplemented by new criteria derived from all of the biochemical processes previously discussed.

## 1 Old criteria for the classification of humus
As they continue to be of value, a summary account of them follows.

**Morphology.** The thickness of litters, the degree of mixing of organic and mineral materials, the presence or absence of an incompletely transformed A0 horizon, are in general a good reflection of overall biological activity and in particular of the speed of decomposition of fresh organic matter. These criteria have already been used at the beginning of this chapter to define *mull*, *moder* and *mor* (the three fundamental types of temperate climate forest humus) in well aerated conditions. Under badly aerated, more or less hydromorphic conditions, the decomposition of fresh organic material is generally slowed down, but it remains very variable, depending on the degree to which anaerobic conditions occur in the surface horizons.

The general classification suggested remains particularly valid as far as temperate forest humus is concerned (Table 2.1).

**Note** that, apart from the peats, which are described in Chapter 11, the humus types produced under anaerobic conditions are **hydromor** and **anmoor**, which should not be confused. Hydromor, as for all mors, is characterised by a completely organic A0 horizon (>30% organic matter), both very acid and saturated with water for most of the year. Anmoor is characteristic of a more biologically active environment, being an intimate mixture of generally well humified organic matter and of mineral material (<30% organic matter) in which poor aeration is reflected in a coherent and plastic structure.

**Table 2.1** Classification of temperate forest humus.

| | | *Aerated conditions* | *Humid conditions more or less aerated* | *Temporary waterlogging (fluctuating water table)* | *Permanently waterlogged (surface water table)* |
|---|---|---|---|---|---|
| | incorporated humus (stable clay–humus complex) | mull | hydromull | | |
| | incompletely mixed (clay–humus complex often unstable) | moder | hydromoder | anmoor | |
| *decreasing biological activity* | superposed humus (thick A0 horizon) | mor | hydromor | | peat |

The structure of clay–humus aggregates forming crumbs, the size and stability of which decreases within the A1 horizon from mull to mor, is also an index of biological activity, not only microbiological, but also of soil fauna (the role of worms as agents of stable structure formation in forest mulls is well known).

**Biochemical parameters.** The two main parameters used are the carbon : nitrogen ratio (C : N) and the amount of mineralisation either of carbon (mineralised C : total C) or nitrogen (mineralised N : total N) measured in the field or by laboratory incubation. But to use both of these criteria as they should be used is a difficult matter of interpretation. For instance, it has long been accepted that there is a correlation between the two such that the greater the nitrogen mineralisation, the lower the C : N ratio. However, Zöttl (1965) showed that this correlation is valid only for organic matter that has been only slightly changed and not incorporated with mineral material to any extent, in particular litters, or A0 horizons of moder and mor. In this respect it should be remembered that the C : N ratio allows *active mor* (or, more exactly, *activatable*, for example by moderate heating or increasing the insolation in the forest) with a ratio of less than or equal to 25 to be distinguished from *inactive mor* with a ratio equal to or greater than 30. But there is no valid correlation between the C : N ratio and the amount of mineralisation of the A1 (or Ah) mixed horizons.

*C : N ratio.* While this ratio has a very wide range as far as fresh organic material and litters are concerned, in general it decreases during decomposition and reaches a fairly constant figure in the A1 horizon which is reasonably characteristic of the type of humus: 10–15 for mull, 15–25 for moder and more than 25 for mor. It should be noted that this ratio is a little lower (10–12) in slightly acid and neutral environments than in acid conditions (acid mull: 12–15).

### The development of C : N ratios during humification

Recent fieldwork, involving the study of litter decomposition in nylon bags, and various laboratory experiments (Viro 1956, Bocock *et al.* 1960, Zöttl 1960) have given a much more detailed insight into the variations of the C : N ratio as fresh organic material develops into humus.

It is known that species with nitrogen-rich leaves and with a relatively low C : N ratio (alder, ash and leguminous trees) free $CO_2$ and mineralised nitrogen at about the same rate, so that the C : N ratio remains constant or is very slightly lowered. The humus formed is an active mull producing much mineral nitrogen. When the C : N ratio of the plant debris is higher, the decomposition slows down, as does the mineralisation of nitrogen compared to that of $CO_2$ which causes the C : N ratio of the humus formed to be lowered; in general, it is clearly less in the A1 horizon than in the litter (forest mull). For debris with a very high C : N ratio (acidifying species), nitrogen mineralisation tends towards zero (mor) while that of carbon still occurs. However, the C : N ratio of the A1 horizon is lowered less than in the case of mull because of the loss of soluble organic nitrogen by leaching.

In mineral horizons (A2, (B), B), the C : N ratio varies according to the soil type. In biologically active soils (brown forest and lessived soils), it is very low (often less than 10) which, according to Stevenson *et al.* (1958), is owing to the fixation of non-exchangeable $NH_4$ by the mineral horizons. Conversely, in soils of low biological activity (podzolic soils) in which soluble organic matter migrates and accumulates in the B horizon, the C : N ratio remains high – equal to or greater than 20 in this horizon. This is one of the essential characteristics for distinguishing lessived from podzolic soils (Stevenson *et al*. 1958, Schnitzer 1981).

### Development of C : N ratios in cultivated soils

The situation is different in cultivated soils where the C : N ratio is determined by the microflora and, where this is active, it is generally 8 to 10. If plant debris such as straw (with a very high C : N ratio) is incorporated into a soil, micro-organisms free the excess carbon as $CO_2$ and fix mineral nitrogen in the organic form (Barbier & Boischot 1954), so that there is a nitrogen deficiency but a reasonable degree of humification. Conversely, if organic matter with a low C : N ratio is added to the soil (for example, dried blood) there is a rapid mineralisation of the excess nitrogen, in the ammonium or nitrate form, but humification remains very restricted. In both cases the humic compounds produced have an equilibrium ratio (for example, 10) characteristic of the soil.

Many authors (for example Henin & Turc 1950, Kononova 1961) have tried to determine the **isohumic coefficient** ('K') – the number by which it is necessary to multiply the weight of fresh organic material to obtain the weight of stable humus formed. This coefficient varies with the nature of the organic material and, in particular, with its C : N ratio. It is very low for debris with low C : N ratios, such as dried blood (C : N = 3.4); it increases for organic matter with moderate C : N ratios, such as dung; and is lowered again for debris with very high C : N ratio, such as straw which changes only slowly into stable humus. Some *K* values are: dung 0.40–0.50; lucerne 0.20–0.30; straw 0.10–0.20. In reality, these *K* values are only approximations as they are dependent to a large extent on other environmental factors such as texture and aeration (in very aerated conditions *K* decreases, while it increases under poor aeration).

*Amount of annual nitrogen mineralisation.* The ideal would be to measure the annual nitrogen mineralisation in the field itself. Unfortunately, this is not really possible despite the numerous more or less ingenious methods developed by research workers. The net amount of nitrogen mineralised in one year depends on the balance between the gross amount mineralised and the amount that is re-incorporated. These opposed processes rapidly alternate, depending on pedoclimatic variations, and their relative importance varies according to the presence or absence of plants that absorb mineral nitrogen or fix atmospheric nitrogen within the rhizosphere (Balandreau 1975). In addition, there are losses by leaching (nitrate particularly), or in the gaseous state ($NH_3$) which are always difficult to evaluate. This indicates why the figures for annual nitrogen mineralisation (by van Praag 1971 and Lemee 1973, 1975) obtained by summing the monthly figures can be no more than tentative and are very probably in excess of the true figure. In this situation, most authors prefer to use laboratory incubation methods where it is possible to say that there is an approximate equivalence between the amount of mineralisation measured by incubation over six months and the net maximum mineralisation in a growing season (Foguelman 1966, Lemee 1967).

As already stated, there is no such thing as a valid correlation between the C : N ratio and the annual amount of mineralisation of A1 mixed horizons. Here are some average figures for the amount of mineralisation obtained (Belgrand 1978):

inactive mor: almost nil.
active mor: 2–3%.

forest mull (plains): 3–5%.
forest mull (mountains): 1%.
carbonate mull (rendzina): 0.5–1%.
steppe mull (chernozem): 0.5–1 per thousand.

These considerable differences are due, on the one hand, to the proportion of labile humic compounds compared to the humic compounds stabilised by prolonged maturation and, on the other, to the very varied efficiency of mineral stabilisers, the iron oxides bound to clays being, from this point of view, much less efficient than active calcium carbonate.

These considerations plainly demonstrate the necessity of using more fundamental physicochemical criteria, based on methods of fractionation which have been shown to be of value in the classification of humus. The fundamental morphological criterion, which aims to measure the degree of incorporation of organic matter in mineral material as an index of the degree of development and thus of humification, is itself limited; carbonate mulls, for example, characterised by a very high biological activity, have, as already noted, organic matter that is only slightly transformed.

## 2 New criteria for humus classification

These criteria are based on the nature and the proportion of the different humus fractions similar to those defined previously (see also Duchaufour & Jacquin 1975, Duchaufour 1976). The origin of these different fractions must be interpreted as a function of the different environmental factors, which thus gives meaningful information on the ecological development of the humus, and also on soil development as a whole, by the relative degree of development of the different kinds of organomineral complex that it controls.

**Degree of change in fresh organic material (degree of humus transformation).** *Slightly transformed forms of humus* are characterised by an abundance of humus compounds that are only slightly altered compared to primary plant material. These compounds are of two types: (i) insoluble compounds, consisting of membranes and inherited humin, the bonding of which to clay is not very strong; and (ii) soluble phenolic precursors produced by the litter that are insolubilised at a variable rate (depending on the nature of the mineral fraction) to form generally persistent AF compounds that do not change to AH. *If insolubilisation is slow, as in acid conditions, migration of AF compounds generally occurs causing podzolisation.*

Paradoxically, the two types of humus that can be placed in this category differ completely in their pH and biological activity, for they are the humus forms of carbonate mulls and mors (or even, to a lesser degree, of moder). In the first case it is the presence of active $CaCO_3$ that prevents humification, while in the second it is the low biological activity. In both cases, the young fractions – inherited humin and fulvic acids – are in fact dominant, but there the similarities stop. The differences are indeed fundamental, for in moder and mors the inherited humin and fulvic acids are rapidly separated in space by movement of the latter. On the other hand, in calcareous mulls, the high level of animal activity (earthworms) causes a great deal of mixing of all the elements of the soil that leads to an intimate mixture of all the humic compounds (no matter what their nature or origin) with the mineral compounds which engenders a stable, aerated crumb structure.

In contrast to these slightly developed forms of humus, there are (i) transformed humus with biological development dominant (temperate climate mull), and (ii) transformed humus characterised by a slow maturation (steppe mull, vertisol humus and, to a lesser degree, the forest mull of a boreal or continental climate).

All other things being equal, hydromorphic humus is generally less transformed than humus developed under aerated conditions, but intermediate forms exist which are usually the result of a lowering of the water table, such as in the case of some very old anmoors (Schaefer 1967) which have a high degree of transformation and even of maturation because of markedly contrasted pedoclimatic phases of long duration.

**Effectiveness of mineral stabilisers.** This has already been the subject of a considerable discussion which it will be sufficient to summarise here. Iron bound to illitic clays has very little stabilising effect compared to amorphous alumina, which is involved particularly with insolubilised compounds, and calcareous materials which react with fresh organic matter and some fulvic acids. Swelling clays behave as strong stabilisers of a humic fraction of the chernozems and vertisols.

**Resistance of the different fractions to extractants.** It is this property that gives the most precise information on the bonding between organic and mineral material and about the nature of the organomineral compounds formed.

The method developed uses a series of reagents in order of increasing strength and effectiveness, each one of which isolates an organomineral complex with well defined properties (on this subject see Vol. 2: *The constituents and properties of soils*, Chs 6 & 9). The way in which reagents act has already

**Table 2.2**  Fractionation of organomineral complexes.

| Reagents | Presumed nature of the complex |
|---|---|
| 1. NaOH solution (tetraborate at 9.7) | mobile complexes: O.M. Al–Fe (podzolic or cryptopodzolic soils) |
| 2. Na-pyrophosphate (pH 9.8) | immobile complexes and bound to clay by $Fe(OH)_n$ and $Al(OH)_n$: (linking cations) (temperate brown soils) |
| 3. NaOH, 0.1N (pH 11–12) | alumina–AH adsorption complexes (andosols) |
| 4. NaOH, 0.1N, after dithionate and HCl–HF | extractable immobilised humin after treatment ($H_1$) |
| 5. non-extractable residue | immobilized humin ($H_2$) (separated by ultrasonics from inherited humin, $H_3$) |

been discussed and at this point it is necessary only to mention in addition the reagent used by Bruckert (Bruckert & Metche 1972; Bruckert & Souchier 1975) which is a NaOH solution buffered at 9.7 that extracts recently insolubilised mobile complexes, characteristic of the accumulative horizons of podzolic soils (or ochric (B) horizons of soils with a podzolic tendency).

In each fraction, AF and AH compounds are generally separated from one another. The AF compounds isolated in the first extraction (NaOH solution buffered at 9.7, preceded by dilute HCl if a decalcification is necessary) are the youngest and least polymerised types, and are thus those that more nearly approach the soluble precursors. As previously stated, these AF compounds are particularly abundant in slightly transformed humus. Sometimes, as a result of migration, they have disappeared from the A1 of mor (or moder) and are then to be found in the spodic B horizons (see Ch. 3).

**Degree of climatic maturation.** Climatic maturation leads to the slow formation of humic compounds with polycondensed nuclei, very resistant to microbial biodegradation and to normal reagents used in extraction. In a chernozem, for example, the amount of AF and AH hardly ever exceeds 50%. However, a small amount of extractable compounds is not necessarily an indication of marked maturity, for certain minerals, such as iron oxides, cause a gross irreversible insolubilisation of the precursors during biological humification and thus there is a risk of confusion between *insolubilised humin* (strictly speaking) and *developed humin* resulting from its slow maturation.

In fact, climatic maturation really can only be appreciated with the aid of complicated biochemical methods, particularly the measurement of molecular volume by a variety of methods, such as IR spectrograph, etc. The simplest method is the application of paper electrophoresis after treatment with HCl–HF to eliminate all traces of silicates, when the very condensed nuclei of grey AH migrate little if at all (Calvez 1970) compared to the slightly condensed nuclei of brown AH which have not been subject to maturation.

Theoretically, strong maturation is also accompanied by a marked reduction in the amount of AF. For example, this is already low in chernozems and it becomes very low indeed in vertisols. Nevertheless, there are exceptions, which is the case in the forest humus of a boreal or continental climate (grey forest soil) where an undoubted maturation (even though not as strong as in the case of chernozems) is reflected in the abundance of grey AH, and which is accompanied by a high production of mobile AF and AH which migrate to give rise to accumulative horizons of a particular type (Ponomareva & Pletnikova 1968, 1975).

**The effect on the nitrogen cycle.** The amount of humus nitrogen mineralisation, discussed previously, is closely related to the plant tissue proteins and their degree of integration into the various humus fractions. The peptides are, in fact, distributed among three fractions (Sinha 1972): (i) a very labile, often soluble fraction which is rapidly mineralised; (ii) a polyphenol-protein fraction which is integrated into the insolubilised compounds and subject to a very variable development depending on the type of humus; (iii) a fraction bound to lignin within the inherited humin which can be important in calcareous mulls.

*In acid mulls*, fractions 1 and 2 are dominant and this means a rapid cycling of nitrogen. *Moders* and *mors* are for the most part inactive because phenol tanning prevents nitrogen mineralisation. There are some exceptions where materials are sufficiently rich in nitrogen and the tanning effect is limited. *Calcareous mulls* have a lower level of mineralisation than acid mulls, peptide compounds being retained for the most part within the inherited humin where they are sequestered by the active calcium carbonate. Finally, strong maturation is always accompanied by a considerable lowering of the amount of mineralisation which is to be seen in the deep humus-rich horizons of chernozems.

## 3   Biochemical classification of natural humus (Fig. 2.8, Table 2.3)

Table 2.3 gives the details of an environmental and biochemical classification based on previously defined criteria. As for soils, *it is the biochemical characters themselves of the humus that are taken into consideration, in that they are the best reflection of the effect of environmental factors on their formation*. In this table, only the main types of humus are dealt with; as will be seen when soil classification is discussed, this allows the basic biochemical processes characteristic of certain classes of soil to be defined. However, there are other types of humus, usually associated with temperate mull, moder or mor, that are controlled by local environmental factors, particularly the pedoclimate, which are not included in the table and will be dealt with at this point.

*Xerophilic or hygrophilic types* are defined in terms of their mean pedoclimatic water balance.

Very dry conditions slow down the decomposition of fresh organic matter and a **xeromoder** or **xeromor** is formed. In contrast, the hygrophilic types have a permanently high water content which, in the case of the **hydromull**, does not prevent good aeration and strong biological activity. It is different in the more acid conditions of an **hydromoder** where the decomposition of fresh organic matter is slowed down and a poorly structured A1 horizon is formed. It is particularly well developed at high altitudes (**alpine hydromoder**). Under extreme conditions of high humidity and often poor aeration, a **hydromor** is formed in which the holorganic A0 horizon (>30% organic matter by weight) is made up of a fine, very black, plastic humus. This differs from a peat in having a greater degree of humification (as is seen in the presence of inherited humin) and often a certain amount of extractable compounds.

**Intermediate types: calcic humus of mountainous areas.** As in the case of soils, intermediate types are numerous and difficult to classify. One of the most difficult problems to resolve is that of humus developed at high altitude on calcareous materials. The humus is always slightly transformed, according to the criteria given previously, since the effect of the low temperature of the mountain climate is added to that of the calcareous material to slow down the process of humification, hence the inherited humin and the slightly altered fresh organic matter are always very abundant and give the soil a dark colour. The amount of organic matter extracted by normal reagents is generally low. There is, in fact, a whole range of calcic humus types of mountainous areas, the naming and classification of which are still uncertain. Using Kubiena's

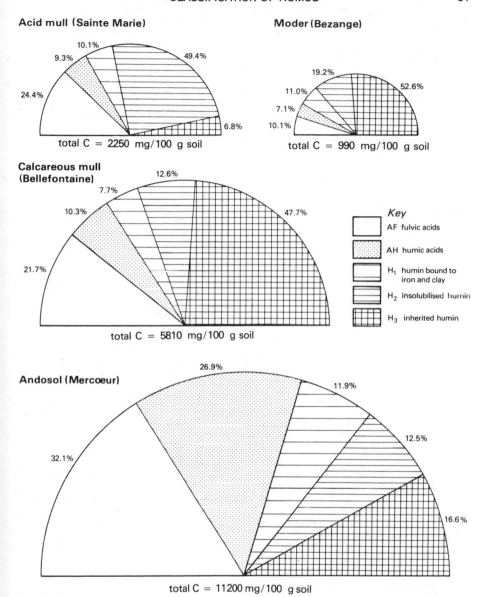

**Figure 2.8** Fractional composition of some temperate forest humus forms (A1 horizons, bound fractions). Note that area of semicircles is proportional to the total amount of humic compounds in per cent.

(1953) definitions provisionally, it appears possible to classify some of them as carbonate mulls, others as moders and yet others as mors. This differentiation is dependent essentially on the way that organic material is incorporated with mineral material (Duchaufour 1978; $V_{3-5}$,* & Ch. 7).

*The roman numerals refer to the captions in Duchaufour (1978).

**Table 2.3** Biochemical classification of humus.

---

*Slightly transformed humus*
Large proportion of organic material with organised structure and
inherited humin; production of fulvic acid with very varied mobility:

(1) weak incorporation of slightly transformed organic matter; maximum     mor
    mobility of AF;
(2) moderate incorporation of slightly transformed organic matter;     moder
    moderate mobility of AF;
(3) massive incorporation of slightly transformed O.M.; immediate     carbonate mull
    insolubilisation of AF by $CaCO_3$.

*Transformed humus – biological humification dominant*
Low proportion of O.M. slightly transformed or inherited humin;
dominance of insolubilised, slightly mobile humic compounds:

| | | |
|---|---|---|
| (1) rapid turnover of humic compounds | acid conditions: insolubilised humin dominant; | acid mull |
| (2) moderately rapid turnover of humic compounds | neutral conditions (saturated insolubilised compounds of all kinds); | eutrophic mull |
| (3) slow turnover of humic compounds | environment rich in allophane, high amounts of extractable insolubilised compounds. | andosol mull |

*Transformed humus with bioclimatic maturation*
Presence of humus compunds with polycondensed nuclei (grey AH,
developed humin), or stabilised by swelling clays:

(1) co-existence of labile and stable ⎰ and mobility of AF and AH     boreal forest mull
    polycondensed compounds ⎱ and AF slightly mobile     chernozemic mull
(2) stabilised and polycondensed compound dominant.     vertisol humus

*Humus development modified by hydromorphic conditions*
As a whole, large proportion of OM relatively unchanged: this
proportion decreasing from peats to anmoor:

(1) very weak development;     fibrous peats,
    (acid, mesotrophic and eutrophic kinds)     (fibrist)
(2) moderately developed: formation of inherited humin, marked     muck or saprist
    increase of extractable AF and AH;
(3) strong development due to pedoclimatic alternations of long duration:     anmoor
    marked increase of all humic compounds often accompanied by a
    certain maturation.

---

### Calcareous mull of high altitude (humocalcareous soils)

This very thick humus, exceptionally rich in organic matter, is formed of crumbs made up almost entirely of carbonates and organic matter, the silicate minerals being present in only small amounts.

### Calcic mull–moder

This type develops from the preceding one by decarbonation and enrichment in soluble compounds from the litter which are precipitated by calcium ions (Bottner 1971). This humus keeps its stable crumb structure although carbonate has been removed; the low amounts of silicates have caused it to be called mull–moder, which is not a very good reflection of its degree of development.

**Calcic mor or tangel**
This is a mor formed on calcareous material at high altitudes, in a cold site and under a general acidifying vegetation. The mixture of organic and mineral material is not as intimate as in the preceding case and a thick, more or less fibrous A0 is formed, the character of which reflects conditions that are more favourable to biological activity than that of an acid mor. Thus the C : N ratio is lower while the pH and S : T are higher, and numerous arthropod faecal pellets are present.

# VI Conclusion

A general conclusion emerges from this study of humification: this is the contrast which exists between the effects of general bioclimatic factors and local factors (site factors such as mineral material) within the whole process of humification. This contrast will be seen again later when the model of soil development is studied.

For the most part, general bioclimatic factors play a determinitive role in the processes of humification. It should be remembered that a particularly good example of this is provided by the zonation in the USSR where, from north to south, there occur mors (podzols), mull–moder (dernovo-podzolic soils), boreal forest mull (grey forest soils) and thick chernozemic mull (chernozem).

Soluble precursors are highly active and abundant in mor soils and, as they are insolubilised slowly and only after migration, this makes them effective agents of podzolisation. The mobility of these soluble precursors decreases towards the south, and their insolubilisation occurs at an earlier and earlier stage in the same direction, so that the intensity of podzolisation decreases until the steppe (chernozem) is reached, when insolubilisation is total and immediate and an additional complication is that of a marked *maturation* of humic compounds.

In an atlantic climate which is less cold, more humid and with less seasonal contrast, the process of bioclimatic humification is very different, for generally no matter what the parent materials are, provided that there is sufficient free iron and fine clay a more or less acid mull is formed, dominated by processes of insolubilisation plus a slight stabilisation of the humus compounds so-formed, in which maturation is involved to only a very slight degree.

Thus, there is a *convergence of humus development* to climatic conditions (and of climatically controlled vegetation) which is relatively independent of the mineral material. It is only on particular parent materials, which are most often exceptional in a given climatic zone, that a *divergent development of humus occurs*. This divergent development is controlled by a composition of the mineral material that can be considered as abnormal in so far as it is characterised by a large excess of a compound which has a decisive influence on humification or, alternatively, to a very marked lack of one or two other materials essential to climatic humification, such as free iron and clay, in the case of a temperate mull. In an atlantic climate, the formation of a forest mull is in fact prevented (i) on material that is almost exclusively quartzitic, (ii) on material that is exceptionally rich in active calcium carbonate compared to silicates, and (iii) on volcanic material rich in allophane. Mor (or moder), calcareous mull, and finally andosol humus correspond respectively to these three special materials, the special properties of which have already been emphasised.

# References

Anderson, D. W. 1979. *J. Soil. Sci.* **30** (1), 77–84.

Andreux, F. 1978. *Etude des etapes initiales de la stabilisation physico-chimique et biochimique d'acides humiques modéles.* State doct. thesis. Univ. Nancy.

Andreux, F., M. Metche and F. Jacquin 1971a. *C.R. Acad. Sci. Paris* **272D**, 2729–31.

Andreux, F., M. Metche and F. Jacquin 1971b. *C.R. Acad. Sci. Paris* **272D**, 2832–5.

Balandreau, J. 1975. *Activité nitrogénasique dans la rhizosphère de quelques graminées.* State doct. thesis. Univ. Nancy I.

Barbier, G. and P. Boischot 1954. *C.R. Acad. Agr.* **1**, 43–8.

Belgrand, M. 1978. *Etude pédologique de quelques stations de la forêt de Marly-le-Rol (78).* DEA Institut Agronomique Paris-Grignon.

Bernier, B. 1960. *Fonds de Rech. Forest.* **5**. Québec: Univ. Laval.

Berthelin, J. and F. Toutain (eds) 1982. Soil biology. In *Pedology*, Vol. 2. *Soil constituents and properties.* London: Academic Press.

Birch, H. F. 1958. *Plant and soil* **10**, 9–31.

Bloomfield, C. 1975. *Soil Biol. Biochem.* **7**, 313–17.

Bocock, K. L., O. Gilbert, C. K. Capstick, D. C. Twinn, J. S. Waid and M. J. Woodmann 1960. *J. Soil Sci.* **11** (1), 1–9.

Bottner, P. 1971. *Evolution des sols en milieu carbonaté. La pédogénèse sur roches calcaires dans une séquence bioclimatique méditerranéo-alpine du sud de la France.* State doct. thesis, Montpellier; *Bull. Sci. Geol.*, no. 37 (1972).

Brachet, J. 1975. *Phytochem.* **14**, 2727–8.

Bruckert, S. 1970. *Influence des composés organiques solubles sur la pédogénèse en milieu acide.* State doct. thesis, Univ. Nancy; *Ann. Agron.* **21** (4), 421–52 and **21** (6), 725–57.

Bruckert, S. and M. Metche 1972. *Bull. ENSAIA Nancy* **XIV** (2), 263–75.

Bruckert, S. and B. Souchier 1975. *C.R. Acad. Sci. Paris* **280D**, 1361–4.

Bruckert, S., F. Toutain, J. Tchikaya and F. Jacquin 1971. *Ecol. Plant.* **6** (4), 329–39.

Brun, J. J. 1978. *Etude de quelques humus forestiers aérés acides de l'Est de la France.* Spec. doct. thesis. Univ. Nancy.

Calvez, H. 1970. *Contribution à l'étude des processus d'extraction et de caractérisation des composés humiques.* Spec. thesis. Univ. Nancy I.

Carballas, T., F. Andreux and M. Metche 1972. *Bull. ENSAIA Nancy* **XIV** (2), 245–61.

Chakrabarty, S. K., H. O. Kretschmer and S. Cherwonka 1974. *Soil Sci.* **117** (6), 318–22.

Chouliaras, N., J. C. Védy and F. Jacquin 1975. *Bull. ENSAIA Nancy* **XVII** (1), 65–74.

Cloos, P. 1981. In *Migrations organo-minérales dans les sols tempérés.* Coll. intern., CNRS Nancy, 1979, 251–8.

Coulson, C. B., R. I. Davies and D. A. Lewis 1960. *J. Soil Sci.* **11** (1), 20–30.

Dell'Agnola, G. and G. Ferrari 1979. *Soil Sci.* **128** (2), 105–9.

Dommergues, Y. and F. Mangenot 1970. *Ecologie microbienne du sol.* Paris: Masson.

Duchaufour, Ph. 1973. *Science du sol* **3**, 151–63.

Duchaufour, Ph. 1976. *Geoderma* **15** (1), 31–40.

Duchaufour, Ph. 1978. *Ecological atlas of soils of the world.* New York: Masson.

Duchaufour, Ph. and Y. Dommergues 1963. *Sols Africains* **VIII** (1), 5–39.

Duchaufour, Ph. and F. Jacquin 1975. *Science du Sol* **1**, 29–36.

Duchaufour, Ph., J. Balandreau and D. Quelen 1971. *Bull. ENSAN* **XIII** (1), 3–6.

Duncan, C. G. 1959. *16th Gen. Meet. Dev. Indust. Microbiol*. 146–56.

Dupuis, Th. and C. Cheverry 1973. *Cah. ORSTOM, Séd. Pédologie* XI (3–4), 215–25.

Filip, Z., K. Haider, H. Beutelspacher and J. P. Martin 1974. *Geoderma* 11 (1), 37–52.

Foguelman, D. 1966. *Etude de l'activité biologique, en particulier de la minéralisation de l'azote, de quelques sols du Languedoc et du Massif de l'Aigoual*. Spec. thesis. Univ. Montpellier.

Fustec-Mathon, E., P. Jambu, P. Bilong, A. Ambles and R. Jacquesy 1981. In *Migrations organo-minérales dans les sols tempérés*. Coll. Intern., CNRS Nancy, 1979, 215–27.

Gerassimov, I. P. 1973. *Soil Sci*. 116 (3), 202–10.

Goh, K. M. and F. J. Stevenson 1971. *Soil Sci*. 112 (6), 392–400.

Grabbe, K. and K. Haider 1971. *Z. Pflanzener. Bodenk*. 129 (3), 202–16.

Guckert, A. 1973. *Contribution à l'etude des polysaccharides dans les sols et de leur rôle dans les mécanismes d'agrégation*. State doct. thesis. Univ. Nancy I.

Guillet, B., J. Rouiller and B. Souchier 1975. *Geoderma* 14 (3), 223–45.

Guittet, J. 1967. *Œcol. Plant*. 2 (1), 43–62.

Gury, M. 1976. *Evolution des sols en milieu acide et hydromorphe sur terrasses alluviales de la Meurthe*. Spec. thesis. Univ. Nancy I.

Haider, K. and J. P. Martin 1967. *Soil Sci. Soc. Am. Proc*. 31 (6), 766–72.

Handley, W. R. 1954. *Forestry Commission Bulletin* 23. London: HMSO.

Handley, W. R. 1961. *Plant and Soil* 15 (1), 37–73.

Henin, S. and L. Turc 1950. *4th Congr. ISSS*. Amsterdam 1, 152–4.

Herbauts, J. 1974. Evaluation de la disponibilité potentielle en azote minéral dans différents types forestiers lorrains. *D.E.A. Pédologie*. Univ. Nancy I.

Hetier, J. M. 1975. *Formation et évolution des andosols en climat tempéré*. State doct. thesis. Univ. Nancy I.

Huang, P. M., T. S. C. Wang, M. K. Wang, M. H. Wu and N. W. Hsu 1977. *Soil Sci*. 123 (4), 213–19.

Jacquin, F. 1963. *Contribution à l'étude des processus de formation et d'évolution de divers composés humiques*. State doct. thesis. Fac. Sci. Nancy.

Jagnow, G. 1973. *Z. Pflanzener. Bodenk*. 134 (1), 20–22.

Jambu, P. 1971. *Contribution à l'étude de l'humification dans les sols hydromorphes calciques*. State doct. thesis. Poitiers.

Jambu, P. and D. Righi 1973. *Sci. du Sol* 3, 207–19.

Jambu, P., T. Dupuis and D. Righi 1981. In *Migrations organo-minérales dans les sols tempérés*. Coll. Intern., CNRS Nancy, 1979, 297–304.

Kilbertus, G. 1970. *Etude écologique de la stratemuscinale dans une pinède sur calcaire lusitanien en Lorraine*. State doct. thesis. Univ. Nancy I.

Kononova, M. M. 1961. *Soil organic matter. Its nature, its role in soil formation*. Oxford: Pergamon.

Kononova, M. M. and I. V. Alexandrova 1973. *Geoderma* 9 (3), 157–64.

Kubiena, W. L. 1953. *The soils of Europe*. London: Thomas Murby.

Laatsch, W. 1963. *Bodenfruchtbarkeit und Nadelholzanbau*. Munich: BVL.

Lemee, G. 1967. *Œcol. Plant*. 2 (4), 284–324.

Lemee, G. 1973. *Œcol. Plant*. 8 (2), 143–74.

Lemee, G. 1975. *Rev. Ecol. Biol. Sol*. 12 (1), 157–67.

Le Tacon, F. 1976. *La présence de calcaire dans le sol: influence sur le comportement de l'épicéa et du pin noir d'Autriche*. State doct. thesis. Univ. Nancy I.

Lineres, M. 1977. *Contribution de l'ion calcium à la stabilisation biologique de la matiére organique des sols*. Spec. doct. thesis. Univ. Bordeaux.

Mangenot, F. 1974. Rapport Premier Coll. Intern. Univ. Nancy I. *Biodégradation et Humification* 1–14.

Mangenot, F., F. Jacquin and M. Metche 1966. *Œcol. Plant*. 1 (1), 79–102.

Manil, G. 1961. *Landbouwhogeschool en de Opzoekingstation*, Ghent 1, 50–83.

Manil, G., F. Delecour, G. Forget and A. El Attar 1963. *Bull. Inst. Agron. Gembloux* 31 (1–2), 1–114.

Martin, J. P. and K. Haider 1971. *Soil Sci*. 111 (1), 54–63.

Martin, J. P., K. Haider and D. Wolf 1972. *Soil Sci. Soc. Am. Proc*. 36 (2), 311–15.

Martin, J. P., S. J. Richards and K. Haider 1967. *Soil Sci. Soc. Am. Proc.* **31** (5), 657–62.

Martinez, F. and J. L. Perez-Rodriguez 1969. *Z. Pflanzener. Bodenk.* **124** (1), 52–7.

Mayaudon, J., M. El Halfawi and L. Batistic 1973. *J. Soil Sci.* **24** (2), 182–92.

Menut, G. 1974. *Recherches écologique sur l'évolution de la matière organique des sols tourbeux.* Spec. thesis. Univ. Nancy I.

Monnier, G. 1965. *Action des matières organiques sur la stabilité structurale des sols.* State doct. thesis. Fac. Sci. Paris.

Nguyen Kha 1973. *Recherches sur l'évolution des sols à texture argileuse en conditions témperées et tropicales.* State doct. thesis. Univ. Nancy I.

Orlov, D. S., I. A. Divovarova and N. I. Gorbunov 1973. *Agrokhimiya* **9**, 140–53.

Paul, E. A. and J. A. van Veen 1978. *11th Congr. ISSS*, Edmonton **3**, 61–102.

Perraud, A. 1971. *Etude de la matière organique des sols forestiers de la Côte-d'Ivoire. Relations sol-végétation-climat.* State doct. thesis. Fac. Sci. Nancy.

Ponomareva, V. V. and T. Pletnikova 1968. *Pochovedeniye* **11**, 104–18.

Ponomareva, V. V. and T. A. Pletnikova 1975. *Pochovedeniye* **9**, 63–73. (*Soviet Soil Sci.* **5**, 565.)

Praag, J. J. van 1971. *Contribution à l'étude de la disponibilité de l'azote et du soufre dans les sols forestiers de l'Ardenne.* State doct. thesis, Gembloux.

Prevot, A. R. 1970. *Humus. Biogénèse, biochimie, biologie.* Saint-Mandé: Tourelle.

Satoh, T. and I. Yamane 1972. *J. Sci. Soil Manure* **43** (1), 41–5.

Schaefer, R. 1967. *Caractères et évolution des activités microbiennes dans une chaîne de sols hydromorphes de la plaine d'Alsace.* State doct. thesis, Fac. Sci., Orsay; *Rev. Ecol. Biol. Sol* **4** (3), 385–437 and **4** (4), 567–92.

Scharpenseel, H. W. 1967. *Trans Comms II and IV, ISSS* 1966, 40–52.

Scharpenseel, H. W. 1972. *Z. Pflanzener. Bodenk.* **133** (3), 241–63.

Scheffer, F. and H. Schluter 1959. *Z. Pflanzener. Bodenk.* **84** (1–3), 184–93.

Schnitzer, M. 1981. In *Migrations organo-minérales dans les sols tempérés*, Coll. Intern., CNRS Nancy, 1979, 229–34.

Schnitzer, M. and S. U. Khan 1972. *Humic substances in the environment.* New York: Dekker.

Schvartz, Ch. 1975. *Evolution des hydrosolubles de litières de Callune et de Hêtre an cours des processus d'humification.* Engng doct. thesis. Univ. Nancy I.

Schwertmann, U. 1969. *Proc. Intern. Clay Conf.* **1**, 683.

Selmi, M. 1975. *Contribution à l'étude de l'humification des litières de Hêtre dans l'Est de la France.* Spec. thesis. Univ. Nancy I.

Sinha, M. K. 1972. *Plant and Soil* **37** (2), 265–71, 273–81.

Souchier, B. 1971. *Evolution des sols sur roches cristallines à l'étage montagnard (Vosges).* State doct. thesis. Univ. Nancy I.

Stevenson, F. J., A. P. Dharival and M. B. Choudhri 1958. *Soil Sci.* **85** (1), 42–6.

Theng, B. K. G. and H. W. Scharpenseel 1975. *Proc. Intern. clay conf.*, 643–53. Willmette, Ill., USA.

Tiurin, I. V. 1951. *Trav. Inst. Sols Dokuchaev* **XXXVIII**.

Toutain, F. 1974. *Etude écologique de l'humification dans les hétraies acidiphiles.* State doct. thesis. Univ. Nancy I.

Toutain, F. and J. C. Védy 1975. *Rev. Ecol. Biol. Sol.* **12** (1), 375–82.

Turenne, J. F. 1975. *Modes d'humification et différenciation podzolique dans deux toposéquences guyanaises.* State doct. thesis. Univ. Nancy I.

Védy, J. C. 1973. *Relations entre le cycle biogéochimique des cations et l'humification en milieu acide.* State doct. thesis. Univ. Nancy I.

Viro, P. J. 1956. *Comm. Inst. Forest Fenniae* **45**, 65 pp.

Wittich, W. 1961. *Der Stickstoff: seine Bedeutung für die Landwirtschaft*, 335–69. Oldenburg: G. Stalling.

Young, L. J. and G. Spycher 1979. *Soil Sci. Soc. Am. J.* **43** (2), 324–8 and 328–32.

Zöttl, H. 1960. *Z. Pflanzener. Bodenk.* **81** (1), 35–50.

Zöttl, H. 1965. *Bericht. Deutsch. Bot. Ges.* **78** (4), 167.

# Chapter 3

# The movement of material within soils

## I  Introduction and definitions

The water that circulates in the soil pores (**gravitational water**) carries with it certain entities either in solution or suspension and is responsible for their general *downward movement*. A great amount of the material thus mobilised can be removed completely from the profile and such an overall loss can be calculated by use of the method of mineralogical balance sheets (see Ch. 1). In contrast, another part of the mobilised material is deposited at a lower level in the profile, i.e. is *redistributed*, which enables two main horizons to be differentiated:

(a)  A horizons that are in general impoverished – **eluvial horizons**;
(b)  B horizons that are in general enriched – **illuvial horizons**.

   Although most movement occurs *vertically*, particularly where permeable materials and level topography are involved, *oblique* or *lateral movement* occurs very frequently on slopes, in hilly topography and where less permeable materials are involved. Thus, *profiles high in the landscape are generally impoverished while those in depressions are enriched*. In certain climates (tropical and mediterranean) and at certain sites, this lateral movement can cause selective erosion, where fine particles as a whole are removed, even on *very gentle* slopes: this is the process of **impoverishment** (Servat 1966, Roose 1970, Roose & Godefroy 1977).
   *Upward movement* of material also occurs, but more rarely as particular circumstances are necessary (to be discussed later). In this case it is the biological factors that play an essential part in the movement of certain elements to the surface, by way of the biogeochemical cycles, which compensates to a certain extent for the processes of downward transport by percolating water.

### 1   Processes of transport (or eluviation): definitions
There is considerable confusion in the nomenclature used to describe the movement of materials in soils. For example, the word *lessivage* can be used in different ways, either very generally to mean the movement of all materials both in solution and in suspension or, on the contrary, in a more restricted

sense as meaning the movement of particles (clays) in suspension only. In fact, it would seem essential to use different names to distinguish between materials that are moving either as soluble salts or as pseudo-soluble organometal complexes or as suspended particles, and in this last case to differentiate further according to the direction of movement and the importance of the processes of redistribution.

**Lixiviation: migration of soluble salts.** This process is concerned mainly with the most mobile cations, those that are capable of forming soluble salts at the pH of the soil: this means essentially the alkali and alkaline earth cations ($Na^+$, $K^+$, $Mg^{2+}$, $Ca^{2+}$) which occur in soil solutions in equilibrium with the *exchangeable* cations retained by the absorbent complex. For example:

$$soil\ K \longleftrightarrow K^+\ soluble$$

The anions that migrate may be in the inorganic form, for example nitrates or carbonates, or organic such as lactates. The heavy polyvalent cations rarely migrate as salts, except $Mn^{2+}$ and $Fe^{2+}$ ions in reducing conditions and sometimes $Al^{3+}$ in very acid conditions.

The gradual movement of alkali and alkaline earth cations generally leads to their replacement on the absorbent complex by $H^+$ or $Al^{3+}$ ions, which results in an *acidification* of non-calcareous profiles, reflected in a *desaturation of the complex* (lowering of the S : T ratio or $V\%$, which represents, it should be remembered, the percentage base saturation expressed in milliequivalents). This loss of cations by lixiviation affects not only the upper part of the profile (A horizon) but often the profile as a whole. *Re-absorption* of cations in the B horizon can occur, but the general balance indicates a deficit, particularly in a humid climate with a strong element of *climatically controlled drainage* (for example an atlantic climate). *The movement of generally very mobile soluble salts favours the processes of subtraction from the whole of the profile (that is to say losses by drainage) rather than redistribution between the A and B horizons*; this is a fundamental aspect of lixiviation.

Soils containing carbonates (calcareous and dolomitic) are subject to a particular kind of lixiviation – **decarbonation** – which generally (but not exclusively) occurs as a result of the action of dissolved $CO_2$:

$$CaCO_3 + CO_2 + H_2O \longrightarrow Ca(HCO_3)_2\ soluble$$

Here again the loss by drainage is general in a humid climate, but in a drier climate with a high potential evapotranspiration (PET), precipitation of the calcium bicarbonate occurs at a certain depth as a very particular kind of illuvial horizon (**calcic** or **ca** horizon). More rarely, in rather more arid areas the loss of calcium from the A horizon occurs as gypsum ($CaSO_4$); then a **gypsic** horizon can form at a certain depth.

**Cheluviation: movement of organometal complexes.** This involves the movement of the heavy cations $Al^{3+}$, $Fe^{3+}$ (occasionally also some alkaline earths derived from litter, such as $Ca^{2+}$) as *organometal complexes* or *chelates* (Pedro & Lubin 1968). This process is generally associated with strong weathering of primary (or secondary) minerals by complexing organic acid (such as oxalic, citric or phenolic acids) which has been called complexolysis in the first chapter; these two processes of weathering and movement as complexes are evidently strictly complementary.

The organomineral complexes are in a pseudo-soluble form and in certain conditions are almost as mobile as soluble salts, *but their solubility is strictly dependent upon the environmental conditions such as Eh and pH, the ionic composition and the concentration of the soil solutions. Because of this these complexes do not remain in solution for as long as the salts in true solution; in the majority of cases they are insolubilised again in the B horizons which, on the one hand, restricts the overall losses from the profile by drainage and, on the other, increases the importance of redistribution of materials within the profile*. In this way **spodic** horizons are formed in podzolic soils by the precipitation of aluminium and iron complexes; these horizons are particularly enriched in amorphous materials, first of all as insoluble complexes which gradually become free hydroxides. Generally, the symbol Bs (or Bfe) is used for these horizons when they are rich in sesquioxides, particularly iron, or Bh if they also contain a great deal of dark-coloured humic acids.

**Pervection (lessivage): movement of particles in suspension** (Paton 1978). This term will be used to describe the process of mechanical movement of clay-size particles, which causes *lessived* soils to form. When this movement is directed *vertically*, as is generally the case, the fine clay particles that are transported are deposited in the B horizon, on the walls of voids or around structural units, thus forming coverings of orientated clay platelets called **cutans** (or more exactly **argillans**, clay being the main component), clearly visible in micromorphological thin sections. Such horizons characterised by clay accumulation (and the iron oxides with which they are intimately bound) are called **argillic** in the American classification, in contrast to the spodic horizons which are enriched in amorphous compounds; in terms of the international system of horizon nomenclature, they are Bt horizons (t = *Ton*, clay in German).

In certain cases, such an accumulation of clay in the Bt horizon fills up the pores in the deeper horizons. If the non-capillary porosity is insufficient to begin with, this can cause a gradual increase in the impermeability of this horizon which thus develops signs of waterlogging, such as the segregation of iron in better aerated zones as rusty patches or concretions (hydromorphic argillic horizons, often symbolised as Btg).

The loss of clay transported to depth by drainage waters is limited; on the other hand, lateral loss of fine clay is common and all the more so as the Bt or Btg horizons are choked with clay, which no longer allows gravitational drainage, so perched water tables are formed and lateral drainage occurs. In certain climates, which are characterised by torrential rain in certain seasons, the water cannot be removed by vertical drainage and a complete waterlogging of the surface horizons results; this causes a *selective erosion* of fine particles by surface run-off which is the *impoverishment* characteristic of tropical soils, planosols and certain mediterranean soils (Servat 1966, Roose 1970, Roose & Godefroy 1977).

## 2   The compensatory processes of return to the surface

Although downward movement of materials is dominant (where the climate is sufficiently humid and the drainage good enough), it is only in exceptional circumstances, even where conditions are very favourable (very humid climate, very porous and inert soil materials), that a particular material is completely eliminated from the upper horizons. This is because of *compensatory* processes of *upward movement* that return at least a part of the previously transported materials towards the surface horizons; these processes are sometimes of a physicochemical nature but more often they are biological.

**Physicochemical processes of upward movement.** These processes involve the upward movement of certain salts in solution, by capillary rise from a waterlogged zone at shallow depth and subject to strong surface evaporation; an upward-moving capillary current is established with surface precipitation of the dissolved salts. Particular examples will be given of three cations that often occur in the soluble state – $Ca^{2+}$, $Na^+$ and $Fe^{2+}$.

The upward movement of calcium bicarbonate occurs in some fersiallitic soils subject to strong seasonal contrasts: in the changeover from wet to dry seasons a rising capillary current is established which causes a resaturation of the surface horizons (Bottner 1971, Lamouroux 1971).

The upward movement of sodium salts (NaCl or $Na_2CO_3$) in an arid (or semi-arid) climate forms saline crusts by evaporation at the surface; in the second case ($Na_2CO_3$), organic matter dissolved in the alkaline conditions also migrates and gives the surface crust a black colour.

The upward movement of ferrous bicarbonate occurs in all climates in hydromorphic soils with a permanent water table and strongly reducing conditions (gley); near to the surface the ferrous ions change to ferric and are precipitated as rusty patches or concretions.

**Biological processes of upward movement: biogeochemical cycles.** These biological processes are rarely of a mechanical nature. However, an important exception is provided by the upward movement of clays caused by earthworms which are responsible, together with all other burrowing species, for a thorough mixing of soil materials from different horizons.

More often it is a case of indirect action by vegetation in which roots remove nutritive elements from depth and return them to the surface by way of the litter: this is **biological mobilisation** as defined by Juste (1965). The elements thus deposited at the surface are gradually re-incorporated in the soil as a result of humification; when the alkaline earth ions ($Ca^{2+}$ and $Mg^{2+}$) are involved, a gradual resaturation of the absorbent complex of the surface horizons results, but the effectiveness of surface immobilisation of these elements is very variable, depending upon the nature and the speed of humification, and this will be discussed in detail later.

## II Mechanisms of migration: eluviation and illuviation

It has already been shown that the mobility of transported materials is dependent on their state, i.e. as soluble salts, complexes in pseudo-solution or in suspension. Therefore, it can be anticipated that processes of movement are also closely related to the state in which materials occur.

### 1 The transport of soluble salts (lixiviation)

Apart from silica, which moves in the soluble state and which will be dealt with separately, this kind of migration is concerned essentially with *alkali and alkaline earth cations*, $Na^+$, $K^+$, $Ca^{2+}$, $Mg^{2+}$. As discussed previously, it causes the gradual desaturation of the absorbent complex and thus the *acidification* of profiles. An important difference between the monovalent ($Na^+$ and $K^+$) and divalent ($Ca^{2+}$ and $Mg^{2+}$) cations needs to be emphasised at the outset: as shown by Védy (1973), monovalent ions are removed as soluble salts both from decomposing litters and from mineral horizons, where most of these ions are in the exchangeable state. On the other hand, the bivalent cations, $Ca^{2+}$ and $Mg^{2+}$, generally occur in litter as complexes, so that the process of movement is different depending on whether they come from litter (complex dominant) or from mineral horizons (salt dominant). *This explains the great mobility of the monovalent cations supplied by litter (for example $K^+$ ions) compared to bivalent cations (particularly $Ca^{2+}$).* This difference also occurs in mineral horizons, but it is not as marked.

There are numerous experimental results which confirm this difference in the behaviour of the $Ca^{2+}$ and $K^+$ ions of litter. Samoylova (1962) showed that within one year deciduous litter loses 80–90% of its potassium as against 20% of its calcium: as the total mass of organic material decreases by the partial mineralisation of the carbon, it results in calcium enrichment, while conversely there is a rapid impoverishment in potassium. Neshatayev *et al.* (1966), in their study of the biogeochemical cycle of the deciduous forest of central Russia, demonstrated an annual addition of 64 kg of calcium as against 55 kg of potassium (per hectare). But after decomposition the litter contains much more calcium (96 kg in total) than potassium (20 kg only). Védy (1973) confirmed these results.

Before making a detailed study of these processes, it is necessary to compare the behaviour of soils without active calcareous material with those soils that contain it; in the first case gradual desaturation of the complex occurs immediately, while in the second, *the occurrence of a more or less abundant reserve of calcium carbonate means that the complex remains saturated in calcium and prevents all processes of acidification; this only begins after the complete decarbonation of the horizon under consideration*.

The mechanism of this phenomenon is easily explicable: in periods of weak biological activity and great humidity, a certain calcium desaturation of the complex can occur; but, in the springtime spurt of biological activity, $CO_2$ pressure increases considerably, which allows a mobilisation of a part of the carbonate reserve, as soluble bicarbonate; the $Ca^{2+}$ is then energetically reabsorbed by the complex, of which it is generally the dominant ion.

Therefore, the movement of cations within calcareous soils has two phases: (i) decarbonation with the complex remaining sensibly saturated, and (ii) desaturation of the complex, which will occur all the earlier as the first is achieved more quickly. Thus a separate section dealing with decarbonation is necessary.

**Desaturation of the complex (non-calcareous conditions).** This process occurs in all environments provided there is sufficient humidity; but it is very gradual in active biological conditions (mull) and it is, in this case, generally balanced by the intervention of the biogeochemical cycle, a point that will be discussed later. In environments with weak biological activity (moder or mor-type humus), the production of soluble organic acids increases considerably, which accelerates the transport of bases; the compensatory processes of upward movement become insufficient and acidification is rapid (as yet it has not been demonstrated, but it is possible that some of the $Ca^{2+}$ and $Mg^{2+}$ ions move as complexes, in the same way as those coming directly from the litter).

As basic cations are removed they are replaced by *acidic* ions – $H^+$ or $Al^{3+}$ – the $H^+$ ions are directly related to the presence of acid organic matter, which has carboxyl groups with a high ionisation constant. For this reason these $H^+$ ions are responsible for strong acidity (pH less than 4), but when the organic compounds are less abundant or absent (for instance in mineral horizons which have been subject to a very slow desaturation of long duration (old loams)) then it is the $Al^{3+}$ ion, gradually freed by the degradation of the clays, which takes the place of the basic ions. In this case the acidity is always much less marked, with the pH rarely less than 5 (Lefevre-Drouet 1966); at pHs above 5 (between 5 and 6), the acidity takes the form of complex aluminous ions – $Al(OH)^{2+}$ or $Al(OH)_2^+$ – which are much less easily exchangeable. This explains how at the same percentage saturation the pH can be different, dependent upon whether the dominant ion is $H^+$, $Al^{3+}$ or even $Al(OH)_n$ under slightly acid conditions. The desaturated horizons, rich in organic material (A0 horizons of mor), are very acid (pH 3 to 3.5) owing to the high concentration of $H^+$ ions (Schwertmann & Veith 1966); the A1 horizons of moder have a *mixed* acidity (30 to 50% $H^+$ ions, 50% to 70% $Al^{3+}$ ions; Delecour *et al.* 1974) and their pH is about 4.5. Finally, essentially mineral horizons which have been subject to pervection (lessived soils) on acid loams have an acidity caused exclusively by $Al^{3+}$ ions and their pH remains at about 5 (Duchaufour & Souchier 1980).

*Comparison of the behaviour of monovalent and bivalent cations.* Comparing these ions with each other, the behaviour of the monovalent cations differs profoundly from the bivalent cations: the first are more mobile in a humid climate and are removed first, except for a small proportion of the $K^+$ ions which are fixed by the clays; they are almost all removed from the profile.

Conversely, bivalent cations are less mobile, their transport is slower and, in addition, a certain proportion of them is retained at depth within the Bt horizon by the absorbent complex.

*The sodium ion* is the most mobile of all the basic cations; even on parent rock rich in sodium minerals, it is rapidly eliminated from the profile. It is only in arid climates, where climatically controlled drainage is practically nil, that the $Na^+$ ion makes up an important part of the absorbent complex (salsodiques soils: see Ch. 13).

*The potassium ion* is also very mobile and subject to rapid movement, but differing from the $Na^+$ ion in that a large amount is retained by the micaceous clays in the fixed state between the silicate sheets (Védy 1973), which allows a potassium reservoir to be maintained within mineral horizons. Under certain conditions of pedoclimatic variation this fixed ion can revert to an exchangeable form.

*The magnesium ion* is more mobile than the $Ca^{2+}$ ion and is removed before it (Durand & Dutil 1971); in addition it is less completely and less rapidly reabsorbed than calcium in the Bt horizons of lessived soils (Duchaufour & Bonneau 1961).

Thus, the ionic composition of the absorbent complex, in a humid climate, where the basic cations are generally present in decreasing amounts in the order $Ca^{2+}$, $Mg^{2+}$, $K^+$ and $Na^+$, reflects not only the decreasing energy of absorption by the complex (stronger for higher valencies), but also the increasing mobility of ions, both properties being closely interrelated.

These experimental results agree completely with the basic theory. Gapon's laws concerning the ionic equilibrium in the system, absorbent complex-solution, explain the effects of dilution (wet season) or concentration (dry season) on the processes of monovalent–divalent exchange. Dilution increases the activity of bivalent ions compared to the monovalent ions: it favours the absorption of the first and the solution of the second; the opposite occurs in periods of strong evaporation, when the resulting high solution concentration causes the exchangeable bivalent ions to go into solution and the monovalent ions of the solution to be absorbed. Scheffer and Schachtschabel (1970) confirmed this in an experiment in which a 1 to 10 dilution caused a very marked exchange between exchangeable $NH_4^+$ and $Ca^{2+}$ in solution (Fig. 3.1).

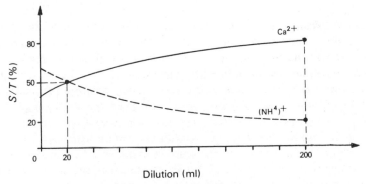

**Figure 3.1** The effect of dilution on the exchange between $NH_4^+$ and $Ca^{2+}$ (montmorillonite) (after Scheffer & Schachtschabel 1970).

This explains why monovalent ions are preferentially leached in a humid climate, since they go into solution at precisely those periods when movement is intense. Conversely, in dry periods the solutions are considerably enriched in bivalent ions but movement is not able to occur; therefore it is possible to explain why monovalent ions are never present in more than small amounts in a humid climate, even if considerable quantities are freed by the weathering of parent material.

**Decarbonation of calcareous soils.** This decarbonation occurs, in sufficiently humid climates, by the removal of calcium bicarbonate (sometimes also $Ca(NO_3)_2$), formed as a result of the gradual solution of carbonates by rain water containing more or less dissolved $CO_2$ (possibly $NO_3^-$ ions): the speed of the process is first of all dependent upon the humidity of the climate, particularly of *climatically dependent drainage*, which will be discussed again in Section IV of this chapter. The average speed of decarbonation in a temperate climate with an atlantic tendency has been evaluated by several authors: according to Scheffer *et al.* (1962), it is rapid to begin with, slowing down gradually (see Fig. 7.3). In 4000 years the decarbonation of a calcareous marl is almost complete; only 2% of active calcareous material remains, but it will take, according to these authors, several more millennia for decarbonation to be complete. According to Wilke (1975), decarbonation is more rapid on terraces of Würm age, composed of more permeable materials; the removal of carbonates is at a rate of about 200 $g/m^2/yr$ and the change from a rendzina to a calcic brown soil will take about 3000 years. In contrast, in a continental climate (in the Berlin region) the rate is much slower (according to Schwertmann 1968) because of the great decrease in the climatically controlled drainage (8 $g/m^2/yr$).

In a mediterranean climate, with a marked dry season, another process occurs which seems to prevent decarbonation: this is the capillary rise of solutions rich in $Ca(HCO_3)_2$, which occurs during the dry season (Bottner 1971). However, this is partially balanced by an active nitrification which occurs in this climate, the abundant formation of very mobile calcium nitrates activating the decomposition and movement of carbonates in spring (Gras 1975). As a whole, fersiallitic profiles on calcareous materials finish by being decarbonated, but the absorbent complex generally remains saturated (see Ch. 12).

*Role of biological factors.* These estimates of the speed of decarbonation are only averages based essentially on climatic factors, which are certainly important but are not the only factors involved. Biological factors, such as the type of humus, also play a very important part and they are able to alter the speed of decarbonation to a considerable extent, even within a particular climatic region. From this point of view, three cases can be distinguished with an increasing speed of decarbonation: (i) bare soil or one with scattered vegetation, poor in organic material: soil solutions contain little $CO_2$ or $NO_3^-$ ions, decarbonation operates slowly; (ii) soil with active mull covered by low, dense vegetation (prairie or steppe): high $CO_2$ production, owing to mineralisation and rhizosphere activity and strong nitrification, considerably increase the speed of the process (Molina 1965, Crahet 1967, Gras 1975); (iii) forest soil with abundant litter: the production of soluble acid organic compounds increases considerably and all the more so as the litter decomposes more slowly (for example, a coniferous moder). These soluble organic compounds, particularly the *phenolic precursors*, appear to have a twofold action: on the one hand they dissolve some of the calcareous material as the soluble bicarbonate salt; on the other they are immobilised as *calcium fulvates* (in which the calcium seems to be partially in the form of a complex; Dupuis & Jambu 1973). It should be recalled that the fulvic acids immobilised by the calcium are particularly abundant in a calcareous environment (see Ch. 2). This is the reason why climatically determined forest soils in a temperate humid climate are completely decarbonated brown soils, no matter what the initial amount of calcareous material in the parent rock. On the other hand, under agricultural conditions the decarbonation is counterbalanced by the frequent cultivation (see Ch. 7).

*Accumulation of calcium: formation of calcic horizons*. The greater part of the bicarbonate formed in the process of decarbonation in a humid climate is removed from the profile by drainage to depth. It is different in a dry climate (dry continental) or even in climates with only a marked dry season (mediterranean) where a calcic horizon is formed at depth (indicated by the symbol 'ca') by the opposite reaction to that which causes its mobilisation:

$$Ca(HCO_3)_2 \longrightarrow CaCO_3\downarrow + H_2O + CO_2$$

This reaction occurs in biologically very active soils (chernozems and some rendzinas) as a result of a decrease with depth in the $CO_2$ concentration, owing to the weaker biological activity away from the humus-rich surface horizons. This accumulation starts as discrete, weakly concretionary forms, either as pulverescent white patches, irregular floury dustings, or as **pseudomycelium**. It should be noted that this precipitation of calcium often leaves the $Mg^{2+}$ ion in solution, which causes the pH to rise, often as far as 9 (reddish chestnut and chestnut soils; Ruellan 1970).

In more arid climates, evaporation plays a more active role than does the reduction of biological activity. In such a case bicarbonate is changed into carbonate as a result of solution concentration and even by the drying out of the profile; the calcareous accumulations are then more massive, and sometimes the calcium carbonate precipitate seems to replace other minerals by a process of epigenesis (Ruellan 1976, Millot 1979). In these conditions more or less indurated calcareous crusts or encrustations are formed, frequently called **petrocalcic horizons**; these encrustations can contain up to 80–90% of $CaCO_3$, and will be described in detail in Chapter 8.

**Movement of silica.** Silica also migrates in the soluble form (monosilicic acid). Its maximum solubility is low (about 100 ppm) and relatively independent of the pH; any considerable increase in soil solution concentration, as a result of the profile drying out, causes silica precipitation as a polymerised gel. However, when the profile is well provided with water, the silica concentration is much less than the maximum figure, and variations in this concentration are to a large extent related to environmental factors, particularly the pH. In addition, it depends also on the *source* of the silica which may be either biological (litter) or geochemical (weathering of silicates or even very slow dissolution of quartz). Even though in slightly weathered materials this second source is not negligible, its relative importance diminishes in very developed and weathered soils, where the biochemical additions by way of litter are the main source of soluble silica. It should be noted, however, that the silica content of litter can vary by a factor of 300 and, in addition, the solution of this silica occurs in an extremely irregular manner which is markedly more rapid in the case of deciduous and grass litter than for coniferous litter (Erhart 1973, Geis 1973, Wilding & Drees 1974, Bartoli & Guillet 1977, Bartoli & Souchier 1978); in the very weathered podzols of the USSR, Belousova (1974) showed that most of the silica present in soil solutions (which can reach 25 ppm) comes from the litter. Unfortunately, as yet there is little information on the forms of silica present in litter and the causes of its mobilisation.

In fact, this silica from litter does not necessarily remain in the soluble state as it moves down into the soil; if the conclusions of Chapter 1 regarding the differential solubility of alumina and silica are agreed with, it can be appreciated that the pH and the solubility of alumina indirectly control the solubility of silica, for any precipitation of alumina in mineral systems causes a certain adsorption of soluble silica by this element (Acquaye & Tinsley 1965). As this adsorption of silica by alumina is the necessary preliminary to the neoformation of clay, *the quantity of silica remaining in solution, in any way, is inversely related to the quantity of neoformed clay in the environment*. It should be remembered, however, that the affinity of alumina for silica is at a maximum in conditions of impeded drainage rich in alkaline earth cations.

In addition, the complex-forming soluble organic matter is also involved indirectly in the silica cycle, for, as previously mentioned, by forming complexes with Al, this organic matter prevents all clay neoformation. Hetier (1975) showed that in andosols the formation of insoluble humus–Al complexes results in processes that favour a continuous and gradual desilication of the profile: *the organic matter, in forming complexes with Al, replaces all the silica of the mixed alumina–silica gels formed in a purely mineral environment*.

From this discussion it is possible to summarise the conditions in which silica movement is of considerable importance.

(a) In an environment with deep weathering, but without acid organic matter, and characterised by a very strong circulation of water causing the removal of monosilicic acid and bases before clay neoformation can occur, which is the situation in the weathering horizons of ferrallitic soils in a humid equatorial climate. It should be remembered, on the other hand, that in the semi-impeded acid horizons which overlie this strongly leaching weathered horizon, a considerable neoformation of kaolinitic clay occurs which uses the residual silica.

(b) In very acid, very permeable conditions, rich in complex-forming organic matter, the silica (coming this time mostly from the litter) is not retained. It moves downwards and is partially immobilised as an amorphous material (allophane compounds) in spodic type horizons (Duchaufour 1954). This is the kind of movement that occurs in podzols and to a lesser extent in andosols.

**Note** that, according to some authors (Nalovic *et al*. 1973), silica is capable of forming stable complexes with iron and thus of favouring its movement in environments lacking in soluble organic matter; this process has been satisfactorily proved by experiment. But in fact it requires a much higher concentration of soluble silica than occurs in soil solutions (Tran Vinh-An & Herbillon 1966). Obviously this process does not occur in ferrallitic weathering horizons (since in these iron is practically immobile), nor in podzolisation where the iron migrates as an organometal complex. In other environments that produce brunification and fersiallitisation, the movement of iron is also reduced to a minimum. Even though it is theoretically possible in these conditions for the movement of iron–silica complexes to occur, it can at most only be of slight importance in soils.

## 2 Movements of organometal complexes (cheluviation)

Many cations, but more particularly the heavy ones – $Mn^{2+}$, $Al^{3+}$, $Fe^{2+}$, $Fe^{3+}$ – move as *organometal pseudo-soluble complexes* often called **chelates**. The alkaline earth cations coming from the litter also move in this form. Heavy cations such as $Al^{3+}$ and $Fe^{3+}$ are present in only small amounts in litter (apart from some exceptions that will be examined); most of them are the result of *biochemical weathering* of primary (and sometimes even secondary) minerals which, as seen in the first chapter, is accelerated in the presence of complexing agents (*complexolysis*).

As stated above, most of the alkaline earth cations derived from litter are in a complexed form; the development of these cations is closely related to their biogeochemical cycles and they will be studied when these cycles are considered. At this point, only the solution and immobilisation of iron, aluminium and manganese complexes resulting from the weathering of minerals will be considered.

**Nature and properties of organometal complexes.** With the exception of the bivalent forms ($Mn^{2+}$, $Fe^{2+}$), the heavy cations produced by weathering are in an insoluble state in most soils with normal pHs: $Fe^{3+}$ is insoluble above a pH of 2.5 – it is then a non-exchangeable complex ion, $Fe(OH)_n$; $Al^{3+}$ at less than a pH of about 5 is in an exchangeable or soluble form, but its concentration even in acid solutions is very low. In these conditions, movement as a salt is reduced or nil for both iron and aluminium; only $Fe^{2+}$, which occurs in a reducing environment, is mobile as the salt $Fe(HCO_3)_2$ below about pH 6.

*As far as the $Al^{3+}$ and $Fe^{3+}$ ions are concerned, the formation of complexes with an organic agent causes the range of solubility to increase greatly towards the higher pHs*: thus, iron in the complex form is mobile even at pH 5; in the case of aluminium, its mobility in the soil is considerably increased by complexing agents: *in comparable situations it has, in aerated conditions, a mobility greater than that of iron*.

*Formation of organometal complexes*. The complexing organic compounds can be of various types: aliphatic acids, phenolic acids, polyphenols and even phenolic polymers, which are already polycondensed and still soluble; even some amino acids have very slight complexing properties. According to the work of Kaurichev and Nozdrunova (1962), Schnitzer (1969) and Jambu *et al*. (1981), the complexing ability is related to the presence in the organic molecule of two juxtaposed functional groups (either two COOH, or one OH and one COOH, or two OH). This gives rise to three sorts of complexes of decreasing solubility (Fig. 3.2), which have been defined by Tan *et al*. (1971), Schnitzer (1981) and Stevenson (1977), and in which the cation : anion ratio becomes very high.

*The true complexes*, in which the cation charge is masked and entirely effaced, are stable. They dissociate with difficulty even if they are subject to considerable pH variations; the complexed aluminium is in the $Al^{3+}$ form and iron as $Fe^{3+}$, or more frequently even in aerated conditions, as $Fe^{2+}$. Lossaint (1959) and Dupuis *et al*. (1970) showed that the reducing ability of some complexing agents (polymers) facilitates complex formation: at a pH of about 3 or 4 (which is normal in strongly complexing environments) practically all of the iron is complexed in the ferrous form.

*Salt complexes* are less stable and these properties are nearer to those of a *salt*; in particular they are sensitive to pH variations which cause their dissociation and precipitation; the metals are mostly in the form of hydroxides – $Al(OH)^{2+}$ or $Fe(OH)^{2+}$ in the case of the most mobile complexes with low metal : anion ratios and $Al(OH)_2^+$ or $Fe(OH)_2^+$ for the less mobile complexes with high metal : anion ratios (Schnitzer 1969, Griffith & Schnitzer 1975).

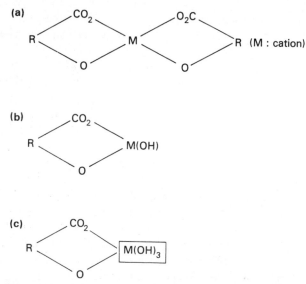

**Figure 3.2**   (a) True complex; (b) salt complex; (c) adsorption complex.

The salt complexes are fixed by clays if there are any present in the environ-ment (brown soils).

*Adsorption complexes* are those in which the complexing organic matter is adsorbed by insoluble inorganic gels.

*As was also emphasised by Schnitzer, the cation : anion ratio is a very impor-tant factor with regard to the solubility of a complex at a given pH; the solubil-ity is all the greater as this ratio is lower* (see Fig. 10.3). *Adsorption complexes are always insoluble.*

### Experimental studies of the properties of complexes

The number of studies on organometal complexes is considerable. Only an outline of this work will be given here. Bruckert (1970) made a comparative study of the complexing ability of aliphatic anions and phenolic compounds extracted from a mor A0 horizon; he showed that the phenolic acids present in the litter have not a great deal of complexing ability. According to this author, the aliphatic acids are responsible for the movement of most of the cations in podzolic soils (from 2 : 3 to 3 : 4); polymers only being involved in the movement of the remainder (1 : 3 to 1 : 4).

Muir *et al*. (1964) have concentrated their attention on the complexes formed by citric acid, one of the most active of complexing agents (together with oxalic acid). The true complexes with iron as the base (molecular ratio of iron : anion, less than 2 : 1) are the most stable and insensi-tive to pH variations; salt complexes are different where the cation : anion ratios can reach 6 : 1 and precipitate as a result of slight variations in pH.

Schnitzer and Skinner (1962, 1965) dissolved the fulvic acids of certain spodic horizons by freeing them of their metal charge and they have studied their complexing ability and solubility at different pHs. For iron, this solubility is stronger at pH 3 than at pH 5. If the pH remains constant, the solubility decreases as the cation : anion ratio increases; this is the case for both iron and aluminium. The solubility of the complex is complete when the molecular ratio is 1 : 1; it decreases in moving towards 3 : 1 and precipitation occurs when the metal : anion ratio is about 5 : 1 to 6 : 1.

Finally, all of the authors cited, as well as Lossaint (1959), Jambu (1971) and Jambu *et al*. (1981), noted the extreme sensitivity of all the types of complex studied to the effect of *calcium* in solution. According to Lossaint, the complexing ability of soil solutions decreases strongly if the percentage of dissolved calcium increases, which causes an increase in the pH (Fig. 3.3). Accord-ing to Jambu, an addition of calcium of about 50–60 ppm to a solution of an organometal complex causes its immediate precipitation.

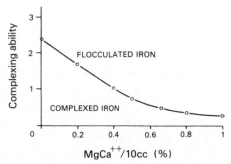

**Figure 3.3** Relation between the complexing ability of a litter extract (g iron complexed/100 g litter) and the amount of calcium and magnesium (after Lossaint 1959).

**Solubilisation and insolubilisation of complexes in the soil.** Judging from the results of experimental pedological research, the essential factor controlling the solubility of complexes is the *cation : anion molecular ratio*; secondly, account has to be taken of adsorption, increase in pH and finally, for certain ions only (iron and manganese), the redox potential (Eh).

Thus, in an aerated acid environment it is essentially the cation : anion ratio which controls complex mobility. The principle of cation mobilisation is simple: the cation : anion ratio is low in moder or mor horizons, which produce a great amount of complexing anions; as downward movement occurs the ratio gradually increases; for, on the one hand, there is a general *sweeping up* of the cations freed by weathering, which increases the amount of complexed metal, while on the other hand the complexing ability of the anions is decreased by biodegradation. As Bruckert (1970) showed, this last process particularly affects the COOH groups which are degraded to $CO_2$; the decarboxylation is sometimes relatively active in spodic horizons (Dommergues & Duchaufour 1965, see Fig. 10.4).

*The increase in the cation : anion ratio which results from this twofold effect causes the precipitation of complexes within spodic horizons – the less stable salt complexes being immobilised more rapidly and before the true complexes* (see Fig. 10.3).

In fact, the mobilisation of cations in the humus-rich surface horizons is only possible when two conditions occur together: humus with little biological activity (moder or mor) and small amounts of free iron and aluminium. When these cations are relatively abundant compared to the complexing anions, surface *insolubilisation* is immediate (see Ch. 2) – this occurs in temperate brown soils with mull, in andosols and, to a certain extent, in cryptopodzolic rankers.

*The processes of adsorption* play a significant complementary role either at the surface or at depth, depending on the conditions. If *clays* are present in sufficient quantities, they adsorb the salt complexes thus forming clay–humus complexes (**brunification**) in Ah horizons. Because of the high biological activity of such an environment, these complexes are present only in small quantities – rapid turnover of mull organic matter (Bruckert 1970, Toutain 1974).

Where little clay is present, precipitation occurs only after the movement of the mobile complexes into the B horizon (podzolisation); this precipitation is accelerated by adsorption on the surface of previously precipitated hydroxides $(Fe(OH)_3, Al(OH)_3)$ – *formation of adsorption complexes*.

*The pH trend and the amount of calcium in solution* play an important part, particularly in calcareous soils or where the adsorbent complex is saturated. Under these conditions iron and aluminium complexes are not formed and the mobility of these elements is reduced to a minimum; the pH of some profiles is variable, being low in the upper part and becoming gradually higher with depth; in this situation the Al and Fe complexes mobilised near the surface are immobilised at depth because of the flocculating action of the $Ca^{2+}$ ions which are generally responsible for the rise in pH.

*The redox potential* is of fundamental importance in the mobility of two metal cations – iron and manganese. This has been known for a long time and was investigated by Betrémieux (1951), Bloomfield (1953, 1955), King and Bloomfield (1966, 1968) and Lossaint (1959). These cations can occur in several forms, either as reduced, relatively mobile bivalent ions ($Fe^{2+}$, $Mn^{2+}$), or as oxidised, much less mobile polyvalent ions ($Fe^{3+}$, $Mn^{3+}$, $Mn^{4+}$), dependent on the Eh.

As mentioned previously, these reduced bivalent cations are much more easily complexed, which further increases the range of their mobility, particularly with regard to the pH. In addition, their occurrence in the reduced state is controlled not only by the Eh but also by the pH, for *with greater acidity these two reduced cations can occur at a higher Eh*. Many investigations have been made of the equilibrium between reduced and oxidised forms of iron and manganese as a function of Eh and pH, and Figure 3.4 is a simplified diagram according to Greene (1963) and Collins and Buol (1970). This shows that the equilibrium curve for manganese is above that of iron and thus, for a given pH or Eh, manganese is more easily reduced, and so more mobile, than iron.

These equilibrium curves explain completely the behaviour of these two cations in the soil. First of all they show why complex formation, in the reduced state, is easier in the moder of podzolic soils than in the mull of acid brown soils. Toutain (1974) investigated this problem in terms of the seasonal variation of Eh and pH of these two types of humus. Rather unexpectedly, the Eh was found to be the same, 300–600 mV in both types of humus on forest soils on sandstone, with minimal values occurring after heavy summer rains.

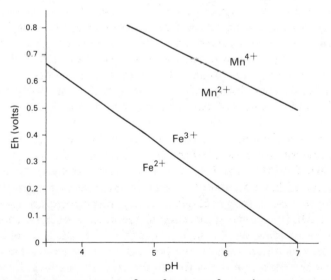

**Figure 3.4** Equilibrium curves for $Fe^{2+}$–$Fe^{3+}$ and $Mn^{2+}$–$Mn^{4+}$ in relation to Eh and pH (after Greene 1963 and Collins & Buol 1970).

However, generally the pH was found to be one unit lower in the moder than the mull, and at certain seasons it fell as low as a value of 3. As these low values are explicable in terms of the occurrence of a great amount of complexing organic matter, it follows that it is at such times that complex formation and mobilisation of ferrous iron can occur, but only in the moder, despite the fact that this environment cannot be considered as being truly reducing. Kaurichev and Mallii (1973) arrived at similar conclusions for the dernovo-podzolic soils of the USSR.

However, the pH and Eh increase with depth in all aerated soils (it is only in some badly aerated, hydromorphic soils that a decrease occurs, particularly in the Eh, because of waterlogging at depth), so that it can be understood why iron changes to the ferric state in deep horizons, which causes the complexes to become insoluble and precipitate.

**Development of the insolubilised complexes.** After the initial phase of insolubilisation, the complexes develop by ageing related to seasonal wetting and drying; the complexing anions develop by *polycondensation*, forming gradually larger molecules: for example, andosols, the oldest complexes, found at the centre of the organo–mineral aggregates, are composed of humic acids, while the recent complexes, in a peripheral situation, are younger, slightly condensed fulvic acids (Hetier 1975). The complexed cations are changed by polymerisation and gradually lose their charge (Schwertmann 1969):

$$Al^{3+} \longrightarrow Al(OH)^{2+} \longrightarrow Al(OH)_2^+ \longrightarrow Al(OH)_3$$

This change is greatly aided by *biological decarboxylation* of the complex-ion anions, which occurs at the same time (Bruckert 1970). Thus, *the gradual condensation of complexing anions is accompanied by a decrease in their complexing properties*. In these conditions, the complexing organic molecules and the complexed cations have a greater and greater tendency to become separated from one another which can be represented thus: *true complex* → *salt complex* → *adsorption complex*. However, as seen in Chapter 2, some of the cations retain their complexed form in the clay–humus aggregates (in both the A1 horizons of mull soils and in spodic horizons).

## 3   Movements of particles in suspension: pervection (Paton 1978)

**Pervection** refers to the process of the mechanical movement of clay occurring in the *dispersed* state – that is to say, as isolated particles; the inclusion of these particles within flocculated aggregates (generally of a reasonable size) prevents the process. Strong biological activity, responsible for a stable crumb structure (mull), is necessarily opposed to the process of pervection, without completely preventing it. In fact, the pervection of clay is a complex phenomenon which depends, on the one hand, on the environmental conditions and, on the other, on the nature of the clays themselves. In addition, as will be seen later, pervection is dependent upon certain pedoclimatic conditions, particularly those concerned with climatically controlled drainage.

Many assessments, both qualitative and quantitative, have been made of pervection. The' presence in the Bt horizon of *oriented* clay skins or *cutans* (*argillan* is a better term) around structural units is accepted by most pedologists as a valid index of pervection. However, some do not accept this in view of their evidence of argillans formed by the transformation of micas and their movement over short distances within a particular horizon (Laves 1970, Smith & Wilding 1972). Some Russian authors (Rode 1964, Zaidelman 1974) think that cutans can be neoformed in place from silica and alumina moved in a soluble or complexed state from the A horizon.

Undoubted proof of pervection has been obtained by the use of soil columns in determining the *quantitative balance* between loss of clay from the A horizon and gains in the B horizon for many soils. Several methods have been used in these precise determinations, which will be considered in the chapter on leached brown soils (Zöttl & Kussmaul 1967, Schwertmann 1968, Smith & Wilding 1972). In addition, it should be noted that several other authors have reproduced these processes experimentally by percolating water through soil columns (Melnikova & Kovenya 1971, Souchier 1971).

**Influence of the environment on pervection.** Pervection is influenced particularly by the pH, the heavy cations $Fe^{3+}$ and $Al^{3+}$, and finally the presence of soluble organic compounds.

Pervection is practically absent in the presence of active calcareous material, for this causes the total immobilisation of clays within very stable crumbs: but as decarbonation occurs, a certain quantity of fine clay is freed by the solution of carbonates and moves into the voids left by their dissolution (Blume 1964, Burnham 1964). In decarbonated, neutral, biologically active soils pervection, as far as the coarsest clays are concerned, is largely prevented by the flocculating action of $Ca^{2+}$ and $Fe^{3+}$ ions, that are involved in stable clay–humus aggregate formation. Only a part of the finest clay, especially montmorillonite, is able to move, together with the ferric iron to which it is closely bound, to form a Bt horizon with ochreous ferriargillans, such as happens in the case of some lessived brown soils on loams.

In the most acid environments (acid brown soils), the role of heavy cations, $Al^{3+}$ and particularly $Fe^{3+}$, is clearly dominant: they form a bridge between the humus and the micaceous clays (in active mull) and behave as energetic flocculating agents; clay movement is reduced to a minimum; in an aerated environment it is particularly the ferric iron which acts in this preventive role (Melnikova & Kovenya 1971, Pedro & Chauvel 1973). In less well aerated environments it is the $Al^{3+}$ ion which takes over from iron, as shown by Gombeer and d'Hoore (1971), Schwertmann (1969) and Guillet *et al.* (1981); this last group of authors emphasise particularly the effectiveness of the *aluminous complex ions* formed by the *ageing* of acid clays, as agents of aggregate formation and hence as a factor in preventing pervection.

From this discussion it is to be expected that there will be a gradual decrease leading to a complete stoppage of clay movement as the acidity increases; although this is effectively the case in environments poor in organic matter, the opposite occurs where there are sufficient amounts of soluble organic matter, for here pervection remains active despite the acidity. This phenomenon has been noted by several authors who attribute it to a twofold effect of the soluble organic matter: *indirect* by the complexing of the flocculating cations $Al^{3+}$ and $Fe^{3+}$, which effectively neutralises their action (Gombeer & d'Hoore 1971); *direct* by the formation around isolated clay particles of a hydrophilic envelope which increases the negative charge and forms a protective barrier against the positive charge of the flocculating cations (Souchier 1971, Dixit *et al.* 1975). Therefore, a relatively important quantity of clay can be moved in podzols.

Thus, in an acid environment two cases can be differentiated. (i) In lessived acid soils, poor in organic matter and at certain seasons badly aerated, there is a partial reduction of the iron causing the breakdown of certain aggregates which frees small quantities of clay, but the aluminous ions slow down their movement: thus the pervection of **deferrified** clay in these conditions is always limited. (ii) In podzolic soils, the more abundant soluble organic matter is involved in neutralising the flocculant action of the heavy cations: thus clay movement in these conditions is more rapid.

Proof of the three kinds of movement that have been described is provided by the different kinds of B horizon accumulation which occur in the slightly acid lessived brown soils, acid lessived soils and podzolic soils (see pp. 85–6).

**Influence of type of clay.** The resistance of clays to movement by water varies according to the type: fundamentally important is particle size and the amount of negative charge.

*Clays of the smallest particle size*, generally *montmorillonites*, are those which are most easily transported, as noted by several authors (Blume 1964, Smith & Wilding 1972). These are practically the only ones that migrate in slightly acid conditions, together with the iron oxides that cover them like a skin, such as in the case of lessived brown soil with mull.

*Micaceous clays*, illites and vermiculites are reasonably mobile, particularly in soils that are being acidified (Guillet *et al.* 1975). However, such movement is prevented, as previously mentioned, by ferric and aluminous ions. Thus the transformation of vermiculites into aluminous vermiculites and even eventually into secondary chlorites, which decreases the negative charges, causes either a blockage of all movement or a blockage that is restricted to the acid and hydromorphic Bg horizons, where this process of aggradation is favoured.

*Finally, the kaolinites* are only slightly mobile: particles are generally large and the charge very low; when these charges are neutralised by ferric ions, no movement at all can take place (Pedro & Chauvel 1973). The experiments of Dixit *et al.* (1975) showed that the kaolinites are practically immobile and resistant to all processes of movement in acid conditions. This has been confirmed by mineralogical balance determinations on soils or parent materials containing kaolinite, in all climates. In a boreal climate on old loams, the inherited kaolinites do not move but even so a *relative accumulation* can occur in surface horizons (because of the movement or the destruction by acid hydrolysis) of other types of clay more susceptible to these processes (dernovo-podzolic soils, Targulian *et al*. 1974). In a temperate humid climate on a Vosgian sandstone containing kaolinite and micaceous clays, only the latter migrate within the podzolic horizons (Guillet *et al*. 1975). Finally, in tropical and equatorial acid soils, where kaolinite becomes predominant, the phenomenon of movement is borne preferentially by the small amount of micaceous clays so that the amount of pervection is greatly decreased. The argillans of the B horizon are often weakly developed in ferruginous tropical soils, and are even less apparent in ferrallitic soils where only kaolinite is present. However, some authors (Chauvel 1977, Eswaran & Sys 1979) have shown the possibility of *destabilisation* of the finest kaolinites, in certain conditions (see Ch. 12).

**Processes and manner of clay accumulation.** While biochemical processes play a major part in the accumulation of complexes, in contrast, the movement of clay being mechanical, its accumulation is controlled particularly by physical factors; the percolation experiments of Melnikova and Kovenya (1971) showed that the speed of movement of clay particles mechanically transported by water is proportional to, but clearly less than, the speed of the water flow: thus any decrease in the speed of water flow leads to a deposition of clay, at least of the coarser particles, which are the basis for the formation of argillans. Now, as far as the summer rains of a temperate humid climate are concerned, a slowing down of water flow normally occurs at a fairly constant depth, which is closely related to the level of maximum density of absorbent roots (Runge 1973). This explains why the depth of the Bt horizon is related to the vegetation and is deeper under forest than under herbaceous vegetation with more superficial root development.

Thus it is the summer rains, together with the vegetation, which initiates the formation of Bt horizons which as it develops causes a gradual slowing down at this level of movement of winter rains which tend to wet the soil more deeply; the saturation of the B horizon in winter by capillary water, effectively stops the downward flow, so that only some very fine clays are able to move to a lower level or even out of the profile completely.

*For well-drained soils in a temperate climate, the limit of summer rain, characteristic of the soil and of its vegetation, roughly coincides with the Bt horizon.*

As the pervection increases, the thickening of the argillans causes a filling up of the voids of the Bt horizon which gradually becomes more impermeable; above this horizon temporary water tables have a tendency to form and the horizon takes on a *hydromorphic* appearance with diffuse rusty patches (Btg). According to Blume (1967), the direction of water flow is modified at this level and becomes lateral, for even on the slightest of slopes an *impoverishment* of the upper horizons of the profile can occur (Duchaufour & Lelong 1967).

Even though the physico-chemical causes of clay precipitation are only of secondary importance, they are, however, significant in two cases: (i) within an acid and slightly permeable Btg horizon, the aggradation of vermiculites into Al-chlorites lowers the negative charge and causes the deposition of clays which are in suspension in this horizon; (ii) the fine clays which are transported by winter rains to a greater depth than the coarser grained clays, can encounter a calcareous horizon and are immediately precipitated; they more or less mix with the decarbonated clay resulting from the weathering of underlying calcareous material: thus, so-called $\beta$-horizons are formed (Ducloux 1970, 1978; Runge 1973; Robin & Coninck 1975).

**Conclusion.** By relating morphology to both eluvial and illuvial processes, it is possible to classify clay-enriched horizons. Some of these, the so-called *argillic* or Bt horizons, in which organic matter is never important, owe their characters almost exclusively to the deposition of transported clays as argillans, that as a result of the parallel orientation of the clay particles can be recognised in thin sections under the polarising microscope.

In slightly acid, biologically active, well-drained conditions the Bt horizon consists of ochreous ferriargillans, made up exclusively of fine-grained, well-crystallised clay particles with an iron skin. Under generally more acid conditions the Bt horizon (or Btg when rusty patches occur as a result of hydromorphism) is formed by the deposition of coarser micaceous clays; in some cases the argillans are decolourised as a result of the clay particles being deferrified by reduction before their mobilisation. In very acid conditions a *degraded* Bt horizon develops, in which the crystal lattice of the clays is considerably altered, so that a vermiculite or Al-vermiculite is dominant (Jamagne 1973, Bullock *et al.* 1974, see Ch. 9).

The origin of two other horizons in which clays have accumulated is more complex, for they contain not only illuviated clay but also clay of a different origin. Thus in the $\beta$-horizons that overlie calcareous materials, the other type of clay is the result of *in situ* decarbonation. In the case of spodic horizons, the B horizon clays are intimately associated with amorphous organo-mineral complexes as a result of the concomitant transport and deposition of both these materials (Belousova *et al.* 1973).

## III  Biogeochemical cycles and biological processes of upward movement

If one excludes the rare examples of physico-chemical upward movement, which have been discussed in the first chapter, it is essentially to the *biogeochemical cycles* that the processes of upward movement should be attributed, which in some profiles reverse the processes of downward transport.

The biogeochemical cycle of the mineral elements (turnover, or **kreislauf**) is now a well-known process. It includes the annual return to the soil surface, by means of litter, of most of the elements previously removed from depth by the roots: this regular supply compensates for the losses by downward transportation and thus maintains the profile in a state of equilibrium. This cycle is completed by the mechanical upward movement caused by the soil animals – earthworms (temperate soils) and termites (tropical soils).

Recent work has elucidated three neglected or little known aspects of these biogeochemical cycles. (i) First of all, the very important part played by roots which is particularly difficult to investigate. In grass-dominated vegetation (steppe), the mass of roots decomposing annually *in situ* can be more important than the aerial vegetative parts that decompose on the surface: the cycle is thus for the most part subterranean and transport of elements is reduced to a minimum – this is the case for **isohumic** soils. (ii) The part played by forest substories, and indeed by herbaceous vegetation generally, appears to be much more important than had been thought up to now; in a fir wood with fescue grass in the Vosges, this grass alone is responsible for the annual return of 25–50% of the litter constituents. (iii) Finally, rainfall contains material derived from air-borne dust and the washings from living leaves which, on occasions, add to the surface something like the amount added by the litter itself. In Belgian oak woods, rainfall adds to the soil annually 42 kg of calcium/hectare and 31 kg of potassium – which is equivalent to and sometimes greater than the quantity of these elements supplied by the litter at the time of leaf fall (Mina 1965, Denaeyer de Smet 1966).

The biogeochemical cycles are responsible for what may be called *biological mobilisation* (Juste 1965), which is very different from the *chemical mobilisation* resulting from the processes of weathering by soluble organic compounds (referred to previously as *acidolysis* or even more frequently as *complexolysis*) that almost always involves transport by cheluviation, so that there is a loss of certain elements from the surface horizons and thus the balance for these horizons is always negative (Védy 1973).

In contrast, the effect of biological mobilisation on general element balance in the profile can be very different and, depending on the vegetation, the form of the element and the type of humification, is either *negative* or *positive*. In the first case, the profile continues to be impoverished despite additions by the litter; while in the second case, elements returned by the litter are generally immobilised in the humic surface horizons which are thus enriched in these elements compared to the underlying mineral horizons (Védy 1973, see also Toutain 1974 and Fig. 3.6).

## 1   Importance and effectiveness of the biogeochemical cycle

The fundamental question concerning the fate of the elements added by the litter is thus the following: what are the factors which favour the accumulation of elements in the surface humic horizons and what are those, on the contrary, which allow their transport and removal from the profile or, in certain cases, their accumulation in the B horizon? An answer will allow a decision to be made on the *effectiveness* of the cycle, which is considered to be high in the first case and low or zero in the second.

These factors will be considered as being either quantitative (direct effects of vegetation) or qualitative concerning the effects of both the type of humification and the kind of humus (indirect effect).

**Quantitative aspects: direct effects.** The simplest and most immediate idea concerning the effectiveness of the cycles is quantitative, for biochemists and pedologists have thought for a long time that the more important the litter supply the more the cycle would have a positive effect. Thus it was assumed that ameliorating species added more nutrient elements to the soil by way of litter than acidifying species, particularly with regard to basic cations, such as $Ca^{2+}$ and $Mg^{2+}$, that are capable of raising the pH and which, as stated previously, are preferentially retained in the litter during humification. In fact, recent work has shown this point of view to be too superficial and in need of revision. Numerous studies have generally established one essential fact: *the weight per hectare of elements added annually for the two types of litter does not differ greatly for the two cations $Ca^{2+}$ and $Mg^{2+}$; only the amount of nitrogen is clearly different, it is more important for the ameliorating litters* (Védy 1973, Toutain 1974, Messenger 1975).

But, as already emphasised in Chapter 2, the nitrogen content of litter has a considerable effect on the process of humification, so that it can be said that high amounts of nitrogen in litter have an *indirectly* favourable effect on the retention of alkaline earth cations.

Védy (1973) compared the additions (in terms of kg/ha/yr) of two litters of coniferous forest growing on Vosgian Triassic sandstone – a fir wood with beech and fescue grass (*Festuca silvatica*) on a brown acid soil with mull, and pine forest with heather on a podzolic soil with moder. The addition of nitrogen is considerably more important in the first type of forest (47 kg as against 37); on the other hand, the addition of calcium and magnesium is more important in the second type of forest (16.5 kg as against 12.7 kg). However, the amount added to the soil is the opposite to that observed in the litter, for the exchangeable calcium is two to five times greater in the A1 horizon of the acid brown soil with mull than in the podzolic soil with moder.

The sub-storey grasses (here fescue grass of the forests) play a very important part in the cycle, as shown by Lemée (1975). They are responsible for the increased addition of nitrogen and also potassium, which is always abundant in grass litters; but grasses contain less calcium and magnesium than tree leaves (Denaeyer de Smet 1966) but, as will be shown, potassium is rapidly removed from surface horizons and is never involved in humification – in which, as has been emphasised, only nitrogen plays an important part.

The effectiveness of the biogeochemical cycle of *potassium* is low despite the great amount added by grasses and its basic character, because of its great mobility; it should be remembered that this mobility is generally greater for those elements occurring as salts (potassium) than as complexes (calcium) in litters. It has been shown that potassium is transported very rapidly from litters and A1 mixed horizons in humid climates. However, it should be noted that silica, which is also transported in the soluble form (monosilicic acid), can be stored, sometimes in great quantities, in the humic horizons, for it occurs in the litter in a more or less condensed, possibly even crystalline form (cristobalite; Wilding & Drees 1974, Bartoli & Souchier 1978).

A purely quantitative comparison of the amount added by various litters is not sufficient to obtain an idea of the direct effect of litter addition. It is necessary to compare the amounts added with the pre-existing reserves in the soil which are either directly assimilable or easily freed by weathering: this approach was taken by Védy (1973). If, for a given cation, little comes from the litter compared to the soil reserves, the complexing agents will have a tendency to enrich themselves in the cation as a result of weathering. Insolubilisation at the surface, if it occurs, can be responsible only for a negligible gain, while the transport of the cation in a soluble form from the mineral soil can lead to a substantial loss. On the other hand, if annual additions by litter are important, immobilisation at the surface by complex formation can lead to spectacular gains in the surface compared to the underlying horizons.

The detailed quantitative investigations of Védy (1973) show that the addition of heavy cations from the litter – iron and aluminium (but not however manganese and certain other minor elements) – is for the most part negligible, compared to the readily weatherable reserves of these elements in the soil: amounts vary from 0.6/1000 to 1/1000 for Vosgian sandstone and granites (to a depth of 20 cm): this explains why for these two elements biological mobilisation is of no importance: sometimes there are exceptions which will be examined.

The same kind of studies applied to the alkaline earths show that annual additions are relatively much more important, but much more variable, compared to parent rock composition. Excluding consideration of calcareous material where the ratio is very low (Toutain 1974) and hence the biogeochemical cycle is totally ineffective, it is sufficient to make a simple comparison between a Triassic sandstone, very poor in calcium, and a calc-alkaline granite which is much richer. An estimate of the ratio of annual addition : soil reserves (to a depth of 20 cm) is about 25–30% for the sandstone and less than 1% for the granite; as will be seen, this explains why the cycle is effective in the first case but not in the second.

**Indirect effect of humification.** It has been shown that ameliorating litter contains the same (or even less) calcium than acidifying litter, while the humic A1 horizons with which they are associated is much richer in calcium in the first case than the second. Therefore, another (indirect) mechanism must be involved that can only be associated with the process of humification.

Humification has practically no influence on the cycle of salts in solution (potassium), which remain very mobile: on the other hand, *it plays an essential part in the cycle of elements which are freed from litter as pseudo-soluble organo-metal complexes, such as aluminium, iron, manganese, calcium* and to a lesser degree magnesium (Védy 1973). The subsequent development of these complexes can follow three possible routes: they can be mineralised, or become insoluble by polycondensation, either rapidly near the top of the A1 horizon, or more gradually after movement into the B horizon. In the first case, the complexed cations become free or exchangeable; in the second they tend to accumulate in the A1 horizon; in both situations the geochemical balance is positive. In the third case redistribution causes the cations to accumulate in the B horizon (important for podzolic soils) and the geochemical balance is negative for the surface horizons that have been impoverished.

Obviously these kinds of cation development are restricted to those involving pseudo-soluble complexes. While this is the most important group of reactions, some cations are associated, at least temporarily, with insoluble membranes and when decomposition is very slow, as in the case of a mor humus, these membranes form an A0 horizon of incompletely decomposed organic matter; in addition the A1 horizon contains abundant inherited humin. So a surface accumulation of cations occurs which is characteristic of slightly active mor or moder humus. All changes that cause the decomposition of this surface accumulation of organic matter at the same time free the elements thus stored in a more assimilable form, particularly nitrogen and calcium. Therefore, methods used to improve raw humus soils should not involve the *removal* of these horizons, but their activation by chemical and mechanical means.

*Thus there is a strong correlation between the final distribution of cations supplied by the biogeochemical cycle and processes of humification.* It should be recalled that these processes are reflected by the relative importance of the various humus fractions, which have been defined in Chapter 2, and the cations are distributed in variable amounts between the different fractions, depending upon the kind of development to which the organic matter in the profile has been subject. To illustrate this point, reference will yet again be made to the contrast between a brown forest soil with mull and a podzolic soil with moder or mor, in a temperate humid climate. (i) In a mull soil the cations are immobilised at the surface within either the extractable humic fraction or the two types of insolubilised humin ($H_1 + H_2$). (ii) In podzolic soils with moder or mor, some of the cations accumulate in an insoluble, unassimilable form, within both undecomposed fresh organic matter and inherited humin ($H_3$) and remain at the surface; other cations are transported as pseudo-soluble complexes which accumulate (at least partially) in this same state in the B horizon.

## 2 Examples of biogeochemical cycles

To illustrate the preceding discussion, a more detailed study will be made of three cations (or groups of cations) which represent three different cases: (i) *calcium*, that stresses the importance of the indirect effect of humification in which the overall balance can be positive or negative depending upon the calcium reserves of the parent material; (ii) *iron and aluminium*, where the amount added by litter is low and is generally removed by cheluviation causing the balance to be negative (however, certain exceptions are of interest); (iii) *manganese*, where in acid conditions biological mobilisation is very great and accumulation in acid mulls is the result of both direct additions and indirect humification.

**Biogeochemical cycle of calcium.** It should be noted first of all that the Mg cycle is similar to that of Ca, with one important difference: the Mg complexes, essentially of the chlorophyll types, are more labile than those of Ca. Because of their rapid biodegradation they are not incorporated to any great extent in the humic compounds, which causes this element to have a mobility intermediate between that of Ca and K (Duchaufour & Bonneau 1961, Védy 1973).

To return to Ca, Toutain and Duchaufour (1970), then Védy (1973), have emphasised different aspects of the biogeochemical cycle of this element, the first by analytical work on the brown and podzolic soils of beech forests in Lorraine, the second by using modified lysimetry to study the *in situ* decomposition of both acidifying and ameliorating litters on two kinds of parent material (which has been discussed in Ch. 2).

These two pieces of research work gave complementary results which mutually support and verify each other and can be summarised as follows:

(a) The biogeochemical cycle is much more effective, if the total balance of calcium is considered, on parent rocks poor in this element (Triassic sandstone) compared to that where rocks are calcium rich (deep granite regolith containing plagioclase); certainly, in active mulls the immobilisation of calcium of biological origin, by humic compounds, occurs on both types of material, but on sandstone the accumulation of biological calcium is reflected in a largely positive balance of total calcium; on the other hand, on granite this accumulation does not compensate for the losses by weathering and the balance remained negative under the experimental conditions. This is illustrated in Figure 3.5 which shows the different forms of calcium, expressed in mEq/100 g in the two cases.

(b) In addition, the authors cited have produced evidence at the same time of the fundamental role of humification which, in mulls, favours the indirect accumulation of calcium; in fact, the biological cycles in mull and moder are completely different. In mull the original litter complexes are rapidly insolubilised by the *humic compounds of insolubilisation* – extractable compounds (AF and AH) and particularly humin ($H_1$ and $H_2$). These compounds, being fairly rapidly biodegradable, free their $Ca^{2+}$ in an exchangeable form, which are thus rapidly assimilable. The cycle is very different in a podzolic soil with moder: a part of the calcium is stored in the A0 and A1 in an unassimilable form (in fact liberated at a very slow rate); another fraction migrates and is immobilised in the B horizon, particularly in the humin form; finally a third part is lost in deep drainage. *The result is an impoverishment of the surface in assimilable calcium reserves and thus a more or less marked acidification results.*

More exact figures can be obtained from the work of Toutain and Védy: the first measured the ratio of exchangeable and extractable calcium in the A and B horizons (ratio CaA : CaB, related to 1 kg of soil). This ratio is 30 in mull, and no more than 10 in moder. In the A1 the non-extractable calcium is distributed differently in the two types of humus: it is concentrated in the inherited humin ($H_3$) of moder, while in contrast there are four times as much in the mull within the insolubilised humin. The same differences have been confirmed experimentally by Védy in his modified lysimetry work on columns of Triassic sandstone. He showed that the biogeochemical cycle is established much more effectively under an ameliorating litter (fir and beech forest with fescue grass) than under an acidifying litter (pine forest with mor). In five years, the gain to the mull in total calcium is 18 mg/100 g as against 9 mg/100 g for the moder. Again it should be emphasised that the gains do not have the same significance since the distribution into the two humic fractions is very different: relatively labile insolubilised humin in the mull and more slowly decomposing inherited humin in the moder.

The experiments of Afanasyeva (1966), in which the biogeochemical cycles

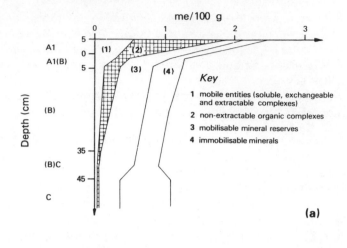

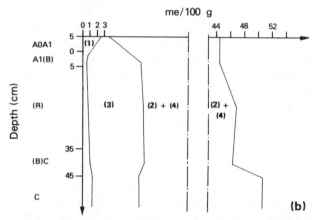

**Figure 3.5** Biochemical balance of calcium, by a modified lysimetry experiment over five years (Védy 1973): (a) biological mobilisation and accumulation on Triassic sandstone (ameliorating litter); (b) chemical weathering and mobilisation on granite regolith (acidifying litter). (Note that fraction 2 cannot be represented as it has a negative balance.)

of calcium under forest or steppe in a continental climate were compared, show that those results obtained for the humid temperate forests can be extrapolated to very different climates. In this case the forest supplies annually to the soil about three times more bases (particularly calcium) than the steppe vegetation; the mull of the steppe, however, has a higher percentage saturation. This is attributed by the author to a more intense and rapid humification.

**Cycles of aluminium and iron.** As previously stated, the biological additions of iron and aluminium are generally negligible, compared to the reserves of iron and aluminium contained in the soil in the free or easily weatherable state: therefore the influence of biological additions on the mobilisation or immobilisation of these elements in the soil is minimal. In active mulls these additions are incorporated with pre-existing reserves, but the gains can never be detected, while, in contrast, in mor- or moder-type humus the chemical mobilisation by complexolysis becomes very significant and the losses are important.

However, this rule does not hold when the parent material is very poor in one or the other of these two elements (or both together), as is the case of quartz sands, poor in iron or weatherable minerals: *under these conditions, the annual supply by litter can become greater than the amounts freed by weathering*. But it is evident that in these conditions, the effectiveness of the cycle can be at most very little; in fact it was shown in Chapter 2 that in the absence of free, active iron (or of aluminium), humification can only lead to the formation of moder (or mor) not of a mull. The development of the soil is towards podzolization not brunification. In these conditions, the biogeochemical cycle is comparable to that observed for calcium on podzolic soils with moder: iron and aluminium of biological origin are necessarily transported to depth and they accumulate in spodic horizons, in the same way as the iron and aluminium supplied by weathering.

Depending upon the composition of the parent material or of the soil, it is sometimes aluminium sometimes iron that is biologically mobilised. On dune sand, richer in iron than aluminium, the biogeochemical cycle of aluminium has a particular importance; in contrast, for certain well developed podzols, where the upper horizons have been entirely stripped of their iron by previous cheluviation, it is the iron which is primarily involved in biological mobilisation.

*The aluminium cycle* has been investigated by Juste (1965) for pine forests growing on sand dunes in Gasgony, and Messenger et al. (1978) for coniferous forests on sands in Michigan. Both studies demonstrated *the particular ability of coniferous trees to concentrate in their leaves more aluminium than iron, in such environments*.

Juste showed that in pine needles the Al : Fe ratio is 8 to 10 while in the sand dune the same ratio is clearly less than 1. Messenger showed that, on the same site, the amount of Al in the leaves is clearly greater for coniferous than for deciduous trees. This aluminium, when freed by mineralisation of organic matter within the soil, tends to displace $Ca^{2+}$ ions and hence, at least in part, to explain the acidifying influence of coniferous trees.

In addition, aluminium present in litter is mobilised rapidly, which is the opposite of what happens to iron (Cheshire et al. 1981).

*As far as iron is concerned*, it is Belousova (1974), in his lysimetry experiments on boreal podzols, who has drawn attention to the effectiveness of the biogeochemical cycle. In such conditions, the upper horizons are impoverished in iron to such a point that almost all of the iron in solution comes from the litter, which supplies between 4 and 40 kg/hectare, and for the most part this iron is retained in the spodic horizon (losses being of the order of 1 to 7 kg/hectare); aluminium is also freed from the litter, but in lesser amounts. In addition, the very slow weathering of certain resistant minerals (alkali feldspars, clays) still continues to provide a certain amount of aluminium in these very impoverished soils; for this element, the two kinds of mobilisation – biological and chemical – appear to occur simultaneously. From his work on the podzolic pseudogleys of Lorraine, Gury (1976) has arrived at very similar conclusions.

**Manganese cycle.** This particularly interesting cycle has been studied by Rousseau (1959), Vallée (1966) and Védy (1973) in the coniferous forests of eastern France, on acid parent materials. It is a remarkable example of a very effective biogeochemical cycle, because the results of both the *direct effect*, which is responsible for a variable concentration of manganese in the litter, and the *indirect effect*, resulting from differences in the processes of humification similar to those affecting calcium, are additive.

It should be recalled first of all that the effectiveness of the cycle is limited to an acid environment (pH equal to or less than 5). At a higher pH, particularly in a calcareous environment, the manganese is immobilised in an insoluble form ($MnO_2$) and it only participates to a slight extent in the biogeochemical cycles. Thus vegetation growing on calcareous materials may show a definite deficiency in this element.

In coniferous forests on acid materials, the biological concentration of manganese in leaves and litters is a common occurrence, although the mineral soil materials (for example, Triassic sandstone or granite in the Vosges) often contains it only in traces. The importance of this foliar accumulation varies according to the type of soil or humus; in the coniferous litters it is eight times higher on acid brown soils with mull than on podzolic soils with moder; this manganese occurs in the leaves in a very mobile complexed form (Védy 1973). Analytical comparisons show that the differences between the humic (A1) horizons of these mull and moder soil types are even greater, for, *on similar parent materials and under comparable environmental conditions the concentration of extractable manganese (as AF and AH compounds) is 100 to 400 times greater in the acid mull than in the moder*.

In an acid mull, the extractable manganese (mobile) can reach 80 mg/100 g compared to 0.14 mg/100 g in a moder, on comparable sites (Védy 1973). For total manganese, the differences are less but remain considerable. Thus on Rhaetic sandstone, the A1 horizons of acid brown soils can contain 2 g/kg of manganese of biological origin, compared to 0.1 g/kg for a podzolic soil with moder (Toutain 1974).

*The differences as a result of litter composition are thus considerably increased by the indirect effect of humification*. According to Védy, the cycle of mobile complexed manganese (as bivalent $Mn^{2+}$) is very similar to the calcium cycle. However, the cycles of these two elements differ in two ways. (i) The considerable accumulation of manganese in the leaves of fir trees growing on mull soils is the result of a greatly increased absorption of manganese in these conditions, which in certain circumstances can cause one-year-old fir and spruce seedlings to be poisoned (Rousseau 1959). This process is unique and there is no equivalent, as will be remembered, in the calcium cycle. (ii) It is possible for manganese to be stored in conditions of good aeration (mull) in an insoluble mineral form ($Mn_2O_3$), which because of its ease of reduction is readily mobilisable.

The ways in which manganese is mobilised and immobilised have been specified by Védy (1973) from the results of his modified lysimetry experiments, which have been discussed several times previously in this book, and by Vallée (1966) from his laboratory incubation experiments. (i) *In mulls*, manganese, as does calcium, accumulates in the insolubilised *humic compounds*, both as extractable mobile forms (AF and AH) and as non-extractable immobilised humins ($H_1 + H_2$). In so far as these compounds are subject to biodegradation, Mn, just like calcium, is freed first of all in the exchangeable form, $Mn^{2+}$. However, at this stage, the manganese cycle behaves rather differently because of its sensitivity to Eh variations, for a part of the mineralised manganese is stored in an easily reducible state, probably as the trivalent oxide $Mn_2O_3$ and, depending upon the seasonal Eh variations, an equilibrium is established ($Mn^{2+} \leftrightarrow Mn^{3+}$) which explains why the coniferous seedlings are able to absorb an excessive quantity of $Mn^{2+}$ at certain periods and become poisoned. (ii) *In moders*, in contrast, Mn remains in a mobile soluble, or exchangeable complexed, state, as $Mn^{2+}$, and is rapidly transported towards lower horizons before accumulating in the spodic B horizon in an insolubilised oxidised state together with insolubilised humic compounds (AF, AH and humin).

It should be emphasised that, in its turn, the mineral manganese stored in the A1 in an easily reducible state, is able to act as a catalyst in the oxidation and polycondensation of humic nuclei. Thus there should be, in fact, as there is for iron, a reciprocal effect between the biological processes on the cycling of the cation on the one hand and of the cation on the activation of biological processes on the other (Duchaufour & Jacquin 1974).

## 3   General conclusions concerning the cycles

Valid conclusions can be made for all the cations, but more especially for those that are particularly involved in pedogenesis: calcium, magnesium, iron, aluminium and manganese.

Climatically determined deciduous forests, such as those in atlantic or continental climates, are characterised by a humus of strong biological activity, of the mull type. For this humus, biological mobilisation of cations is general; in the case of calcium and manganese, it causes marked enrichment of the surface humic horizons, when parent materials are poor in these elements. The chemical mobilisation which causes losses in calcium, iron, aluminium and manganese is reduced to a minimum; in these conditions, the cationic composition of the humic horizons is comparable, no matter what the initial composition of the rock. It is this which has been called *convergent evolution of climatically determined humus* (see also Chs 2 & 4).

It is only on certain parent materials poor in iron and weatherable minerals, or when a particular vegetation is substituted for the climatic-determined one by man, that the biological cycles lose their effectiveness. As a result of a slowing down of microbial activity, a moder or a mor of slight biological activity is formed at the surface; the humus development diverges from that of the general climatically determined development. Under its influence, the chemical mobilisation by weathering, or even cheluviation, prevails over the processes of biological movement towards the surface, the upper part of the

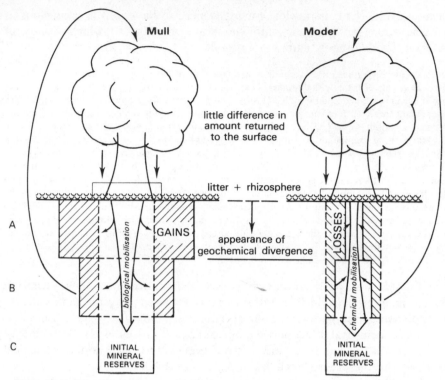

**Figure 3.6**  Divergent biogeochemical cycles under mull and moder (Toutain 1974).

profile is impoverished in elements such as Ca, Mg, Fe, Al and Mn, in this process of podzolisation. Figure 3.6, taken from Toutain (1974), provides a good illustration of these two fundamental kinds of development.

## IV   Influence of environmental factors on the transport of material within soils

In the same way as for weathering and humification, the transport of material within soils is dependent upon two groups of environmental factors which are involved at very different levels: (i) overall bioclimatic factors; and (ii) local site factors.

### 1   Effect of general bioclimatic factors

Climate exerts a basic influence, first of all *directly*, in so far as it controls the pedoclimate and thus to a large degree the movement of water in soils; then *indirectly* through the climatic control of vegetation. Climate and vegetation control the whole direction taken by humification which, as shown above, profoundly influences the biogeochemical cycles.

**Direct effect of climate: drainage controlled by climate.** In general, this effect is best seen in conditions of good drainage where materials are sufficiently permeable to allow the soil solution to percolate *vertically downward*. Those situations in which particular processes occur because of poor local drainage conditions will be discussed in the next section.

As far as general climate is concerned, it is possible to evaluate the expected effects of percolation in terms of the *climatically controlled drainage*, which is the key element of transport processes and which it is essential not to confuse, despite the similarity of terms, with *local drainage* which is a site factor controlled by rock permeability and topography.

Climatically controlled drainage is a climatic factor completely independent of the local site factors. The annual climatically controlled drainage is defined as the difference between the precipitation ($P$ mm) and the potential evapotranspiration ($PET$ mm), in which the temperature is involved indirectly only in so far as it influences the $PET$. Various authors have suggested formulae for the calculation of $PET$ and, as an example, that by Turc (1961) will be considered.

However, it is not sufficient to consider the climatically controlled drainage on an annual basis, i.e. the difference between mean annual $P$ and $PET$, as this can lead to totally false interpretations, because it is necessary in addition to take account of *dry periods*, during which $PET$ is greater than the rainfall and hence no through drainage can occur. Therefore, the climatically controlled drainage should be determined on a monthly basis, and only those months in which $P$ exceeds $PET$ and drainage can occur (i.e. the balance is positive) should be used in the calculation:

$$D_{monthly} = P - PET$$

$$D_{annual} = \Sigma(\text{monthly drainage}) - 100 \text{ mm}$$

(100 mm is an approximate figure for the soil water reserves at the beginning of the rainy season).

Figure 3.7 clearly shows the errors that can arise by using the mean annual values for $P - PET$ in the case of results obtained in mediterranean and continental climates. The first, despite a very high $PET$ in the summer dry season, is still associated with a considerable transport of materials within the soil as most drainage is connected with the dominant winter rains. The second case, on the other hand, does not favour transport processes because the rains coincide with the high $PET$ of summer and only very limited drainage occurs.

Generally in atlantic climates, where precipitation is well distributed throughout the year while $PET$ is on average clearly lower than that which characterises continental climates, conditions are more favourable to the processes of transport than in continental climates.

The use of these formulae makes it possible to define climates in terms of their increasing aridity, with the amount of material transported in soils decreasing in the same direction from the more humid climates to the more arid. In this respect, reference should be made to the classifications of Emberger (1939) and Ganssen (1972) into perhumid, humid, subhumid, semi-arid and arid climates.

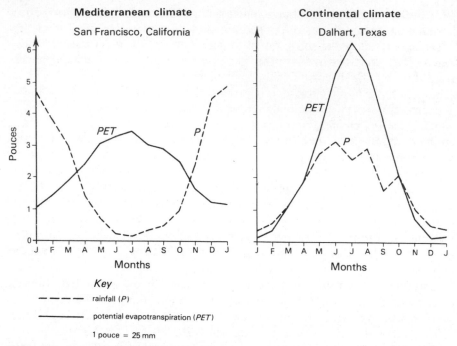

**Figure 3.7** Rainfall and evapotranspiration in mediterranean and continental climates (after Thornthwaite, quoted by Emberger 1939).

Arkley (1967) has made a detailed classification of soils in terms of climate, by which it is possible to define the relationships that exist between climatically controlled drainage and the importance of the processes of transport which is *valid in well-drained conditions*: a summary of his conclusions follows:

(a) *Arid climate with practically no drainage (PET always greater than P)*: no material, even the most mobile ion ($Na^+$), can be removed from the profile – this is the domain of the salsodic soils.

(b) *Weak annual drainage:* removal of $Na^+$ ions and incomplete decarbonation of the upper horizons (chestnut soils of the steppe).

(c) *Climatically controlled drainage less than 150–200 mm:* complete decarbonation of the surface horizons, accumulation of $CaCO_3$ within calcic horizons – these were the soils formerly called 'pedocals' by Marbut (1928) – there is no pervection of clay nor transport of sesquioxides.

(d) *Climatically controlled drainage between 150–200 and 400 mm:* moderate pervection of clays accompanied by a weak to moderate acidification of the absorbent complex; no calcic horizon is formed, the calcium being removed from the profile.

(e) *Climatically controlled drainage greater than 400 mm:* considerable pervection of clays accompanied in cold climatic zones (or at sites with particular characteristics) by a strong acidification with cheluviation;

migration of Al and Fe as complexes. These last two zones correspond to a large extent to that of Marbut's 'pedalfers'.

Note that it should be emphasised that dry seasons not only prevent all drainage, but are even capable, in certain circumstances, of having the opposite effect. They cause the rise towards the surface of certain cations $(Ca^{2+})$ in a soluble form that leads to a seasonal resaturation of the absorbent complex – which is known to be characteristic of certain fersiallitic soils. But these processes of capillary rise have no effect on complexes or clay suspensions.

**Indirect effect: vegetation and climatically controlled humification.** The combined effect of climate and vegetation allows bioclimatic soil zones to be defined, where each is characterised by a definite kind of humification, which has a determinative effect on the processes of transport by its control on the biogeochemical cycles. The succession of climatically controlled humus types characterising the *zones* of the USSR defined by Dokuchaev is a good example:

(a) The low temperatures of the boreal climate combined with the presence of an acidifying litter (coniferous forests, ericaceous species) favour the formation of mor, accompanied by the persistence of active soluble compounds which cause the transport by cheluviation of iron and aluminium (*climatically determined podzolisation*).
(b) Further south, the temperatures are not so low, the ameliorating litter of deciduous trees and the debris of herbaceous plants allows a more and more active humification (mull–moder, then mull); the process of cheluviation is gradually replaced by pervection. Over a considerable area (that of the dernovopodzolic soils) it seems that the two processes co-exist (Targulian *et al*. 1974).
(c) Finally, in the steppe zone and that of the forest–steppe, pervection itself ceases; only decarbonation occurs. The combined effects of vegetation and climate cause not only an active humification, but also a marked *maturation* of the humic compounds: this results in a considerable immobilisation of the organo–mineral complexes produced by the litter within the humus-rich horizons. *The alkaline earth ions, $Ca^{2+}$ and $Mg^{2+}$, are thus stored in large quantities in a complexed form (and in part in the exchangeable form) in the A1 horizons of chernozems.*

There are important differences between soils of the continental deciduous forest zone (grey forest soils) and those of the atlantic deciduous forests (brunified soils), despite the fact that the humus is a mull in both cases. In general, the humic compounds of the atlantic mull are much more labile than those of the continental mull, which is associated with a lesser degree of maturation. This is reflected in the more rapid turnover of organic matter, the smaller quantity of cations stored in the A1 horizons, the lower percentage saturation of the absorbent complex and the higher acidity of the atlantic mull soils compared to the continental mull soils (Ponomareva & Plotnikova 1976).

Finally, it should be noted that very similar comparisons have been made by Perraud (1971), in the climatic zones of tropical Africa, who showed that humification and particularly polycondensation of the humic compounds (and thus, as a consequence, of the organo-mineral complexes of which they form part) increases as the intensity of the dry seasons increases; *this results in a greater and greater accumulation of bases in the humus from the south towards the north*. This allowed three zones to be defined, essentially based on the percentage saturation of the absorbent complex, increasing from south to north.

There is no doubt that, in these regions, the more or less high percentage saturation is a factor of biological origin and thus relatively recent and independent of the general processes of pedogenesis which are much older and slower (see Ch. 4, Sec. II(2) and Ch. 12).

## 2   Local effects of site factors

Site factors of three kinds: (i) the nature of the soil material, its permeability and its mineral composition; (ii) topographic factors; and (iii) local conditions of humification. However, topography and the permeability of the soil material are involved in controlling the *water balance*, which will be discussed first of all. With regard to local conditions of humification, they depend to a large extent on the two other site factors, to which it is necessary to add that of human influence.

**Water balance: local drainage.** The climatically controlled drainage discussed previously is, as seen there, an overall environmental factor that gives general indications and allows the dominant evolutionary tendency in a particular zone to be defined. It is applicable to sites that are perfectly horizontal and where the material is reasonably permeable, for then downward percolation is dominant. However, this ideal situation is disturbed by local site factors, such as when slightly permeable soil materials prevent percolation and when even the slightest of slopes modifies water flow so that it is oblique or lateral (Blume 1973). More generally, sites higher in the topography are impoverished by eluviation, while in contrast depression sites are considerably enriched, particularly if they have been previously silted up by fine-grained deposits. Thus *local drainage*, as determined by site factors, can often modify the effects of the climatically controlled drainage.

In fact, the presence of water in the profile acts in two very different ways: first of all *mechanically*, by more or less preventing water percolation or making it move in a different direction; secondly, it can have an indirect physico-chemical effect, by modifying the *Eh* and hence the process of oxidation/reduction.

*Mechanical effects*. Mechanical effects are the result, on the one hand, of the permeability of the soil material (favouring vertical movement) and, on the other, of the slope (favouring lateral movements and lateral or oblique transport of solutions or suspensions). Slightly permeable soil material prevents vertical pervection but often favours the lateral movement of clay: *generally, this happens where the soil materials are poorly structured and sorted*, such as sandy material in which the pores have been filled up by a small quantity of silt, poorly structured loams, and finally very clayey materials.

Paradoxically, the pervection of fine clays is greater and more generalised in strongly structured uniform loams, with considerable non-capillary porosity, than in humus-poor sands, where the movement of clay, often in a lateral direction, is limited to short distances, for the grain size discontinuities within the sand cause the speed of the suspensions to slow down or stop, accompanied by at least a partial precipitation of the clay. In this case, the accumulative horizon has a banded appearance because of the episodic nature of the pervection, which is restricted to humid periods (Vandamme & de Leenheer 1968, Legros 1975, Gile 1979).

As far as permeable materials at the surface are concerned, in which pervection of clay is active, the accumulation of the clay in the Bt horizon frequently causes a gradual blocking up of the deeper horizons which can lead to completely hydromorphic conditions: this favours, in certain cases, lateral movement with losses from the profile (impoverishment). In other cases, if the vertical movement continues to be important, the horizon of accumulation increases in thickness upward, deposition occurring near and nearer the surface (upward movement of the B horizon).

*Effect of the water balance on the Eh.* The indirect effect of the water balance on the redox potential complements the direct mechanical effects. In slightly permeable conditions, gravitational water tends to stagnate, causing either a waterlogging of the pores or the formation of a perched water table near the surface. A lowering of the Eh results, which is particularly marked in warm periods and is accompanied by the reduction of certain cations (iron and manganese), causing an increase in their mobility.

In terms of mobility, the lowering of the Eh has three important consequences. (i) The reduced iron becomes more mobile than the aluminium, while the reverse is the case in good aeration. (ii) The reduced iron is separated from the clay; the two entities migrate separately and no longer simultaneously as in well-aerated conditions. (iii) Soluble organic compounds (of which certain are complexing with regard to the cations) are biodegraded less rapidly; these organic compounds favour the simultaneous movement of clays and cations. In very acid conditions, they become very active and accelerate the degradation of clay by complexolysis (see Ch. 1), freeing their constituents in a soluble or pseudo-soluble form (hydromorphic pseudopodzolisation).

Two situations can be distinguished: (i) moderately reducing conditions without a water table, i.e. materials that are very compacted and weakly porous (acid loams); and (ii) very reducing conditions with a water table present.

MODERATELY REDUCING CONDITIONS. Usually, such conditions are to be found in very weathered old loams with weak porosity: the iron is partially separated from the clay and moves on its own, sometimes in the complexed state; but in these conditions the biodegradation of the complexing anion occurs fairly quickly when the profile drys out, so that the iron reprecipitates in the ferric form at a lower level. With regard to the clays, their movement follows but, as previously noted, more slowly. In addition, they can be subject to considerable *degradation*: a degraded Bt horizon is then formed where the argillans are decolourised, often badly crystallised, and accompanied by a bleached and pulverescent silica deposit (Jamagne 1973, Bullock *et al.* 1974, see also Ch. 9).

VERY REDUCED CONDITIONS: PRESENCE OF A TEMPORARY OR PERMANENT WATER TABLE (HYDROMORPHIC SOILS). The more or less complete destruction of fine aggregates by iron reduction favours the simultaneous mobilisation of iron and clay, but their mobility depends largely on the pH and the percentage calcium saturation – low in gleys with a saturated complex it increases greatly in pseudogleys and acid stagnogleys. The consequence is either a *segregation of iron* (rusty patches and concretions) or *an elimination of iron and clay* by lateral movement (bleached or *albic* horizon). The acid water table subject to a slow lateral flow mobilises the iron and clay in an increasingly spectacular manner as the organic matter develops towards a hydromoder or a hydromor, and complex formation increases. The reduced iron can be transported a great distance and precipitated in the ferric state in better aerated zones, in the form of large concretions, or even as an indurated cuirasse. When the clays are partially hydrolysed (in the case of podzolic hydromorphic soils), both silica and alumina are liberated and mobilised – pseudopodzolisation (see Ch. 9, Gury 1976, Sokolova and Targulian 1977, Duchaufour 1978a).

**Chemical properties of soil material.** As already noted with regard to the biogeochemical cycles (see Conclusion, p. 104), the chemical properties of soil material in well drained conditions only play a secondary part, compared to the bioclimatic influence, in the final equilibrium between the cations (particularly $Ca^{2+}$ and $Mg^{2+}$) which is characteristic of profiles. This fundamental idea will be developed in Chapter 4 in the concept of the climatically determined soil in which the upper horizons have similar properties no matter what the initial composition of the soil material and are referred to as **analogous** soils. This is the case, for example, of the *brunified soils* in an atlantic climate.

*In contrast, the process of pervection of clay depends more directly on the initial composition of the soil material than on the cycling of cations.* It is more marked on unconsolidated sedimentary materials (loams and marls) than on deep regoliths of crystalline rocks or on indurated acid rocks (sandstone, schist); in the first case an argillic horizon is visible in the lessived brown soil that results, which is not the case in the second situation where an acid brown soil develops. These two soils are good examples of brunified analogous soils in an atlantic climate (Ch. 9).

However, in certain cases where the parent material has a very special composition, soil development takes a completely different route from that of the climatically determined equilibrium. This introduces the concept of an **environmental threshold** that will be discussed in Chapter 4 and used to characterise a second kind of equilibrium, that of site.

As far as the composition of the soil material is concerned, it is the amount of iron and fine clays that is particularly important. Three examples will be considered:

(a) *Very calcareous material*, for example chalk, or soft calcareous materials, without silicate impurities. If, in fact, the calcareous material contains sufficient impurities, decarbonation allows them to accumulate on the surface, the usual processes, particularly pervection, can then intervene if the residual material is thick enough. But soft calcareous material with 98% or 99% $CaCO_3$, even if it is protected against erosion and subject to an acidifying forest litter, cannot decarbonate at a sufficient rate to counteract the mechanical mixing of the soil by animals: the profile remains with carbonate in the surface, which prevents all processes of pervection (Duchaufour 1978b: $V_2$).

(b) *Very quartzose materials* (or poor in weatherable minerals). For the reason given in Chapter 2, mull is not able to form owing to a lack of sufficient clay and active iron; the humus can only be a moder or mor no matter what the vegetation. As in a boreal climate, the transport of iron and aluminium by cheluviation replaces pervection so that development is directed towards podzolisation that is all the more marked as the environment is poor in iron (Souchier 1971). The accumulation is in the form of a *spodic* B horizon and is no longer *argillic*.

(c) *Volcanic material producing allophanes by weathering*. Lacking clay, the clay–humus complex cannot be formed. The humus–Al and humus–Fe complexes are formed in considerable quantities, but because of the great amount of complexed cations, as well as the absence of a quartzose framework, the immobilisation of the complexes is immediate. Only silica and bases are transported (Hetier 1975, see Ch. 6).

**Local conditions of humification.** It should be recalled that these local conditions are closely related to the chemical composition of the soil material. The three examples which have just been given correspond to a divergent development of humus (carbonate mull, acid mor and andosol mull), compared to the formation of a climatically determined mull. But humus depends also on other factors – Eh and hydromorphic conditions – which have already been discussed, and finally a biologically important factor, the vegetation, which is itself dependent upon human action. *An artificial change of vegetation on certain soil materials is able to modify the type of humus and, consequently, the nature of the processes of transport within soils*. This is so for cases of *degradation* under human influence, which will be returned to at greater length in Chapter 4. For example, the replacement of a deciduous forest by an ericaceous heath on certain loamy sands or sands leads to a transformation of humus from mull to mor; moderate acidification and leaching are replaced by marked acidification and a podzolisation characterised by a large-scale movement of organo-metal complexes. A secondary spodic horizon is formed, generally above the argillic horizon formed previously under the forest; the superposed horizons of accumulation are the evidence of the profile's history (Duchaufour 1948, see Figs 9.6 and 7).

## V  Conclusion

From the discussion of the three fundamental pedogenic processes of weathering, humification and the movement of materials within soils, that have occupied the first three chapters, two conclusions can be made. (i) There is a great similarity of approach, for in each of the chapters the general effect of the bioclimatic factors (climate and climatically determined vegetation) is compared with the more localised effect of the site factors (topography, parent material and hydromorphism) that control pedogenesis on a more restricted scale. This fundamental comparison will be made yet again in Chapter 4 when a model of pedogenesis is discussed. (ii) Humification acts as a pedogenic integrator so that the three fundamental processes are not independent but mutually react with one another. This is seen particularly well in temperate and cold climates where the humus type not only defines the process of humification but also that of the biogeochemical cycling of bases and thus of acidity. It is the same in the case of weathering and the movement of materials where the humus type controls whether the hydrolysis is neutral or acid, gradual and of moderate intensity or involves the much more effective and closely linked processes of complexolysis and cheluviation. It is only in regions with a hot climate, where there is intense biological activity and a great depth of soil, that the surface humification is dissociated from pedogenesis occurring at depth. As will be seen, in humid equatorial climates there is a complete difference between processes occurring at depth completely independent of organic matter and those near the surface that are absolutely controlled by organic matter. This is particularly so with regard to the movement of materials where, at depth under neutral and well-drained conditions, silica and bases are mobile, while under the more acid, and sometimes hydromorphic, conditions of the surface it is the aluminium and iron that are affected (see Ch. 12, Sec. C).

## References

Acquaye, D. K. and J. Tinsley 1965. In *Experimental pedology*, E. G. Hallsworth and D. V. Crawford (eds), 126–39. London: Butterworth.
Afanasyeva, Y. A. 1966. *Pochvovedeniye* 6, 1–10. (*Soviet Soil Sci*. 6, 615–25).
Arkley, R. J. 1967. *Soil Sci*. 103(6), 389–400.
Bartoli, F. and B. Guillet 1966. *C.R. Acad. Sci. Paris* 284D, 353–6.

Bartoli, F. and B. Souchier 1978. *Ann. Sci. Forest* **35**(3), 187–202.

Belousova, N. I. 1974. *Pochvovedeniye* **12**, 55–69. (*Soviet Soil Sci.* **6**, 694–708.)

Belousova, N. I., T. A. Sokolova and N. A. Tyapkina 1973. *Pochvovedeniye* **11**, 116–32. (*Soviet Soil Sci.* **6**, 692–708.)

Betrémieux, R. 1951. *Etude expérimentale de l'évolution du fer et du manganèse dans les sols*. Engng doct. thesis, Univ. Paris; *Ann. Agron.* **8**, 193–295.

Bloomfield, C. 1953. *J. Soil Sci.* **4**(1), 5–17.

Bloomfield, C. 1955. *J. Soil Sci.* **6**(2), 284–92.

Blume, H. P. 1964. *8th Congr. ISSS*, Bucharest **V**, 715–22.

Blume, H. P. 1967. *Isotopes and radiation techniques*. Vienna.

Blume, H. P. 1973. *Pseudogley and gley. Trans Comms V and VI ISSS*, E. Schlichting and U. Schwertmann (eds), 187–94. Weinheim: Chemie.

Bottner, P. 1971. *Evolution des sols en milieu carbonaté. La pédogénèse sur roches mères calcaires dans une séquence bioclimatique méditerranéo-alpine du sud de la France*. State doct. thesis. Univ. Montpellier, Sciences géologiques, Mém. no. 37, 1972.

Bruckert, S. 1970. *Influence des composés organiques solubles sur la pédogénèse en milieu acide*. State doct. thesis, Univ. Nancy; *Ann. Agron.* **21**(4), 421–52 and **21**(6), 725–57.

Bullock, P., M. H. Milford and M. G. Cline 1974. *Soil Sci. Soc. Am. Proc.* **38**(4), 621–8.

Burnham, C. P. 1964. *8th Congr. ISSS*, Bucharest **VII**, 132.

Chauvel, A. 1977. *Recherches sur la transformation des sols ferrallitiques dans la Zone Tropicale à saisons contrasteés*. Documents ORSTOM, no. 62, Bondy and Paris.

Cheshire, M. V., M. L. Berrow, B. A. Goodman and C. M. Mundie 1981. In *Migrations organo-minérales dans les sol tempérés*, Coll. Intern., CNRS, Nancy, 1979, 241–6.

Collins, F. J. and S. W. Buol 1970. *Soil Sci.* **110**(2), 111–19.

Crahet, M. 1967. *Bull. AFES* **4**, 17–34.

Delecour, F., H. Van Praag and G. Cherduville 1974. *Pédologie*, Ghent **XXIV**(3), 216–37.

Denaeyer de Smet, S. 1966. *Bull. Soc. R. Bot. Belgique* **99**, 345–75.

Dixit, S., R. Gombeer and J. d'Hoore 1975. *Geoderma* **13**(4), 325–30.

Dommergues, Y. and Ph. Duchaufour 1965. *Science du Sol* **1**, 43–59.

Duchaufour, Ph. 1948. *Recherches écologiques sur la chênaie atlantique française*. State doct. thesis. Fac. Sci. Montpellier.

Duchaufour, Ph. 1954. *Bull. AFES* **60**, 234–6.

Duchaufour, Ph. 1978a. *Science du Sol. Bull. AFES* **4**, 215–27.

Duchaufour, Ph. 1978b. *Ecological atlas of soils of the world*. New York: Masson.

Duchaufour, Ph. and M. Bonneau 1961. *Rev. Forest. Fr.* **12**, 793–9.

Duchaufour, Ph. and F. Jacquin 1974. *10th Congr. ISSS*, Moscow **VI**, 84–90.

Duchaufour, Ph. and F. Lelong 1967. *C.R. Acad. Sci. Paris* **264D**, 2884–7.

Duchaufour, Ph. and B. Souchier 1980. *C.R. Acad. Agric. Paris* **4**, 391–9.

Ducloux, J. 1970. *Bull. AFES* **3**, 15–27.

Ducloux, J. 1978. *Contribution à l'étude des sols lessivés sous climat atlantique*. State doct. thesis. Univ. Poitiers.

Dupuis, T. and P. Jambu 1973. *C.R. Acad. Sci. Paris* **276D**, 489–91.

Dupuis, T., P. Jambu and J. Dupuis 1970. *C.R. Acad. Sci. Paris* **270D**, 2264–7.

Durand, R. and P. Dutil 1971. *Science du Sol* **1**, 65–78.

Emberger, L. 1939. *Mém. Soc. Sci. Nat. Maroc. Inst. Rubel. Zurich* **14**, 40–157.

Erhart, H. 1973. *Itinéraires géochimiques et cycle géologique de la silice*. Paris: Doin.

Eswaran, H. and C. Sys 1979. *Pédologie*, Ghent **XXIX**(2), 175–90.

Ganssen, R. 1972. *Bodengeographie*. Stuttgart: Koehler.

Geis, J. W. 1973. *Soil Sci.* **116**(2), 113–31.

Gile, L. H. 1979. *Soil Sci. Soc. Am. J.* **43**(5), 994–1003.

Gombeer, R. and J. d'Hoore 1971. *Pédologie*, Ghent **XXI**(3), 311–42.

Gras, F. 1975. *Lessols très calcaires du Liban Sud: évolution et mise en valeur*. Engng doct. thesis. Univ. Strasbourg.

Greene, H. 1963. *J. Soil Sci.* **14**(1), 1–11.

Griffith, S. M. and M. Schnitzer 1975. *Can. J. Soil Sci.* **55**(3), 251–67.

Guillet, B., M. Gury and B. Souchier 1975. *Geoderma* **14**(3), 223–45.

Guillet, B., J. Rouiller and J. C. Védy 1981. In *Migrations organo-minérales dans les sols tempérés*. Coll. Intern., CNRS Nancy, 1979, 49–56.

Gury, M. 1976. *Evolution des sols en milieu acide et hydromorphe sur terrasses alluviales de la Meurthe*. Spec. thesis. Univ. Nancy I.

Hetier, J. M. 1975. *Formation et évolution des andosols en climat tempéré*. State doct. thesis. Univ. Nancy I.

Jamagne, M. 1973. *Contribution à l'étude pédogénétique des formations loessiques du Nord de la France*. Fac. Agr. Gembloux.

Jambu, P. 1971. *Contribution à l'étude de l'humification dans les sols hydromorphes calciques*. State doct. thesis. Univ. Poitiers.

Jambu, P., T. Dupuis and D. Righi 1981. In *Migrations organo-minèrales dans les sols tempérés*. Coll. Intern., CNRS, Nancy, 1979, 297–304.

Juste, C. 1965. *Contribution à la dynamique de l'aluminium dans les sols acides du Sud-Ouest atlantique*. Engng doct. thesis. Fac. Sci. Nancy.

Kaurichev, I. S. and N. N. Mallii 1973. *Pochvovedeniye* **7**, 19–23. (*Soils Fertil.*, 1973 **36**(2), 447.)

Kaurichev, I. S. and E. M. Nozdrunova 1962. *Izv. Timizyazev Akad*. **5**, 91.

King, H. G. and C. Bloomfield 1966. *J. Sci. Fd Agric*. **17**, 39–43.

King, H. G. and C. Bloomfield 1968. *J. Soil Sci*. **19**(1), 67–76.

Lamouroux, M. 1971. *Etude des sol formés sur roches carbonatées. Pédogénèse fersiallitique au Liban*. State doct. thesis. Univ. Strasbourg.

Laves, D. 1970. *Ber. Deutsch. Ges. Wiss. B. Miner. Lagerstettenf.* **15** (3–4), 363–81.

Lefevre-Drouet, E. 1966. *Ann. Agron*. **17**(5), 553–70.

Legros, J. P. 1975. *C.R. Acad. Sci. Paris* **281**, 1817–20.

Lemée, G. 1975. *Rev. Ecol. Biol. Sol* **12**(1), 157–67.

Lossaint, P. 1959. *Etude expérimentale de la mobilisation du fer des sols sous l'influence des litières forestières*. State doct. thesis. Fac. Sci. Strasbourg.

Marbut, C. F. 1928. *Proc. 1st. Intern. Congr. Soil Sci*. **31**, 57–77.

Melnikova, M. and S. V. Kovenya 1971. *Pochvovedeniye* **10**, 42–9. (*Soviet Soil Sci*. **3**(5), 611–18.)

Messenger, A. S. 1975. *Soil Sci. Soc. Am. Proc*. **39**(4), 698–702.

Messenger, A. S., J. R. Kline and D. Wilderotter 1978. *Plant and Soil* **49**(3), 703–9.

Millot, G. 1979. In *Alteration des roches cristallines en milieu superficiel, INRA Seminar, Bull. AFES* nos. 2 and 3, 259–61.

Mina, V. N. 1965. *Pochvovedeniye* **6**, 7–17. (*Soviet Soil Sci*. **6**, 603–8.)

Molina, J. S. 1965. *Reuta Fac. Agron. Vet. Univ. Buenos Aires* **16**, 3.

Muir, J. W., R. I. Morisson, C. J. Bown and J. Logan 1964. *J. Soil Sci*. **15**(2), 220–25.

Nalovic, L., S. Henin and J. Trichet 1973. *C.R. Acad. Sci. Paris* **276D**, 3005–6.

Neshatayev, Y. N., O. G. Rastvorova, L. S. Schastnaya, I. A. Tereshenkova and V. P. Tsyplenkov 1966. *Pochvovedeniye* **12**, 31–9. (*Soviet Soil Sci*. **12**, 1372–9.)

Paton, T. R. 1978. *The formation of soil material*. London: George Allen & Unwin.

Pedro, G. and A. Chauvel 1973. *C.R. Acad. Sci. Paris* **277D**, 1133–6.

Pedro, G. and J. C. Lubin 1968. *Ann. Agron*. **19**(3), 294–347.

Perraud, A. 1971. *Etude de la matière organique des sols forestiers de la Côte-d'Ivoire. Relations sol-végétation-climat*. State doct. thesis. Univ. Nancy I.

Ponomareva, V. V. and T. A. Plotnikova 1976. *Pochvovedeniye* **1**, 33–40.

Robin, A.-M. and F. de Coninck 1975. *Science du Sol* **3**, 213–28.

Rode, A. A. 1964. *Pochvovedeniye* **7**, 9–22. (*Soviet Soil Sci*. **7**, 660–71.)

Roose, E. J. 1970. *Cah. ORSTOM. Sér. Péd*. **VIII**(4), 469–82.

Roose, E. J. and J. Godefroy 1977. *Cah. ORSTOM, Sér. Péd*. **XV**(4), 409–36.

Rousseau, L. Z. 1959. *De l'influence du type d'humus sur le développement des plantules du Sapin dans les Vosges*. Engng doct. thesis. Fac. Sci. Nancy.

Ruellan, A. 1970. *Les sols à profil calcaire différencié des plaines de la Basse Moulouya*. State doct. thesis. Univ. Strasbourg; *Mém. ORSTOM*, no. 54.

Ruellan, A. 1976. *Bull. Soc. Géol. Fr*. (7) **XVIII**, 1, 41–4.

Runge, E. C. 1973. *Soil Sci.* **115**(3), 183–93.

Samoylova, Y. M. 1962. *Pochvovedeniye* **3**, 96–194. (*Soviet Soil Sci.* **3**, 315–23.)

Scheffer, F. and P. Schachtschabel 1970. *Lehrbuch der Bodenkunde*. Stuttgart: Enke.

Scheffer, F., E. Welte and B. Meyer 1962. *Z. Pflanzener. Bodenk.* **91**(1), 1–17.

Schnitzer, M. 1969. *Soil Sci. Soc. Am. Proc.* **33**(1), 75–81.

Schnitzer, M. 1981. In *Migrations organo-minerales dans les sols tempérés*, Coll Intern., CNRS Nancy, 1979, 229–34.

Schnitzer, M. and S. I. Skinner 1962. *Soil Sci.* **96**(3), 181.

Schnitzer, M. and S. I. Skinner 1965. *Soil Sci.* **99**(4), 278.

Schwertmann, U. 1968. *Sizungsber. Ges. Naturforsch. Freunde zu Berlin* **8**(1), 16.

Schwertmann, U. 1969. *Proc. Intern. Clay Conf.* **1**, 683.

Schwertmann, U. and J. Veith 1966. *Z. Pflanzener. Bodenk.* **113**(3), 226–36.

Servat, E. 1966. *Conf. Pédologie Médit.*, Madrid, 406–11.

Smith, H. and L. P. Wilding 1972. *Soil Sci. Soc. Am. Proc.* **36**(5), 808–15.

Sokolova, T. A. and V. O. Targulian 1977. In *Problems of soil science*, 479–92. Moscow: Nauka.

Souchier, B. 1971. *Evolution des sols sur roches cristallines à l'étage montagnard (Vosges)*. State doct. thesis. Univ. Nancy I.

Stevenson, F. J. 1977. *Soil Sci.* **123**(1), 10–17.

Tan, K., L. King and H. Morris 1971. *Soil Sci. Soc. Am. Proc.* **35**(5), 748–51.

Targulian, V. O., A. G. Birina, A. V. Kukizov, T. A. Sokolova and L. Tselisheva 1974. *Guidebook to dernovo-podzolic soils*. Moscow: 10th Congr. ISSS.

Toutain, F. 1974. *Etude écologique de l'humification dans les hêtraies acidiphiles*. State doct. thesis. Univ. Nancy I.

Toutain, F. and Ph. Duchaufour 1970. *Ann. Sci. Forest* **27**(1), 39–61.

Tran Vinh An, J. and A. J. Herbillon 1966. *C.R. Conf. Péd. méditerranéenne*, Madrid, 255–64.

Turc, L. 1961. *Ann. Agron.* **12**(1), 13–49.

Vallée, G. 1966. *Nouvelles contributions à l'étude du rôle du manganèse dans la régénération de la sapinière vosgienne*. Engng doct. thesis. Fac. Sci. Nancy.

Vandamme, J. and L. de Leenheer 1968. *Pédologie*, Ghent **XVIII** (3), 374–406.

Védy, J. C. 1973. *Relations entre le cycle biochimique des cations et l'humification en milieu acide*. State doct. thesis. Univ. Nancy I.

Wilding, L. and L. Drees 1974. *Clays and Clay Minerals* **22**, 295–306.

Wilke, B. M. 1975. *Z. Pflanzerner. Bodenk.* **2**, 153–71.

Zaidelman, F. R. 1974. *Podzolisation and gleyification*. Moscow: Nauka.

Zöttl, H. and H. Kussmaul 1967. *Anales Edafol. Agrobiol.* **XXVI** (1–4), 381.

# Chapter 4

# General principles of the origin and development of soils

## I Soil development cycles: definitions

When mineral material is exposed at the surface, it is gradually colonised by vegetation – herbaceous plants first of all, then shrubs, and finally, in a humid temperate climate, trees. At the same time, the soil forms and develops: first of all a humus-rich horizon is formed on the surface (AC type profile) then a (B) or B horizon gradually appears and thickens little by little: *thus a succession of profiles of increasing development occurs in parallel with the succession of plant communities*. This twofold development leads to a stable equilibrium that characterises both vegetation and soil, and which ecologists refer to as the **climax**; such a development towards the climax can be referred to as *progressive*.

### 1 Phases of the cycle of progressive development

Various authors have investigated many examples of such simultaneous development of soil and vegetation on a variety of parent materials, such as dune sands, alluvium and particularly moraines exposed by the glacial retreat of the last century, which form an ideal material for study, as the age of soils on the surfaces successively uncovered by glacial retreat is easily determined. Thus, Ludi (1945) investigated one such cycle on the Aletsch glacial moraines which are at an elevation of 2000 m and, in their initial state, are calcareous. Such materials are first colonised by a calcicolous herbaceous vegetation but then, as a result of decarbonation, an acid ranker develops with acidophilic shrubs (green alder) and finally, on a moraine exposed for almost a century, an ericaceous heath (*Rhododendron* and *Vaccinium*) is established on an, as yet, slightly developed podzol which is, of course, still far from being in a state of stable equilibrium.

The close relationship that exists between *the phases* of plant colonisation and those of soil development in this example will be noticed (Table 4.1). *Such a succession of phases leading to a stable profile make up what is often called a cycle of development*. The dynamics of the profiles concerned involve the three fundamental processes discussed in the first three chapters: (i) the incorporation of organic matter to form the humus-rich surface horizon; (ii)

**Table 4.1** Progressive development on a glacial moraine.

| Vegetation | Mosses, dwarf willows | → | green alder | → | ericaceous heath |
| Soils | initial eutrophic brown soil | → | mor ranker | → | young podzol |
| Age | (5 years) | | (30–40 years) | | (90 years) |

the weathering of minerals; and (iii) the movement of material within the soil in the soluble or pseudo-soluble state, or as suspensions, all of which, it will be remembered, are closely related to the environment. *It is the interaction of these three fundamental processes, as controlled by the environmental conditions, which allows profile development as a whole to be explained*. But in the first three chapters the *time factor* was not considered, at least in any detail. This will now have to be done in the synthesis which is the subject of the present chapter.

First, we must note that the three processes proceed at different rates; therefore they are only coincident in part and the equilibrium state will develop more or less sooner for one than the other, which inevitably causes a time lag between one process and another. Thus it has been established that humus-rich A1 horizons are formed more rapidly than mineral (B) or B horizons. Some (B) or B horizons are more closely related to the vegetation- and humus-rich horizons than others and so develop more rapidly. As far as illuvial B horizons are concerned, there are two possible origins: either they are formed directly by the movement of materials from the A horizon or, in contrast, a previously formed weathered (B) horizon is enriched by movement of materials. This can be represented as shown in Figure 4.1.

However, it must be stressed that the final phase of progressive development is not necessarily of the ABC type. Thus development can stop at the AC stage, as in the case of a rendzina, or at the A(B)C stage, as in the case of an acid brown soil, as well as at the ABC stage, where the B horizon can be *argillic* or *spodic* depending on the vegetation or the environmental conditions.

## 2 Regressive development

Thus vegetation has a fundamental role in soil development and the maintenance of its equilibrium state, both in terms of biological equilibrium, resulting from the microflora and fauna, and biochemical equilibrium, resulting from humification and the biogeochemical cycle. In addition, it prevents profile destruction by its protective effect against erosion, by both rain (surface runoff) and wind. In this respect it should be pointed out that forests give

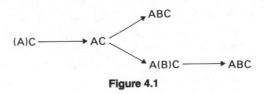

**Figure 4.1**

greater protection than herbaceous covers (steppe or prairie) and therefore the most pedologically developed profiles are to be found under forests.

*The destruction of such a stable equilibrium (or climax) as that described above is generally the result of a fairly sudden removal of vegetation – particularly forest – caused by a geological process, such as a natural catastrophe or, more simply, by man.* The sudden increase in erosion results in either the complete removal of the profile, exposing the C or R horizon or the opposite – its burial by new alluvial material deposited by surface water flow. In both cases, mineral materials in their initial states are again exposed; hence the name **regressive development** is often given to this type of reverse development (Pallmann 1947, Pallmann *et al*. 1949).

Very frequently, however, this erosive regression is incomplete, resulting in a **truncation** in which the upper horizons of the profile are removed. This exposes, for example, a more compact or more indurated B horizon which can subsequently act as parent material for new development that is often different from the first, if the climatic conditions have changed. This second development will thus give rise to *polycyclic or complex soils*, the interpretation of which is difficult and will be examined in more detail later.

Such cycles of progressive and regressive development can succeed one another on very different time scales, which may be either at relatively short intervals as a result of natural or man-made catastrophes, such as avalanches, floods and forest clearing, or at geologically long time intervals, when they are characterised by a succession of climates which are sometimes favourable to vegetation (**phases of biostasis**) or, in contrast, favourable to erosion (**phases of rhexistasis**, Erhart 1967).

This regressive development can be compared to profile rejuvenation. In most cases, particularly when it occurs on steep, mountain slopes, rejuvenation occurs suddenly because of an external cause such as clearing avalanches etc.; however, there is also another kind of rejuvenation which is *continuous* and which occurs in certain circumstances even under protective near-climax vegetation. In this case, more moderate erosion limits the progressive development to rapidly formed horizons, such as those that are humus rich, while the more slowly developed (B) or B horizons are not able to form. Even on almost level sites, certain profiles on unindurated rocks always remain young and of the AC type, which is the case for initial rendzinas on chalk and of some pelosols on unindurated shales. On more indurated rocks, AC profiles mainly occur where steeper slopes increase the amount of erosion (see Ch. 6, Sec. II/2).

## 3 Short and long cycles

It has already been mentioned that (B) or B horizons form more slowly than A horizons; even so, as will be seen, the time taken for the complete development of B horizons is very variable, depending upon the conditions. This can be less than 1000 years, for certain spodic horizons of podzols, to several hundred thousand years for those formed in a hot climate, such as the oxic horizon of a ferrallitic soil. It is obvious that the concept of soil–vegetation equilibria cannot be the same in both of these cases, for where (B) or B horizons are formed rapidly a *short cycle* is involved in which *biochemical weathering is dominant*; while where their formation is slow, a *long cycle* is involved in which *geochemical weathering is dominant*.

From the conclusions reached in the three preceding chapters, it can be said that *while short cycles, characteristic of temperate or cold climates, are completely dependent on the development of organic matter, in contrast, long cycles which are dominant in hot climates are relatively independent of the processes of humification*. This statement may appear paradoxical and contrary to the law of the effect of temperature on reaction speed but, in fact, it is explicable by the differences between temperate and tropical weathering which were emphasised in the first chapter. Temperate weathering is not complete and, in terms of clay formation, never goes beyond the phase of bisiallitisation. Tropical weathering is much stronger and all minerals, resistant or not, are weathered and silica loss is much greater, leading to the formation of kaolinite or even gibbsite.

Thus it appears as if weathering consists of two phases, where the first is rapid and affects the most susceptible of minerals and leads to bisiallitisation and the second is much slower in which all the remaining minerals are attacked, including the 2 : 1 clays formed in the first phase. Now, this slow phase is well developed in hot climates (Spaargaren 1979) but is practically absent in temperate climates (Novikoff *et al.* 1972).

Short cycles are absolutely characterised in terms of the simultaneous and gradual development of soil and vegetation to produce an *environmentally determined humus type*, such as those described in Chapter 2. *These cycles have been completed within a period of geological time in which there have been no great climatic changes* and, although the vegetation may have been subject to certain periodic changes (which will be discussed later), the main direction of pedogenesis has not been greatly modified. It is man's destructive actions that have caused important processes of degradation of the soil–vegetation equilibrium, which will be examined in a later section.

Soil formation in the atlantic temperate zone illustrates this particularly well, for short cycles of development are most characteristic, with durations not exceeding the 10 000 to 12 000 years of the postglacial period, during which there has been little climatic variation. However, this does not exclude the occurrence of some *long cycles* operating on very old materials, such as pre-Würmian loams, where things become more complicated, for these are polycyclic soils in which the present-day cycle, which alone characterises the soil–vegetation equilibrium, is superposed on the generally pre-Würmian long cycle.

In hot, humid climates, although some short cycles involved in the formation of humus-rich horizons are in operation, pedologists do not consider them as being characteristic of climatically controlled development, for the effect of these cycles is restricted to the surface, while generally profiles are up to several metres deep. The deeper horizons are outside the influence of surface organic matter and they form slowly as a result of physicochemical processes controlled exclusively by the pedoclimatic conditions at depth. As will be seen, vegetation is still closely related to pedoclimatic conditions, but not directly with pedogenesis at depth, as it is in the much shallower profiles of the temperate zone.

In the case of long cycles, the idea of gradual development and of climax is therefore rather vaguer than in the situation where short cycles are dominant. The law of climatically determined pedogenesis is nevertheless applicable, but because of the slowness of the pedogenic cycle the evidence is more difficult to obtain because of various disturbing factors. (i) Over such a long period it is rare for considerable climatic variations not to have occurred. (ii) The amount and even the kind of soil development are more closely related to the age of outcrop, so that the final stage of soil development occurs only on materials that have been exposed for a very long time.

## II  The time factor: determination of soil age

Methods of soil age determination are different according to whether *short cycles*, in which humification is the dominant process, are being dealt with, or *long cycles*, that are relatively independent of humification. In the case of short cycles, methods are concerned with determining the speed of biological processes in terms of the carbon cycle, the relationships between vegetation and humification and the formation and development of organomineral complexes. The results, while not being absolute, are sufficiently accurate so that, by cross checking between the results of the different methods, the main stages of soil formation and their relative duration can be determined. In contrast, methods for determining the length of long cycles are fewer and the results obtained much less reliable.

### 1  Investigation of the length of short cycles in temperate climates
A brief review of the methods used will be given before considering the results obtained from a comparison of these methods.

**Methods of investigation.** There are a series of methods that go from the simplest, such as the direct observation of processes after colonisation of a new parent material, to the more complex, such as those of palynology, archaeology, calculation of the speed of a process and extrapolating it over time and $^{14}C$ dating. In general, apart from the first, these methods involve calculations and interpretations which are often difficult, and hence results are not obtained directly.

*Observation of a process after colonisation of a new parent material.* This is the simplest method which was used by Ludi and referred to previously. It is well suited to investigating the formation of humus-rich surface horizons, at least as far as biological humification is concerned. Humus-rich horizons reach a state of equilibrium when the annual addition of carbon by litter and root decomposition is exactly balanced by the amount lost in mineralisation. The time required to reach this equilibrium varies according to the climate, the initial parent material colonised and the colonising vegetation. Of the many

**Table 4.2** Climatic phases and associated vegetation since the last glaciation.

| | Age | Vegetation |
|---|---|---|
| last glaciation | More than 8500 BC | tundra |
| pre-boreal | 8500–7500 BC | pine–birch |
| boreal | 7500–5500 BC | pine–hazel |
| atlantic | 5500–3000 BC | mixed oak forest and hazel |
| sub-boreal | 3000–800 BC | beech–fir (montane) |
| sub-atlantic | 800 BC–AD 2000 | beech–fir–birch (montane) |

authors who have done this kind of research, consideration of results will be restricted here to those of Laatsch (1963).

*Palynology*. The study of peat pollen is a very old-established technique. The gradual upward growth of peat and the associated stratification of the pollen grains allow the late Quaternary climatic sequence to be determined, which is important to pedologists in evaluating the vegetation associated with this succession of climates which have been involved in the formation of temperate soils. In summary, this classification from oldest to youngest should be recalled (Table 4.2). In addition, it is to be noted that in the Bronze Age (at about 1000 BC) large-scale forest clearing by man started.

But, the investigation of phases of soil development has been pursued much further by modern palynological techniques. Thus, several authors (for example, Dimbleby 1961 and Munaut 1967) have shown that pollen is *transported* within some very porous sandy *podzolic* soils, so that there is a stratification in their mineral horizons comparable to that found in peats. This makes it possible to determine the climatic phases that have controlled soil development, which, as they have been dated by other means, gives the age of the soil; in addition, the successive vegetation types responsible for the process of podzolisation can be characterised. However, this method is inapplicable to acid brown soils with mull, which have very great biological activity, for the mechanical mixing of the horizons by soil fauna appear not to allow the development of a regular pollen stratification corresponding to successive plant communities.

*Prehistory and archaeology*. Investigation of tumuli dated by traces of ancient civilisations has provided important information to pedologists (Dimbleby 1962). A good example is provided by the work of Sachse (1965), from the Halle region, who compared the soil developed on a tumulus 4500 years old with the soil buried under the tumulus. A brunified chernozem with signs of pervection (index of clay movements of 1 : 1.5) is developed on the tumulus, while the soil under the tumulus shows no sign of pervection. Thus, in this time interval the climate has become more favourable to pervection which, as it only occurs in the soil on the tumulus, has taken 4500 years to develop.

*Extrapolation over time of a process of known speed*. This is done by measuring a particular biochemical process over a relatively short time period, either in the field or the laboratory. The speed of the process thus being known, it is then possible, by the detailed investigation of the present state of a profile, to determine the minimum time required for its formation. This method has been applied to a wide range of processes.

(a) *Solution and transport of carbonates* under a given climate and vegetation, such as was determined by Arkley (1963, 1967) in New Mexico. In this regard see also the results of various authors, referred to in Chapter 3, in the discussion of the decarbonation of calcareous soils.

(b) *Pervection of clay*. In Chapter 3, a comparison was made between the rapid pervection of fine clays in calcium-rich, aerated conditions and the slow pervection in acid, badly aerated conditions. It can be seen that the first is characteristic of short and the second of long cycles. Of these, only the first has been measured with reasonable accuracy, by German workers, for a uniform loess under various kinds of vegetation, by making a very exact determination of clay balances on soil columns in terms of losses from the A horizon and gains in the B, which has enabled the speed of pervection in the Postglacial period to be determined (Zöttl & Kussmaul 1967). These determinations will be looked at in greater detail in the study of brunification and pervection (Ch. 9).

(c) *Podzolisation*. Stone and McFee (1965) have provided a good illustration of the measurement of the speed of podzolisation, from the time of its initiation, in very favourable conditions, with coniferous vegetation on sandy parent materials in the Adirondacks. Determination of the fire history enabled the authors to establish an exact correlation between three factors: the age of the vegetation, the thickness of the surface A0 horizon and the quantity of organic matter immobilised in the Bh after migration. The correlation between the thickness of the A0 horizon and the quantity of organic matter in the B is linear and, as this organic matter accumulates at a rate of almost 100 kg/hectare per annum, it takes only a few centuries for the podzol characteristics to develop. However, as yet the profile equilibrium state has not been reached, nor has the rate of development slowed down. Nevertheless, this work and that of Ludi (1945) show that the initial stages of podzolisation are exceptionally rapid in favourable conditions.

*Dating of organic matter by* $^{14}C$. This method requires very careful interpretation for it does not give the age of the soil directly and therefore the results should be compared with those obtained by other methods. Its basis is that in living plant materials there is a very low but constant amount of the radioactive isotope $^{14}C$ (about $1.85 \times 10^{-10}$ of total percentage of carbon). When plant material is dead (fossilised and humified), the $^{14}C$ decreases as it is no longer replaced and, as the half-life of this isotope is about 5500 years, it is possible to calculate the age of this fossilised or humified organic matter. By this means Guillet (1965) was able to determine the age of successive layers of peat in the Vosges and to compare the results with those obtained from palynology. The humic material of the peat is sufficiently inert for the age of the carbon to correspond well with its date of formation. Unfortunately, this is not the case for soil humus, which is constantly renewed by the addition of recently formed organic matter and rejuvenated by the mineralisation of older material at a very variable rate. *Dating by* $^{14}C$, *therefore, does not give the real age of the soil, but rather what may be called the mean residence time (MRT) of the carbon, which increases as the rate of organic matter turnover decreases*. Thus, as will be seen, the biological humus of the brown atlantic soils has a much lower MRT than that of the climatically matured, stabilised chernozem humus. In conditions of rapid organic matter turnover, the MRT is not necessarily related to the time at which pedological processes involving this organic matter were initiated, which is the case for some atlantic forest podzols (Tamm & Holmen 1967, Guillet 1972). In contrast, there is a high correlation between the MRT, measured in the Bh, and the age of degraded podzols, formed under an ericaceous vegetation, where the organic matter is almost inert (see below, Sec. III/4).

**Results: the length of short cycles.** Comparing all available results, it can be said that, in a temperate climate, all horizons characteristic of short cycles have been formed within the Postglacial period and that the equilibrium state of a forest profile is reached in less than 10 000 years.

*A1 humus-rich horizons*, resulting from biological humification, are those which form most rapidly. Laatsch (1963) showed that 600 to 1500 years is required to reach equilibrium, depending on the kind of vegetation and the chemical richness of the parent material. The MRT, as determined by $^{14}C$, is very low and is never greater than 300–400 years (Paul *et al.* 1964).[*] In contrast, the *andosol mulls*, in which mineralisation is blocked by allophane, have a much higher MRT that can be up to 4000 years for the most resistant fraction (Hetier 1975). Such a prevention of mineralisation is of physico-chemical origin, but in the case of the *chernozems* it is the result of *maturation* which affects a part of the humus (see Ch. 2). The MRT of a chernozem often reaches 1000 years at the surface (Paul *et al.* 1964), where the most labile part of the organic matter is particularly concentrated, and this increases with depth, frequently reaching 4000–5000 years at the base of A1 (Scharpenseel 1972).

*Weathered (B) horizons of the cambic type* (brunified soils with mull) are evidently formed at a much slower rate than those of corresponding A1 horizons and this is largely dependent on the nature of the parent materials. It is relatively rapid – 3000–5000 years (Wilke 1975) – if the soil is formed by decarbonation of calcareous material, but on granitic material it is slower and,

---

[*]It is to be noted that a higher value (1250–1450 years) was found in a cultivated, Ap, horizon at Rothamsted (Jenkinson & Rayner 1977). This is explicable by the fact that even atlantic mulls have a minor (15%) stabilised fraction with a high MRT (2000 years). After cultivation, the importance of this fraction increases because of the increased speed of turnover of the labile fraction.

according to Birkeland (1974), a minimum of 10 000 years is necessary for the complete development of a brown soil (B) horizon.

In this respect, it is to be noted that there is often confusion in the case of the development of acid brown soils on granitic materials between the strictly pedogenic *brunification* and the much slower prepedogenic *deep regolith formation*, which seems never to have been influenced by organic matter. Souchier (1971) has clearly distinguished the two processes in his work on these soils in the Vosges, with brunification being assigned to a short cycle and deep regolith formation to a long cycle in which, additionally, the stages are not well known. The great differences that exist between these two processes, such as the neoformation of kaolinite only occurring in deep regolith formation, have been referred to in the first chapter.

*Illuvial horizons* are formed as rapidly as, and sometimes more rapidly than, weathered (B) horizons. In favourable conditions, podzolisation starts and continues, at least in its initial phases, at a very rapid rate. The two processes of acidolysis and complexolysis are involved from the start, with a spodic B horizon being formed in parallel with an A0 surface horizon. In exceptionally favourable conditions (humid climate, coniferous forest, sandy parent materials), complete podzol development can occur in a few centuries – such as in the case of **egg-cup** or **funnel-shaped** podzols, that are seen under certain isolated conifers, where they are, in fact, the same age as the trees (Duchaufour 1978: $I_4$).

However, the speed of podzolisation varies considerably depending upon the environmental conditions and the kind of podzol. Thus, podzolic soils under deciduous forest on parent materials with little sand have, according to palynological studies, developed relatively slowly and, according to Dimbleby (1962), Munaut (1967) and Guillet (1972), belong to the *atlantic* period. In contrast, the *podzols of degradation* under *Calluna* are formed more quickly because of the more active and less biodegradable organic matter. On the atlantic plains this amounts to 2000–3000 years and about 1000–2000 years for the Triassic sandstone of the Vosges (Guillet 1972). Note that even when a relative equilibrium state has been reached, as reflected in the thickness of the ashy A2 horizon, podzol development still takes place very slowly by a gradual thickening of the spodic B horizon by a mechanism which will be investigated in Chapter 10 (Franzmeier & Whiteside 1963, Harris 1970). If the initiation of a spodic horizon is relatively sudden, that of pervection under mull on sedimentary rocks is much more gradual. This process has been followed by many authors working on calcareous loess, who showed that, while calcium carbonate is present, fine clays are freed very gradually, increasing as decarbonation increases. Thus pervection starts slowly, then it increases in speed, to slow down again when conditions become acid. In well aerated and slightly acid conditions, the graph of pervection is S-shaped (Zöttl & Kussmaul 1967) and the development of the equilibrium profile on loess takes 6000–8000 years. The full development of lessived brown soils, which are characteristic of the atlantic deciduous forests, has taken about 10 000 years, or practically the whole of the Postglacial period (Meyer 1960, Wilke 1975).

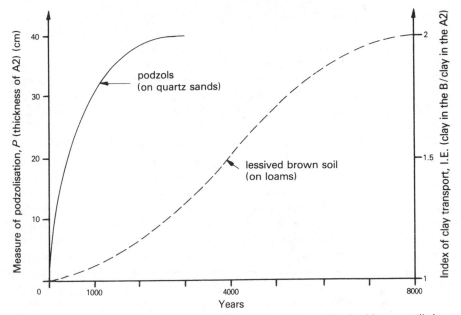

**Figure 4.2** Comparative speed of development of podzols and lessived brown soils in an atlantic climate.

In Figure 4.2 a comparison is made between the relative speeds of podzolization and pervection in an atlantic climate.

## 2  Investigation of long cycles: time of development

The investigation of a long cycle is much more difficult than that of short cycles and the results obtained are less precise and more equivocal. First of all, as in the case of the short cycles, the methods employed will be reviewed, to be followed by a more detailed examination of the results in a separate section.

**Methods of investigating long cycles.** There are no more than two, of which the first is practically the same as that discussed in the case of the short cycles where a physicochemical process measured over a short time period is extrapolated. This was used by Leneuf and Aubert (1960) to determine the speed of ferrallitisation in terms of the amount of silica removed in drainage waters in a year in relation to the amount of silica in the fresh bedrock. The results can be no more than very approximate, for, as will be seen, the time involved is very long and the amount of silica in solution decreases as the weathering progresses. However, it does give an order of magnitude of the time required for ferrallitisation in particular conditions. Similarly, account should be taken of the tentative conclusions of Spaargaren (1979), that in southern Italy it takes 500 000 years for a metre thickness of terra rossa to form on an indurated limestone – about the same time as is needed for the slow change of the illites of the limestone to kaolinites.

The second and simplest method is also that which has been most used (Begon & Jamagne 1973, Icole 1973, Jamagne 1973, Mückenhausen 1973a,b, Hubschmann 1975, Bornand 1978, Daniels *et al.* 1978) and *involves a comparison of the kind (and the degree) of development to which materials of similar composition but of geomorphologically different ages have been subject*. Classical studies of this type have been made on alluvial terraces, the age of which increases with height, and also on moraines and periglacial aeolian loams.

Unfortunately, there are many difficulties and possible errors, of which the main ones are that the dating of pre-Würmian deposits is often very inaccurate and that the materials of the different terraces of a particular valley are not always comparable, as there is frequent contamination by (generally aeolian) loam which has not been precisely evaluated, either in terms of its amount or age, compared to the subjacent material. In certain cases, such as those studied by Hubschmann (1975) on the terraces of the Garonne Basin, these loams are thick enough for a soil to be formed which is of a much more recent age than that of the terrace itself. However, generally the mixture between the loams and the older alluvial materials is so intimate that valid comparisons are possible from one terrace level to another. To be valid, these investigations must be of a precise and detailed nature that alone are capable of improving the quality of the information obtained, as the following examples illustrate: (i) *the mineralogical balances* of Hemme (1970) to assess the relative importance of an intimate mixture of aeolian loams and decarbonated clays in a terra fusca; (ii) *microstructural investigations* of Begon & Jamagne (1973) and Jamagne (1973) of various types of clays of different pedogenic phases to which materials have been subjected; (iii) *investigation of the weathering of the cortex of pebbles* (Icole 1973), in which the pebbles are selected so as to have a comparable structure and mineralogy from one terrace level to the other (Duchaufour 1978: $I_6$).

**Results: duration of long cycles.** It would seem that the minimum time required for long cycles is something rather less than 100 000 years. According to the work of Leneuf and Aubert (1960), Segalen (1973), Birkeland (1974), Yaalon (1975) and Daniels *et al.* (1978), the time period can extend from about 100 000 years (unsaturated ferruginous soils of the **ultisol** type) to a million years (ferrallitic soils), for hot and humid climates.

As far as cold temperate climates are concerned, if these figures are compared with the length of the climatic phases of the late Quaternary, it can be seen that, as the Würm glaciation lasted 80 000 to 90 000 years, some temperate soils affected by long cycles must go back at least to the Riss—Würm interglacial and they could be much older. Some temperate soils that show signs of fersiallitic or ferruginous (or even ferrallitic) processes, certainly belong to the early Quaternary or even Tertiary. But if, as previously stated, pedogenesis could continue through certain phases of the Würm period, it has been markedly restricted or interrupted during the glacial phases. At the same time, the processes of mechanical disturbance of the upper part of soils (by cryoturbation or solifluction) checked the biochemical or physicochemical processes of pedogenesis. In these circumstances, most of the indications of long cycles in temperate regions that do occur, such as strong weathering, marked desaturation of the absorbent complex and some aspects of clay pervection, must have been acquired prior to the Würm glaciation. This topic will be discussed later and the conclusions to be drawn from it examined.

When warmer regions (from the Mediterranean to the Equator) which have not been subject to glaciations are considered, the pedogeneic implications of long cycles are totally different, for although pedogenesis has had phases of slowing down and speeding up in line with climatic variations, it has not been interrupted. Thus, it is not possible to differentiate two phases of profile development, as in a cold temperate climate, and it also explains why weathering profiles are much deeper in warmer climates.

If we assume, as has been demonstrated, that organic matter is involved only in short cycles, it can readily be seen that, in the case of tropical soils, the recent formation of surface organic horizons often has no connection with the obviously older (B) or B mineral horizons. However, frequently in the case of ferrallitic soils, as will be seen in Chapter 12, a short cycle controlled by organic matter affects the upper part of the profile and is superposed on the long cycle that affects the whole of the profile. On the other hand, as already stated, the correspondence between pedogenesis and climate is less evident for long than for short cycles, as a very long time is required for an equilibrium state to be reached, while many parent materials have been deposited, or uncovered by erosion, at too recent a date for profiles to achieve this state. Thus, several soil types at various stages of development coexist in the same climatic zone; this obviously complicates the environmental study of tropical pedogenesis.

## III  Effect of environmental factors on development cycles

As the short cycles of temperate or cold climates are simpler and better known than the long cycles, they will form the basis of the discussion to be given here. It will then be seen to what extent they are applicable to the long cycles. Short cycles are closely related to vegetation, which is evidently an essential environmental factor. This vegetation contributes a specific kind of organic matter to the mineral material, which, as was emphasised in the first three chapters, profoundly influences pedogenesis by its effect on weathering, the formation of organomineral complexes and the biogeochemical cycling of elements. *Thus humus occurs as the connecting link between the living and the mineral worlds, where it integrates all environmental factors*, controls the direction of pedogenic development and gives the soil its major properties.

Therefore, to understand soil formation it is necessary to understand completely the reciprocal effect of natural vegetation on soil and of soil on vegetation or, in other words, of investigating the equilibria of *natural ecosystems*. Only stable ecosystems which have been long established and not disturbed by man (*the climax*) permit the soil–vegetation equilibrium characteristic of a particular environment to be determined. Some of these ecosystems are to a large extent controlled by the general climate (*climatic climax*), while others are in contrast controlled by local site factors (*site climax*).

## 1 Definitions: environmental factors and pedogenesis

In the first three chapters, dealing with the three fundamental processes of pedogenesis (weathering, humification and the transport of materials within soils), a basic distinction was made between two major groups of environmental factors: *general bioclimatic factors* and *site factors*, which will be used again in the present synthesis.

The general bioclimatic factors delineate the *climatically controlled zones* of soil and vegetation, considered on a world scale, such as those described by the Russian school, and in particular Dokutchaev at the end of the last century. The climatically controlled plant formations (or **biomes**) running parallel to lines of latitude are evidently relatively independent of local factors, particularly that of parent rock, while each climatic zone is characterised by the same type of **zonal** soil. From the beginning, however, certain exceptions have been recognised within these climatic zones and these are **intrazonal** soils, which are more closely related to local site conditions. Generally, these intrazonal soils occur as patches of restricted extent within a particular zone, characterised by a *specialised* vegetation which is different from the general climatic plant association of the zone, such as peats within forest zones and halophytic vegetation within the steppe zone, etc.

These basic ideas remain generally valid and several modern authors have adopted them – such as Lemée (1967), Ganssen (1972) and Boulaine (1975) who distinguish 12 major climatic zones on a world scale. In addition, they have been able to be more exact and recognise more subtle differences as a result of recent work in three areas:

(a)  the controlling part played by the type of humus;
(b)  the idea of a threshold effect;
(c)  the possibility of a break in the natural equilibrium causing degradation as a result of human intervention.

**The part played by the type of humus (climatic climax).** The recognition of the controlling influence of humification on pedogenesis has allowed the concept of zonal soils to be completed and made more precise. In a given zone the climatically controlled plant formation may be relatively uniform – which is not the case for the soils, at least as far as the characters of the deep mineral horizons are concerned, for they bear the imprint of the parent material only. The humic horizons and part of the A2, (B) and B mineral horizons, in as far as they are affected by climate and vegetation, are subject to a *convergent development*. On this basis Pallmann (1947) and Pallmann *et al.* (1949) have defined *analogous soils* as being soils formed by the same climatically controlled vegetation on different parent materials, which thus have a similar type of humus and pedogenesis while, in contrast, the lower part of the profile reflects the properties of the parent materials.

**The idea of a threshold effect leading to a site climax.** Even though regional climatically controlled plant associations (Duvigneaud 1974) are relatively independent of local environmental conditions, this is not so everywhere, for in some extreme cases there is a threshold effect which prevents their establishment and leads to the development of a *site association* (Duvigneaud 1974), or even a *specialised association* (Favarger & Roger 1956), where humification is different from that of the climatically controlled plant association. *Thus it is possible to speak of a divergent humus development and, as a consequence, of divergent soils* which at equilibrium give rise to a *site climax*. Thresholds may result from various causes: mineralogical, physicochemical (i.e. the nature of the parent material) or pedoclimatic, such as the water balance, which is often controlled by the topography (marsh peats being a particularly good example).

**Degradation by human intervention.** Frequently man disturbs the natural equilibrium by modifying the climatically determined plant association and causing it to be replaced by a new *secondary association*. This in turn leads to a change in the humus type which causes the original profile to develop in a totally different way compared to the normal progressive development. It is a question of *degradation*, in which the new artificial equilibrium is sometimes relatively stable and at other times, in contrast, is very temporary. It is possible for the original equilibrium to be re-established by a restoration of the vegetation, either naturally or artificially. This process of climax restoration is sometimes referred to as **regradation**.

**The case of long cycles**
It is evident that the preceding considerations are fully applicable to short cycles which are so characteristic of boreal and temperate pedogenesis where there is such a close genetic connection between vegetation, humification and pedogenesis. However, to what extent can these considerations be applied to the hot climate zones where long cycles are dominant? In fact, even though in these regions vegetation and organic matter no longer have a controlling influence on soil formation, it is still possible to compare the overall effect of the general climate, which is involved, particularly in terms of variations in moisture status, with that of local site factors. *Thus the idea of the climatically controlled soil and the site-controlled soil is of interest in all climatic zones.* However, the characteristic equilibria of long cycles have certain peculiarities. First of all, the relationships between environmental factors and soil formation are undoubtedly more difficult to establish because of the slowness of development. Then, when there is evidence of a soil–humus–vegetation correlation, this is not necessarily the result of a direct reciprocal reaction between the three terms, as it is in the case of the short cycles, but rather *to a parallel development of the biological factors and pedogenesis, which are both controlled by the climate* or possibly by an adaptation of site-controlled plant associations to soil properties controlled by the site. Thus, despite the absence of strong genetic connections between the biological factors of vegetation and humus and the soil, they remain important as soil indicators.

## 2 The effect of general bioclimatic factors
The general climate and climatically controlled vegetation jointly control pedogenesis in climatic zones, being responsible for the convergent humifica-

tion on a variety of parent materials, characteristic of *analogous soils*. The climatic zonation of humus has been discussed several times in the preceding chapters and particularly at the end of Chapter 2.

A very good example of analogous soils is provided by the zone of boreal podzols of the USSR (*Taiga*). Comparing the soils on two types of parent materials (aeolian loams and morainic sands), the upper horizons are the same consisting of a very thick A0 and an ashy A2 characterised by strong clay hydrolysis. In contrast, the B horizons are markedly different (see Ch. 10).

In 1949, Pallmann *et al*. gave other examples of analogous soils associated with the altitudinally controlled plant communities of the Swiss Alps. The succession from the lowest to the highest altitude is comparable to that in going from south to north in the USSR. Mull in the coniferous–deciduous mixed forests of the lower montane zone, thick mull–moder of the upper montane zone, moder in the coniferous forests with *Vaccinium* in the lower sub-alpine zone and mor in the coniferous forests with *Rhododendron* in the upper sub-alpine zone. The soils of this last zone provide a particularly good example of analogous soils, for a thick acid A0 mor horizon occurs on both silica-rich parent materials (such as granites, schists and sandstones) and calcareous materials (Bottner 1972).

**Intrazonal soils**

Most of these soils can be considered as having a *mixed* origin, for their development requires the joint effects of the overall bioclimatic conditions and local site conditions, such as parent rock, topography etc. Thus such soils are more or less restricted to particular sites, *but only within particular climatic zones* (Table 4.3).

**Sequences of climatically controlled soil development in non-mountainous areas** (Table 4.4). The best way to understand fully the influence of general climate on soil development is to compare a sequence of zones that differ from one another in only one climatic factor, such as the mean temperature or moisture status involving the climatically controlled drainage. Table 4.4, derived partially from Ganssen (1972), gives three sequences established by this procedure. Two of them are related to variations in moisture status, one in a cold and the other in a hot climate, while the third is characteristic of humid climates where the temperature varies. In addition, Figure 4.3 shows the distribution of the different zones on a world scale.

**Table 4.3**  Distribution of the main intrazonal soils

| Soils | Climatic conditions | Site conditions |
|---|---|---|
| vertisols | hot climate, with marked seasonal contrast | impeded depressions |
| planosols | tropical or subtropical climate with marked seasonal contrast | temporary surface water table |
| salsodic soils | arid climate (hot or cold) | parent material rich in sodium |
| andosols | very humid climate | volcanic parent material |
| stagnogleys | very humid climate (mountains) | hydromorphic |

**Table 4.4** Climatically determined soil sequences.

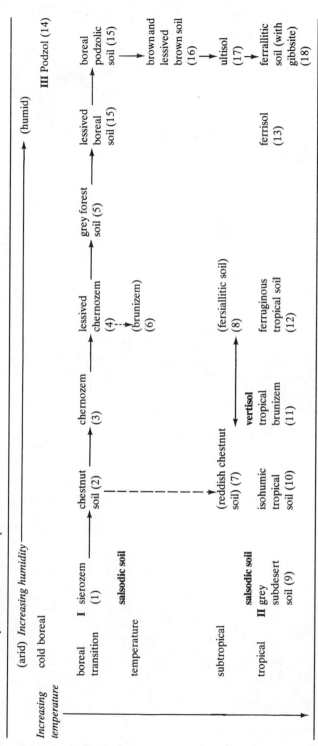

*Increasing temperature* ↓    (arid) *Increasing humidity* → (humid)    **III** Podzol (14)

| | | | | | | | |
|---|---|---|---|---|---|---|---|
| boreal transition | **I** sierozem (1) · **salsodic soil** | chestnut soil (2) → | chernozem (3) → | lessived chernozem (4) → | grey forest soil (5) → | lessived boreal soil (15) → | boreal podzolic soil (15) |
| temperature | | | | (brunizem) (6) | | | brown and lessived brown soil (16) |
| subtropical | | (reddish chestnut soil) (7) | | (fersiallitic soil) (8) | | | ultisol (17) |
| tropical | **II** grey subdesert soil (9) · **salsodic soil** | isohumic tropical soil (10) | **vertisol** tropical brunizem (11) | ferruginous tropical soil (12) | ferrisol (13) | | ferrallitic soil (with gibbsite) (18) |

**I**, cold climate sequence with decreasing aridity; **II**, hot climate sequence with decreasing aridity; **III**, humid climate sequence with mean temperature increasing. Intrazonal soils shown in bold type also.

Plant associations: (1) cold semi-desert, (2) sparsely vegetated steppe, (3) densely vegetated steppe, (4) forest–steppe, (5) boreal continental deciduous forest, (6) prairie, (7) garrique or maquis, (8) *Sclerophyll* forest, (9) hot semi-desert, (10) sparsely vegetated savanna, cacti and grasses, (11) thornbush savanna, (12) tropical savanna with sparse trees, (13) humid tropical forest, (14) boreal coniferous forest and ericacae, (15) mixed boreal forest, (16) atlantic deciduous forest, (17) humid subtropical forest, (18) very humid equatorial forest.

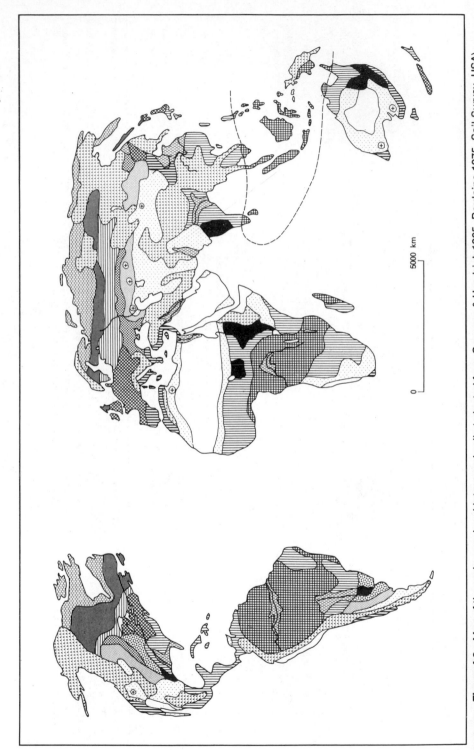

**Figure 4.3** Map of the main zonal and intrazonal soils (adapted from Ganssen & Haedrich 1965, Boulaine 1975, Soil Survey, USA).

5000 km

0

Three important climatically determined soil groups, which are outside these three sequences, have been placed in brackets with a dashed arrow to indicate the climatically determined group to which they are most closely related.

In addition, certain intrazonal soils with mixed development, such as sal-sodic soils and vertisols, have been put in boxes near the zonal soils to which they are most closely related. However, the planosols, andosols and stagnogleys are not included in the table, for they are not restricted to any one zone.

*From Figure 4.3 it can be seen that the vertisols are important on outcrops of basic rocks in the dry Tropics.*

*Sequence I: succession of soils in cold climates with decreasing aridity.* A preliminary comment is required as this sequence, which is represented horizontally in Table 4.4, actually occurs in the USSR from south to north and a decrease of temperature in this direction may, quite legitimately, be expected. This temperature gradient does occur but it is not pedologically significant: in winter, freezing temperatures are common throughout the sequence as a result of the extreme continentality of the Russian climate, and in summer it has its greatest effect indirectly on the moisture status through *PET*. Thus, where steppe is dominant, *PET* is greater than precipitation, while forest becomes dominant when precipitation exceeds *PET* and in between is the contested forest–steppe zone (with thick chernozems) where the two vegetation types coexist or alternate with one another in space and time.

In all of this sequence except at the two extremities *humification*, often together with *maturation*, is favoured both by the strongly seasonal climates and in part of the sequence by the kind of vegetation. In the *steppe* zone the abundant supply of organic matter both from litter and the rhizosphere is incorporated to a considerable depth in the soil (**isohumism**). This humification

---

mountains — various types of altitudinal sequences

deserts

semi-deserts: cold (*sierozem*), hot (*grey and brown sub arid soils, reddish chestnut soils, tropical isohumic soils, etc.*)

sparsely vegetated steppe (*chestnut soils, burozems*)

densely vegetated steppe (*chernozems*)

prairies (*brunizems*)

temperate deciduous forest (*brunified soils*)

deciduous and mixed forest of the boreal transition zone (*polzolic and boreal lessived soils, grey forest soils*)

boreal zone, coniferous forests (*podzols*)

tundra and boreal peaty soils

subtropical and mediterranean zone with a dry season (*fersiallitic soils* dominant)

humid subtropical zone (*ultisols*) and tropical zone with dry season (*ferruginous soils and ferrisols*)

humid equitorial zone, dense forest (*ferrallitic* and *ferrisols* dominant)

hydromorphic intrazonal soils (*alluvium, gleys, planosols*)

*vertic,* intrazonal soils

*salodic,* intrazonal soils

decreases towards both extremities of the sequence. Thus, in the zone of maximum aridity the much lower density of the steppe vegetation results in a smaller amount of organic matter added to the soil, while in the forest zone much more organic matter comes from litter than from the rhizosphere, which causes a decrease in the thickness of the A1 mixed horizons. In addition, maturation is less marked, acidity increases and the forest mull becomes moder when the temperature is lowered and the processes of complexolysis and cheluviation appear, which are intensified in the podzol zone (Fig. 4.4).

Studies of the movement of materials within the soils of this sequence have been dealt with in the environmental section of Chapter 3. It should be remembered that the maximum of humification of the densely vegetated steppe is accompanied by a biological accumulation of calcium in the A1 and, as far as the movement of material is concerned, the three basic processes of *decarbonation*, *pervection* and *cheluviation* replace one another as the climatically determined drainage increases.

Some groups of climatically determined soils are more or less related to the soils of this sequence, although in fact they do not belong to it. Thus the *reddish chestnut soils* are related to the chestnut steppe soils by their isohumic character but, in addition, the hotter climate causes a greater development of iron oxides which produces similarities to the mediterranean fersiallitic soils. In the same way the **brunizems** replace the lessived chernozems in a less cold climate.

*Sequence II: succession of soils in tropical climates with decreasing aridity.* In spite of the difficulties caused by the dominance of long cycles in tropical climates, it is still possible to recognise a climatically determined sequence of soils in areas where disturbing effects, such as those resulting from climatic variations over time, parent materials of different ages and termite and human activity, are limited. These disturbing factors are much less important in South America, in particular Colombia, than in tropical Africa, although in this case the sequence is not oriented east–west but is determined by the great variations in precipitation, from a few hundred millimetres up to 10 m, caused by the great topographic barrier encountered by the humid winds from the Pacific. In this environment the great variations in precipitations suppress even the reduced effects of disturbance that do occur.

Organic matter is involved to only a slight extent in the most humid part of the sequence. In contrast, the strong seasonal alternations in the more arid

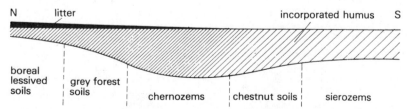

**Figure 4.4** Climatically determined humification: sequence I, USSR (the density of the shading is proportional to the amount of incorporated organic matter).

part of the sequence favour isohumic development which assumes a particular character at sites where vertisols tend to be formed.

Soil differentiation is mainly the result of climatically controlled weathering, as this is reflected in the movement of material within soils. Thus with increasing humidity there is a greater immobilisation of iron and aluminium while the loss of silica in drainage water increases. At the same time the acidity of the profiles increases and clay neoformation, which is dominant in hot climates, becomes less rich in silica, according to the sequence:

$$2 : 1 \text{ clay} \longrightarrow 1 : 1 \text{ clay (kaolinite)} \longrightarrow 1 : 1 \text{ clay} + \text{gibbsite}$$

This succession is very clearly seen where the parent material is the same, and precipitation increases as the importance of the dry season decreases. It is to be noted that in the same direction the amount of clay pervection decreases.

**Note** that the fersiallitic soils, which are characteristic of mediterranean climates, can be considered as the homologues, in a cooler climate, of the ferruginous tropical soils, with the main differences being that 2 : 1 rather than 1 : 1 clays are dominant.

*Sequence III: succession of soils in humid climates with increasing average temperature*. This differs from sequence I in that the rise in mean temperature from north to south does not cause an increase in aridity, for in the same direction the precipitation increases so that $P - PET$ is always clearly positive. *Throughout this sequence, the processes of transport are at a maximum, but the materials that are transported vary according to whether the temperature is high, medium or low*. Such a sequence is well developed on the east coast of America from Canada in the north to the Brazilian coast in the south. In this sequence, once again, humification is the dominant pedogenic factor in the boreal and temperate zones, and it can be said that *within these two zones the types of humus absolutely control the kind of weathering, the cycling of the bases and the processes of transport*. In contrast, in the equatorial zone, humification no longer accounts for the acidification, and weathering takes the form of a total hydrolysis which, as mentioned previously, favours the loss of silica and bases.

The relationships between different climates and different processes are given in Table 4.5.

**Note** that the soils of this sequence that are not shown in Table 4.5 are intergrades: (i) *boreal–temperate intergrades*, brown podzolic or humic ochric soils, (ii) *temperate–humid tropical intergrades*, subtropical ferruginous soils (ultisols of the USA classification).

With regard to the three major elements (aluminium, iron and silica) it can be said in overall terms for sequence III that the loss of the first two decreases from north to south, while silica losses increase.

Figure 4.5, taken from Volobuyev (1962), summarises all the previous discussion. It shows the geochemical composition of climatically determined soils in terms of silica, iron and aluminium

**Table 4.5**  Soil sequence of humid climates.

|  | *Boreal zone* | *Temperate zone* | *Equatorial zone* |
|---|---|---|---|
| type of humus | mor | mull | tropical acid mull |
| base cycle | storage in A0 (inactive forms) acidification | maximum effectiveness of the biogeochemical cycle (slight or moderate acidity) | cycle of little importance, acidification |
| processes of weathering and movement | complexolysis, cheluviation (podzolisation) | gradual transformation and moderate pervection (brunification) | total hydrolysis, neoformation of kaolinite (ferrallitisation) |
| transported entities | Al–Fe (complexes) | clay (and iron) | silica (+bases) |

with reference to the two main climatic factors – humidity (abscissa) and mean temperature (ordinate). *A* represents an arid climate where the accumulation of bases occurs under a steppe vegetation; *B* is the area of temperate brown forest soil in which the loss of all elements (Al, Fe, Si and bases) is moderate and in equilibrium; *C* is the area of podzolisation where the accumulation of residual silica (as quartz) is great, while all the other elements and particularly the sesquioxides are mobilised; finally *D* is the area of ferrallitisation in which silica and bases are removed and sesquioxides are immobilised within the profile.

**Altitudinal sequences: vegetation and soil zones.** With altitude, the rapid decrease of mean temperature (about 0.5°C per 100 m) and growth season are responsible for a *sequence of vegetation and soil zones*, which in temperate regions is comparable to the succession through zones of latitude. As the

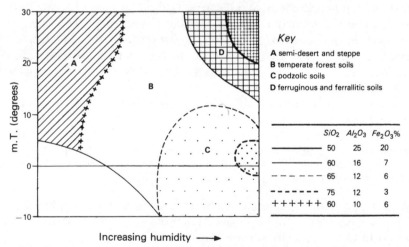

**Figure 4.5**   The average composition of soils in terms of $SiO_2$, $Al_2O_3$, $Fe_2O_3$ in relation to humidity and mean temperature (after Volobuyuv 1962). Note that the difference from 100% represents other elements, particularly the bases.

altitude increases there is a general slowing down of organic matter decomposition, so that gradually profiles become more humic; however, as long as there is a certain amount of deciduous vegetation mixed with the coniferous, the biological activity is sufficient to guarantee the formation of a mull or thick mull–moder (montane zone). At a greater altitude in the sub-alpine zone, acidifying species (conifers and Ericaceae) become dominant and there is a change in soil development towards the podzolic – first brown podzolics and then podzols (Fig. 4.6).

In contrast, the *alpine meadows* above the tree line have a relatively active humus of a moder or mull–moder type. The climatically determined soils are much less well developed and are rankers (on silicate rocks) or humic lithocalcic soils (on limestone).

On granitic material the altitudinal succession is as follows:

*acid brown soil → humic ochric brown soil → brown podzolic soil → podzol → alpine ranker*

However, it must be noted that it is only on the colder slopes that this succession is well developed (Bartoli 1966), for on the warmer slopes there is a compensation effect with elevation between the lowering of the air temperature and an increase in the intensity of the solar radiation. The radiation raises the soil temperature and thus the speed of organic matter biodegradation, which causes the climatic belts to be much less clear. This is additional proof of the control that organic matter has on pedogenesis in these climates.

It is to be remembered that Pallmann found the best examples of *analogous soils* in mountainous areas where at the sub-alpine level there is a considerable accumulation of acid organic matter as moder or mor on practically all types of parent material. Thus a mor is developed on a lithocalcic soil on indurated limestone, an andopodzolic soil on lavas, a podzolic soil or a podzol on marls or moraines (Duchaufour 1978, $I_1$ & $I_2$).

*In tropical areas*, an altitudinal succession comparable to that in temperate

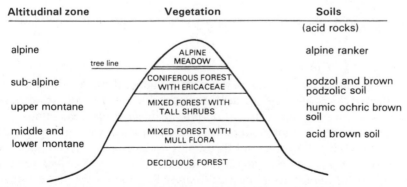

| Altitudinal zone | Vegetation | Soils |
|---|---|---|
| | | (acid rocks) |
| alpine | ALPINE MEADOW | alpine ranker |
| *tree line* | | |
| sub-alpine | CONIFEROUS FOREST WITH ERICACEAE | podzol and brown podzolic soil |
| upper montane | MIXED FOREST WITH TALL SHRUBS | humic ochric brown soil |
| middle and lower montane | MIXED FOREST WITH MULL FLORA | acid brown soil |
| | DECIDUOUS FOREST | |

**Figure 4.6** Altitudinal zones of vegetation and soils (on granite in the temperate zone).

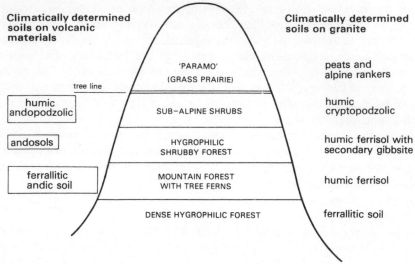

**Figure 4.7** Altitudinal zones of vegetation and soils (on volcanic materials and granite in the tropical zone, Colombia).

zone mountains can be recognised, with the important difference that there is an absence of seasons. The organic matter, the effect of which on lowland soil development is regarded as negligible, rapidly becomes more important and in very wet conditions a peaty or hydromor horizon is formed at the surface. To begin with, a surface podzolisation of a particular type is superposed on the ferrallitisation, which persists in an attenuated form (altitudinal humic ferrisol). Then the profile contracts in thickness and becomes entirely impregnated by the very acid peaty humus (Fig. 4.7). In Colombia, the succession on granite from low to high altitude is as follows:

*ferrallitic soils → humic ferrisols → (podzolised) humic ferrisols → humic cryptopodzolic soils → peaty rankers.*

**Note** that in Figure 4.7 the *analogous soils* formed on volcanic materials at equivalent elevations are also given.

### 3   Effect of site factors
This involves those factors, independent of the general climate, that have only *local* effects which, in general, are hardly visible on a world scale but are readily apparent in detailed studies and the resulting large-scale maps. These factors cause either a slight *modulation* of the climatically determined equilibrium or, if certain environmental thresholds are crossed, to a *complete reversal* of the direction of climatically controlled development. *The lack or insufficiency of local drainage, as defined in Chapter 3, is one of the site factors most frequently involved and is responsible for all those soils classed as hydro-*

*morphic, such as the peat and gley soils characteristic of sites with a permanent water table, to be found in all climatic zones* (see Ch. 11).

Here, a more detailed investigation will be made of two environmental factors – mineral material and topography – that not only play an essential part in site equilibria, but also are generally closely interrelated.

**The role of mineral material.** In addition to its effects on the moisture status and drainage which have been discussed at length in Section IV of Chapter 3, the mineral material of soils also plays an essential part in terms of its mineralogy and geochemistry. In a temperate atlantic climate, *brunification* and mull formation are characteristic of a great range of materials and thus give rise to a considerable number of analogous soils. Nevertheless, an environmental threshold is involved, which has been discussed in Chapters 2 and 3, in allowing these processes to operate. Thus after decarbonation, which is general in a humid climate, for brunification to occur there must be present sufficient clay (e.g. 7–8% in the lower Vosges) and active iron bound to this clay (about 0.5% free iron in the Liassic or Triassic Vosgesian sandstone). All materials that do not satisfy these criteria, no matter whether they are quartzose and are very poor in clay and active iron, or contain excessive amounts of carbonates or, finally, where the clay minerals have been replaced by amorphous alumina and allophane, they characteristically have different forms of pedogenesis, podzolisation, carbonation (rendzinas) and andosol formation respectively. Thus these three soil types (atlantic podzol, rendzina and andosol) are good examples of an equilibrium determined by site, in the same way as hydromorphic or peat soils (see Ch. 3, Sec. IV).

However, soil mineral material is not always involved in such an extreme manner and often it merely slows down or accelerates a given process. In this way it is possible to explain the formation of intermediate soils, often called *intergrades*, between two classes or groups. Such profiles, which occur at the junction of different classes (or subclasses), pose a classificatory problem. It is only by the establishment of a developmental sequence that they can be located in relation to other soils, as can be seen in several cases in Duchaufour (1978).

At this point consideration will be restricted to three typical examples from the temperate zone:

(i) *Ochric brown soils* (or simply *ochric soils*) are intergrades between brunified and podzolised soils. As mentioned previously, they are of common occurrence in the upper montane altitudinal zone, but at lower altitudes their formation is controlled by an insufficiency in the amount of iron in the parent material. Souchier (1971) has shown in the investigation of some 100 soils on Vosgesian granite that there is a close relationship between the amount of iron and the degree of podzolisation: greater than 4% acid brown soil, 2.5–4% ochric soil, 1.5–2.5% brown podzolic soil, less than 1.5% podzolic soil ($Fe_2O_3$%, total iron).

(ii) *Calcareous brown soils* on marls are incompletely decarbonated as a result of the intervention of a retarding environmental factor, such as slope, vegetation, or a particular moisture regime and are thus an intergrade between the brunified and the calcimagnesian soils (rendzinas).

(iii) *Pélosols* are a particular type of soil formed on materials very rich in semi-swelling clays which have little pedological development as the unconsolidated clays erode very rapidly and, in addition, because of their impermeability, are often hydromorphic. When they are well protected by a dense forest cover against erosion and excessive surface water-logging, a climatically determined brunification can occur, which makes such brown pélosols a good example of an analogous soil.

**Role of topography: catenas.** Topography is one of the most important site factors, as slopes are fundamentally involved in pedogenesis for they modify the way that water behaves so that lateral surface run-off increases at the expense of infiltration which in turn increases the erosive effect. As was seen in Chapter 3, this has important consequences both mechanically, in the transport of all kinds of material, and physicochemically by affecting the water balance and hence the Eh.

The importance of topography in pedogenesis becomes marked in regions of broken relief and to an even greater extent in truly mountainous areas, where soils vary so rapidly from one point to another that they form a veritable mosaic which, even at a large scale, is often difficult to map except as soil associations. In mountains the multiplicity and diversity of site-controlled soil–vegetation equilibria tend to make less apparent the climatically determined altitudinal zones, which has allowed the concept to be contested by various authors despite its undeniable reality.

The lateral movement of water which occurs down slope, even where the gradient is slight and not readily apparent, is responsible for pedogenic differences depending on slope position and for the fact that the soils are genetically related to one another. Thus, sites at the top end of slopes tend to be impoverished, to the advantage of sites on lower slopes and particularly those of depressions that are enriched in materials from all the higher positions. Such a regular succession of soils, identical along a given contour line but varying in a continuous fashion down slope, has been called a *chain of soils* or *catena*.

There is a great variety of catenas depending upon the kind of processes activated by the topography. Other environmental factors that are involved, apart from the main one of slope, are those of general climate and parent material. This complex of processes can be classified into four groups, in terms of the decreasing intensity of the forces involved, as follows: (i) powerful mechanical action resulting in the transport of solid particles of varied dimensions, i.e. *erosion*; (ii) less powerful mechanical action which causes the selective transport of the finer particles, i.e. *lateral or oblique pervection of clays*; (iii) transport in the form of both pseudo- and true solutions; (iv) local changes in the moisture status and of the redox potential.

By considering these four groups of processes, different kinds of catenas can be described; however, there is considerable overlap between them and in many (ii), (iii) and (iv) are involved simultaneously. It is possible to distinguish three kinds of catenas in which the processes of mechanical transport gradually give way to those which involve changes in hydromorphic conditions.

*Catenas involving erosion*. Erosion removes practically all the fine earth leaving behind only pebbles and gravels, if any are present, as a surface accumulation. In some cases this causes a *general rejuvenation* by removing all the soil and exposing the C horizon. Even when such removal is very quick, which is often the case, it is still possible for humus-rich horizons to develop in the stable intervals between periods of erosion, causing, for instance, the development on mountain slopes on fresh igneous rocks of erosional rankers with AC profiles. In other cases, erosion is not complete and *truncation* only occurs, in which a more indurated B horizon is exposed. Erosion is controlled not only directly by the slope but indirectly by the *materials involved*. Thus unconsolidated and fine-grained materials, such as silts, are particularly erosion prone; furthermore, discontinuities of grain size or lithology within the profile considerably increase the effects of erosion, by abruptly changing the direction of water movement from vertical to lateral. Equally, however, the general climate is important, for the sudden and violent rainfall of mediterranean and tropical climates considerably increases the erosive power and even on gentle slopes surface wash can selectively remove all the silt and clay, which Servat (1966) and Roose (1970) have called **impoverishment** (see Chs 11 & 12).

As an example of a catena controlled by erosion, that which occurs on the side slopes of dry valleys which are inset into the Jurassic limestone plateaux of northeastern France will be considered. In this case rejuvenation is of a particular kind for it involves very old materials that have been subject to a long cycle of pedogenesis, which will be considered in greater detail in the next section. This terra fusca, which is a residual, decarbonated clay, is often contaminated by aeolian loams that may even form a uniform overlying layer. At the edge of the plateau the terra fusca is partially truncated by erosion and often exposed to a certain degree of *secondary recarbonation* and develops either towards a *brunified rendzina of erosion* or a *calcic brown soil*. The crest of the valley-side slope is stripped and results in a *lithosol*, while at the foot of the slope a colluvium occurs made up of a mixture of terra fusca, loams and limestone gravels on which a more or less gravelly *colluvial rendzina* or *secondary brunified rendzina* is formed. The valley floor has a great thickness of the finer materials (such as clays, loams and carbonates) and a *colluvial calcareous brown soil* is formed (Duchaufour 1978: $IX_6$ plateau soils, $V_1$ & $VI_6$ lower slope soils, Fig. 4.8).

Note that in this example rejuvenation leads to entirely different results on upper and lower slopes, with the more or less complete removal of the profile from the upper slope and recarbonation as a result of intense mechanical mixing on the lower slopes.

*Catenas involving transport of material in solution or in fine suspension*. In this type of catena the coarser particles of sand and silt remain in place, transport by lateral drainage is confined exclusively to fine clays, salts and pseudo-soluble complexes that are very gradually moved down slope, so that the whole of the upper part of the slope is impoverished and acidified while the lower part of the slope is generally enriched and the pH rises. When, as a

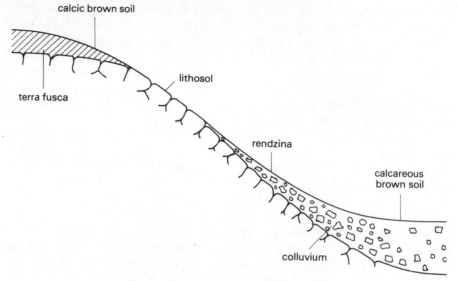

**Figure 4.8**  Soil catena on hard limestone.

result of a general vegetational change, degradation resulting in secondary podzolisation occurs, this always starts at the top of the slope because the lower parts of the slope which are less acid and richer in clay are more strongly resistant to degradation.

*As a whole, such catenas reflect the general impact of climatically controlled pedogenesis,* so that there is a marked difference between the catenas of temperate, or cold, climates and those of the Tropics. The first are largely controlled by the kind of humification taking place in the upper part of the catena while, in contrast, this is of little importance in the Tropics, provided that sufficient bases are present and the pedoclimate is seasonally dry.

*In temperate climates*, two cases can be distinguished: (i) in which an acid brown soil with mull occurs in the upper part of the catena which, owing to lateral pervection of clay and the movement of bases, is responsible for the formation of a more clayey and less acid soil downslope; (ii) in contrast, where a podzol or a ranker with a mor horizon occurs in the upper part of the catena, it is responsible for the large-scale movement of organomineral complexes that accumulate down slope forming a *brown podzolic soil of slopes* with a thick spodic B horizon which can completely obliterate the A2 horizon.

*In a tropical climate with a dry season* oblique pervection of clay is less common, for it is the 2 : 1 micaceous clays that are particularly affected by this movement and they are present in only small amounts in this climate. As complete weathering occurs, it is the soluble freed constituents that are most easily moved. It should be recalled that in slightly acid conditions iron and aluminium hardly move compared to silica and bases and these are concentrated in the lower part of the catenas. Therefore, the upper parts of these catenas are impoverished in silica and calcium which makes them suitable sites for the formation of a moderate amount of kaolinite. In contrast, down slope, where the whole of the profile is enriched in silica and bases, the most important process is the large-scale formation of montmorillonite (vertisolisation).

Two examples of catenas will be given, one from a podzolising humid mountain climate and the other from a tropical dry climate around an inselberg.

### Mountain climate (Fig. 4.9)

As described by Schweikle (1971), the catena occurs at an elevation of 700 m in a cold humid climate ($P$ = 200 mm; mean temperature 6–7°C) at Grönbach in the Black Forest on mottled Triassic sandstones. Under these climatic conditions the general tendency is to produce a moder or a fairly peaty mor throughout the catena, which consists of an upper reasonably well drained zone (profile I), then a flatter zone where lateral drainage is not so good (profile II) and then again into a more steeply sloping area (profile III).

*Profile I* is a *humic ochric soil* in which, as it is well aerated and rich enough in iron, the organomineral complexes are immobilised, so that the processes of lateral transport are reduced to a minimum and only some of the bases migrate. Despite the formation of a moder–mor surface, brunification still occurs but is accompanied by a certain podzolic tendency.

*Profile II* is a *podzolic stagnogley* in which almost permanent waterlogging, together with strong acidity, causes a large-scale reduction of iron and, probably, an acid hydrolysis of some of the clays. All the entities freed by these processes are moved laterally by way of the water table, both organomineral complexes with iron and aluminium, as well as some clays in suspension, which causes the strongly bleached A2 horizon to be extremely impoverished.

*Profile III* inherits those mobile entities coming from profile II. Accumulation appears to be initiated by the better aeration which causes the $Fe^{2+}$ to be precipitated as $Fe^{3+}$. The flocculated nature of this precipitate then traps all the other mobile entities, such as clay, organic matter and even free aluminium. This massive accumulation starts at the surface and goes to a depth of 40 cm, which gives the whole profile a spodic appearance, but the absence of differentiated horizons causes this soil to be classed as a *humic brown podzolic soil* with a fluffy structure

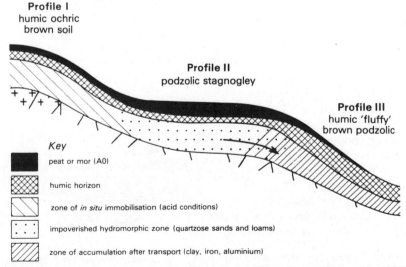

Profile I
humic ochric
brown soil

Profile II
podzolic stagnogley

Profile III
humic 'fluffy'
brown podzolic

Key

peat or mor (A0)

humic horizon

zone of *in situ* immobilisation (acid conditions)

impoverished hydromorphic zone (quartzose sands and loams)

zone of accumulation after transport (clay, iron, aluminium)

**Figure 4.9**  Catena on Triassic sandstone in the Black Forest (Schweikle 1971).

('lockerbraunerde' of Schlichting). This catena shows how acid hydromorphism acts as an accelerator of weathering and mobilisation, for the whole of profile II can be regarded as an A horizon and profile III as a B horizon.

### Dry tropical climate (Fig. 4.10)

This is provided by the Kosselili inselberg in Chad which was described by Paquet (1969) and Bocquier (1973). In contrast to the preceding examples, organic matter is involved little, if at all, and the mobile elements are no longer aluminium and iron but silica and bases ($Ca^{2+}$ and $Mg^{2+}$) which are concentrated in the lower part of the catena as the water flow gradually decreases. This causes a *complete reversal of pedogenesis in downslope areas compared to up slope*, for neoformation of montmorillonite completely replaces that of kaolinite. Microstructural studies show that, as a result of the continual influx of silica and bases from up slope, gradually montmorillonite formation occurs further and further up slope, replacing former A2 horizons. The sequence from top to bottom is the following:

*ferruginous tropical soil* → *planosol* → *solodised-solonetz* → *vertisol*

Here again, the vertisol at the base receives everything transported from an inselberg, but these entities do not accumulate as such as they are incorporated within the neoformed clays.

*Catenas with Eh variations in hydromorphic conditions*. The catena concept can be extended to very gently sloping plains and plateaux that characteristically have a water table which may be either temporary (perched) or permanent. The water involved is rarely completely immobile and very slight differences in level (from some tens to a few centimetres) are sufficient to cause its slow movement which, while it is obviously insufficient to cause mechanical transport, is responsible for solution transport. In these conditions, $Fe^{2+}$ is transported down slope to the point where the water table intersects the surface, where it is oxidised and precipitated as ferric oxide. It is in this way that the very hard ironpan of the Landes of Gasgony and certain ironstones of tropical regions are formed. Profiles situated up slope from such sites, on the contrary, lose almost all of their free iron (see Fig. 10.8).

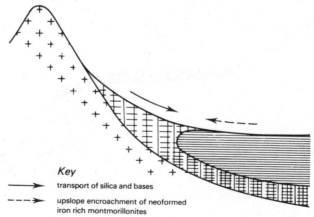

Key

⟶ transport of silica and bases

--→ upslope encroachment of neoformed iron rich montmorillonites

**Figure 4.10** Catena around an inselberg in the Tropics (after Paquet 1969, Bocquier 1973). (For symbols, see general key, p. ix.)

When a badly drained plateau is adjacent to a depression, the movement of the two water tables (on the surface and at depth) is linked so that a *pseudogley* is formed on the plateau and a *gley* in depression (see Fig. 11.1).

It can be concluded that topography is of great importance as a site factor in pedogenesis both in temperate climates, where it either prevents the complete development of the profile by slowing down processes or modifies the climatically controlled development as a whole by affecting lateral movement, and in the Tropics, where it can completely reverse the climatically controlled processes of weathering.

## 4 The effect of man: process of degradation

Man is able to modify the natural development of soils either very suddenly and directly as a result of cultivation or more slowly and indirectly by replacing the original (climax) by secondary vegetation. These changes are responsible for a modification of the type of humus and hence for a new kind of pedogenesis which is reflected in profile changes referred to as degradation (Pallmann *et al*. 1949). Such degradation is entirely Postglacial and thus has an important impact on the short cycles of pedogenesis.

Note that the term degradation is used here in an environmental and biological sense; it has a very different meaning when the degradation of an argillic (Bt) horizon is considered or of the clays themselves, where the physicochemical process involved is acting only on a particular horizon or even on particular clay minerals.

**Direct effects as a result of cultivation**. Cultivation to a generally standard depth results in a homogenised brown or black upper horizon, with a clear lower limit, in which the structure has been modified and the original humus type is no longer recognisable. In profile descriptions it is referred to as an Ap horizon. In contrast, except for **agric** horizons (see below), the lower (B) or B horizons are only slightly modified by cultivation and *it is possible to follow the history of the development of cultivated soils by making a detailed study of these (B) or B horizons and comparing them with profiles developed under natural vegetation*.

In certain sandy soils, very long-term cultivation has been responsible for the development of some characteristic horizons that are used as diagnostic horizons in the American classification.

An **anthropic** horizon is one that is black or brown, humus rich, with a very high phosphorous content (more than 250 ppm $P_2O_5$). Such a horizon occurs in the **postpodzols** on the Flandrian sands of the Campine and Sologne where the soil has been fertilised by a mixture of dung and the A0 horizon of a podzol (previously used as animal bedding). This has given rise to a thick uniformly black A1 horizon without a distinct A2, which at depth gives way to a zone of fairly soft, ochreous pan (Ameryckx 1960). In places where less podzolised soils have been fertilised by dung mixed with straw, this anthropic horizon has a browner colour.

**Plaggen** horizons are slightly different as they have been formed by the addition of large amounts of organic matter at irregular intervals over a period of centuries. The resulting black soil is markedly raised above surrounding areas and has parallel bedding planes (**plaggenboden**).

Finally, over a long period even deeper horizons can be modified by cultivation to produce an *agric* horizon enriched in clay and humus, situated immediately below the anthropic (Ap) horizon. The fact that *black coverings* of humus occur in fissures, old root channels and worm holes is evidence that certain cultural practices, such as the addition of mineral fertilisers and intensive cultivation, cause the mobilisation of not only a part of the clay but even of the organic matter, both of which are deposited in B horizon voids at the base of the cultivated layer.

**Indirect effects: degradation.** The degradation of vegetation and soils is the frequent result of long established practices, particularly with regard to forests, such as excessive clearing, grazing, fire, collecting of litter etc., which have exerted pressure on natural vegetation throughout history. These practices interrupt the biogeochemical cycle of nutritive elements and cause the original vegetation to be replaced by a different *secondary* vegetation which changes the process of humification. In turn, these changes cause a new soil type to develop which is very different from the original that characterised the climax. Three examples of degradation will be considered, two of which occur in the atlantic forest zone, both on plainlands and mountains, on two different parent materials, while the third deals with sub-alpine forest and its replacement by grassland.

*Atlantic heathlands* have replaced oak forests on certain parent materials as a result of clearing of the atlantic plainlands in the Bronze Age, 2000–3000 years ago. To begin with, it must be emphasised that this process is restricted to those parent materials that are most acid and poorest in clay and free iron (the preventive part played by these two entities in the process of podzolisation was discussed in Ch. 2). Such materials as residual clay with flints, that have a sandy surface layer, and sand or sandstone outcrops of all kinds are particularly susceptible. This process has been investigated by various authors – Dimbleby (1952a) in Yorkshire, Galoux (1954) then Coninck and Herbillon (1969) in the Belgian Campine. By the use of palynology, Munaut (1967) has given much more detailed information on the stages of development and the dates of formation of these atlantic heaths.

All of this work shows that the original forest soil was either a *lessived acid soil*, where there was sufficient clay and iron in the parent material, or a *podzolic soil* on quartzose sands, which are poorest in these two constituents. On the first soil the podzolisation caused by vegetation change can only occur after the prior pervection of the clay and iron, the presence of which would prevent it. This form of podzolisation, preceded by an increase in the pervection, is known as *indirect* (see Ch. 10).

In contrast, on very quartzose materials, the soil at a site climax under forest is a podzolic one which is, however, not too well developed, for the A2 is not markedly bleached and there is little humus in the B horizon. The rapid invasion of cleared areas by *Calluna* causes a well differentiated *iron–humus podzol of degradation* to develop, in which the A2 horizon is strongly bleached and there is a characteristically black (Bh) horizon.

Very similar conclusions were reached as a result of work on Triassic sandstone in the lower Vosges (Guillet 1972), where again the soils of the original mixed fir–beech woods were brown podzolics or podzolics, with an iron B horizon poor in organic matter. The effect of *Calluna* is essentially the same as in the oak forests of the plain and a strongly differentiated iron–humus podzol of degradation with a black Bh horizon is formed. Additionally, the use of $^{14}$C has allowed the mechanism of formation of the Bh horizon to be determined. It was shown that the complexing organic compounds of the primary forest litter remain relatively biodegradable and hence they cannot accumulate in the B horizon, while those of the *Calluna* litter are resistant and are able to accumulate as a characteristic black Bh horizon.

Vegetation history, and hence the role of *Calluna* in degradation, can be determined by pollen stratigraphy. In forest podzolic soils on sandstone, the pollen is exclusively of forest species and it reflects only a beech–fir alternation, while in contrast the iron–humus podzols have a concentration of *Calluna* pollen in the B horizon (Fig. 4.11). This *Calluna* invaded the areas cleared by man for temporary cereal cultivation, of which traces have also been found. Pine woods with bracken and bilberries occurring today are a recent association created by man that has excluded *Calluna*, so that pine pollen is concentrated in the surface.

An exact date for the time of *Calluna* dominance has been obtained by the use of **indicator pollens** carried to the site in small quantities by the wind from neighbouring forested areas, which makes it possible to decide if the *Calluna* phase preceded or followed the appearance or disappearance of certain well dated forest species.

In addition, the mean residence time of the organic matter of the Bh of the podzols of degradation, determined by $^{14}$C, is *about half the total age of these soils; that is to say, the time of the* Calluna *invasion*. The fact that the MRT has not been reduced in the period of much slower podzolisation since the disappearance of the *Calluna* points to the conclusion that the humic precursors from the *Calluna*, which are immobilised in the Bh horizon, are so exceptionally resistant to biodegradation as to be practically inert; this is further supported by the results of incubation experiments. In contrast, the MRT of the B horizons of forest podzolic soils is relatively low and is not related to the age of the podzolisation, which may amount to several thousand years and be referrable to the atlantic period.

*Wet heathlands with Molinia* have a kind of degradation characteristic of only slightly permeable situations, that are rich enough in fine-grained materials to resist podzolisation. Materials favourable to this kind of degradation are old loams subject to a long cycle of weathering, often Preglacial (to be discussed in the next section). In this case the degradation is towards an *increase in the processes of hydromorphism*.

Degradation was initiated only in the last few centuries, or up to 1000 years ago at most, by the destruction of the primary forest in patches or large clearings. Becker (1971) has taken as an example of this the loam-covered old terraces of the Charmes forest of Lorraine. The original forest soil is

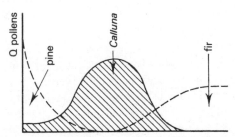

**Figure 4.11** Simplified pollen diagram for a humus–iron podzol on Vosgesian Triassic sandstone; successive layers of pollen with depth (Guillet 1972).

polycyclic and pervected, and its Bt horizon is almost impermeable owing to the tight packing of the pervected clay in the pores, but hydromorphic conditions are moderated by the high *PET* maintained by the closed forest which enables the upper soil horizons to be reasonably aerated and the humus to remain an acid mull. Forest clearing causes a marked reduction in *PET* which leads to the rise to the surface in winter of the temporary water table, causing the humus-rich horizons to be deprived of oxygen and to be changed into a hydromoder or hydromor. At the same time, the large-scale reduction and mobilisation of iron causes the soil to develop in the direction of a *pseudogley* (see Ch. 11).

*Sub-alpine meadow*, or meadow–larch woodland soils, provides another example of an environmentally controlled degradation. The original association of pine forest with ericaceous species (*Vaccinium* and *Rhododendron*) was gradually destroyed by intensive grazing and replaced by a dominantly gramineous secondary association. Under the influence of this ameliorating vegetation, the original mor changed into a mull and the characteristic A2 and B horizons of the podzol disappeared. The soil developed towards a very shallow brown meadow soil. It should be noted that at this altitude the result of degradation is the opposite of that which occurs in an atlantic climate, for here the climax soil is a podzol while the degraded soil is a brown soil.

*Re-establishment of climatic equilibria.* When the destructive action of man stops, it is possible for the initial equilibrium to be gradually re-established, but this is not always easy as the equilibria resulting from degradation can be relatively stable and regarded as a **paraclimax**, such as the old heathlands in the atlantic climatic zone. In general, modern silvicultural practices, such as soil cultivation and the use of fertilisers, accelerate the speed at which the climax is re-established.

Dimbleby (1952b) showed the favourable influence of certain transitory, and but slightly demanding, ameliorating species (such as birch) that are capable of colonising the old atlantic heaths and preparing the way for oak by gradually changing the podzol of degradation to a brown soil with mull. Some deep roots are capable of penetrating the indurated B (iron pan) horizon and of obtaining bases from the parent material that is being weathered. By the action of the biogeochemical cycle, these bases are concentrated at the surface, the pH increases, as does the speed of litter decomposition which allows earthworms to move in and cause a mixing of the podzol horizons initiating brunification.

5   *Conclusion: the environmental control of soil–vegetation equilibria*
In conclusion, a summary can be made of the preceding discussion on the environmental control of short cycles of pedogenesis.

The soil–vegetation equilibria that have been examined can be considered as belonging to one of four main types: either climatic or site climaxes, in both of which development is approaching completion and with which man has not interfered, or those in which development is clearly incomplete, being at a juvenile stage as a result of erosion, and finally those degraded by man.

It is to be noted that the same soil, such as a podzol, can belong to each of these four types:

*juvenile soil:* shallow podzol on slopes;
*climatic climax:* boreal or sub-alpine podzol;
*site climax:* atlantic forest podzol on quartzose sand;
*degradation:* iron–humus podzol (of heathland).

It also needs to be emphasised that an even better example of a site podzol than that of the atlantic forest podzol is provided by the *tropical hydromorphic podzol*, developed on sandy coastal plains with a phreatic water table in equatorial regions (Turenne 1975). Very particular site conditions, such as acid water and very poor quartzose materials, are evidently necessary for its formation which is so radically different from the normal climatically controlled process of ferrallitisation (see Ch. 10).

## IV Study of long cycles

Long cycles of pedogenesis can have a duration of 100 000 years or more and are particularly associated with the soils of tropical regions. It should be remembered that in these regions organic matter is no longer of prime importance in pedogenesis, and correlations between organic matter and soil are no longer as close as those in temperate climates, indeed there is no necessary genetic connection between them.

Many temperate soils formed on very old parent materials have also been subject to a development of very long duration, in the course of which they have acquired certain characteristics which are similar to those of present-day tropical soils. Thus these characteristics are the result of a long cycle, prior to the Postglacial short cycle, which alone reflects the current soil–humus–vegetation equilibrium. In such a polycyclic (or polygenetic) soil the short cycle affects only the upper part of the profile while, in contrast, the deeper horizons reflect the older development. *It is to be noted that such a soil can be considered as analogous with soils formed on younger parent material.*

However, the evidence for differentiating an older and a more recent cycle of pedogenesis is not always clear. This is the case in temperate climates where many soils occur on parent materials of early Würmian age. As their development started before the end of the Würm period, it must be measured in tens of thousands of years and is not limited to Postglacial times. But, as will be seen later, these soils are not fundamentally different from Postglacial soils, with development differing only in degree not in kind, such as having a more marked acidity. These soils will be referred to as being *old*.

In tropical climates there are also many soils of intermediate age, i.e. only thousands to tens of thousands of years old, but they differ from the old temperate soils (discussed above) in still having certain youthful characteristics compared with the fully developed climatically determined profiles. An example of this is provided by certain *ferrisols* on relatively young parent materials, which are incompletely weathered ferrallitic soils.

In addition, certain tropical soils, formed on slopes or recent alluvium, are in the initial stage of their development and thus have been subject to only a *short cycle* of pedogenesis, for example the *eutrophic brown soils* on basic rocks.

## 1 Definitions (Figs 4.12 & 13)

To start the study of this particularly difficult and complex question correctly, the terms used need to be defined exactly, for there is considerable confusion about them in the literature. This will be followed by a discussion of all the varied current theories before considering particular examples in a detailed manner.

**Palaeosols.** These are old soils formed under different environmental conditions (in particular climate and vegetation) from those of the present day. The term is a very general one and does not refer to the subsequent development of the palaeosol, which may have been buried under a more recent deposit or, in contrast, have been subject to a second cycle of development.

**Fossil soils.** These are generally considered as being palaeosols buried under more recent, usually thick, deposits, which prevent all subsequent pedogenesis. It is different if the deposit is not sufficiently thick, for then the palaeosol will be affected in some way by more recent pedogenesis to produce a *compound* or *complex soil*.

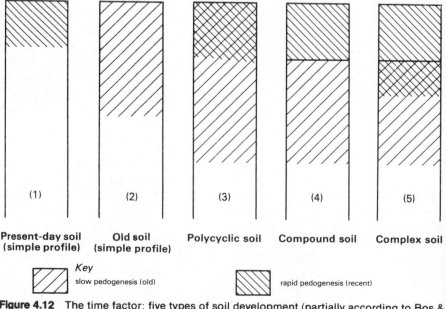

**Figure 4.12** The time factor: five types of soil development (partially according to Bos & Sevink 1975).

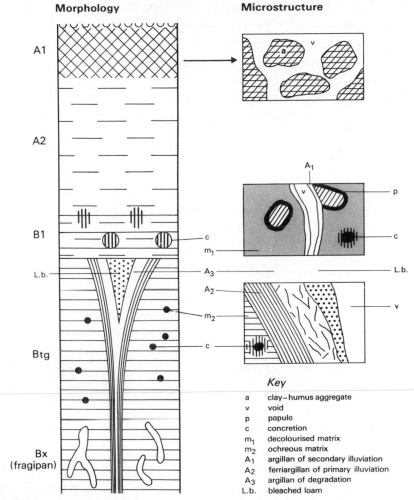

**Figure 4.13** Morphology and microfabrics of glossic lessived polycyclic soil on Rissian loam, Lorraine.

The deeply buried palaeosols, which are thus inaccessible to roots, do not directly interest the pedologist, for they do not play any part in the formation of environmental equilibria nor in various pedological problems. On the other hand, they provide in an indirect manner important information on their age and the origin of the parent material.

Thus, the study of soils buried by **loess** of periglacial aerolian origin has been made by a number of authors, for example Lieberoth (1963) in Saxony, and Jamagne (1969) in the north of France. The addition of loess occurred intermittently so that the study of the successively buried soils gives information on the climate and vegetation at the time that the deposits were made. By comparing these results with those from polycyclic or complex soils formed on contemporary loams, it is possible to interpret profile genesis and to separate the present-day characters from those which have been inherited.

**Polycyclic (or polygenetic) soils.** As stated previously, these are soils that have been subject to two cycles of development under different climatic conditions, in which the present-day characters are markedly different from the older *inherited* characters. In most cases there has been a fairly long pedogenic break between the two cycles, for example, as a result of a prolonged glacial phase or even a phase of rhexistasis in Erhart's sense (i.e. a phase of climatically controlled erosion which interrupts all biochemical processes of soil formation). Of course the profile shows signs of this interruption, thus, when glaciation is involved, *cryoturbation* to a certain depth can occur or, when intensely eroded, profile *truncation* can result. In both cases the material of the present-day profile has been modified as cryoturbation mixes the old surface horizons of the palaeosol while erosion often exposes an old B horizon (for example a Bt horizon). Microstructural investigations often provide information on these successive phases, which it is very necessary to take account of in an interpretation of the profile.

**Old soils.** Despite its lack of precision, this term will be used, as suggested previously, for certain temperate or warm temperate soils of an intermediate age that have developed over a period of more than 10 000 years, during which there may have been phases of increased or decreased pedogenesis but where the fundamental process has remained practically unchanged. In this category would be placed the lessived acid soils on the early Würmian loams in France and the fersiallitic soils of the Midi where the Würmian period is marked only by an increase in precipitation and no interruption of pedogenesis. This use of the term 'old soil' has also been advocated by Ruellan (1970).

**Compound and complex soils.** These terms are used in the sense of Bos and Sevink (1975), for all profiles formed from two geologically distinct layers in which the underlying layer is generally a palaeosol. Thus they are somewhat similar to polycyclic soils except that in this case it is only the deeper layer that is affected by the long cycle of pedogenesis.

The term **compound soil** is used when the two phases of pedogenesis have acted separately on each one of the two layers without any interaction between them, so that the palaeosol buried under the more recent deposit is not involved in the recent phase of pedogenesis. In so far as the palaeosol occurs at such a shallow depth, it can be reached by some roots, and it can be considered as forming part of the profile.

In contrast, in **complex soils** the recent phase of pedogenesis affects not only the surface layer but also the deeper palaeosol, which has thus been affected by two successive cycles of pedogenesis: for example, a long cycle involving weathering and a first phase of pervection, then a short cycle characterised by another phase of clay accumulation owing to pervection from the more recently deposited surface layer.

**Note: extension of the polycyclic idea**

When there has been, during Postglacial times, a considerable change in the vegetation and humification, it can result in such important pedogenic changes that traces of both successive stages are to be seen in the profile; *this is, however, not the result of a long cycle followed by a short cycle, but rather of two successive short cycles both dependent on biological processes entirely confined to the Postglacial period*. Those profiles resulting from *degradation*, such as the secondary podzols, are a good example of polycyclic soils resulting from differences in the process of humification. It should be recalled that these profiles are often characterised by a spodic B horizon (resulting from the secondary heath) overlying an argillic Bt horizon (evidence of an old forest cycle). Another example will be considered in detail in Section 3.

It is possible for the same kind of soil, depending upon its recent history, to belong to each of the types specified above. For example, in the case of a terra rossa, a rubified fersiallitic soil formed on decalcified clays on indurated limestone, it is possible to distinguish it as:

(a) *a present-day soil* (short cycle) in the mountains of Lebanon (Lamouroux 1971) where, owing to exceptionally favourable conditions, it can form in less than 10 000 years (see Ch. 12);

(b) *an old soil* in the Midi of France where development has been slower because of less favourable climatic conditions, but where, nevertheless, pedogenesis has been virtually continuous so that it can be considered to be an old soil;

(c) *a fossil soil* in Perigord or Charente, where the climate is markedly colder, which can be considered to be a palaeosol which is sometimes buried to such a depth by more recent deposits, such as an aeolian loam, that it is no longer subject to pedogenesis and is indeed fossilised;

(d) *a complex soil* when the surface aeolian loam is only some decimetres thick, so that the terra rossa is involved in recent pedogenesis, such as pervection of fine clays from the surface loam;

(e) *a polycyclic soil* in dolines on karst plateaus, where the *residual* terra rossa is mixed with limestone debris and subject to secondary recarbonation to form a calcareous brown soil on rubified fersiallitic material. In this case the terra rossa is a parent material, not a soil.

## 2   Interpretation of long cycles: prolonged or successive development

A discussion is necessary at this point as the preceding definitions are not accepted by all pedologists. As far as long cycles are concerned, two theories are involved and at present a definitive decision about this very difficult problem cannot be made. Discussion will be restricted to posing this problem and suggesting the most reasonable provisional solution, pending more facts in the area of Quaternary geology and palaeoclimatology that will allow a more definite decision to be made.

The key question is whether the climatically controlled pedogenesis of the long cycles has been continuous or discontinuous (or even completely different). According to Novak *et al.* (1970), Icole (1973) and Hubschmann (1975), there has not been a succession of different kinds of climatically controlled pedogenesis within the Quaternary, but only *one kind of pedogenesis* involving an increasing degree of acidification, weathering and pervection which are affected by climatic variations only to the extent of being accelerated or slowed down, or even momentarily interrupted.

With regard to this, Hubschmann has proposed restricting the term palaeosol to soils that have been subject to a climatically controlled pedogenesis clearly different from that of the present day, and of using the term **archeosol** for temperate soils on old parent materials that have been subject to one and the same climatically controlled pedogenesis over a long period of time which, according to Hubschmann, is the more common case.

This new idea is in marked contrast to the classical idea of successive kinds of pedogenesis resulting from climatic variations of sufficient significance over time to cause the development of particular physicochemical processes. In this theory, the time factor is not solely responsible, but other environmental factors are involved and in particular those of a general bioclimatic nature that have been subject to important variations in the Quaternary. This is the theory of Kubiena (1953) and Mückenhausen (1973a, 1974), which has been adopted as the basis for the organisation of this whole chapter. These two authors classify all those Quaternary palaeosols which they see as having developed under a subtropical or even tropical climate, and thus very different from the climate of the present day, into one or a few special groups, such as the **plastosols** on acid materials and **terrae calxis** on calcareous material.

To begin with, it can be said that the problem is very different depending on whether temperate or tropical regions are being considered. In the temperate zone, successive glacial phases have interrupted the continuity of pedogenesis, while in the Tropics, because climatic variations have been of smaller amplitude, continuity of pedogenesis is more likely. Therefore, these two cases will be considered separately.

**The problem of long cycles in a temperate climate.** A summary will be given of a number of examples which are favourable to the idea of successive rather than continuous pedogenesis in temperate conditions, but none of them is absolutely decisive.

The climatic variations in the Quaternary are now well known and, even though previously somewhat exaggerated, they are by no means negligible. Thus, from the biological and palaeontological evidence, the Riss–Würm interglacial had a mean temperature which was 3°C higher than the present day (Flint 1971). Although not spectacular, this difference is important as it is the kind of present-day temperature contrast between the Midi and the plains of Central France which, as far as climatically controlled pedogenesis is concerned, is across an important boundary.

Palaeosols of the same age and origin have different characteristics according to whether they are situated in one or other of these two zones, so that the terra fusca of the Jurassic of Lorraine is homologous to the terra rossa of the Jurassic of Provence, which has been subject to a much more intense rubification. As these differences cannot be related to either parent material or time of development, a climatic factor would appear to be involved.

A final line of reasoning would seem to have an even greater importance. Most authors who have studied Quaternary palaeosols, even those who are proponents of pedogenic continuity, have noted the occurrence of a discontinuity of the weathering process which is characteristic of the soils on either side of a particular time period in the Quaternary. This points to an abrupt change in pedogenesis at a certain time in the Quaternary that without doubt can be ascribed to a no less abrupt change of climate. It is true that there is disagreement about the date of this change between Jamagne (1969), who has suggested the Riss–Würm interglacial for northern France and Belgium,

and Icole (1973) and Hubschmann (1975), who have suggested the Mindel
–Riss interglacial for the Pyrenees. These differences are not surprising if
account is taken of the difficulties of precisely dating terraces, on which an
important disturbing factor is the rejuvenation by aeolian addition subsequent to terrace formation, and also that such a time difference is not
unreasonable between two such markedly different geographical and climatic
provinces.

The nature of the palaeosols so defined is very characteristic, with a large
amount of neoformed kaolin, a high degree of weathering, considerable
rubification and the occurrence of deep tongues of material in which the clays
are strongly degraded. None of these characters occurs even in the oldest of
the soils that postdate this climatic change.

Icole (1973), Novak et al. (1970) and Bornand (1978) have given convinc-
ing details. Thus the best contrast between the two periods of pedogenesis is
provided by the *weathering of feldspars, for in the oldest materials it is responsible for a massive neoformation of kaolinite, while in the younger materials
practically no clay is produced; thus there is a contrast in weathering between
all or nothing, which was discussed in the first chapter in terms of tropical and
temperate weathering respectively*. It is obvious that time alone is not sufficient
to explain such a contrast of weathering processes. This is confirmed from a
different angle by the study of soils that are old but have all been formed
subsequent to the change in climate, for on this side of the climatic threshold
nothing other than gradual weathering changes are to be seen. For example,
between the different parts of the Würm period pedogenesis is continuous
and all processes, such as acidification, amount and depth of weathering,
transformation or degradation of 2 : 1 clays, which are dominant, vary gradually in intensity in relation to one factor – that of time.

From this discussion it can be seen that for northern Europe the term
*polycyclic soil should be kept for the first type of soils, with a pedogenic
discontinuity, formed on loams that are of Rissian age or older, and the term
old soil should be restricted to the second type of soil that has a continuity of
pedogenesis and is formed on materials of early Würm age at the most.*

**The problem of long cycles in tropical climates.** In so far as tropical soils which
have been subject to a long cycle of pedogenesis are concerned, the answer to
the question of whether prolonged or successive pedogenesis has occurred is
very different. Obviously no glacial phases have periodically interrupted
pedogenesis and, additionally, if climatic variation has often been important
in the Quaternary, it would have had a greater effect on precipitation than on
temperature, which means that the dominance throughout of clay genesis by
neoformation can be assumed. *From this it is to be expected that continuity of
pedogenesis would occur in hot rather than in temperate climates, which has
been confirmed by most workers.*

Thus *time* can be seen as a fundamental environmental factor, for in a given
climatic region the degree of weathering and the soil type can differ according
to the age of the parent material.

From the conclusions of the first chapter with regard to weathering, it can be seen for well drained sites in hot climates that the dominant neoformed clays gradually lose their silica as weathering and acidification increase. Thus the three types of weathering – (i) fersiallitisation (2 : 1 clays dominant); (ii) ferrugination (kaolinite and 2 : 1 clays); and (iii) ferrallitisation (kaolinite and gibbsite) – in fact do not belong to different cycles but to three phases of the same cycle, the end stage of which is ferrallitisation. In subtropical climates with a marked dry season stage (i) is rarely exceeded; in a dry tropical climate development stops at stage (ii); it is only in humid equatorial climates that stage (iii) is reached. The increase in the speed of pedogenesis, owing to rise of temperature on the one hand and increase of precipitation on the other, is fundamental to the final equilibrium state that is reached.

Parent material composition is also involved, for basic materials such as basalt and dolerite develop from stage (i) to stage (iii) more rapidly than acid materials, rich in silica and quartz. Thus it is easy to understand that the interaction of the three factors of climate, composition and age of parent materials produces a pattern of extreme complexity, which explains the difficulties of defining the climatic zones of the hot regions, compared to those that characterise the temperate or cold zones. Even in the equatorial humid zone (the one most favourable for ferrallitisation) ferrallitic soils (the end result of development) are not really dominant for they are restricted to the oldest parent materials, which have not been rejuvenated, or to parent materials that can develop rapidly.

## 3 Examples of palaeosols and polycyclic soils

Soils formed on Tertiary and early Quaternary parent materials have all the characters of a tropical development that have been discussed and are undoubtedly palaeosols. In agreement with most authors, those soils of northwestern Europe formed on loams of Riss age are acceptable as palaeosols, in contrast to those formed on loams of early Würm age, that are considered as old soils and in which there has been continuity of pedogenic process, with the time factor being the only one involved.

**Palaeosols and polycyclic soils of the Tertiary and early Quaternary.** All these palaeosols show signs of the imprint of a hot climate, of the tropical kind: kaolinite is the dominant clay; weathering is of the *ferruginous* type, if a certain amount of micaceous minerals persist, or of the *ferrallitic* type, where they are present in no more than trace amounts and gibbsite is present.

Examples will be considered of weathered materials on calcareous and aluminosilicate rocks.

*Weathered material on limestones.* This deals with material produced by the decarbonation of indurated limestones, such as those of the middle and upper Jurassic, very often referred to by the general term **terra fusca** when rubification at most is slight, and **terra rossa** when it is marked. These two materials, according to Mückenhausen (1973b, 1974), are homologous and of varying age. The oldest material in which kaolin is dominant goes back as far as the Tertiary or earliest Quaternary, while the more recent have a marked dominance of illites and were formed in various interglacial phases. *Terrae fuscae* were formed in cooler, more humid climates than that required for the genesis of *terrae rossae*, which additionally required a dry season. These two palaeosols have served as parent materials for numerous types of polycyclic soils, the recent development of which, in terms of short cycles, is connected with present-day conditions of either climate or site. This has resulted in brunification in a temperate climate and isohumism, of a reddish chestnut soil, in a dry steppe climate. Mechanical effects owing to topography have caused a secondary recarbonation of these materials, giving rise to *brunified rendzinas* or *polycyclic calcareous brown soils*. Chapter 7 (on calcimagnesium soils) will consider these polycyclic soils in greater detail.

When they are very old, these decarbonated materials can be very thick, in excess of 1 m, and have the characteristics of a very acid ferruginous or ferrallitic palaeosol, in which kaolinite is the dominant clay. This is the case for some materials in certain areas of the Swabian Jura studied by Professor Schlichting's group (Hemme 1970, Alaily & Schlichting 1975). However, because almost always there has been an addition of much more recent aeolian loams to the surface of these materials, a necessary preliminary has been its assessment by very precise mineralogical balances, to arrive at an accurate idea of what can be ascribed to weathering. These mineralogical balances show that illites have been added in the loams, while kaolinite is almost the only clay of the weathered material.

Recent pedogenesis on this material is in complete contrast to the older development, for under a markedly cold, humid, mountain climate, almost always a surface mor occurs and, depending on local drainage conditions, either a podzol or podzolic stagnogley is formed. In both soils the A2 is completely bleached and they differ only in the thickness and the nature of the B horizons. Clays disappear, particularly the 2 : 1 clays, either by lateral pervection or acid hydrolysis. The total profile thickness does not exceed 30–40 cm, with the ferrallitic material reappearing beneath the B horizon.

*Weathered materials of the old massifs*. These have been investigated in the Ardennes and the Rhine schist massive – particularly by Kerpen (1960) and Mückenhausen (1973b) – and involve *grey loams* variegated with ferruginous patches that occur on the oldest weathering surface of the Rhine schist massive. Both authors regard this material as a **plastosol** of Kubiena, which is a kaolinite-rich palaeosol with a very massive structure, of Tertiary age. Certain *siderolithic* outcrops, which occur on the southwestern border of the French Massif Central, can be considered as having the same origin (Callot 1977).

It is to be noted that these examples of palaeosols show a *convergence of climatic development*, for no matter whether limestone or aluminosilicate rocks are involved, they both have characteristics of marked ferrallitisation in an acid medium. This is additional proof of the reality of climatically controlled development, no matter whether recent or older pedogenesis is considered.

The recent Postglacial development of the plastosols is very variable, but generally their great impermeability is responsible for a hydromorphic development. Under the best of conditions a pervected hydromorphic soil is formed, but very impermeable conditions cause forests to be unstable and generally degraded, and the rise of the water table causes a pseudogley to form (Mückenhausen 1973a).

**Palaeosols and polycyclic soils formed on middle to late Quaternary loams** (Fig. 4.13). The polycyclic development on these older loams of Lorraine, investigated by a group from CNRS in collaboration with CNRF, will be taken as an example (Duchaufour *et al*. 1973). The older pedogenic history has been determined by micromorphological analysis, particularly of the lower horizons, while the recent pedogenesis has been determined in Becker's (1971) investigation of the hydromorphic state of the near-surface horizons. The result of these investigations shows that the loams have been subject to three main phases of development: (i) pre-Würm subtropical, (ii) Würm surface cryoturbation, (iii) post-Würm, brunification and hydromorphism.

*The first phase of pre-Würm age* is displayed by the deep Bg horizon, which is *in situ*, for the original structure is completely preserved and undisturbed. It is an old Bt horizon that has the character of a *plastosol* or *brown loam* of an ochreous, slightly rubified colour and with a massive structure. When this horizon has not been truncated, there is often preserved a bell- or funnel-shaped, tongue-like (**glossic**) projection, the central part of which generally consists of a bleached quartzose loam, that is the remnant of an A2. More often, this *glossic* horizon has been removed by cryoturbation and only a deeper fragipan horizon is preserved, with branching white veins against an ochreous background. These *glossic* structures, together with the tight packing of the soil material, have been ascribed to the first phase of glaciation subsequent to the subtropical weathering. Microfabric analysis gives evidence of intense clay pervection within the glossic structures leading to the deposition of thick ochreous ferri-argillans along contraction cracks – primary illuviation argillans of de Coninck and Herbillon (1969) and Fedoroff (1970) – which have been degraded and bleached in the central part by the circulation of acid waters, that are also responsible for the bleached network associated with the lower fragipan (argillans of degradation). Where the clay involved in these processes is dominantly an open vermiculite, its degree of crystallinity has been much decreased (see also Ch. 9).

*The second phase at the end of the Würm is of cryoturbation* which affected the upper part of the profile including all the A2 and the upper part of the Bg horizon of the lessived glossic soil of Riss age. The funnel-shaped glossic structures are generally truncated and reworked. A B1 horizon is formed corresponding to the upper part of the Bg horizon, which is proved by the occurrence within the plasma of papules, or dislocated fragments of ferriargillans. The surface cryoturbated layer was then exposed to a new pedogenesis, under very acid conditions that caused pervection to decrease, which is reflected in the occurrence of bleached very thin argillans, and as they are *in situ* they must be subsequent to the cryoturbation (secondary illuviated argillans).

*The final post-Würm phase* corresponds to the current short cycle which is related to the forest vegetation and has affected only the upper part of the profile. The relative impermeability of the B1 and Bg horizons favours hydromorphic development, which under closed forest and the resulting high *PET* causes hydromorphic conditions to remain at some depth and a lessived hydromorphic soil is formed, which is an *analogue* of profiles on other parent materials. However, man-induced *degradation* is common, causing the water table to rise and a pseudogley or even a podzolic pseudogley to form. This type of hydromorphic degradation has been dealt with previously.

Many examples of soils with multiple pedogenesis on loams have been described in France by Ducloux and Ranger (1975) and Ducloux (1978), and in Great Britain by Avery (1973), Bullock and Murphy (1979) and Catt (1979). The deep *glossic* horizons are often called **palaeoargillic**, while the brunified upper horizon is often formed from recent loess. Such soils formed from two materials of different ages can be considered as being both polycyclic and complex.

**Post-Würm polycyclic soils of biological origin.** In this case the successive pedogenic cycles are the result of variations in vegetation and humification entirely within the post-Würm period. The type of humus characteristic of the first cycle is different from that which forms in the second and in certain cases traces of it are preserved.

A good example is provided by the palaeochernozems on central European loess (Austrian–Saxony) which have been investigated by Kopp (1965). The formation of these soils under a steppe vegetation with artemisia goes back to the last glacial period, as shown by palynology and $^{14}$C, and the incorporation to considerable depth of strongly polymerised humic acids has given the whole profile a dark colour (the extraordinary stability of this organic matter which has persisted for 7000 years is to be noted). The more recent phase dates from the beginning of the atlantic period when forest replaced steppe vegetation causing a surface brunification, together with moderate pervection, which has produced a Bt horizon with a polyhedral structure and cutans.

In France the black earths of Limagne have had a similar development (Duchaufour 1978: II$_2$).

## 4   Examples of complex profiles

Complex profiles are made up of two layers of different ages, the lower one of which has been affected by the two successive cycles of pedogenesis. This is different, it should be remembered, from the case of a compound profile, where the lower layer has been affected only by the older cycle of pedogenesis (see Fig. 4.12).

Undoubtedly, the truncation by erosion of the oldest profile has caused difficulties of interpretation which have made it necessary to use precise mineralogical balances and microfabric analyses to unravel the frequently complicated history of this type of profile.

The best example of this is provided by the deposition of a variable thickness of aeolian loam (often reworked by surface water) over older surface materials, which is common at the present day in the Mediterranean zone and even in central Europe because of the addition of Saharan dust (Yaalon & Ganor 1973). The influence on pedogenesis of this addition of dust varies according to the way it is added. Thus if this has been moderate, gradual and continuous over a long period, the dust is mixed with the *in-situ* material (mechanical or biological reworking) and so a second layer is not formed but rejuvenation of the whole profile occurs which makes its investigation and dating difficult. Hubschmann (1975) showed this to be the case on terraces in the Pyrennean foothills, and this has already been pointed out with regard to the weathered material of the limestones of the Swabian Jura.

In contrast, if the addition of loam occurs in greater quantities and is more concentrated in time, a distinct upper layer is formed which buries the previously formed profile and, if it is thick enough, an undoubted *fossilisation* of a palaeosol results, i.e. a fossil soil is formed. If, on the contrary, this layer does not exceed some tens of centimetres, the palaeosol is involved in the new pedogenesis and a complex profile results, such as in the case of many soils of the Paris basin and Lorraine.

One of the classic examples of a complex profile is that where loams overlie a thin bed of decarbonated clay on indurated limestone. Among the many investigations of such materials, the following can be cited: Ducloux (1970, 1978), Mathieu (1975), Robin and de Coninck (1975) and Catt (1979). In fact, despite their apparent simplicity, the interpretation of these profiles requires care, for two questions need to be answered:

*Question 1.* Is the surface loam younger than the deeper layer of decarbonated clay, which is thus a buried palaeosol of the terra fusca type, or have the two layers been formed simultaneously? The second case is the same as that in covered karst (Bonte 1963, Mathieu 1975) where the surface cover overlying the limestone acts as a reservoir for water, regulating infiltration, which thus increases the amount of dissolved $CO_2$ and the corrosive effect of water; also it protects the clay residual from decarbonation against erosion. It is often difficult to differentiate between these two possibilities, both of which could be valid according to the situation. Mathieu (1975) investigated typical cases of the formation of decarbonated clay at depth while, in contrast, Robin and de Coninck (1975) described a complex profile which seemed rather to be made up of a palaeosol of the terra fusca type buried under *aeolian* loamy sands.

*Question 2.* Is there a genetic connection between the two layers? This can only be in the form of illuvial clay coming from the surface loam being deposited in the lower clay layer which is thus of mixed origin. The operation of this process, determined by microfabric and mineralogical analyses, has been shown to occur in most of the examples investigated and is responsible for the 'beta' horizon, referred to in Chapter 2.

No matter what the genetic interpretation of these complex profiles, their morphological identification is generally easy, as the marked heterogeneity presented by the relatively well differentiated overlying loam and the lower clay (beta) horizons is plainly visible. The interpretation of other complex profiles made up of the same materials but which have been mechanically disturbed by cryoturbation is not so easy, particularly if detailed mineralogical and microfabric analyses are not available. In such profiles, the lower clay layer has been mixed with some of the loam by cryoturbation and solifluction, the layer thus homogenised was then covered by a new layer of loam. Finally, pervection of clay from the surface loam occurs and gives the lower layer all the appearance of a Bt horizon. Such a profile has been analysed and described in detail in Duchaufour (1978) profile $IX_6$, microstructure $XX_8$ and $XX_9$ (see also Fig. 7.4b).

## V  Conclusion

At the end of this general chapter on soil development two important questions can be asked. They represent only one aspect of the long-term differences between biologists and ecologists on the one hand and geologists and geomorphologists on the other, as to whether pedology is a biological or an earth science. These questions are concerned first with whether the idea of *climax* can be applied to soil and secondly with the use of biological processes in the genetic classification of soils.

The idea of climax is defended only by those pedologists who have biological interests and is disputed by the others. Does a stable equilibrium exist reflecting the interaction of soil and vegetation? From the preceding discussion it can be said that the concept of climax can be applied very well to the *short cycles* characteristic of temperate or cold climates, for it has been shown that organic matter (*humus*) is itself an active intermediary between vegetation and soil. When a new parent material is colonised, there is a strict parallelism in development between vegetation and soil in which, in both cases, the ultimate state is called the *climax* and thus it is an idea that is as valid for soil as for vegetation.

It is evidently not the same for soils with slow development, characteristic of hot climates, which are the products of a *long cycle*. In this second case, if the development of vegetation and humification are always relatively rapid processes, profile development is in contrast slow. *The same vegetation and the same type of humus can occur on different soils, depending on the age and kind of parent materials*. In these circumstances there is a correlation only between vegetation and humification but not between vegetation and soil, and hence *the idea of climax should be limited to the processes of humification*.

As far as the problem of soil classification is concerned, which will be dealt with in the second part of the book, logically it can be said that an environmentally based classification should depend whenever possible on the idea of the climax – that is to say, in those situations where a reciprocal soil–vegetation equilibrium exists. Consequently, when dealing with a polycyclic soil which has been formed by the successive action of a long and a short cycle, classificatory priority should always be given to the short cycle, and the long cycle should only be considered at a lower level of classification, such as at that of the group. However, this presupposes that *the short and long cycles are identifiable*, or in other words, that it is possible to differentiate between current and inherited profile features, but in many cases this is not certain. In particular, this is so for many of the soils described by Ruellan (1970) as old soils, which do not show any differences between older and more recent characters and where development, even if it is of long duration, is apparently continuous. Old and recent soils have the same type of development, only differing in degree, and it is evident that these soils should be placed in the same class.

## References

Alaily, F. A. and E. Schlichting 1975. *Mitteilg. Deutsch. Bodenk. Ges.* **22**, 621–4.
Ameryckx, J. 1960. *Pédologie*, Ghent **X** (1), 124–90.
Arkley, R. J. 1963. *Soil Sci.* **96** (4), 239–48.

Arkley, R. J. 1967. *Soil Sci.* **103** (6), 382–400.

Avery, B. W. 1973. *J. Soil Sci.* **24** (3), 324–38.

Bartoli, C. 1966. *Etudes écologiques sur les associations forestières de la Haute Maurienne.* Engng doct. thesis. Fac. Sci. Montpellier.

Becker, M. 1971. *Etude des relations sol-végétation en conditions d'hydromorphie dans une forêt de la plaine lorraine.* State doct. thesis. Fac. Sci. Nancy.

Begon, J. C. and M. Jamagne 1973. In *Pseudogley and gley. Trans Comms. V and VI ISSS*, E. Schlichting and U. Schwertmann (eds), 307–17. Weinheim: Chemie.

Birkeland, P. 1974. *Pedology weathering and geomorphological research.* Oxford: Oxford University Press.

Bocquier, G. 1973. *Genèse et évolution de deux toposéquences de sols tropicaux de Tchad. Interprétation biogéodynamique.* State doct. thesis. Univ. Strasbourg.

Bonte, A. 1963. *Sedimentology* **2** (4), 333–40.

Bornand, M. 1978. *Alteration des materiaux fluvio-glaciaires. Genese et evolution des sols sur terrasses quaternaires, dans la moyenne vallee du Rhone.* State doct. thesis. Univ. Montpellier.

Bos, R. H. G. and J. Sevink 1975. *J. Soil Sci.* **26** (3), 223–33.

Bottner, P. 1972. *Evolution des sols en milieu carbonaté.* State doct. thesis. Univ. Montpellier; *Mém. Bull. Sci. Géol.* no. 37.

Boulaine, J. 1975. *Geographie des sols.* Paris: PUF.

Bullock, P. and C. P. Murphy 1979. *Geoderma* **22**, 225–52.

Callot, G. 1977. *Science du Sol* **4**, 189.

Catt, J. A. 1979. Soils and Quaternary geology in Britain. *J. Soil Sci.* **30** (4), 607–42.

Coninck, F. de and A. Herbillon 1969. *Pédologie*, Ghent **XIX** (2), 159–272.

Daniels, R. B., E. E. Gamble and W. H. Wheeler 1978. *Soil Sci. Soc. Am. J.* **42** (1), 98–105.

Dimbleby, G. W. 1952a. *J. Ecol.* **40** (2), 331–41.

Dimbleby, G. W. 1952b. *Plant and Soil* **4**, 141–53.

Dimbleby, G. W. 1961. *J. Soil Sci.* **12** (1), 12–22.

Dimbleby, G. W. 1962. *Oxford Forestry Memoirs* **23**.

Duchaufour, Ph. 1978. *Ecological atlas of soils of the world.* New York: Masson.

Duchaufour, Ph., F. Le Tacon, M. Becker and J. M. Hettier 1973. *Pseudogley and Gley. Trans Comms V and VI ISSS*, E. Schlichting and U. Schwertmann (eds), 287–93. Weinheim: Chemie.

Ducloux, J. 1970. *Bull. AFES* **3**, 15.

Ducloux, J. 1978. *Contribution a l'étude des sols lessivés sous climat atlantique.* State doct. thesis. Univ. Poitiers.

Ducloux, J. and J. Ranger 1975. *Ann. Soc. Sci. Nat. Charente-Maritime* **VI** (2), 116–32.

Duvigneaud, P. 1974. *La synthese ecologique.* Paris: Doin.

Erhart, H. 1967. *La genese des sols en tant que phenomene geologique.* Paris: Masson.

Favarger, G. and P. A. Roger 1956. *Flore et vegetation des Alpes.* Neuchâtel: Delachaux et Niestlé.

Fedoroff, N. 1970. *Colloque Micromorphologie des sols.* Grignon: ENSA.

Flint, R. F. 1971. *Glacial and Quaternary geology.* New York: Wiley.

Franzmeier, D. P. and E. P. Whiteside 1963. *Q. Bull.* **46** (1), 2.

Galoux, A. 1954. *La Chênaie sessiliflore de Haute Campine. Essai de biosociologie.* Stat. Rech. Groenendaal.

Ganssen, R. 1972. *Bodengeographie.* Stuttgart: Koehler.

Ganssen, R. and F. Haedrich 1965. *Atlas zur Bodenkunde.* Mannheim: Bibliograph Institut.

Guillet, B. 1965. *La méthode de datation par le carbone 14: application à la détermination de l'âge de la tourbière du Beillard (Vosges).* 3rd yr thesis. Fac. Sci. Nancy.

Guillet, B. 1972. *Relations entre l'histoire de la végétation et la podzolisation dans les Vosges.* State doct. thesis. Univ. Nancy.

Harris, S. A. 1970. In *Paleopedology*, D. H. Yaalon (ed.), 191–203. Jerusalem: International Society of Soil Science/Israel University Press.

Hemme, H. 1970. *Die Stellung der lessivierten Terra fusca in der Bodengesellschaft der Schwabischen Alb.* Thesis. Univ. Hohenheim.

Hetier, J. M. 1975. *Formation et évolution des andosols en climat tempéré*. State doct. thesis. Univ. Nancy.

Hubschmann, J. 1975. *Morphogénèse et pédogénèse quaternaires dans le piémont des Pyrénées garonnaises et ariégeoises*. State doct. thesis. Univ. Toulouse.

Icole, M. 1973. *Géochimie des altérations dans les nappes d'alluvions du piémont occidental nord pyrénéen*. State doct. thesis. Univ. Paris VI.

Jamagne, M. 1969. *VIIIth INQUA Congr*. Paris, 359–72. Versailles: INRA.

Jamagne, M. 1973. *Contribution à l'étude pédogénétique des formations loessiques du nord de la France*. State doct. thesis. Inst. Agron. Gembloux.

Jenkinson, D. S. and J. H. Rayner 1977. *Soil Sci*. **123** (5), 298–305.

Kerpen, W. 1960. *Die Boden des Versuchsgebiet Rengen*. Berichte Univ. Bonn (5).

Kopp, V. E. 1965. *Eiszeitalter und Gegenwart* **17**, 97.

Kubiena, W. L. 1953. *The soils of Europe*. London: Thomas Murby.

Laatsch, W. 1963. *Bodenfruchtbarkeit und Nadelholzanbau*. Munich: BVL.

Lamouroux, M. 1971. *Etude des sols formés sur roches carbonatées. Pédogénèse fersiallitique au Liban*. State doct. thesis. Univ. Strasbourg.

Lemée, G. 1967. *Précis de Biogéographie*. Paris: Masson.

Leneuf, N. and G. Aubert 1960. *7th Congr. ISSS Madison* **IV** (V31), 225–8.

Lieberoth, I. 1963. *Geologie* **12** (2), 149–87.

Ludi, W. 1945. *Forschungsinst. Rübel*. Zurich: Berichthaus.

Mathieu, C. 1975. *Science du Sol* **3**, 183–206.

Meyer, B. 1960. *7th Congr. ISSS Madison* **IV** (V25), 177–83.

Mückenhausen, E. 1973a. *Pseudogley and gley. Trans Comms V and VI ISSS*, E. Schlichting and U. Schwertmann (eds), 147–57. Weinheim: Chemie.

Mückenhausen, E. 1973b. *Anal. Edaf. y Agrobiol*. **XXXII** (1–2), 1–20.

Mückenhausen, E. 1974. *Bodenkunde*. Frankfurt-am-Main: DLG.

Munaut, A. V. 1967. *Recherches paléo-écologiques en basse et moyenne Belgique*. Univ. Louvain.

Novak, R. J., H. L. Motto and L. A. Douglas 1970. In *Paleopedology*, D. H. Yaalon (ed.), 211–24. Jerusalem: International Society of Soil Science/Israel University Press.

Novikoff, A., G. Tsaw Lassou, J. Gac, B. Bourgeat and Y. Tardy 1972. *Bull. Sci. Geol*. **25** (4), 287–305.

Paquet, H. 1969. *Evolution géochimique des minéraux argileux dans les altérations et les sols des climats méditerranean à saisons contrastées*. State doct. thesis. Univ. Strasbourg.

Pallmann, H. 1947. *Congr. Intern. Ped. Medit*., Montpellier, 1–36.

Pallmann, H., F. Richard and R. Bach 1949. *10th Congr. IUFRO*, Zurich, 57–95.

Paul, E. A., C. A. Campbell, E. A. Rennie and K. J. McCallum 1964. *8th Congr. ISSS Bucharest* **III** (II21), 201–8.

Robin, A.-M. and F. de Coninck 1975. *Science du Sol* **3**, 213–28.

Roose, E. J. 1970. *Cah. ORSTOM, Sér. Pédol*. **VIII** (4), 469.

Ruellan, A. 1970. *Symposium on age of parent materials and soils*. Amsterdam.

Sachse, M. F. 1965. *Deutsch. Akad. Landwirtschaft. Berlin* **9** (10), 867–80.

Scharpenseel, H. W. 1972. *Z. Pflanzener. Bodenk*. **133** (3), 241–63.

Schweikle, V. 1971. *Die Stellung der Stagnogleye in der Bodengesellschaft der Schwartzwaldhoch-flache*. Univ. Stuttgart-Hohenheim, Sulz Kreis Horb.

Segalen, P. 1973. *L'aluminium dans les sols*. Document ORSTOM, no. 22.

Servat, E. 1966. *Conf. Pédologie Medit*., Madrid, 406–11.

Souchier, B. 1971. *Evolution des sols sur roches cristallines à l'étage montagnard (Vosges)*. State doct. thesis. Univ. Nancy I.

Spaargaren, O. C. 1979. *Weathering and soil formation in a limestone area near Pastena (Italy)*. Publ. no. 30. Doct. thesis. Univ. Amsterdam.

Stone, E. L. and W. W. McFee 1965. *Soil Sci. Soc. Am. Proc*. **29** (4), 432–6.

Tamm, C. O. and H. Holmen 1967. *Saertryk Meddel. fradet Norske Skogsforsoksvesen* **XXIII**, 69–85.

Turenne, J. F. 1975. *Modes d'humification et différenciation podzolique dans deux toposéquences guyanaises*. State doct. thesis. Univ. Nancy I.

Volobuyev, V. R. 1962. *Pochvovedeniye* **5**, 73–82.
Wilke, B. M. 1975. *Z. Pflanzener. Bodenk* **2**, 153–71.
Yaalon, D. H. 1975. *Geoderma* **14** (3), 189–205.
Yaalon, D. H. and H. Ganor 1973. *Soil Sci.* **116** (3), 146–55.
Zöttl, H. W. and H. Küssmaul 1967. *Anal. Edaf. y Agrobiol.* **XXVI** (1–4), 381.

*Part II*

*PEDOGENESIS: THE BASIS OF SOIL CLASSIFICATION*

# Chapter 5

# Soil classification

## I  General introduction

Soil classification has to solve a twofold problem:

(a) It has to classify the higher units, to group the major soil types of the world according to the genesis of their basic properties and thus provide a framework, of some kind, to act as a basis for the science of pedology;
(b) It has to provide a means of making large-scale maps for practical purposes, such as in agronomy, which often necessitates the use of detailed characteristics, that are of local importance only, in defining and naming units. It is obvious that this classification of lower units involves entirely different problems from that involved in dealing with the higher units.

The final aim, however, is to produce a single hierarchical classification which includes all types of units. Such an overall classification must have a pyramidal form where the relatively few higher units make up the summit and the lower units, increasingly subdivided and numerous, make up the base.

In several countries, soil classification has generally been attempted simultaneously at these two levels − scientific classification of higher units and detailed locally applicable classifications for large-scale soil mapping. Then it is necessary to integrate the two levels by fitting the lower units into the framework of the higher units.

In this book the problem of the classification of the lower units (series) will not be dealt with and this chapter is concerned only with the higher units. It is generally agreed that this classification must entail an abstraction of detailed soil properties, so that it is based absolutely on the fundamental processes of profile development, i.e. pedogenesis, and not on factorial properties, hence the term **genetic** which is commonly used for this kind of classification.

## 1  Development of ideas in soil classification

Very generally, it is possible to distinguish three main phases in the history of soil classification. At first, pedologists, under the influence of the Russian school, used environmental factors as a basis and in particular the most important of them, that of climate, which resulted in the soils being divided into three major classes: **zonal**, that is to say essentially dependent on the factors of climate and climatically determined vegetation; **intrazonal**, dependent upon local peculiarities of site; and the **azonal** immature soils such as those on slopes rejuvenated by erosion. Properties of the soils themselves, which were moreover not well known, were only used at a lower level.

As knowledge of the physics and chemistry of the soil has increased, particularly with regard to weathering, it has been used as a major factor in the definition of the higher units of soil classification. Thus pedologists have defined soils as a function of the processes of weathering, of the kind of weathering complex and its degree of saturation and also of the processes of movement. For example, Pallmann (1947) suggested an hierarchical classification in which an insoluble, practically immobile, **filter** fraction is distinguished from a **percolate** fraction comprising entities transported in solution or suspension, the direction of movement, the chemical make-up of the filter and the properties of the percolate being used to define the successive hierarchical levels of classes, orders, associations and types.

More recently, the history of soil classification has been marked by the development of two tendencies. *Intrinsic soil characteristics* have been taken as a basis for classification in preference to the external factors, such as climate, that are responsible for their formation. However, pedogenesis is not ignored for *the characters that are selected are those that reflect the phases of soil formation and development*, so that the basis remains *genetic*; that is to say, it emphasises soil development. In addition, *environmental syntheses* are used in modern classifications, which involve an evaluation of all factors, environmental, morphological, physicochemical and biological. The selected characters are not independent of one another but are part of an interacting unified whole. Such an approach has allowed *classes* and *orders*, that is to say the higher units that appear in most classifications, to be differentiated.

## 2   The principles of modern classifications: similarities and differences

Although most modern classifications emphasise intrinsic soil characteristics, this is not done at random, for *they are chosen and ranked in so far as they are the best reflection of one or several developmental processes, which are themselves controlled by the environment*.

In other words, soil history, or the stages of soil development as related to the environment, must not be ignored. In this respect it should be remembered that Schröder (1973) defined the trilogy: *environment → process → soil characters*, a concept restated in a slightly different form by Gerassimov (1974), who defined *elementary pedological processes* related to the environment.

Comparing the main present-day world classifications of western and eastern Europe and also the USA and FAO, it is possible to say that, despite the apparent great differences, they are all basically the same and they differ only in their minor details.

*The Russian classification* has retained its original environmental divisions, and classes remain by definition *zonal* or *climatic*. (For example, the different podzols – boreal, atlantic, tropical – are each classed within their own climatic zone, even though they are subject to the same developmental process.) Within the zonal classes, soils are grouped according to whether development has been controlled by the parent material (lithogenesis) or by hydromorphic conditions (biohydrogenesis). Developmental processes are taken into account, however, in defining the lower units or types (Rozov & Ivanova 1968, Labova 1977, 1978).

*The American and FAO classifications*, in contrast, emphasise profile characteristics, defined in terms of analytical criteria, independent of process and environmental conditions. However, in a revision of the American scheme (Soil Survey Staff 1975), an attempt has been made to emphasise genetic processes and even the environmental conditions within particular units, which has been successful in certain cases but less so in others.

*The western European schemes*, particularly those of the German Federal Republic (Mückenhausen 1962, 1965), Great Britain (Avery 1973, 1980) and France (CPCS 1967), are different, for in them the choice of characters and their hierarchical level cannot be separated from the study of processes and the environmental conditions: thus *the three panels of the triptych, environment, process and characters, are considered simultaneously*. This third method gives a better result because it achieves the proposed objective, that of showing the *genetic relationship* of the various taxonomic units.

## II  Criteria used in modern classifications

The first attempts at classification considered profile characteristics independently of one another and then placed them in order of decreasing importance so as to differentiate the units at successive levels. In contrast, because of the numerous difficulties caused by this approach, more recent classifications have attempted to relate the basic characters to one another so as to differentiate a certain number of families with a particular genetic parentage (orders or classes), which thus expresses the synthetic nature of these classifications much more clearly.

### 1  The first criteria used in genetic classifications
It is possible to consider the basic criteria of the various present-day genetic classifications under the following headings:

**The amount of profile development, together with the degree of development of the parent material**

(A)C profile  :  raw mineral soils;
AC profile    :  little-differentiated soils containing organic matter;
A(B)C profile:  soils developed as a result of weathering with a weathered (B) horizon;
ABC profile  :  soils developed as a result of weathering and movement of material, with an illuvial B horizon.

These criteria have not been used at the same level in all classifications, but most of them have differentiated an (A)C class of profiles and one or several AC profile classes, that have humic profiles without a distinct (B) or B mineral horizon. A problem that is very difficult to resolve is created by the common occurrence of intermediate profiles, such as certain humic AC soils that are very weathered and developed (some rankers and andosols).

**Climatically controlled weathering.** This, as related to clay formation, is taken as being fundamental in all present-day classifications. The basis of this has been discussed in Chapter 1. A typical example is provided by the American classification that differentiates two types of (B) horizon – the cambic (B) with moderate and incomplete weathering and the oxic (B) with a maximum of weathering. Two types of Bt horizon are also differentiated – those of the alfisols (lessived temperate soils) and those of the ultisols (lessived subtropical soils), that differ in their degree of weathering. A similar distinction is found in the French classification where the classes *fersiallitic*, *ferruginous* and *ferrallitic* are characterised by increasing weathering.

**The movement of material within soils.** This includes, as will be remembered, both downward and upward movement (biogeochemical cycles) which controls the equilibrium of bases and thus the pH. All classifications take these factors into account but at different levels.

Generally, *cheluviation* (the movement of mobile organometal complexes) is used in all classifications to define the class of podzols. In contrast, *pervection of clay* is used at various levels depending upon the classification. In the American classification it is regarded as fundamental and the presence of an *argillic* horizon (i.e. one enriched in clay by pervection) is used to characterise two classes – alfisols and ultisols – while the French classification uses it only at the subclass level, e.g. to differentiate the lessived soils within the brunified class of soils. Generally, *percentage base saturation* is only used at a lower level, but in the American classification it is regarded as an important criterion in distinguishing alfisols (slightly developed lessived soils) and ultisols (very developed lessived soils).

**Pedoclimate.** This deals with the internal soil climate, characterised by both seasonal temperature variations and the amount of water present. The latter factor takes on a particular importance when at certain seasons it becomes so excessive as to cause an oxygen deficit and greatly affect the redox potential, for this allows the *hydromorphic* class of soils to be differentiated in the various western European classifications. In contrast, the American classification uses this criterion only at the sub-order level and 8 of the 10 orders described have *hydromorphic* sub-orders, while the *histosol* order, or that of peaty soils, is regarded as entirely *hydromorphic*.

For the non-hydromorphic soils, that is to say those that are normally drained, at least in the surface, most present-day classifications have taken into account the pedoclimatic characteristics at the subclass level. The American classification uses them in a very precise manner to define the seasonal limits of temperature and humidity, which in terms of current practice are difficult to apply. In this respect, the importance given in certain classifications to pedoclimatic characteristics to differentiate subclasses within well drained soils appears to be excessive. Their use can also be criticised on the grounds that pedoclimate is too intimately linked to the *external factors* of the environment, particularly the general climate, to be considered as a soil property in itself.

## 2  Present-day tendencies: taking account of new criteria

Recent progress in pedology in two important areas will not fail to have a profound influence on soil classification in the near future. These areas are concerned with the *time* factor and the *integrating* role of organic matter in soil dynamics.

**The utilisation of the time factor.** As already stated in Chapter 4, time can be considered as a basic environmental factor which it is impossible to ignore in classification. It is necessary to recall only that some soils, particularly those of temperate or cold climates, characteristically have a short cycle of development while, in contrast, others develop very slowly as a result of long cycles. In addition, *old* soils can be considered as occupying an intermediate position.

Short cycles are not a major classificatory problem, as they are completely reflected in the current soil–vegetation equilibria. The problem is more difficult to resolve when it is a case of soils affected by long or intermediate cycles. First, when pedogenesis, although slow, has been continuously progressive, the old soil can be classified in the same taxonomic unit as those soils subject to a comparable pedogenesis during a short cycle. In this situation old and recent soils differing only in their degree of development belong to the same class and are differentiated at the group, or even subgroup, level. For example, the *acid lessived or degraded soils*, that are generally restricted to old parent materials, should be placed within the class of brunified soils, together with the *lessived brown soils* subject only to a short cycle.

Secondly, in the case of polycyclic soils (Ch. 4) that have been subject to discontinuous development in which two successive very different types of pedogenesis have been involved, the problem is different. Thus, the profile has two groups of characters that are incompatible with continuous development, from an environmental point of view. In fact, to the characters inherited from the old development have been added the present-day characters, acquired under the influence of current bioclimatic factors and these latter characters, according to the author, must take precedence over the former: in other words, *the palaeosol must be considered as the parent material of the present-day soil*. The previously quoted case of podzols in the Swabian Jura provide an example, developed on old ferruginous or ferrallitic parent materials. It would seem obvious that such soils should be classified as podzolised and not as ferruginous or ferrallitic. This particular approach will allow soils developed on palaeosols to be distinguished as being polycyclic, compared to the corresponding monocyclic group.

This approach is used in the French classification of 1967, and the

American classification is very similar as it differentiates such soils from those of the present day by the addition of the prefix *pale* (e.g. paleudults, palexeroll etc.). In contrast, in the classification of Kubiena (1953) and Mückenhausen (1962, 1965) greater emphasis is given to the older rather than more recent developments and the palaeosol is used to differentiate particular groups, e.g. *terrae calcis* (palaeosols formed of decarbonated clay on indurated limestone) and *plastosols* (palaeosols with kaolinite on siliceous materials).

**The integrating role of organic matter.** Recent work has shown the importance of organic matter in soils, at least as far as the rapid development associated with short cycles is concerned. This role, which has been discussed at length in Part I and particularly in Chapter 2, is in fact a double one, *for not only does soil organic matter integrate all of the environmental factors but it is also the motive force in pedogenesis, including weathering, because it controls the formation of specific organomineral complexes on which all soil properties depend.*

In these circumstances, the solution to the problem of the choice and ranking of classificatory criteria must be facilitated by emphasising the biochemical properties of the organic matter, in so far as they are better known than formerly. This can only be the case for soils affected by short cycles which alone have a pedogenesis that is dependent on the organic cycle. For such soils the classification of humus types according to their biochemical properties (Table 2.3) can serve as a basis for the differentiation of several important classes of temperate soils. The simple characterisation of soil organic matter implies, as was stated in Part I, a similar characterisation of all the fundamental processes of weathering, movement of material, and base cycling, as well as explaining the main morphological or physicochemical properties acquired by the soil during its development

However, until now few classifications have recognised this fundamental role of organic matter in pedogenesis or given it the place it should have in the differentiation of higher units because, first, it is considered that the properties of organic matter and its biochemical development are not well enough known and, secondly, organic matter is a relatively labile soil constituent that is capable of rapid change under man's influence, particularly as a result of cultivation. This second reason has caused the Americans to use, in preference to the natural humus that is not resistant to cultivation, four to five **epipedons** (the upper part of the profile) whose properties supposedly survive cultivation. Apart from the histosols (peats), which are unique, only one order in the American classification, the **mollisols**, has been defined in terms of organic matter, i.e. that of a **mollic epipedon**, a diagnostic horizon to be defined later. Even this restricted use of organic matter, however, is unsatisfactory, for the definition of the mollic horizon is based neither on biochemical properties of the humus nor on the environmental factors that have controlled its formation.

So that organic matter can be taken into consideration effectively as a basic classificatory element, it is necessary that the two reasons given above for its lack of use should be resolved. To do this, two avenues of research require investigation:

(a) A better characterisation of organic matter will deal with the first point.
(b) The second difficulty requires the selection of a method that will allow the use of *natural humus* as a reference, even to classify soils modified by cultivation. *Cultivated soils would then be classified by comparing them with similar profiles developed under permanent vegetation.*

## III   Definition and hierarchical position of taxonomic units: horizons

An ideal classification should be hierarchical, that is, of pyramidal form with at most two or three divisions at the summit. But, from the preceding discussion it can be seen to be easier, initially, to differentiate a certain number of soil families with a common genesis, that correspond to natural *groups*. This is the solution adopted in most classifications where 10 to 15 basic *classes* or *orders* have been differentiated.

An additional major difficulty in soil classification is the absence of a clear boundary between soils, for soil types are not well defined entities, as are plants and animals, but form part of a *continuous spectrum* where all transitions occur. However, the aim of all classifications is precisely this, that of placing soils in separate compartments by establishing more or less artificial boundaries between soil groups.

### 1   Attempts at hierarchical classifications
It must be admitted that most attempts at a hierarchical classification have not been successful because it is impossible to select a *criterion* (or a *group of criteria*) that is fundamental enough to differentiate satisfactorily two or three major groups at the highest level.

In attempting to do this, several classifications have used the criterion of hydromorphism; for example Avery (1956) distinguished two major initial categories, hydromorphic and non-hydromorphic soils, Kubiena (1953) and Mückenhausen (1965) divided soils into three major groups – terrestrial, semi-terrestrial and sub-aqueous. However, from the start the artificial nature of this basic division is apparent. Thus, according to Kubiena, pseudogleys are terrestrial soils while gleys are semi-terrestrial despite the great similarity in their development and characters. In the French classification both of them are in the hydromorphic class.

Going back to the conclusions of the first part of this book, and in particular those of Chapter 4, a new possibility for an initial two-way division is apparent, by considering the contrast between soils whose development is controlled by organic matter (short cycles) and those whose development is relatively independent of organic matter (long cycles). Perhaps this solution will be considered as valid in some future classifications. However, there is no need to disguise the fact that there would be difficulties. Thus, several soils of hot climates controlled by long cycles are controlled in the upper parts of their profiles by organic matter, such as the more or less podzolised ferrisols with secondary gibbsite. In addition, certain classes of soil that up to now have been considered to be relatively homogeneous and well differentiated, by the dominance of a key characteristic, would be split into two units belonging to each of these new basic divisions. Such is the case for hydromorphic and sodic soils where certain groups of these two classes are very considerably affected by organic matter. Finally, certain soils of hot climates, such as the vertisols, have characteristic organomineral complexes that result, however, from special site factors

## 2  Problems arising from the continuity of developmental sequences

The solution of this major difficulty lies in establishing lineages or developmental sequences showing the genetic connections that occur between soil groups and the possibilities of them changing from one to another. It is this which has been attempted in Duchaufour (1978), where two kinds of sequences are considered: (i) those based on the degree of development, in which the effect of the same basic process, that depends on the environment on the one hand and the age of the parent material on the other, is varied so that it is possible to distinguish initial, immature and mature soils; (ii) those resulting from the interaction of several basic processes, that are used to define different classes, which allows intermediate soils between the various classes (or *intergrades*) to be dealt with more satisfactorily. The process considered to be the main one is represented by a noun, while the secondary process contributing to the first is represented by an adjective, for example *podzolic* pseudogley. In many cases the secondary process is one of degradation resulting from man's modification of the vegetation, such as in a *podzolic* lessived soil.

This method of presenting sequences does not, of course, replace true classifications, but it complements them and makes them more explicit. The tables and diagrams already presented in Duchaufour (1978) will not be repeated, but the reader will be referred to them when the classification of each major soil group is dealt with.

The whole of this discussion can be summarised as follows:

(a)  Initially, as in most classifications, a limited number of *genetically related groups* or *soil classes* will be differentiated, in terms of a coherent grouping of characters of all kinds. In the case of short cycle temperate soils, *organic matter* composition will be given primary consideration because of its marked integrating role.

(b)  Transitions between classes and genetic lineages within subclasses among their main divisions will be explained as far as possible in the form of linked developmental tables.

(c)  With regard to polycyclic soils, characterised by the successive effects of a long and a short cycle, only the short cycle will be considered as effectively reflecting the current soil–humus–vegetation equilibrium and hence this more recent development will be taken as being more important than the older development, which allows a parent material of a particular type (a palaeosol) to be differentiated.

Two very different kinds of classifications will be discussed: (a) those based on an *analytical grouping of key characteristics*, that only fairly recently have taken any account of environmentally controlled processes and factors. Examples of this approach are the American classification together with that of the FAO, which is based on similar but much simpler criteria; (b) those *in which the environmentally controlled processes and the properties of organic matter* are the bases for the differentiation of the main units, such as the CPCS French classification of 1967, a variant of which will be used in the subsequent chapters of this book.

## 3 International horizon nomenclature

Capital letters are used for the main horizons, while lower-case letters as indices are used for their subdivisions. The letters chosen indicate the nature of the essential constituents, while numerical indices are generally used to indicate quantitative variations in these constituents.

### Main horizons

(The distinction, which is considered to be fundamental, between a weathered (B) and an accumulative B has not been made in the USA and FAO classifications.)

A: surface horizon containing organic matter; often impoverished in fine materials or in iron by pervection.
(B): structural or weathered B, differing from parent rock by its greater degree of weathering (presence of free $Fe_2O_3$) and from the surface A horizon by its different structure (also called Bw).
B: horizon enriched by illuviation of fine-grained or amorphous materials – clay, iron and aluminium oxides and sometimes humus.
C: original material from which are formed A, B or (B) horizons.
G: greenish grey horizon, rich in ferrous iron, with rusty patches formed within or at the upper boundary of a permanent water table.
R: subjacent indurated rock.

### Subdivisions of main horizons

*A horizons*
A00 (or L): litter, identifiable plant debris.
A0(0): organic horizon with original plant structure modified or destroyed (more than 30% organic matter).
A1 (or Ah): mixed horizon containing organic matter (less than 30%) and mineral material.
Ap: cultivated organic horizon, which is thus homogenised and with a clear lower boundary.
A2 (or Ae, E): horizon poor in organic matter, from which clay and sesquioxides have often been removed by pervection, light coloured (often referred to as *eluvial*).
A/B (or B1): transition horizon between eluvial and illuvial, marking the beginning of the accumulation of fine-grained or amorphous materials.

*B and G horizons*
Bt: clay accumulation (T = *Ton*, clay in German).
Bh: humic accumulation.

Bs (or Bfe, Bir): sesquioxide accumulation dominant.
Bb: placic horizon; thin wavy ironpan.
Go: oxidised gley with patches and concretions.
Gr: reduced gley, greenish grey with ferrous iron dominant.
Depending on the conditions, certain indices can be used with A, B or C horizons:
g: pseudogley of temporary hydromorphic conditions, variegated with grey, white and rusty patches, sometimes with black concretions;
ca: horizon enriched in calcium carbonate;
sa: horizon enriched in salts;
x: fragipan.

**Note: special horizons**
Attention needs to be drawn to certain complex horizons that have a particular importance in classification.
A0H: *humified layer* of a mor horizon in which the plant material has been greatly altered compared to the *fermentation layer* A0F.
A1B: horizon with transformed humus from the insolubilisation of certain organomineral complexes, characteristic of certain humic profiles (ranker, andosols, humic ochreous soils).
BtBs: both *argillic* and *spodic* mixed horizon.
Beta ($\beta$): horizon both of weathering and a particular kind of accumulation, rich in fine clay and iron, formed above certain beds of indurated limestone under a loamy or sandy cover.

## IV  The American classification (soil taxonomy)

### 1  General principles: horizons

The American classification resulted from the work of a team of people, the first version of which, published in 1960, was produced by the Soils Cartographic Service under the direction of G. Smith, with the aid of a Belgian team (Tavernier, Ghent). Several revisions and modifications have been made since 1960, and have lead to a definitive version (Soil Survey Staff 1975, Flach 1978).

The concept of the profile, i.e. the soil section, is replaced by that of the pedon, considered as a volume of soil, in which the horizons form its superposed layers. The basis of the classification is the identification of diagnostic horizons, carefully arranged and defined in terms of their physical, morphological and chemical properties as a whole, and very precisely described and quantified. The use of those diagnostic horizons, at least the most important of them, allows the main orders of the classification to be differentiated, except for the *aridisols* and *vertisols*.

*These fundamental diagnostic horizons* are classified into two main groups: (i) those of the *surface*, containing organic matter and called *epipedons and defined so that their properties remain the same, even though they are cultivated*; (ii) those at *depth*, that are essentially mineral and hence corresponding to the (B) or B horizons. There is also a certain number of *secondary diagnostic horizons* defined that are used to differentiate particular groups.

With a few exceptions, orders are divided into sub-orders by the use, in a very precise way, of the *pedoclimatic* characteristics of humidity and temperature. Saturation by water (hydromorphism) is considered as a type of pedo-climate and is only used at the sub-order level, which is different from most western classifications that have one or two hydromorphic classes. Only one order, the *aridisols* of the desert or semi-desert, is differentiated in terms of pedoclimate.

## 2   Complete list of diagnostic horizons

**Fundamental diagnostic horizons: epipedons.** These are surface horizons with more or less humus and include the whole of the A horizon and, some-times, the upper part of the B if it contains humus.

*Ochric horizon:* not strongly coloured and poor in organic matter.
*Mollic horizon:* dark coloured by organic matter (more than 1%); thick (more than 10 cm on hard rock and 25 cm on unindurated loams); friable and non-massive structure; saturation greater than 50%; C : N generally less than 17.
*Umbric horizon:* same in colour, thickness and structure, but saturation less than 50% and C : N generally greater than 17.
*Other kinds of epipedon:* the *histic* (organic, generally peaty), and the *anthropic* and *plaggen*, which result from former cultivation and are defined in Chapter 4.

**Fundamental diagnostic horizons (at depth).** Dominantly mineral and either a (B) horizon of weathering (cambic, oxic) or an illuvial B horizon enriched by migration (argillic, spodic, natric).

*Cambic horizon* results from the incomplete weathering of primary minerals, having a charac-teristic structure and an exchange capacity greater than 16 mEq/100 g clay. If carbonate is present, there is less in the cambic horizon than in the C horizon. For other materials the colour is generally stronger than that of the C horizon and the amount of clay is higher.
*Oxic horizon* results from very strong weathering so that all primary minerals, except for quartz, have almost totally disappeared. Exchange capacity is low and, if measured at pH 7, is always less than 16 mEq/100 g clay, or 10 mEq if measured at the pH of the soil. Colour is generally strong because of the high amounts of free metal oxides (iron oxides and often also aluminium).
*Argillic horizon:* Bt horizon characterised by the presence of illuvial clay, giving a particular microstructure (*argillans* around structural units or, if the texture is sandy, forming *bridges* between sand grains). For loams or clay loams, the index of transportation must be greater than 1 : 1.2. In sandy materials, the increase in the amount of clay in the B horizon compared to the A must be greater than 3% and in clayey materials this figure must increase to 8%.
*Spodic horizons* result from the accumulation of organic and amorphous mineral materials (Bh and Bs) with a particular microstructure (either cemented *en masse* or an unindurated pseudo-silt of the *intergranular-aggregate* type. This horizon should also have certain chemical criteria:

(i)  Al and Fe (pyrophosphate at pH 10)/clay $> 0.2$;
(ii) Al and Fe (pyrophosphate at pH 10)/Al and Fe (citrate–dithionite) $> 1/2$.

Another chemical index is based on the thickness of the horizon and the high exchange capacity of the accumulated amorphous materials.
*Natric horizon* is a form of argillic horizon found in sodic soils subject to pervection, characterised by a columnar structure and high exchangeable sodium (greater than 15%).

**Secondary diagnostic horizons**
*Calcic horizon:* enriched in calcium carbonate (ca).
*Gypsic horizon:* enriched in gypsum ($CaSO_4$).

*Salic horizon:* enriched in NaCl.
*Albic horizon:* eluvial, decolourised, bleached (planosols).
*Placic horizon:* hydromorphic indurated spodic horizon, a kind of thin wavy ironpan.
*Duripan:* indurated non-calcareous horizon.
*Fragipan:* very compacted but *brittle*, loamy horizon.
*Plinthite:* massive clayey horizon of tropical climates, generally hydromorphic with red patches, capable of indurating on drying out.

## 3 Nomenclature: orders, sub-orders, great groups

The nomenclature is based on entirely new concepts and the 10 basic orders are as follows:

(1) *Entisols:* very slightly developed soils without diagnostic horizons.
(2) *Vertisols:* soils with swelling clays,
(3) *Inceptisols:* slightly developed soils in which diagnostic horizons can form rapidly.
(4) *Aridisols:* soils of arid regions.
(5) *Mollisols:* soils with a mollic horizon.
(6) *Spodosols:* soils with a spodic horizon.
(7) *Alfisols:* soils with an argillic horizon, weathering not excessive.
(8) *Ultisols:* soils with an argillic horizon, very strong weathering.
(9) *Oxisols:* soils with an oxic horizon.
(10) *Histosols:* organic hydromorphic soils.

Sub-order names are derived by the use of a two- or three-letter suffix from the name of the order (*ent, ert, ept, id, oll, od, alf, ult, ox, ist*) and a prefix to indicate the character of the sub-order, which is very often pedoclimatic, e.g. *aqu*, hydromorphic; *ust*, of a hot climate; *ud*, of a humid climate; *bor*, of a cold climate; *xer*, of a dry climate; *trop*, of a tropical climate (examples – *aquod*, hydromorphic spodosol; *ustalf*, alfisol of a hot climate).

The names of the great groups are formed by adding to the subgroup name another prefix signifying the presence of a diagnostic horizon, e.g. *frag*udalf, a udalf with a fragipan, or a pedoclimatic term, e.g. *cry*aquept, *therm*aquod.

Finally, the subgroups are named by using the group name together with an adjective indicating the mode of development, e.g. *orthic*, the typical soils; *aquic*, hydromorphic; *udic*, of a humid climate, etc.

A list of orders and sub-orders, with their approximate equivalents to the terms used in this book, is given below. In subsequent chapters, details will be given, wherever possible, of the subdivisions of the great groups.

## 4 List of orders and suborders

### Entisols

Very slightly developed soils without diagnostic horizons (suffix *ent*).

(a) *Aquents:* hydromorphic soils, gleyed alluvial soils.
(b) *Arents:* soils in which the diagnostic horizons have been destroyed by cultivation.
(c) *Psamments:* sandy regosols.
(d) *Orthents:* regosol or lithosol.
(e) *Fluvents:* alluvial soils.

## Vertisols
Soils with swelling clays (suffix *ert*).
(a) *Torrerts:*   vertisols of a very dry climate.
(b) *Uderts:*     vertisols of a humid climate.
(c) *Usterts:*    vertisols of a hot climate with marked seasonal variations.
(d) *Xererts:*    vertisols of a dry climate with marked seasonal variations.

## Inceptisols
Soils with rapidly forming diagnostic horizons – umbric or cambic (B) (suffix *ept*).
(a) *Aquepts:*    surface pseudogley with A(B)g profile.
(b) *Andepts:*    andosols (on volcanic ash).
(c) *Umbrepts:*   ranker or brown soils with an umbric horizon.
(d) *Ochrepts:*   temperate brown soils.
(e) *Plaggepts:*  soils with a plaggen horizon.
(f) *Tropepts:*   brown soils of tropical climates.

## Aridisols
Soils of arid climates – desert soils (suffix *id*).
(a) *Orthids:*    sierozems, brown sierozem soils.
(b) *Argids:*     arid soils subject to pervection and with an argillic B.

## Mollisols
Soils with a mollic A1 horizon (suffix *oll*).
(a) *Rendolls:*   calcareous soils – rendzinas.
(b) *Albolls:*    planosols and solonetz with an albic horizon.
(c) *Aquolls:*    gleyed humic soils, gleyed brunizems.
(d) *Borolls:*    chernozems.
(e) *Udolls:*     brunizems (prairie soils).
(f) *Ustolls:*    southern chernozems.
(g) *Xerolls:*    reddish chestnut and chestnut soils.

## Spodosols
Soils with spodic B – podzols (suffix *od*).
(a) *Aquods:*     hydromorphic podzols – gleyed humic podzols.
(b) *Humods:*     humic podzols (non-hydromorphic).
(c) *Orthods:*    podzolic soils and iron–humus podzols.
(d) *Ferrods:*    iron podzols.

## Alfisols
Soils with an argillic B horizon, weathering not excessive – lessived soils (suffix *alf*).
(a) *Aqualfs:*    lessived pseudogley, planosols.
(b) *Boralfs:*    boreal lessived soil.
(c) *Udalfs:*     temperate lessived soil.
(d) *Ustalfs:*    ferruginous and fersiallitic soils.
(e) *Xeralfs:*    fersiallitic soils of a dry climate.

## Ultisols
Soils with an argillic B, strongly weathered and very unsaturated – ferruginous soils or ferrisols (suffix *ult*).
(a) *Aquults:*    hydromorphic ultisols.
(b) *Udults:*     ultisols of a humid climate.
(c) *Ustults*     ultisols of a hot climate.
(d) *Xerults:*    ultisols of a dry climate.
(e) *Humults:*    humic ultisols.

**Oxisols**

Soils with an oxic horizon rich in sesquioxides – ferrallitic (suffix *ox*).

(a) *Aquox:*      hydromorphic ferrallitic soils.
(b) *Orthox:*     ferrallitic soils of a humid climate.
(c) *Ustox:*      ferrallitic soils of a hot, dry climate.
(d) *Humox:*      humic ferrallitic soils.
(e) *Torrox:*     ferrallitic soils of an arid climate.

**Histosols**

Soils with a histic horizon, generally hydromorphic organic soils – peats (suffix *ist*).

(a) *Fibrists:*   fibrous histosols, organic matter slightly developed.
(b) *Folists:*    drained organic soils of a cold or very humid climate.
(c) *Hemists:*    peaty soils with partially humified organic matter.
(d) *Saprists:*   peaty soils with very humified organic matter.

**Note** that more exact correlations are given in Duchaufour (1978).

## 5   Discussion of the American classification

The principal merit of the American classification is its very great precision. Most of the criteria used are quantitative and, in so far as the necessary facts are available, it is a relatively easy matter to classify and name a particular profile without equivocation. However, it must be realised that this apparent precision is often an illusion, as all the quantitative facts are rarely available. For example, this is the case when pedoclimatic factors are used (particularly those dealing with the moisture status) which, it will be remembered, are used to differentiate the sub-orders. To be able to classify a soil satisfactorily at the sub-order level, it is necessary to know exactly the number of days on which the soil is dry, i.e. when the moisture is below the wilting point, and whether the days are consecutive or not, together with the season and the soil temperature; these conditions have to occur at least 6 years out of 10. As these facts are rarely available, generally regional climatic data is used which is contrary to the aims of the classification; indeed, vague climatic designations are being used which have been rejected in most modern classifications, such as temperate brown soil, tropical brown soil, cold climate chernozem etc. As most sub-orders are differentiated in terms of pedoclimate, confusion is inevitable from this high level of the classification.

The American idea of using diagnostic horizons is very valuable and has been adopted in many classifications. Thus, *argillic*, *spodic*, *natric* and *calcic* horizons, as well as many secondary diagnostic horizons, such as fragipan and plinthite, have been used in the French classification. However, two criticisms can be made of the use of these diagnostic horizons in the American classification. The first concerns their genetic *significance*, for several of the most important of them are defined without taking into account the environmental conditions or the developmental processes involved, which results in soils being classified together that have no genetic similarity. Thus a *mollic horizon* is defined without taking account of its biochemical compositions or the environmental conditions of its humus development, so that calcareous mulls, thick forest mulls and climatically matured chernozems all have mollic horizons which are in fact, as explained in Chapter 2, very different from one another. It is the same in the case of the cambic horizon, which is defined mostly in negative terms, for it classifies together all of the temperate brown soils (ochrepts) and some of the tropical brown soils (tropepts, i.e. those in which the horizon of weathering is not oxic) despite the fact that tropical and temperate weathering are very different (see Ch. 1). This was not overlooked by Ehwald (1965).

The second criticism concerns the often exclusive use of only *one diagnostic horizon* independently of the other horizons of the profile, as they all form part of a whole in which each affects the other during profile development. Therefore, all valid classifications should be based on an overall consideration of the horizons. For example, podzols are defined in terms only of the presence of a spodic horizon, but in the north of the USSR there are soils with no spodic horizons, whose development, both in terms of process and environmental conditions, is obviously of the podzolic type and yet these soils are clearly separated from podzols in the American classification.

As diagnostic horizons are used to define the principal orders, environmental heterogeneity is produced within several of them. For example, the three levels of hot climate weathering – fersiallitic, ferruginous and ferrallitic – that are fundamental to the French classification (see Moormann & van Wambeke 1978, Sys 1967, 1978) are not clearly differentiated, except for the third group which generally corresponds to the oxisols. The fersiallitic soils and a part of the ferruginous soils are classified together with the lessived temperate soils in the alfisol order, while other ferruginous soils are classified as ultisols. This lack of homogeneity in the classification of tropical and subtropical soils can only be deplored when their developments, particularly that of weathering, have so much in common.

# V FAO classification

## 1 General characters

The FAO classification was produced as a basis for the world soil map. As a result of considerable international consultation, several successive approximations were proposed leading to the development of a definitive version in 1974.

This work showed that it was easier to reach international agreement on the differentiation of units at the group level than on the establishment of an hierarchical classification. The FAO group deliberately avoided the latter solution and did not hesitate to increase the number of basic units to 26, each of them being subdivided into 2 to 9 secondary units. The basic units are arranged in a logical order according to an increasing degree of weathering and profile development. Slightly developed soils independent of climatic factors are dealt with first of all, then soils whose development can be considered as moderate, and finally the later units deal with soils of hot climates that are considered as the most developed and weathered.

This classification is similar to the American in so far as it uses the same fundamental diagnostic horizons and hence the same criticisms can be made that insufficient account is taken of the environment and of their conditions of genesis (for example, the cambic horizon). However, this classification also contains amendments of the American classification: (i) it is much more simple; (ii) the mollic horizon, of doubtful environmental significance, loses its importance; (iii) a new diagnostic horizon that takes account of hydromorphic or *gley* properties is defined, which is missing in the American classification, and allows most of the hydromorphic soils to be grouped within the *gleysol* unit; (iv) the pedoclimatic criteria which, as stated above, are a serious fault in the American classification, are only used to differentiate the two units of the *arid regime*; and (v) the nomenclature is considerably simplified and common pedological terms are used; however, only some of the confusing terms have been omitted and yet others have been created for units that are not well known.

The grouping together in the *cambisols* and *luvisols* of soils having very different types of weathering (temperate and tropical) and some of the excessive simplification are to be regretted. In certain cases, soils have been classified together that should have been separated, e.g. the *rendzinas* are too heterogeneous and need subdividing. The *arenosols* classify together all the sandy soils, even those that are well developed, which is a confusion between pedogenic and material classification (the same criticism can be made of the psamments of the American classification).

## 2   List of main units

(1)  Fluvisols:       alluvial and colluvial soils.
(2)  Gleysols:        hydromorphic soils.
(3)  Regosols:        slightly developed soils on unindurated materials.
(4)  Lithosols:       slightly developed soils on hard rock.
(5)  Arenosols:       soils with a sandy texture.
(6)  Rendzinas.
(7)  Rankers.
(8)  Andosols.
(9)  Vertisols.
(10) Solontchaks.
(11) Solonetz.
(12) Yermosols:       arid soils without organic matter.
(13) Xerosols:        arid soils with organic matter.
(14) Kastanozems:     chestnut steppe soils.
(15) Chernozems.
(16) Phaeozems:       brunizems and lessived chernozems.
(17) Greyzems:        grey forest soils.
(18) Cambisols:       temperate or tropical brown soils.
(19) Luvisols:        temperate lessived soils, lessived fersiallitic soils.
(20) Podzols.
(21) Podzoluvisols:   lessived soils or glossic podzolic soils.
(22) Planosols.

(23) Acrisols:          ultisols (USA), lessived acid ferruginous soils.
(24) Nitosols:          tropical ferrisols.
(25) Ferralsols:        ferrallitic soils.
(26) Histosols:         peaty organic soils.

**Note** that the relationship of this terminology with that used in this book is only indicated where they are different. More precise correlations are to be found in Duchaufour (1978).

## VI   The proposed environmental classification

### 1   Objectives

At present the French classification of 1967 is being revised and this work is not yet finished. In this book a version of the old French classification will be used that has been modified and amended to give it a more environmental basis and hence it can be referred to as an *environmental classification*.

However, this by no means implies a return to the old names for classes and subclasses, using environmental terms, particularly those of climate. On the contrary, all such terms as soils of *hot climates*, *cold climates*, *tropical* or *mediterranean* are avoided, both because of their imprecision and because they allude to a cause rather than effect, and soils should be described in terms of their intrinsic properties. However, *the choice of criteria to be used at the different levels is based, as stated previously, on those that are important in terms of the environmentally controlled processes of soil development*. In other words, in this classification there is no separation between developmental processes and soil characters, both being taken into account together, for there cannot be a correct *classification* without an *interpretation*.

Nevertheless, there are exceptions to this rule where certain environmental terms have been retained. For example, the *arid* pedoclimatic regime of certain soils, the nature of hydromorphism and hydromorphic soils, and finally a subclass of the lessived soils has been characterised in terms of general climate (boreal and continental). In the first two cases the use of terms for easily recognisable environmental characters, instead of extended descriptions of the distinctive soil properties, is to be preferred in a classificatory scheme that should be brief and condensed. The third case is caused by a lack of knowledge of these soils which in addition poses, at least for certain pedologists, a serious classificatory problem.

### 2   The main characteristics of the proposed classification

At this point, only the classes and subclasses of this classification will be dealt with (details of the groups will be given in subsequent chapters) and this will conform to the modern classificatory principles given at the beginning of this chapter.

(a) *Soil classification is a process of synthesis*. Even though certain diagnostic horizons, particularly the argillic, spodic, natric, calcic and oxic, are described and used in classification, they are never used individually, for it is the whole of the naturally developed profile that is considered, as each horizon affects all the others in the course of its environmentally controlled development. *It is the genetic link between the horizons that must form the basis of classification.* As stated previously, cultivated pro-files will be classified in terms of an equivalent profile developed under permanent vegetation and the use of a suitable adjective will be enough to specify the modifications caused by cultivation.

(b) *Palaeosols as soil classes have been omitted.* In fact, it is necessary to choose between alternative uses of this term. Either the palaeosol has been modified by recent processes and it is these that are used in classifi-cation, which means that the palaeosol is considered to be a *parent mater-ial*, distinguished at the group level (polycyclic soils); or an undoubtedly old soil is involved but which is still in equilibrium with present-day climate and vegetation, so that it is not possible to consider it to be a true palaeosol.

(c) *Soils that occur between classes, called intergrades*, that have been subject to several basic developmental processes, are important and several sub-classes are differentiated in terms of their intergrade characters.

Intergrades between groups also occur and, as already stated, they are difficult to place within an hierarchical classification in which the units are separated from one another in a very artificial way. Only by the use of developmental sequences, which should be part of all classifications, is it possible to locate these intergrades in terms of the basic groups or subgroups. Examples of developmental sequences are given in Duchaufour (1978) and a certain number of them will be repeated, in summary, in this book.

## 3   Choice of criteria: classes and subclasses

**Classes.** Twelve classes are differentiated and in all cases the basic criteria used are the degree of soil development and profile differentiation, the kind of weathering and clay formation, and the basic physicochemical processes, related in the majority of cases to the nature and properties of the organomineral complexes that occur in the soil. However, the basic criteria used differ according to the general climatic conditions that control soil development and also, for certain classes, the relative importance of site conditions compared to that of the general climate.

To start with, the 12 classes can be split into three major divisions.

*Division I:* pedogenesis controlled more or less intimately by the develop-ment and the effect of organic matter, which is generally the case in tem-perate and cold climates, except for class V which has a very strongly season-ally contrasted climate. Classes II, III, IV, V, VI and VII.

*Division II:* for the greater part of the profile, pedogenesis is relatively inde-pendent of organic matter, which is the case for soils of hot climates. The kind and amount of sesquioxides of iron and aluminium, but particularly iron, are important. Classes IX, X and XI.

*Division III:* pedogenesis controlled by local site conditions where develop-ment is controlled by particular physicochemical conditions. Classes VIII and XII.

However, in spite of their interest, particularly in a theoretical way, these three divisions will not be used in the general classification for the reasons given previously in Section III.

**Division I.** The role of organic matter as an integrator in classes II, III, IV, V, VI and VII will be summarised.

*Class II. Slightly differentiated humic soils:* very rapid insolubilisation of the abundant organomineral complexes, giving a profile uniformly coloured by humus.

*Class III. Calcimagnesium soils:* humification blocked at an early stage by calcium carbonate; considerable amount of slightly transformed humus incorporated in the profile; limited weathering.

*Class IV. Isohumic soils:* organic matter stabilised by a prolonged climatic maturation is incorporated to a considerable depth by biological processes.

*Class V. Vertisols:* very stable, dark coloured complexes incorporated to considerable depths by vertic movements (swelling clays – organic matter); development related to a pedoclimate with marked seasonal contrasts and impeded drainage.

*Class VI. Brunified soils:* mull with rapid turnover and little thickness, resulting from insolubilisation by iron and clay, in which the iron is amorphous or slightly crystalline.

*Class VII. Podzolised soils:* formation of mobile organomineral complexes; processes of complexolysis and cheluviation dominant.

**Division II.** The three classes have in common a particular kind of sesquioxide development (strongly crystalline); each of them is defined by the degree of weathering on the one hand and the process of weathering on the other, this last factor is responsible for both qualitative and quantitative differences in the clays.

*Class IX.* 2 : 1 clays dominant (transformation and neoformation).

*Class X.* 1 : 1 clays dominant, weathering of primary materials still incomplete (neoformation dominant).

*Class XI.* Complete alteration of primary minerals; 1 : 1 clays exclusively neoformed (often also gibbsite present).

**Division III.** Particular physicochemical processes, related to local site conditions.

*Class VIII.* Oxidation/reduction of iron related to hydromorphic conditions.
*Class XII.* Involvement of the sodium ion.

**Note** that class I is characterised by an incomplete development of the profile as a result of either very particular climatic conditions or local conditions of site.

**Subclasses.** These are defined in terms of variants of the general development: nature of the agents of insolubilisation of the humic precursors, iron and aluminium (class II); amount of humus (class III); replacement of a calcic horizon by a (B) or Bt horizon (class IV); intervention of a process considered as secondary, such as clay pervection (class VI); intensity of redox processes and amount of organic matter (class VIII); degree of profile rubification and acidification (class IX); degree of weathering and clay pervection (class X);

relative importance of kaolinite and gibbsite in the weathering complex (class XI); form taken by the Na$^+$, as a salt or exchangeable cation (class XII).

**Note** that many subclasses should be considered as intergrades which form a transition between two classes: class III, subclass 2–class IV, subclasses 2 and 3–class V, subclass 2–class VI, subclass 3–class VII, subclass 2–class IX, subclasses 1 and 3–class X, subclass 2. In these circumstances their systematic position is necessarily a matter of discussion, and the question should remain open dependent upon the future developments in pedology.

## 4   List of classes and subclasses

### I   Slightly developed soils

(1) Slightly developed climatic soils (cryosols and desert soils).
(2) Slightly developed soils resulting from erosion.
(3) Slightly developed soils resulting from deposition (alluvial and colluvial soils).

### II   Desaturated humic soils with little horizon differentiation (AC profile)
*Rapid insolubilisation of the abundant organometal complexes (humic compounds of insolubilisation), forming a dark uniform profile.*

(1) Without, or poor in, allophanes: rankers.
(2) Rich in allophanes: andosols.

### III   Calcimagnesian soils
*Humification blocked at an early stage by calcium carbonate; slightly transformed humus incorporated in large amounts in the profile; limited amount of weathering (inheritance important).*

(1) *Humus rich:* A1C – rendzinas and pararendzinas.
(2) *Humus poor: brunified* intergrade; weathered (B) well developed. Brunified calcimagnesian soils.
(3) *Very humus rich:* profile A0A1C or A1(B)C.

### IV   Isohumic (steppe) soils
*Organic matter stabilised by long-term processes of climatic maturation, incorporated to considerable depth by biological means; dominance of 2 : 1 clays.*

(1) *Saturated complex* A1Cca: chernozems, chestnut soils.
(2) *Desaturated complex*, brunified intergrades A(B)C or ABtC: brunizems.
(3) *Isohumic–fersiallitic intergrade:* reddish chestnut soils, sub-arid soils.
(4) *Arid regime:* sierozems.

### V   Vertisols
*Soils with swelling clays: very stable, dark-coloured organomineral complexes*

*incorporated to considerable depth by vertic movements; pedoclimate of strong seasonal contrasts.*

(1) *Dark vertisols:*
    little development (inherited clays);
    developed (neoformed clays).
(2) *Coloured vertic soils* (intergrades or degraded):
    little sign of vertic characters;
    very marked vertic characters.

## VI   Brunified soils with A(B)C or ABtC profiles
*Humus with rapid turnover resulting from the processes of insolubilisation by iron and clay; moderate biochemical weathering: illites–vermiculites and associated hydroxides.*

(1) *Brown soils with (B) horizon of weathering.*
(2) *Lessived soils with Bt of the argillic type.*
(3) *Continental or boreal lessived soils* (with *pseudopodzolic* development).

## VII   Podzolised soils
*Organic matter slightly transformed, forming mobile organomineral complexes: weathering by complexolysis dominant; marked differentiation of eluvial and illuvial horizons.*

(1) *None or slightly hydromorphic podzolised soils.*
(2) *Hydromorphic podzolised soils* (with water table).

## VIII   Hydromorphic soils
*Soils with local segregation of iron by process of oxidation/reduction.*

(1) Soils with marked oxidation/reduction processes (soils with water tables): pseudogley–stagnogley–gley–peats.
(2) Processes of oxidation/reduction often reduced; hydromorphism results from the absorption of water by very clayey materials beneath a surface horizon impoverished in clays: pélosols–planosols.

## IX   Fersiallitic soils
*Particular kinds of development of iron (rubification): 2 : 1 clays dominant (transformation and neoformation). CEC (of clays) greater than 25 mEq/100 g.*

(1) Incomplete rubification: brown fersiallitic soils.
(2) Complete rubification, saturated or almost saturated complex: red fersiallitic soils.
(3) Complex desaturated and partially degraded: acid fersiallitic soils.

### X  Ferruginous soils
*Abundance of crystalline iron oxides (goethite or hematite); geochemical weathering still incomplete, neoformed 1 : 1 clays dominant. CEC (of clays) 15–25 mEq/100 g.*

(1) Persistence of primary minerals and 2 : 1 clays in all horizons: ferruginous soils, *sensu stricto*.
(2) Complete weathering of primary minerals at least in the upper part of the profile: ferrisols.

### XI  Ferrallitic soils
*Geochemical weathering of primary minerals complete (except for quartz); only 1 : 1 clays present; large amounts of sesquioxides: crystalline iron and aluminium oxides. CEC (of clays) less than 16 mEq/100 g.*

(1) Ferrallitic soils, *sensu stricto*: kaolinite dominant.
(2) Ferrallite: sesquioxides (gibbsite and iron oxides) dominant.
(3) Ferrallitic soils with hydromorphic segregation of iron.

### XII  Salsodic soils
*Development controlled by the Na⁺ ion which can be in two forms.*
(1) Saline form: saline soils.
(2) Exchangeable form: alkaline soils.

## 5  Definitions of groups and subgroups
The criteria for the definition of groups and subgroups vary according to the class; details will be found in subsequent chapters. Here, only an outline will be given with examples.

*Degree of development*, e.g. *brown podzolic soils–podzolic soils–podzols.*
*Intergrade groups*, e.g. *brown soils, andosolic, ochric, vertisolic.*
*Groups with special horizons*, e.g. *brown soils with fragipan, pseudogley etc.*
*The time factor*, e.g. *eutrophic brown soil, monocyclic (on diorite)* or *polycyclic (terra fusca).*
*The effect of man*, anthropic groups, e.g. *anthropic brown rendzinas (of cultivation).*

## References

Avery, B. W. 1956. *6th Congr. ISSS*, Paris **E** (V45), 279–85.
Avery, B. W. 1973. *J. Soil Sci.* **24** (3), 324–38.
Avery, B. W. 1980. *Soil classification for England and Wales*. Technical Monograph No. 14. Harpenden: Soil Survey of England and Wales.
CPCS 1967. *Classification des sols*. Ecole Nationale Supérieure Agronomique Grignon.
Duchaufour, Ph. 1978. *Ecological atlas of soils of the world*. New York: Masson USA Inc.
Ehwald, E. 1965. *Sitzungsberichte* **14**, 12.
FAO–Unesco 1974. *Soil map of the world*. Rome: FAO–Unesco.

Flach, K. W. 1978. *Soil taxonomy; approved amendments and clarification of definitions*. Soil Conservation Service.

Ganssen, R. 1972. *Bodengeographie*. Stuttgart: Koehler.

Gerassimov, I. P. 1974. *10th Congr. ISSS*, Moscow **VI**, 482–8.

Kubiena, W. L. 1953. *The soils of Europe*. London: Thomas Murby.

Lobova, E. V. 1977. The new world soil map, scale 1 : 10,000,000. In *Problems of soil science*, V. Kovda (ed.), 310–20. Moscow: Nauka.

Lobova, E. V. 1978. *Intern. J. Ecol. Environ. Sci.* **4**, 75–82.

Moormann, F. R. and A. van Wambeke 1978. *11th Congr. ISSS*, Edmonton **2**, 272–83.

Mückenhausen, E. 1962. *Entstehung, Eigenschaften und Systematik der Boden der Bundesrepublik Deutschland*. Frankfurt-am-Main: DLG.

Mückenhausen, E. 1965. The soil classification system of the Federal Republic of Germany. *Pedologie*, Ghent, special number **3**, 57–74.

Mückenhausen, E. 1974. *Bodenkunde*. Frankfurt-am-Main: DLG.

Pallmann, H. 1947. *Congr. Intern. Pédol. Méditer.*, Montpellier, 1–36.

Rozov, N. N. and E. N. Ivanova 1968. In *Approaches to soil classification*, 53–77. Rome: FAO–Unesco.

Schröder, D. 1973. *Pseudogley and gley. Trans Comms V and VI ISSS*, E. Schlichting and U. Schwertmann (eds), 413–19. Weinheim: Chemie.

Soil Survey Staff 1975. *Soil taxonomy*. Agriculture Handbook **436**. Washington DC: Soil Conservation Service.

Sys, C. 1967. *Pédologie*, Ghent **XVII** (3), 284–325.

Sys, C. 1978. *Pédologie*, Ghent **XXVIII** (3), 307–35.

*Chapter 6*

# Immature soils and soils with little profile differentiation

In several classifications, soils with an AC profile are grouped in a single class; however, the apparent absence of a weathered (B) horizon, distinct from the dark humic A horizon, is not necessarily an indication of immaturity. Thus, certain very humic rankers or andosols with an AC profile have undergone a considerable amount of development, in terms of weathering, and the weak profile differentiation is the result of very particular environmental conditions.

In a modern classification, it would seem to be essential to separate those AC soils, whose weak profile development is the result of their youth, from those that have been subject to a marked development. Hence AC soils are separated into two distinct classes:

(a)  immature soils;
(b)  soils with little profile differentiation.

## A  IMMATURE SOILS

### I  General characters and classification

Immature soils with an AC profile are characterised by slight weathering of the mineral material and generally small amounts of organic matter in the profile. Several classifications distinguish raw mineral soils, that have practically no organic matter, from immature soils in which a rapidly formed humic horizon is present. But it is thought that such a distinction is unnecessary, for *in these very young soils the rapidly formed organic matter merely overlies the mineral material, without the formation of true organomineral complexes and without it influencing the development of the mineral material.*

However, the third subclass that will be differentiated does contain exceptions, as will be seen, which can be considered as occupying intermediate or intergrade positions with more developed soils. Thus soils formed from materials deposited by water (alluvial or colluvial soils) generally have a phreatic water table with great oscillations in level. This favours strong biological activity which is often responsible for a fairly deep incorporation of organic matter, that is sometimes very transformed. In addition, the soil material is slightly affected by weathering (brunified alluvial soils). However, these exceptions do not justify the splitting of the alluvial soils into several distinct types, for otherwise they are a homogeneous subclass in terms of their properties, environmental conditions and geographical position.

In the case of the first two subclasses, it is either a climatic factor (such as being very cold or very dry) or a topographic factor (slope and erosion) that prevents development. Profiles remain shallow, only more or less surficial humic horizons develop and lie directly on C or R horizons. This basic division has been recognised in several classifications, in particular that of Kubiena (1953) who differentiated two great groups of immature terrestrial soils: **the raw climax soils** (tundras) and the **non-climax raw soils** (also called **syrozems**). As far as alluvial soils are concerned, Kubiena placed them in the semi-terrestrial soils, because of the presence of a water table.

These general considerations allow three subclasses to be distinguished:

(a) *Immature climatic soils.* The organic matter can be relatively abundant, particularly for cold climate soils, but it is only slightly transformed and developed, and thus has no more than a slight effect on the development of the mineral fraction. This subclass can be divided into immature soils of arid regions (desert soils) and immature soils with a frozen horizon (cryosols, tundras).

(b) *Immature erosional soils.* These are soils formed on slopes in all climatic zones. Thus, here, it is a physical cause that prevents development. The kind of parent material, depending on whether it is indurated or not, is particularly important.

(c) *Immature depositional soils.* These involve materials transported and deposited by water and are thus characteristic of alluvial plains. Generally, these soils are of great agricultural and economic importance. Their very particular hydrological conditions (some of them have water tables that fluctuate greatly in level) give them special properties. In several types the rapidly transformed organic matter is incorporated to considerable depth and forms organomineral complexes, which relates them to other classes of soils without a water table and with slower development. In addition, the mineral material can be somewhat weathered, either as a result of current processes or by *inheritance* from the transported material that has been previously weathered.

## II  Immature climatic soils

Apart from pedoclimatic differences which are obviously fundamental, the basic difference between the soils of hot deserts in the Arctic is in their organic matter content, which is very low or non-existent in the first but can be relatively important in the second. Little will be said about the first group as they are of little scientific or economic interest.

## 1   Desert soils (aridisols)

Desert soils have little or no humus and little chemical weathering. In contrast, the processes of physical disaggregation, related to abrupt variations in temperature and humidity, are active and cause a breakdown of materials (often considerable) in the surface of the soil. There is no development of structure, but particular kinds of texture occur: powdery, sandy or even clayey, with a network of polygonal cracks in the dry season (**takyr**). The lack of drainage at all seasons is very characteristic, but after the rare rains the strong evaporation very frequently causes the formation of gypseous–calcareous crust. The following types can be distinguished:

(a)  the *reg* or *hamada:* stony desert where the stones have an iron–manganese skin;
(b)  the *erg:* sandy desert;
(c)  the *takyr:* clay soil with polygonal cracks;
(d)  *arid soils with gypseous–calcareous crust.*

## 2   Cryosols (Kayricheva & Gromiko 1974)

Cryosols have a permanently frozen horizon at a particular depth (permaice or permafrost), and they occur in the *tundra* zone which characteristically has a vegetation of lichens with scattered dwarf bushes. Even though there is little vegetation, the extreme slowness of organic matter decomposition, resulting from the low mean temperature, sometimes causes a considerable accumulation of slightly transformed organic matter, frequently with hydromorphic characters of the anmoor or peaty type.

   **Raw cryosols**, practically without humus, are generally stony with the stones in geometric patterns as a result of freeze–thaw cycles (**arctic polygonal soils**, Fedoroff 1966). **Humic cryosols** are where development has started because vegetation has established itself while weathering remains very limited.

   The presence of a permanently frozen horizon at a particular depth is responsible for the formation of a near-surface water table when the surface thaws. Even though weathering is not great, the small amount of iron freed is in the ferrous form and a greenish-grey, gley horizon develops. In addition, organic matter decomposition is slowed down for two reasons – the low mean temperature and the lack of aeration – and often a peaty or anmoor surface horizon develops. Cryoturbation is obviously a frequent phenomenon and this is responsible for the formation of variously oriented sinuous bands within the profile which are either organic and black or ferruginous and ochreous.

   Taking account of the degree of hydromorphism and the development of surface organic matter, which can either be a mor when well drained or anmoor or peat when poorly drained, three basic types of humic cryosols can be distinguished:

(a)  **gleyed cryosols:**      related to the low humic hydromorphic soils;

(b) **peaty cryosols:**     related to the humic hydromorphic soils;
(c) **cryosols with mor:** related to the rankers.


## III   Immature erosional soils

These are soils poor in organic matter, characteristic of slopes where rejuvenation by erosion is a common occurrence. Generally, the organic matter present is not combined with the mineral material so that true clay–humus complexes are not formed.

These soils are common on unindurated rock, such as chalk, clayey sediments, and loess, where even on relatively gentle slopes all development is prevented. Such immature soils are often called **regosols**. As soon as development becomes a little more important and the profile acquires particular properties, it places them in another class, e.g. the still slightly developed **grey** or **white rendzinas** on chalk and the **pélosols**, often with hydromorphic characters, on marls or clayey materials. In the chapters where these particular soil groups are dealt with, these weakly developed soils will be considered again.

Soils on indurated rocks are generally called **lithosols**, although the boundary with the preceding group is in fact difficult to define. Here again, an increase in the amount of surface organic matter causes the initiation of development which connects the lithosols to the *rankers* (on non-calcareous silicate rocks) that will be discussed in the following section.


## IV   Immature depositional soils

These are the alluvial and colluvial soils, which are distinguished from one another by their geomorphic positions and origins and also by their hydrology. **Alluvial soils** are recent valley deposits near to rivers where they are very often flooded. They are characterised by the presence of a **phreatic water table**, with strong seasonal variations in level. **Colluvial** soils are characteristic of lower hillslopes, where water tables are generally absent, and are made up of material deposited from the erosion of the upper slopes. Both groups have certain characteristics in common, such as the absence of structure, a heterogeneous texture that varies from one point to another, great porosity and good surface aeration, and the absence of profile differentiation.

*Apart from their physical properties and their very special hydrological conditions, these two groups of soils often have particular kinds of profile development involving either organic compounds (humification) or the mineral compounds (weathering), or even the two together. Hence the occurrence of numerous intergrades transitional to other classes such as the calcimagnesian, brunified, hydromorphic, vertic and isohumic soils.*

However, material can be weathered prior to its erosion and transport by water, so that it is relatively rich in iron and coloured brown, but this apparent brunification is in fact inherited, for it is not associated with the structural properties characteristic of the brown soils.

With regard to the organic matter, the physical conditions are exceptionally favourable to vegetation (the depth, aeration and amount of water often being nearly optimal), and to biological activity in general, which often causes an increase in the speed of humification and sometimes even of maturation of the organic matter, which is thus more or less intimately incorporated.

Note that alluvial soils of marine origin are characterised, at least initially, by the presence of a saline water table. However, they can develop by desalinisation and form a particular group known as **polders**, which will be discussed at the same time as the salsodic soils (Ch. 13).

## 1   Alluvial soils

In terms of texture, mineralogical composition and degree of weathering, alluvial soils have the composition and properties of the materials that have been transported. Now, these are extremely variable reflecting, on the one hand, the geomorphological and the geological nature of the regions traversed and, on the other, the conditions of the alluviation, in particular the current speed. Thus alluvial soils can be calcareous or acid, sandy, or even pebbly, loamy or clay rich, slightly weathered (*grey* alluvial soils) or, on the contrary, more weathered and fairly rich in iron (brunified alluvial soils).

However, alluvial soils, despite their heterogeneity, have certain properties in common, related to their hydrological status.

(a) *The constant presence of a permanent but markedly oscillating phreatic water table* (sometimes varying from the surface, in times of flood, to several metres in depth). In these circumstances the water is constantly renewed and is poor in organic matter so that generally it does not cause reduction or segregation of iron and, because of the relatively high amounts of dissolved oxygen, respiration of submerged roots is assured. However, when for one cause or another (distance from the river, very fine materials, reduced amplitude of water-table oscillations, such as when it is able to remain relatively high in summer), the circulation of water slows down, redox phenomena occur at depth to produce rusty patches on an olive–beige background. These are the **gleyed** or **semi-gleyed alluvial soils**, transitional to **low humic gley soils** (Duchaufour 1978: $XIV_1$), in other cases peat can occur: these hydromorphic inter-mediate soils are frequently concentrated in the depressions that run parallel to the bed of the major streams outside the levée banks (Fig. 6.1).

(b) *Frequent heterogeneity of texture and grain size* is seen in very abrupt variations, both laterally and vertically within the profile. Generally fine materials (homogeneous loam or clay loam) of variable thickness overlie a layer of sand or gravel. It is to be noted that, very frequently, roots stop at this coarser layer for it prevents both the downward movement of gravitational water and greatly limits capillary rise from the water table. Consequently, the thickness of the fine-textured surface layer is very

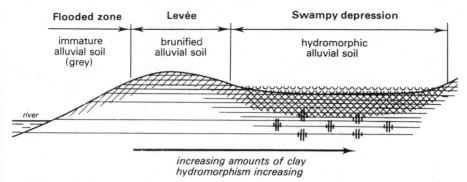

**Figure 6.1** Development of alluvial soils adjacent to a river (see general key p. ix).

important as far as soil fertility and the supply of water to plants are concerned. It is possible to go very rapidly from a point where the vegetation is very hydrophilic to another where it is, on the contrary, xerophilic.

(c) *Humification is generally activated* by the favourable soil moisture status (except when it is too dry) and a mull humus is formed. However, it becomes peaty (anmoor) in more hydromorphic situations. Buried humic horizons (frequently peaty) are a common occurrence under more recent deposits of loam.

When an anmoor has been subject to a long enough phase of desiccation, it can develop by *maturation* towards a **chernozemic** type of humus, or even **vertic**, if the material is very rich in clay. This is the way in which isohumic or vertic alluvial soils or **tchernitza** (Mückenhausen 1975) have developed (Duchaufour 1978: III₃).

Finally, very calcareous alluvial soils are characterised by humification similar to that of rendzinas – **rendzina-like alluvial soils**.

**Brunification of alluvial soils in a temperate climate.** Brunified or brown soils relatively rich in clay and free iron (Duchaufour 1978: III₂) can be of two kinds: (i) those resulting from the beginning of weathering of an initially slightly weathered parent material, for example a sand in a less frequently flooded area such as levée bank (Fig. 6.1), in contrast to the immature alluvial soils of frequently flooded areas (Duchaufour 1978: III₁); and (ii) others owing their colour, richness in clay and free iron to *inheritance* from the kind of material transported (for example, loamy or loessial material). Often it is difficult to distinguish between these two kinds of brunification, but in both cases the absence of structure and the presence of a water table allow them to be differentiated from brown soils, *sensu stricto*, formed on old terraces without a water table.

**Conclusion: proposed classification** (Fig. 6.2)

1 *Slightly hydromorphic alluvial soils*

(a) *Low humic alluvial soils*
Grey alluvial soil (slightly developed) of two varieties – acid or calcareous.
Brunified alluvial soil of two varieties – acid or calcareous.

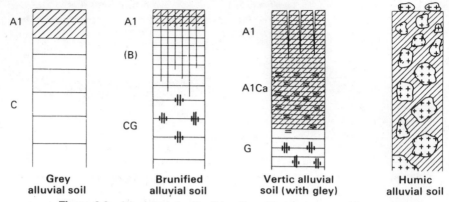

**Figure 6.2**  Alluvial and colluvial soil profiles (see general key, p. ix).

(b) *Humic alluvial soils*
   Rendzina-like alluvial soils, with calcareous mull.
   Chernozemic or vertic alluvial soils (**tchernitza**).

2. *Hydromorphic alluvial soils*

   (a) *Low humic alluvial soils with gley (semi-gley).*
   (b) *Peaty alluvial soils or with anmoor.*

**Utilisation of alluvial soils.** Alluvial soils are generally fertile because they contain great quantities of weathering minerals, in spite of the absence of structural development. In addition, they are well supplied with water, without being waterlogged, as a result of the nature of the water table. However, they do have drawbacks for they are often flooded in winter or in spring, and in hot and dry periods the water table can fall so low that the zone of capillary rise cannot reach the roots of plants, particularly when grasses are involved. The presence of sandy or gravelly layers, which interrupt capillary rise further, aggravates the situation. The best alluvial soils are those with a *thick layer of loam or clay loam* (of at least 1 m) overlying a layer of coarser texture that does not interrupt capillary rise (that is to say, where the thickness above the lowest level of the water table does not exceed 30–40 cm). Because of their deep rooting, poplars are generally well suited to alluvial soils, but even so account must be taken of the fact that gravelly beds prevent such root penetration and an interruption of capillary rise can then result, just as it does in the case of grasslands.

## 2   Colluvial soils

**Colluvial soils** are depositional, generally situated on piedmont areas, with uniform profiles without horizons, very porous, either made up of a mixture of fine and coarse materials (coarse-grained colluvium on lower parts of steep slopes in mountains) or, in contrast, of fine materials (fine colluvium situated a certain distance from slopes which are themselves more gentle). Organic matter content is very variable, sometimes very low and at other times high, which is the case for humic colluvium of mountains.

**Stabilised** colluvium, i.e. where new additions by deposition are infrequent, are generally more or less brunified in a temperate climate. As in the case of alluvial soils, there are two kinds of brunification, one resulting from development *in situ* (for example, sandy colluvial soils), the other when the transported material itself is brown.

Depending upon the type of bedrock, there are two main kinds of colluvial soil: acid and calcareous. Because of their relationship with the rendzinas (certain of which are also colluvial), the calcareous colluvial soils, no matter whether they are humic (humo-calcareous soils and humo-calcic soils) or only slightly humic (colluvial brown calcareous soils), are generally found in most classifications in the **calcimagnesian** class of soils. At this point only the *acid humic colluvial soils* will be discussed (Duchaufour 1978: $III_4$).

**Humic colluvial soils.** At the foot of mountain slopes on crystalline rocks, this type of colluvial soil occurs down slope from the erosional ranker. It benefits from the addition of soluble and insoluble materials from up slope. The humus is a mesotrophic mull, very thick and very active (often 30 cm or more), overlying a C horizon that contains materials of all dimensions which are very little weathered. As a whole, it is very porous and riddled by roots. This soil, even though it is shallow, behaves as though it were deep. As the exchange capacity is high within the humus, the base content can be reasonably high, in spite of the strong acidity (pH 5.3). The perfect root penetration because of the porosity allows reputed neutrophile species (*Mercurialis perennis*), in fact demanding of calcium and nitrogen, to colonise these sites. Mixed forests occur on them (ash, elm, maple, etc.) and they are generally planted with conifers, which grow very well, but all cultivation is prevented by the numerous rocks.

Note that when the exchange capacity of these soils is measured, not at pH 7 but at the pH of the soil, the base saturation appears to be much higher, of the order of 85–90%. In addition, the $Al^{3+}$ ion is practically absent, the $H^+$ ion representing the $T - S$ value. This explains why reputed neutrophile species can possibly occur on these soils (Duchaufour & Souchier 1980).

# B  UNSATURATED HUMIC SOILS
## (with little profile differentiation)

## I  Introduction

The soils of this class, like most of those of the preceding class, have an AC profile (or even A(B)C, but with the (B) remaining only slightly developed), which allows them to be considered as having slightly differentiated profiles. In general, they are very humic and more or less strongly coloured by organic matter, but they differ from calcimagnesian or isohumic soils in being unsaturated and even acid. They occur in almost constantly humid, coastal or mountainous climates, on non-calcareous materials, so that they are very rapidly impoverished in basic cations by their removal from the profile as weathering progresses. The only exceptions are provided by soils that are still young (e.g. eutrophic andosols) and even these are partially unsaturated.

In general, these soils differ from those of the preceding class, except in their initial phases, by having a marked development of a very particular type, both as regards mineral material and organic matter. Weathering can be very strong with great amounts of hydroxides being freed and in which the dominant role is played either by iron hydroxides (cryptopodzolic ranker) or aluminium hydroxides (andosols). This is obviously related to the composition and nature of the parent rock which is generally a crystalline silicate rock, rich in weatherable minerals in the first case and volcanic materials in the second.

This abundance of free hydroxides has profound implications for the process of humification, which is also specific to these soils. When fresh organic material decomposes, the formation of water-soluble humic precursors is favoured by the humid climate: *but these precursors are rapidly insolubilised by the iron or aluminium hydroxides, so that only a limited migration can occur; the humification is characterised by a marked dominance of insolubilised humic compounds, generally occurring as complexes with the polyvalent cations of iron and aluminium.*

These complexes are only subject to a very localised redistribution within the profiles, so that they remain only slightly differentiated and uniformly coloured by organic matter.

**Development and classification.** Two main subclasses can be differentiated – that of rankers on generally crystalline silicate rocks, rich in weatherable minerals, and that of andosols formed on volcanic materials. Each of these two subclasses can be differentiated into groups with varying development, forming developmental sequences from the initial to the most developed stage. Youthful phases are characterised by weak humification and very reduced weathering, while developed phases are the opposite.

*Ranker:*   erosional ranker (formed on slopes) → cryptopodzolic ranker.
*Andosol:*  vitrisol (formed on recent volcanic material) → developed andosol.

Slopes generally prevent development of rankers; the age of the material plays the dominant role in the degree of development of the andosols.

As development occurs characters change. In youthful phases, the organic matter is only slightly transformed and is superposed, or at most juxtaposed, with the mineral material; in contrast, in the developed phases it forms dark-coloured organomineral complexes which are incorporated in the profile, sometimes to a very great depth, thus forming a very particular kind of humic horizon for which the symbol A1B has been used.

## II Rankers

Rankers are unsaturated acid soils with an AC profile (or having the appearance of an AC soil), with a well developed humic horizon often having a clear boundary with the mineral material, particularly if an indurated rock is involved (R, and an AR profile). These soils are formed from generally crystalline silicate rocks, rich in weatherable minerals, characteristic of humid mountains (in certain cases also in coastal regions with an atlantic type of climate which are equally very humid). In extreme climates (high altitudes) the slowness of fresh organic matter decomposition and the low level of weathering of the mineral material must be ascribed to the not very high mean temperature (Alpine ranker).

### 1 The main types of ranker: environment and general characters (Fig. 6.3)

From the environmental point of view, rankers can be classified into three basic groups. Two of them correspond to an equilibrium state, i.e. to a *climax*, and they can be considered as being climatic rankers. This is the case for alpine rankers and cryptopodzolic rankers (of which there are two varieties – sub-alpine and atlantic). The third is immature because of a physical cause and it occurs *at all altitudes on slopes*. This is the erosional (or slope) ranker, which is the least developed of the three; its humus is always weakly humified, being a mor or moder with a coarse, often fibrous, structure.

The two kinds of climatic rankers are very different from one another in terms of the nature and the degree of their development, related in both cases to the climatic zones in which they occur and which differ in their mean temperatures.

The **alpine ranker**, and also the **arctic ranker** which is similar to it, is characteristic of the zones of altitude and latitude beyond the temperature

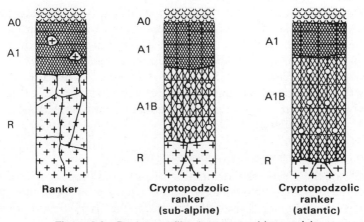

**Figure 6.3** Ranker profiles (see general key, p. ix).

limit for forest vegetation. Thus the mean temperature is very low, which slows down the biological activity as well as the biochemical processes of weathering. Humus changes only very slowly and the profile is thus of the AC or AR type.

The **cryptopodzolic ranker**, in contrast, has been subject to considerable development both as regards humification and weathering. *It is characteristic of zones that are less cold, where the general climax is a forest but where local climatic conditions (wind) prevent forest development which is thus replaced by a lower vegetation, such as heath or grassland.*

However, weathering, humification and the process of redistribution of the surface-formed pseudo-soluble organometal complexes remain limited to the surface horizon of the profile. A dark (A1B) or (AhB) horizon is characteristic of this soil type and the profile is thus of the A0(A1B)C or A0(A1B)R kind.

The whole of this discussion can be summarised as follows:

*Ranker with AC* (or AR) *profile* $\begin{cases} \text{Climatic type: alpine or arctic ranker.} \\ \text{Slope type: erosional ranker.} \end{cases}$

*Developed ranker with A0(A1B)C profile:* cryptopodzolic ranker.

## 2   *Erosional (or slope) ranker* (Duchaufour 1978: III$_5$)

The most frequent type is the erosional ranker with moder, which is characteristic of steep slopes on acid, indurated rocks (sandstone, granites) in mountainous areas. This soil has a very simple profile, being composed of only an A0A1 horizon which is 20–30 cm thick, made up of very acid, coarse humus and rock fragments, all enveloped in a network of roots; frequently, the whole of it is raised up and separated from the indurated rock like a covering. This soil is constantly impoverished in soluble products, which may be complexed, and also in clay, and in addition it is very shallow. It is covered by sparse coniferous forests which are able to take nutritive elements directly from minerals which are being weathered.

Other types of rankers occur that have a different type of humus: **rankers with mor** of a coniferous and ericaceous vegetation which produces the raw humus. **Rankers with mull**, in contrast, are formed under a deciduous forest vegetation with a mixture of species, or more commonly under a grass or prairie vegetation. Generally, the mull is formed of a mixture of organic matter with fine material coming from the breakdown of a very thin surface zone of the indurated rock, which explains the sharp boundary between the A1 and C (or R) horizon.

## 3   *Alpine ranker* (Duchaufour 1978: III$_6$)

The profile of the alpine ranker, as for the erosional ranker, is again of the AC type, but it is the climate of the alpine type, not erosion, that prevents weathering and organic matter decomposition. This type of ranker occurs in all topographic sites, both level and sloping.

The most typical profile has a mor-type humus (or hydromor), 25–30 cm thick overlying, often very sharply, slightly weathered indurated rock. Here again, organic matter decomposition is very slow; however, the amounts of humification and of the organic matter that can be extracted are higher than for erosional rankers, for biological activity is somewhat favoured at certain periods by the richness in nitrogen of plant debris (prairie litter with low C : N) and also because of the intensity of solar radiation reaching the soil during periods of exposure to sunlight.

The great variation in the humification of alpine rankers, often over small distances, is explicable in terms of the *local microclimate*, for at high altitudes the least variation in microrelief and local drainage results in considerable pedoclimatic variations. Thus there can be a hydromor (or hydromoder) at cold and humid sites, or a xeromoder when conditions are very dry, while at cool, well aerated sites even an organic rich mull–moder can form, but with an aggregate-type structure. As a result of the very strong biological activity at these sites, there is a great abundance of coprogenic micro-aggregates.

In certain cases, a beginning of weathering causes the formation of a B horizon, of very little thickness, coloured brown or rusty depending on the type of humus and the pedoclimate. This B horizon can be a brunified horizon (mull–moder type of humus: **brunified ranker**) or, on the contrary, a spodic B if the humus is a moder or a mor. In this last case the beginnings of an A2 horizon can occur in the form of an ashy trace; thus a *dwarf podzol* is formed.

### 4   Cryptopodzolic ranker (Duchaufour 1978: IV₁ and IV₂)

4   *Cryptopodzolic ranker* (Duchaufour 1978: $IV_1$ and $IV_2$)

This profile with an AC appearance is in fact a developed profile: the horizon occurring beneath the litter (A0A1 surface horizon) is uniformly coloured by organic matter and is much thicker than in the preceding case. It is an A1(B) horizon, very strongly humified as a result of insolubilisation by iron and aluminium hydroxides of precursor molecules produced in the upper horizons or *in situ* by the rhizosphere. Under the influence of this abundant organic matter, the processes of weathering by *complexolysis* are relatively important. Therefore, such a soil is a very special one for while there is no doubt that it is related to the immature rankers *sensu stricto* with AC profile, it differs from them in having several podzolic characters. Hence the name *cryptopodzolic*, i.e. a weakly podzolised profile but without visible horizon differentiation.

**Environmental conditions.** The factors of climate and vegetation play an essential role in the genesis of this profile. Thus, while it occurs in regions that are generally forested (coastal plains and the sub-alpine zone of mountains), in fact it occupies particular sites where the forest is replaced by lower vegetation (sub-alpine meadows or ericaceous heaths in mountains, and gorse, broom heathlands on atlantic coasts), because of a local environmental factor – generally the violence of the wind that inhibits tree growth on exposed sea coasts and on windy crests and summits of atlantic zone mountains. However, the local climate also has seasonal phases of strong insolation which play a certain part in the processes of complex insolubilisation and humification.

Finally, richness in free iron and aluminium resulting from weathering is also an essential factor, for on quartzose rocks poor in weatherable minerals the ranker is replaced by a true podzol with well differentiated A2 and Bh horizons.

**Biochemical characters.** The biochemical characters of the A1B horizon are a reflection of these special environmental conditions: the amount of free iron and aluminium is high and it generally exceeds 1%, the distribution being sensibly uniform in all horizons, except sometimes for the aluminium which, being more mobile, has a slight tendency to increase towards the base.

There is a moderate amount of clay (some 8% to 10% at the surface on granite) but this decreases fairly rapidly with depth. The degree of base saturation is always extremely low. As far as the organic matter is concerned, which is abundant (8–12%), extractable compounds (AF and AH) exceed 50%, insolubilised humin also being always abundant. *The biochemical tests of the Soil Taxonomy, as well as those used by French pedologists, show that in fact it is a weakly developed spodic horizon, hence the name* **cryptopodzolic**.

**Origin and development of cryptopodzolic rankers.** This is dependent on the particular development of the organic matter, which is itself very closely related to the particular environmental conditions. Study of the biological activity, and particularly of organic matter mineralisation and humification, by Bonneau (1967) and Foguelman (1966), is important in this regard. In the humid phase (winter in the atlantic climate, time of snowmelt in the sub-alpine zone), there is a massive production of water-soluble organic compounds that, in these climates, causes a limited *complexolysis* and is thus responsible for the moderate *biochemical podzolisation* that is seen. It should be emphasised here that it is a *climatic* type of podzolisation, relatively independent of vegetation conditions, since it occurs even under the influence of grass litter with low C : N (12 to 15 in the A0A1). In the spring, this phase is followed by an active biodegradation (accompanied by a strong mineralisation of C and N) of some of the compounds, but during the greater dryness and insolation of summer there is first of all a total insolubilisation, followed by a polycondensation of the existing complexes. During this phase, a strong re-organisation of the previously freed mineral nitrogen occurs, which is an indication of a very active biological humification.

The great concentration of roots in the surface, as well as the high number of complexing cations, prevents all migration to depth of the organomineral complexes formed in the humid phase: in fact, a limited migration occurs, *but it is difficult to see because of the very special character of the weathering which is concentrated at the surface and decreases very rapidly with depth, the opposite of that which occurs in forest soils with a B horizon.* The constancy in the amount of Fe and Al over a depth of some tens of centimetres is explained by the balance between the rapid decrease of weathering with depth and the limited movement of the freed elements. Only the bases $Ca^{2+}$ and $Mg^{2+}$ are removed from the profile, which acidifies rapidly.

It should be noted that, when a profile of this type is invaded by forest, a profile is rapidly formed in which there is an A0A1 horizon of little thickness and an ochreous mineral (B) horizon (ochric brown soil and brown podzolic soil; Duchaufour 1978: $II_3$). This occurs particularly at the lower altitudinal limit of the sub-alpine meadows in the zone contested with sub-alpine forest, when a decrease or cessation of grazing causes the upper limit of the forest to rise.

**Main types.** While having the same general development, the sub-alpine cryptopodzolic ranker is slightly different from the atlantic ranker of the coastal zone (Franz 1956, Carballas *et al*. 1967, Guitian Ojea & Carballas 1968).

(a) *The sub-alpine ranker* never exceeds 60 cm in thickness, it is moderately humic (less than 10% organic matter in the A1B) and because of the cold climate it has a great amount of weakly polycondensed fulvic acids.
(b) *The atlantic ranker* can exceed 1 m in thickness. The organic matter, often in excess of 10%, is strongly polycondensed and contains mainly humic acids as a result of the hotter and dryer seasonal phases. In addition, this type of soil has a more specifically hydromorphic character, as it occurs to a great extent in talwegs and low-lying areas, whereas the immature erosional rankers occur in higher parts of the landscape (Fig. 6.4).

A fitting conclusion to this section can be made by comparing the cryptopodzolic rankers, developed soils, even though undifferentiated, with erosional or alpine rankers which alone have a true AC profile. For this very particular kind of soil, a cryptopodzolisation that is moderate and limited to the surface horizons (which often causes it to be difficult to recognise) is not incompatible with a biological activity that is often intense in spring and summer. It is a climatic podzolisation that is related to that seen in certain boreal profiles (**dernovopodzolic soils**).

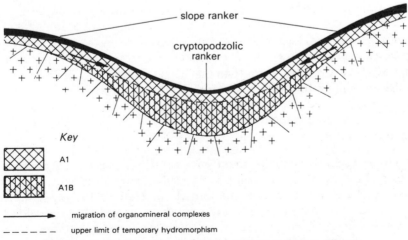

Key

A1

A1B

⟶ migration of organomineral complexes

– – – – – upper limit of temporary hydromorphism

**Figure 6.4** Development of atlantic rankers in Galicia (Spain).

# III  Andosols

## I  Introduction

These dark coloured, often black soils were first described by Japanese workers and by Dudal and Soepraptohardjo (1960) as being characteristic of outcrops of volcanic ash and formed essentially of **allophane–humus** complexes. Subsequently, it was recognised that they were much more widespread and could occur on consolidated eruptive rocks which were somewhat glassy (Mancini 1964), or sometimes even finely crystalline, provided weathering occurs rapidly and frees sufficient quantities of amorphous materials capable of fixing organic matter (Hetier 1971).

**General characteristics.** In the case of andosols, just as in that of podzols, organic matter plays a dominant role in pedogenesis, as shown by Hetier (1975): *it is responsible for weathering, being an acidolysis or even complexolysis* (see Ch. 1), *but differs from podzolisation in that the large amounts of humus–aluminium (and also humus–iron) complexes are formed* in situ, *without being subject to movement, by a moderate polycondensation of both organic and mineral amorphous materials*. The amorphous mineral compounds stabilise the organic matter and protect it against microbial biodegradation, which leads to its accumulation in the profile. This massive formation of amorphous organomineral complexes is independent of vegetation (which is very often forest, the complete opposite of that in the cryptopodzolic rankers). It is controlled solely by climatic factors (constant humidity without a dry season) and parent material that is always volcanic.

The two essential components of andosols are both amorphous (although the second may be poorly crystalline): *humified organic matter*, resulting from the insolubilisation of precursors, and *allophanes*, imperfectly crystallised aluminosilicates, the nature of which is still not exactly known.

The profile, as in the case of the rankers, appears to be an AC but differs in that on certain types of consolidated rock a (B) horizon of little thickness, coloured by iron, can form. Even so, this (B) horizon is not too distinct and the greater part of the humic horizon, where the complexes accumulate, must be considered as a mixed A1B horizon, as was done in the case of the cryptopodzolic rankers. However, the quantity of amorphous organomineral complexes is very much greater than that in the rankers and, in addition, here it is the aluminium, not the iron, which plays the principal role in complex formation.

## 2  Environment

**The parent rock.** This is the main environmental factor; the presence of non-crystalline *glass* in the eruptive rocks has long been considered as a fundamental condition of andosol formation. Hetier (1975) showed that andosols are also found on entirely crystalline parent rock, provided the minerals are fine grained and readily weatherable. *It is the speed of weathering of the parent material that is responsible for andosol formation*; the greater this is, the more characteristic the andosol produced. As the speed of weathering reaches a maximum on basic volcanic ash or very fine scoria, with large surface areas for interaction, it is here that the most typical andosols occur (with little differentiation and without a (B) horizon). In contrast, on indurated rock, development is slower and is concentrated at the surface and a (B) horizon, generally less weathered and poorer in organic material, is formed (andosol with A(B)C profile). Some very acid rocks, even though they are very glassy (some trachytes), weather too slowly to form typical andosols and generally transitional or intergrade soils are formed (andopodzolic soils, andic–ochric soils, see Ch. 9).

**Climate.** The climate plays a no less important role, for andosols are only able to form under a constantly humid climate, without a dry season (or when the *PET* is sufficiently low for the soil to remain constantly moist). It will be seen that certain physical and chemical properties of andosols are profoundly modified when they are dried out. *If the dry periods are long, or frequent, irreversible developments in the amorphous materials occur and the andosol changes to another type of soil*. In particular, mineral gels can only change into clays by crystallisation as a result of wetting and drying cycles, as has been shown in a general way in Chapter 1 (Sherman *et al*. 1964, Sokolov & Karayeva 1965). It follows that andosols *sensu stricto* can only occur under very definite climatic conditions, i.e. very humid mountains, at all latitudes where the andosols form at a characteristic level, the mean altitude of which is a function of the general climate. At higher or lower altitudes andosols *sensu stricto* give way to transitional types, andic rankers at higher altitudes and andic soils of various kinds, richer in clay, at lower altitudes (Fig. 6.5).

**Topography.** The effect of topography is often combined with that of climate for, in general, andosols are soils with a humid pedoclimate, but one that is relatively well drained. On lower slopes and valley bottoms, often with greater insolation, particularly in hot climates, the constant lateral addition of silica and bases favours the formation of vertisols (see Caldas & Salguero 1975, Quantin *et al*. 1977, Fig. 6.6). Similar sequences have been described in humid tropical mountains with soils going from andosols to ferrallitic soils (Martini 1976).

As far as lowlands are concerned, andosols are restricted to the equatorial perhumid zone and once a dry season appears they are replaced by vertic eutrophic brown soils (Siefferman 1969).

**Vegetation.** The vegetation plays no more than a minor role and is generally forest of the ameliorating type which produces a litter that decomposes rapidly. Nevertheless, on acid rocks (trachytes) and under very cold climates, it can be replaced by an acidifying vegetation (*conifers, Vaccinium* sp.), where the production of water-soluble complexing compounds is favoured, while the free cations are in smaller quantities, *which results in a greater mobility of the complexes formed and intergrade, andopodzolic soils develop*.

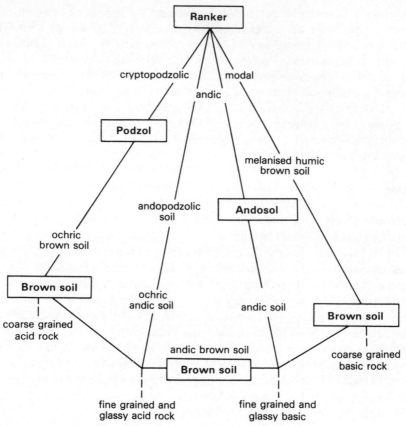

**Figure 6.5**  Soil altitudinal zones in the French Massif Central on different parent rock (according to Hetier 1971).

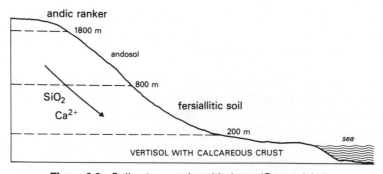

**Figure 6.6**  Soil catena on basaltic lavas (Canary Isles).

## 3   Geochemical and biochemical characters

If the genesis of andosols is complicated, their very specific biochemical characters are relatively constant and uniform, at least for *modal* types. Only the immature soils (the **vitrisols**) and the numerous intergrades are clearly different, for in this last case, to those characters that are truly andosolic must be added those that relate them to other classes.

**Morphology and physicochemical properties.** In those types without profile differentiation, the uniformly dark or black profile has the appearance only of an AC soil; the main horizon which has a strong accumulation of organic and mineral amorphous materials and is thus very strongly humified, must be designated by the symbol A1B (or AhB). It should be noted, however, that often a thin very dark B horizon is formed between the weathered, constantly humid zone and that of slight weathering which is often dryer, particularly on ashes and scoriae.

In those types with a differentiated profile (Quantin 1974) on consolidated materials, where the weathering is less rapid, a lighter and more strongly coloured (brown, ochreous or reddish) (B) horizon, with less organic matter, is formed. Generally, it is not very thick and both its upper and lower boundaries are indistinct. On fine-grained ashes it extends into a (B)C horizon, without structure but partially weathered, as is shown by its abnormally high amount of allophane (Duchaufour 1978: $IV_3$, $IV_4$ & $IV_6$).

The physical properties of andosols have often been described: very low apparent density (less than 0.8 in the modal type, decreasing to 0.5 in the tropical humic hydromorphic types), very high porosity and permeability, structure characteristically of fine aggregates of silt size (called **pseudosilts**) with a particular thixotrophic consistency – that is to say, slimy and soapy to the touch when wet, pulverescent when dry. In developed andosols, sand-size particles, which are generally unweathered glass shards, occur in small quantities and any increase in them is accompanied by a strong decrease in the above properties and indicates a vitric tendency.

The enormous water-holding capacity of andosols has often been commented on (Colmet-Daage *et al.* 1967); it exceeds 100% and can reach 200% in hydromorphic tropical andosols. Wilting point is also very high and it corresponds to an amount of water which is often of the order of half the field capacity and so greatly limits the amount of available water. Experience has shown that prolonged desiccation can greatly lower the field capacity, often irreversibly; from about 100% it can be lowered to at least 40% (Sokolov & Karayeva 1965, Colmet-Daage *et al.* 1967).

The exchange capacity of andosols, the value of which is exceptionally high (of the order of 50 to 100 mEq/100 g), is made up for the most part of charges that vary with the pH. Measured at the pH of the soil, which is generally acid, it can be reduced to a half or even a third of the value obtained at pH 7. This explains why the actual degree of base saturation is often underestimated, and why its correlation with pH ceases to have significance. These particular properties are in fact those of the two main constituents of the complex, the humic compounds and allophane; clay (where permanent charges are dominant) only plays a very subsidiary role from this point of view.

## The mineral constituents of the pseudosilt fabric

(a) *The main mineral component, often called 'allophane'*, is an amorphous or at least poorly crystalline, aluminosilicate, the properties of which are not well known. According to some (de Villiers 1971), allophane is made up of two parts: (1) a siliceous core in which there is some degree of organisation, such as the replacement of 1 in 4 of the silicons by 4-coordinated aluminium, which gives rise to permanent charges; and (2) a purely amorphous peripheral zone, mainly of aluminium complexes (hence 6-coordinated) where there is little amorphous silica and which is responsible for the variable charge. It is known that these variable charges are much more important than the permanent charges in the absorbent complex of andosols. Without completely agreeing with this idea while it is still unproven, Hetier, from his dynamic studies which will be discussed later, considers that the really active part of the allophane is indeed the *outside covering of amorphous alumina*, which is responsible for all of the biochemical properties of the pseudosilts while the amorphous silica has only a secondary role.

It should be noted that the separation and determination of allophanes are particularly difficult. It is necessary to use solvents that are gentle enough not to attack certain clays, particularly halloysite, but sufficiently strong to extract the mineral gels completely; however, the most condensed part, that often has *cryptocrystalline* properties, commonly resists this extraction. After numerous experiments on the kinetics of extraction both on soils themselves and on artificial mixtures of allophane and clay, Hetier recommended the use of a *combined* reagent (Tamm's reagent plus dithionite so that free iron can also be extracted). Yoshinaga's reagent (Yoshinaga & Aomine 1962), citrate–dithionite–bicarbonate, according to Hetier, is only capable of an incomplete extraction. On the other hand, others such as Moinereau (1974a,b) and Wada and Aomine (1973) say that the combined reagent does not extract the most condensed fraction and only an additional extraction with the aid of an alkaline reagent ($Na_2CO_3$ and even dilute NaOH) is necessary. However, the conditions required for the effective use of this reagent need to be very strictly controlled (short duration in the cold) to prevent the dissolution of halloysite, which is very sensitive to this kind of reagent.

Hetier states that in andosols *sensu stricto* there should be at least 10% allophane extracted by these means, while in the case of intergrades to other classes, i.e. the andic soils, there should be 5% to 10% of this allophane.

(b) *The clay fraction*, which is always of very minor importance in the andosols *sensu stricto* and a little more important in the andic soils, is very difficult to study because it is covered by a thick amorphous layer which it is necessary to remove completely to obtain interpretable X-ray diffraction (XRD) traces. Repeated extraction with Na-hypochlorite to dissolve the organic fraction and with combined oxalate–dithionite reagent must be used before all particle size determinations. Study of the clays shows them to be very heterogeneous, a fact which will be explained when the dynamics of these clays are studied. The 1 : 1 clays are dominant and generally they are very poorly crystallised; for example the globular halloysites consisting of concentrically rolled sheets, and imogolites, poorly crystalline fibrous silicates closely related to allophanes; well crystallised kaolinites are also found with the typical hexagonal shape; and finally even some 2 : 1 montmorillonite-type clays. *As a whole, neoformed clays are dominant which, it should be recalled, is a characteristic of volcanic materials rich in glass* (Paton 1978). Nevertheless, in many cases inherited micaceous clays are present, as are those, such as vermiculites and chlorites, that have been transformed only slightly (Miyazawa 1967).

**The organic constituents of the pseudosilt fabric.** The extraction of the organic constituents by the classical methods that have been given in Chapter 2 gives very significant results; for this reason the reader is referred to the results obtained from an andic mull given in Figures 2.1 and 2.2. *The substantial dominance of extractible compounds from the insolubilised soluble precursors is to be noted*. The fractions (AF + AH), extractable at pH 9.8, are almost equal, in terms of percentage, to the amount extracted by sodium hydroxide at pH 12; in the first extraction fulvic acid is dominant, in the second, in contrast, it is the more condensed humic acid; *but study of the humic acid by paper electrophoresis shows that there are no strongly polycondensed fractions of the grey humic acid type*. With regard to humin, always in minor amounts, there are two fractions of nearly equal importance: inherited humin with a still organised structure, and insolubilised humin, of which little can be extracted, that is very probably bound to iron.

The exceptionally high amount of extractable humic compounds at pH 12 shows that here it is aluminium and not iron which is mainly involved in the insolubilisation of the complexes; as a whole, the amount extracted from andosols is extremely high, being of the order of 60%. Only an estimate can be made of the amount of humification (which takes account of the humin insolubilised by the iron) and this is about 75–80% in andosols *sensu stricto*. Overall, the humified organic matter contained in andosols represents the enormous amount of 15–20% of the total mass, which is proof of the stabilising effect of the allophanes (this represents some 500 to 1000 tons of organic matter per hectare).

**The microfabric of an andosol pseudosilt.** These organomineral complexes are not organised in a random way. They give rise to micro-aggregates, the structure of which has been studied in a detailed manner by Hetier *et al.* (1974). These workers have used successive extractions of the organomineral complexes in the way described previously and after each extraction they have examined the residue by stereoscan. For each fraction extracted, mean residence time has been determined by the use of $^{14}$C (see Ch. 3), giving additional information of great interest. Figures 6.7 and 6.8, taken from Hetier's thesis, give some idea of the structure of an aggregate.

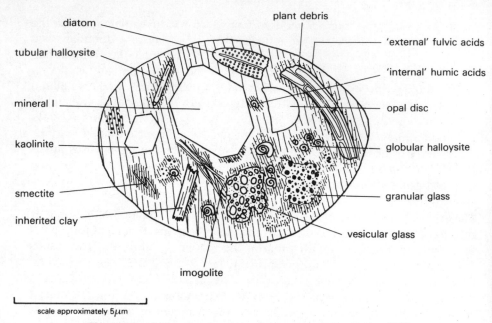

Figure 6.7 Andosols: structure of aggregate (after Hetier 1975).

Labels (Figure 6.7):
- diatom
- tubular halloysite
- mineral I
- kaolinite
- smectite
- inherited clay
- imogolite
- plant debris
- 'external' fulvic acids
- 'internal' humic acids
- opal disc
- globular halloysite
- granular glass
- vesicular glass

scale approximately 5μm

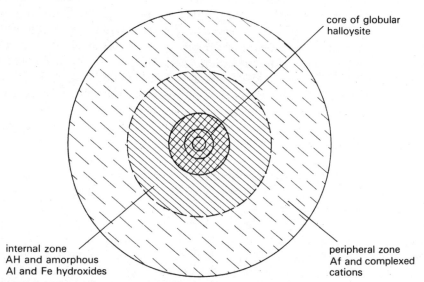

Labels (Figure 6.8):
- core of globular halloysite
- internal zone AH and amorphous Al and Fe hydroxides
- peripheral zone Af and complexed cations

Figure 6.8 Distribution of humified compounds in the organomineral complex (after Hetier 1975).

Figure 6.7 shows the whole of an aggregate formed of an amorphous organic and mineral mass containing inclusions of a varied nature, such as glass, primary minerals, kaolinite, organic debris (= inherited humin); some of these inclusions are formed of more condensed organomineral cements, generally having a core of globular halloysite; these *condensed internal layers*, resistant to the first extractions, are only dissolved at pH 12. Figure 6.8 shows the structure of an amorphous complex, with the slightly condensed external layer and the internal more condensed layer. The outside part is dominantly complexes of fulvic acid (with a relatively rapid turnover, as shown by the fairly low MRT), and $Al^{3+}$ and $Fe^{3+}$ cations. The next more condensed layer is much richer in amorphous minerals that immobilise, by adsorption, humic acids which have begun to polymerise and whose mean age is very high (MRT can reach 4000 years), indicating a certain biological inertia: the core is crystalline, generally globular halloysite.

## 4 Origin and development of andosols

Andosols offer to the geochemist and biochemist a material for investigation of considerable interest. In fact, they are a natural laboratory where reciprocal interaction between amorphous organic and mineral matter reaches an exceptional level, which allows the development of these two constituents to be followed more easily than in other soils. The stages of clay formation and the processes of humification have been studied by many workers of all countries. Here only a brief outline will be given, based mainly on the work of Trichet (1969), Wada and Aomine (1973), Quantin (1972, 1974), Hetier (1975) and Moinereau (1977).

Three successive phases can be distinguished in andosol development (Hetier 1975).

**First stage: prepedologic weathering.** This is the initial weathering of the lavas prior to colonisation by vegetation and incorporation of organic matter, which has been investigated by Trichet (1969), who showed that the glass gives rise by hydration to completely amorphous spheres; then the gels develop by gradual crystallisation into neoformed clays. Hetier thought that this neoformation occurred as a result of the internal reorganisation of the entities freed from rocks without necessarily passing through a solution phase. Under conditions that are still rich in silica and alkaline earth cations, the first neoformed clays are montmorillonites; since the bases and silica are rapidly eliminated, the environment acidifies and clays poorer in silica, such as imogolite and halloysite, appear. Several workers are of the opinion that in this phase imogolites can form a stage between allophanes and certain halloysites (Wada & Inoue 1967, Cortes & Franzmeier 1972, Dudas & Harward 1975):

$$\text{allophanes} \longrightarrow \text{imogolite} \longrightarrow \text{halloysite}$$

This development is accompanied by a loss of silica in the soluble state.

**Second stage: involvement of the organic matter.** The incorporation into the soil of important quantities of water-soluble compounds from the litter modifies the course of weathering. This is the phase of pedogenesis *sensu stricto*. Weathering increases in speed and takes on the characteristic appearance of an acidolysis or even complexolysis. It would seem that first of all the aluminium and iron cations are solubilised as complexes, but because of their liberation *en masse* the cation : anion ratio is very high and in these circumstances the insolubilisation of these complexes occurs immediately, without any migration taking place: *the humic compounds maintain the mineral gels in the amorphous state, preventing all neoformation of clay* (see Ch. 1). If the hydroxides are immobilised, it is not the same for silica and bases which are gradually eliminated by solution. It is probable, in fact, that the incorporation of organic matter increases the speed at which soluble silica is eliminated.

For their part, the amorphous mineral compounds have in their turn an important effect on the development of organic matter, aluminium in this respect playing the major role. The maintenance of a constant humidity increases the speed at which fresh organic matter decomposes, but the amorphous mineral compounds stabilise and protect the humified compounds against all microbial biodegradation so that they are able to accumulate in considerable quantities in the profile.

This twofold effect has been demonstrated by Hetier from his study of the decomposition of labelled maize litter (either total or only the water-soluble phase) by incubation with various mineral materials, such as that from the (B) horizons of andosols themselves, or the (B) horizons of brown soils or, finally, synthetic materials with an allophane base. The insoluble tissues of the labelled maize decomposed more quickly with andosolic-type material than with brown soil material. As for the water-soluble compounds, they are actively mineralised as long as they remain free, but this mineralisation is markedly slowed down from the time that the insolubilisation of complexes occurs. Finally, when the complexes have undergone maturation (by a phase of desiccation) their mineralisation is almost entirely blocked and only goes on at a very slow rate.

Another experiment by the same author has shown the part played by organic matter in the elimination of silica and bases. In these experiments, water-soluble compounds from the litter at various concentrations and pHs were percolated through columns of different materials: scoriae, andosolic (B) horizon materials, and artificial gels. With a very low concentration of carbon in solution and at pH 7, transport of silica and bases is very weak (5 ppm silica). When this concentration is moderate and the environment moderately acid (which are the conditions for andosol formation), the amount transported increases by three to four times. Finally, if the concentration and acidity exceed a certain threshold value, it is no longer only the silica that is mobilised, but also the iron and aluminium, which are the conditions for podzolisation.

As stated by Wada and Aomine (1973), the soluble organic matter and the low pH increase the speed of silica solution, which is characteristic of the phase of andosol formation *sensu stricto*. It can be concluded that the silica initially in the allophanes is not bonded very strongly to the aluminium. It is easily displaced and dissolved as a result of the formation of the much more strongly bonded organo-aluminium complexes.

**Third stage: development by ageing of the mixed gels.** This third stage is marked by a twofold condensation affecting, on the one hand, the organic fraction and, on the other, the mineral fraction of the complexes, described in Chapter 3. Here it may be recalled that the stages are:

true complexes ⟶ salt complexes ⟶ adsorption complexes

In the most condensed zones of the pseudosilts, the organic and mineral constituents develop towards forms which are bound to one another less and less intimately. The aluminium hydroxides are gradually freed as such, the humic acids occurring only in the adsorbed state on their surfaces: in these circumstances the inhibitory effect of humic compounds on the crystallisation of mineral gels ceases; a limited neoformation of clay then becomes possible, which is favoured by the nature of the material (presence of glass), but these new clays of neoformation are different from those formed in the pre-pedologic phase, for between these two periods the environment has been acidified and very greatly impoverished in silica. In these circumstances, it is *kaolinite* that is formed and sometimes even *gibbsite*, when the silica deficit is even more marked, particularly in hot climates. In extreme cases, ferrallitisation is able to follow andosol formation (see Ch. 12, Sec. C). At this stage, the humic acids no longer develop but remain practically inert. All mineralisation is prevented by their internal position within the pseudosilt fabric and being covered by a considerable thickness of insoluble organomineral complexes. The *MRT* by $^{14}C$ of these humic acids is very high and can exceed 4000 years. If the estimates of Wada and Aomine (1973) are accepted, it is a little less than the true age of the andosol. This shows that the humic acids are preserved without transformation. In particular as a result of the constantly very humid pedoclimate, *they are not subject to any form of maturation as is the humus of the chernozem. There is no formation of grey humic acids nor of transformed humin with marked ring condensation*. This is additional proof, in a negative way, of the major role played by phases of desiccation in the process of humus maturation.

In conclusion, it is to be noted that the general process of andosol formation is strictly controlled by the pedoclimate. When the soil never dries out, little of the third phase of development occurs, the processes of clay neoformation and mineralisation of the most developed organic compounds is markedly slowed down. The soil remains an andosol *sensu stricto*. If, on the contrary, the soil dries out at certain seasons, the independent development of organic and mineral constituents is favoured, the overall amount of organic matter decreases at the same time as clay neoformation (brunification) and even gibbsite crystallisation (ferrallitisation in hot climates particularly) occurs. These are intergrades towards brown or ferrallitic soils. In addition, Moinereau (1977) showed that cultivation by increasing the speed of humus turnover also increases the speed of brunification.

## 5   Classification of andosols: main types

The study of the development (Fig. 6.9) and formation of andosols leads to the adoption of two fundamental criteria in their classification: (i) *the degree of development*, which means the separation of young andosols, rich in volcanic glass (abundant sandy fraction) and poor in organomineral complexes (vitrisols), from developed andosols, very rich in organomineral complexes, of allophane and humic compounds; and (ii) *the formation of intergrades towards other classes, called andic soils*, that can be of three kinds: either

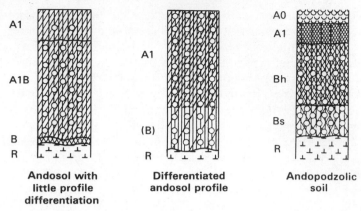

| Andosol with little profile differentiation | Differentiated andosol profile | Andopodzolic soil |

**Figure 6.9** Andosol profiles (see general key, p. ix).

*podzolised andic soils* if the dominant process is a movement of mobile complexes of aluminium, or even of iron; or *brunified andic soils*, if the dominant process is clay neoformation; and finally, *ferrallitic andic soils* if crystalline gibbsite is particularly abundant.

Little will be said about the very sandy vitrisols, which have very slightly weathered glass and are poor in organic matter. If this organic matter is more abundant, it remains independent of the mineral fraction and does not form true organomineral complexes.

(a) Vitrisols: *immature andic soils*.
(b) Andosols *sensu stricto* (or modal):
    (i)  little differentiated, humic AC profile;
    (ii)  differentiated, less humic A(B)C profile;
    (iii)  hydromorphic, very humic AG profile.

**Note** that in the first two categories, two types can be distinguished: *eutrophic* or *oligotrophic* depending upon the degree of saturation.

(c) Andic soils (intergrades):
    andic ranker;
    andopodzolic soil;
    brunified andic soil;
    ferrallitic andic soil.

**Andosols** (*sensu stricto*). According to Hetier, amorphous and allophanic materials exceed 10% in the A1B or (B). In addition, the ratio of allophane to clay is greater than 1.5.

Three great groups correspond to the definition:

(a) *Humic andosol with little profile differentiation* (Duchaufour 1978: IV$_3$).
    This is a very uniform profile with a slight decrease in organic matter with depth. The deep horizon, still coloured by organic matter and rich in organomineral complexes, is in reality an A1B horizon; sometimes, how-

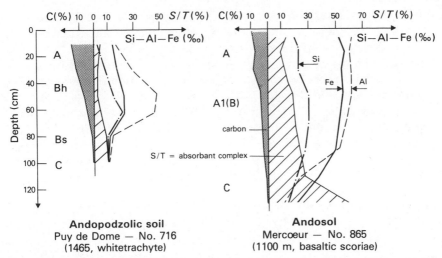

**Figure 6.10** Analytical characters of andopodzolic soils and andosols, Châine des Puys (after Hetier 1973).

ever, there is the beginning of a (B) horizon as a black band at the base of the profile. The amorphous mineral compounds of Al, Fe and Si are uniformly distributed (in Fig. 6.10 these are expressed as Al, Fe and Si not as $Al_2O_3$, $Fe_2O_3$ and $SiO_2$%). This profile is characteristic of finely divided materials (ash and scoriae) with a great surface area of contact.

(b) *Andosols with differentiated profile* (with a B horizon; Duchaufour 1978: IV₄). This profile is characteristic of outcrops of massive, indurated rock with a slower speed of weathering, and where the decrease of organic matter is more marked, for it is less easily incorporated into horizons at depth. In these circumstances, a (B) horizon (called **chromic**, i.e. of brown or ochreous colour) can be distinguished from the clearly darker humic horizon of the surface. This does not prevent the (B) horizon being very weathered and rich in amorphous compounds.

These two types can be divided into two subgroups dependent on their base saturation. Eutrophic andosols have an S : T ratio greater than 50%, oligotrophic andosols less than 50%. In fact, the first are rare, since the complete development of an andosol involves, as stated previously, a considerable removal of bases (particularly Ca and Mg) and silica. This is reflected in the fact that eutrophic andosols are richer and oligotrophic andosols poorer in amorphous silica, which supports the idea of Martini and Palencia (1975) that eutrophic andosols are in fact young soils still at the start of their development, for they occur, according to these authors, on recent volcanic materials and under climates where the humidity is less, which slows down the process of desaturation and elimination of silica. This can be expressed as follows:

vitrisols  ⟶  eutrophic andosol  ⟶  oligotrophic andosol
       (slightly developed)             (developed)

(c) *Hydromorphic very humic andosol* (Duchaufour 1978: $IV_6$). This soil is characteristic of very humid tropical mountains, at an altitude greater than 3000 m (called the zone of mists). The profile is almost constantly soaked with water, which slows down the decomposition of fresh organic matter which takes on a peaty appearance. In addition, in spite of the great porosity and the slope, which assures normal drainage, the great amount of water retained by the humic horizon causes a more or less complete reduction of iron so that the mineral horizon has a blue–grey colour (horizon Cg or even G in extreme cases). In these very developed soils the quantity of amorphous mineral and organic materials can be very high, of the order of 70–80% of the total.

**Andic soils** (intergrade types). At this point the *andic ranker* will not be detailed. This is a soil *analogous* to the cryptopodzolic ranker developed on volcanic rocks and seen in the Massif Central above the limit of forest vegetation (about 1300–1400 m) and is characterised by a profile that is not so deep but richer in incompletely decomposed organic matter than the andosols *sensu stricto*.

Three kinds of intergrade will be discussed which are related to three different soil classes: podzolised soils, brunified soils and ferrallitic soils.

(a) *Andopodzolic soils* (Duchaufour 1978: $IV_5$). These are formed on acid volcanic rocks, poor in bases and with slower weathering (white silica-rich trachyte) and subject to the influence of an acidifying vegetation (*Vaccinium*). These soils have both andic and podzol characters, particularly in that *some of the amorphous silica and alumina, and even the iron, migrates and accumulates at depth to form a true spodic horizon, Bh black, Bs rusty* (see Fig. 6.10). This is a reflection of an abundance of complexing organic acids produced by the litter, while, in contrast, the amount of Fe and Al cations freed by weathering is much less. In these circumstances the mobility of the complexes increases and their insolubilisation is no longer immediate, as in the case of andosols (see Ch. 3 on organometal complexes).

(b) *Brunified andic soils*. These soils are subject to more marked phases of desiccation and they occur in the temperate zone at a lower elevation than that of the andosols. In the Massif Central of France they occur between about 700 m and 1000 m – with *andic brown soils*, more closely related to the brown soils because of their considerable clay formation, below 700 m (Hetier 1971). The humic horizon is more clearly demarcated and confined to the surface. A clay loam (B) horizon, with a marked polyhedral structure, is formed. However, it remains richer in amorphous materials than brown soils on granite at a similar altitude.

(c) *Ferrallitic andic soils*. Characteristic of humid tropical climates, these soils have been subject to a threefold developmental process: a more or less rapid mineralisation of the organic matter which is concentrated within a not too thick surface humic horizon; clay formation (kaolinisation) limited as a result of a marked deficit in silica (eliminated by transport during previous phases); crystallisation of hydroxides such as gibbsite and goethite (or haematite), which become the dominant entities in the profile. This also confirms that in a humid tropical climate, the process of andosol formation on volcanic materials is a starting point that is exceptionally favourable to the subsequent formation of a **ferrallite** very rich in crystalline sesquioxides and poor in clay. This transformation is well known to be caused by a change of climate and the appearance of a dry season (Quantin *et al*. 1978, see Ch. 12, Sec. C).

Note that in the deep horizons a **duripan** can be formed by the precipitation of silica.

In Oceania, Quantin (1974) described fersiallitic andic soils. They seem to result from the development of eutrophic andosols, still rich in silica, towards soils characterised by strong neoformation of 2 : 1 clays, this development being also favoured there by the occurrence of a dry season. Chichester *et al*. (1969) and Moinereau (1977) have also described andic soils with montmorillonites in regions with a mediterranean-type climate.

## 6    Characteristics and suitability for agriculture and forestry
The andosols, being found mainly in mountains and characterised by a high humidity, are generally covered by forest and grasslands. The slightly and moderately unsaturated andosols are very favourable to plant growth because of their great porosity, their aeration, and the ease with which they are penetrated by roots. However, the very high wilting point and the slowness of the rewetting of the allophanes when they have dried out, cause great problems of water supply to plants in periods of exceptional dryness. Their main chemical problem, common to all andosols, is that of an almost irreversible precipitation of phosphate ions by aluminium, which is the reason for the frequent deficit in this element.

Schaefer *et al*. (1969) showed that, despite a very slow mineralisation, the nitrogen cycle is uniform, so that the nitrogen nutrition is assured at a relatively satisfactory level in the andosols.

The very acid andosols and the andopodzolic soils are much less fertile and they become clearly unsuitable for plants, for the presence of free aluminium, toxic to roots of certain plants, is no longer compensated for by a high amount of calcium. In addition, the acidity resulting from the presence of amorphous mineral gels disturbs the nitrogen cycle. The mineralisation of water-soluble compounds is still slowed down, while, in contrast, their insolubilisation – and thus the immobilisation of assimilable nitrogen – is speeded up, which results in a deficit of nitrogen.

## References

Bonneau, M. 1967. *Science du Sol* 1, 49.
Caldas, E. F. and M. L. Salguero 1975. *Andosoles de las islas Canarias*. Santa Cruz de Tenerife: Ed. Confederacion.

Carballas, T., Ph. Duchaufour and F. Jacquin 1967. *Bull. ENSAN* **IX** (1), 20.

Chichester, F. W., C. T. Youngberg and M. E. Harward 1969. *Soil Sci. Soc. Am. Proc.* **33** (1), 115–20.

Colmet-Daage, F., F. Cucalon, M. Delaune, J. and M. Gautheyrou and B. Moreau 1967. *Cah. ORSTOM, sér Pédologie* **V** (1), 3–38.

Cortes, A. and D. P. Franzmeier 1972. *Geoderma* **8** (2–3), 165–76.

Duchaufour, Ph. 1978. *Ecological atlas of soils of the world*. New York: Masson.

Duchaufour, Ph. and B. Souchier 1980. *C.R. Acad. Agric. Paris* **4**, 391–9.

Dudal, R. and M. Soepraptohardjo 1960. *7th Congr. ISSS*, Madison **IV** (V32), 229–37.

Dudas, M. J. and M. E. Harward 1975. *Soil Sci. Soc. Am. Proc.* **39** (3), 561–6 and 571–7.

Fedoroff, N. 1966. Les sols du Spitzberg occidental. *Mem. CNRS*, RCP **42**, 111–228.

Foguelman, D. 1966. *Etude de l'acivité biologique, en particulier de la minéralisation de l'azote, de quelques sols du Languedoc et du Massif de l'Aigoual*. Spec. thesis. Montpellier.

Franz, H. 1956. *6th Congr. ISSS, Paris* **E** (V22), 135–41.

Guitian Ojea, F. and T. Carballas 1968. *Anal. Edafol. y Agron.* **XXVII** (1–2), 57–73.

Hetier, J. M. 1971. *Science du Sol* **2**, 51–82.

Hetier, J. M. 1973. *Science du Sol* **2**, 97–114.

Hetier, J. M. 1975. *Formation et évolution des andosols en climat tempéré*. State doct. thesis. Univ. Nancy I.

Hetier, J. M., F. Guttierez Jerez and S. Bruckert 1974. *C.R. Acad. Sci. Paris* **278D**, 2735–7.

Kayricheva, I. and I. Gromiko 1974. *Atlas of soils of the USSR*. Moscow: Kolos.

Kubiena, W. 1953. *The soils of Europe*. London: Thomas Murby.

Mancini, F. 1964. *Session soil classification*. Florence: FAO.

Martini, J. A. 1976. *Soil Sci. Soc. Am. J.* **40** (6), 895–900.

Martini, J. A. and J. A. Palencia 1975. *Soil Sci.* **120** (4), 278–87.

Miyazawa, K. 1967. *Pedologist* **11** (1), 25.

Moinereau, J. 1974a. *Science du Sol* **3**, 173–93.

Moinereau, J. 1974b. *Science du Sol* **4**, 353–67.

Moinereau, J. 1977. *Alteration des roches: formation et evolution des sols sur basalte, sous climat tempéré humide*. State doct. thesis. Montpellier.

Mückenhausen, E. 1975. *Bodenkunde*. Frankfurt-am-Main: DLG.

Paton, T. R. 1978. *The formation of soil material*. London: George Allen & Unwin.

Quantin, P. 1972. *Cah. ORSTOM, Sér. Pédologie* **X** (2), 123–51 and **X** (3), 273–301.

Quantin, P. 1974. *Cah. ORSTOM, Sér. Pédologie* **XII** (1), 1–12.

Quantin, P., M. L. T. Salguero and E. F. Caldas 1977. *Cah. ORSTOM. Sér. Pédologie* **XV** (4), 391–407.

Quantin, P., M. L. T. Salguero and E. F. Caldas 1978. *Cah. ORSTOM Sér. Pédologie* **XVI** (2), 155–75.

Schaefer, R., A. Urbina De Alcayaga and E. San Martin 1969. *Panel of volcanic ash soils in Latin America*. Costa Rica.

Sherman, G. D., Y. Matsusaka and H. Ikawa 1964. *Agrochimia* **8**, 146.

Siefferman, G. 1969. *Sols de quelques régions volcaniques du Cameroun*. State doct. thesis. Fac. Sci. Strasbourg.

Sokolov, I. A. and Z. S. Karayeva 1965. *Pochvovedeniye* **5**, 12–22. (*Soviet Soil Sci.* **5**, 467–75.)

Trichet, J. 1969. *Contribution à l'étude de l'alteration experimentale de verres volcaniques*. State doct. thesis. Fac. Sci. Paris.

Villiers, J. M. de 1971. *Soil Sci.* **112** (1), 2–8.

Wada, K. and S. Aomine 1973. *Soil Sci.* **116** (3), 170–77.

Wada, K. and A. Inoue 1967. *Soil Sci. Pl. Nutr.* **13** (1), 9–16.

Yoshinaga, N. and S. Aomine 1962. *Soil Sci. Pl. Nutr.* **8** (2), 6–13.

# Chapter 7

# Calcimagnesian soils

## I  Introduction: general characters

When calcium carbonate is particularly abundant in a rock, and it is freed in the *active* form, it has a very particular effect on soil development which is generally opposed to the climatically determined development. From the beginning of pedology, **rendzinas** were differentiated from other soils and classed by the Russian school among typical intrazonal soils. It should be remembered that active carbonate is a curb on the processes of weathering, little iron being freed. Because of the very high biological activity, particularly animal, fresh organic matter is finely divided and incorporated to considerable depth, but humification is slowed down by the effect of calcium carbonate which *stabilises* humic compounds in a slightly transformed state and protects them against biodegradation (**carbonate mull**): *the profile, rich in well incorporated but slightly transformed organic matter, is of the AC type.*

If the presence of calcareous material in the parent rock appears to be a necessary condition for the formation of a rendzina, it is not sufficient by itself, for in many cases, particularly in a humid climate, profile decarbonation occurs more or less rapidly which results in a relative accumulation of silicate materials, the amount of organic matter decreases, and a brown colour resulting from the freeing of iron appears: *the profile is brunified.* However, as long as the A1 horizon is thick and very humic and the (B) horizon is no more than slightly developed, the term *rendzina* generally continues to be used (brunified rendzina), but when the humic horizon decreases in thickness (eutrophic mull), and a brown (B) horizon with a strongly developed polyhedral structure becomes the dominant horizon, this is a special kind of calcimagnesian soil, which has many of the characters of a brunified soil, to which class it is transitional.

The composition of the calcareous material, related to the topography, is evidently the fundamental environmental factor that controls the formation and development of calcimagnesian soils; *depending upon the local conditions it is responsible, to a very variable extent, for the maintenance within the profile of a sufficient quantity of active carbonate in intimate contact with the incorporated organic matter*; thus the l   ι conditions of site, parent material and topography control the formation or, in certain cases, the transformation of calcimagnesian soils.

*The bioclimatic factors* of climate and vegetation play only a secondary role: under the climate of temperate plainlands most rendzinas *sensu stricto* are in fact very shallow and too dry to permit forest growth so that their vegetation is a kind of **edaphic steppe** composed of xerophilic grasses or at best a scrub formation of calcicolous bushes: however, at certain sites where the parent material is fissured or colluvial, it allows the deeper penetration of more powerful woody roots, and so in these particular circumstances a forest is able to establish itself on calcimagnesian soils. It results first of all in an increase in the amount of organic matter (forested black rendzina), generally followed by a decarbonation that takes place at a much more rapid rate than under grassland: *the establishment of forest on a rendzina is very often a factor in its brunification*.

The cold and humid climates of mountains have the same effect, for, compared to the climate of the plainlands, decarbonation increases in speed. But this type of climate has another essential effect on the development of calcimagnesian soils: it slows down organic matter decomposition and causes its accumulation in the profile in quantities that are often considerably greater than in the plains. The calcimagnesian soils of mountains form a particular subclass which is distinguished by its extreme richness in organic matter.

Thus it is possible to distinguish three main subclasses:

(a) humic calcimagnesian soils with an AC profile: *rendzinas*;
(b) very humic soils with an A1C or A0A1C (sometimes A1(B)C) profile: *very humic calcimagnesian soils*;
(c) low humic soils with a developed (B) horizon: *brunified calcimagnesian soils*.

Before consideration is given to the formation and development of calcimagnesian soils under different conditions, the main environmental and biochemical characters of three types of profiles corresponding to these three subclasses will be discussed.


## II   Environment, morphology and biochemistry of the basic types

### 1   Humic calcimagnesian soils: rendzinas

A forested **humic black rendzina** will be taken as an example even though, as stated above, it is not widely distributed, for generally brunification occurs rapidly under forest and a **brunified rendzina** results. It is only under particular site conditions where decarbonation is too slow or incomplete that a rendzina *sensu stricto* is preserved; this is the case for footslope **calcareous colluvium** which is enriched by new additions of carbonate; also on outcrops of unindurated, very pure carbonate rocks where decarbonation produces only a small residue of silicate material; or finally on slopes of calcareous marl where erosion prevents decarbonation. From these examples, an understanding can be gained of the fundamental part played by site conditions, which will be examined in greater detail in the following section.

**The profile of the colluvial rendzina** (Duchaufour 1978: $V_1$). This has a very humic A1 horizon, 30–40 cm thick, with a brown–black colour and a very stable and well aerated crumb structure (angular crumbs of 2 to 5 mm) resulting from the formation of complexes of humus, clay and calcium carbonate. The amount of organic matter is very high and can reach 15% in the surface, but decreases gradually towards the base of the horizon; generally, numerous pebbles occur in all the horizons, their dimensions being dependent on the nature of the colluvium.

Frequently, at depth there is a **calcic** horizon of the pseudo-mycelium type, or forming an irregular white pulverescent deposit within the C horizon, by the reprecipitation of a part of the carbonate dissolved in the A horizon under the influence of the organic matter and biological activity. The amount of $CaCO_3$ is very high throughout the profile (5–10% of active carbonate), but it is less at the top of the A1 than at the base, because of the beginnings of decarbonation at the top of the profile. The amounts of iron and clay are very variable within the A1 horizon, being dependent on the nature and the origin of the colluvium: when it is derived from indurated limestone, it is generally mixed with terra rossa or terra fusca and the amount of free iron can be considerable (0.5% to 1%). Humus can be extracted only to a slight extent, even after decarbonation with dilute HCl. In most cases, the sum of AF + AH does not exceed 25%; again it consists mainly of fulvic acids recently insolubilised by carbonates. The relatively low C : N ratio (less than 15 at the top of the A1) is due to the strong biological activity; it becomes less and is nearer 10 at the bottom of the horizon, which is a reflection of a decrease in the proportion of primary organic matter coming from the litter and an increase in the degree of humification. The exchange capacity, where the saturation by $Ca^{2+}$ and $Mg^{2+}$ reaches 100%, is always high because of the abundance of organic matter and it varies between 30 and 50 mEq/100 g. The type of clay, which for the most part is inherited from the parent material, is not a particular characteristic of the soil; it can be either montmorillonite or illite, although the latter is more frequent, but kaolinite also occurs in rendzinas formed on an old terra fusca colluvium. This clay is preserved without modification within aggregates.

The basic properties of the rendzina *sensu stricto*, such as have been described, are maintained even when certain environmental conditions are changed, e.g. parent material or vegetation. However, reflecting these developmental differences are certain characters that distinguish these profiles from the previously described colluvial rendzinas.

On a pure and fairly unindurated limestone, on a level site (Duchaufour 1978: $V_2$), the mixture of organic and mineral material is the result of biological activity alone and mechanical processes are not involved: this causes the profile to be much shallower and the degree of incorporation of organic matter with the mineral material less complete; often a desaturated moder is formed at the surface. The concentration of organic matter decreases more rapidly with depth; clay and iron, even though there is very little in the chalky weathering C horizon, are concentrated at the surface by relative accumulation; in fact, here, the dissolution of the carbonates in the surface, being no longer compensated for by new additions, is considerable; the soil keeps its rendzina character only because of the very great reserves of $CaCO_3$ of the parent material (99%) compared to the silicate impurities.

Under grassland, the so-called grey rendzina occurs which is less humic (5–8% organic matter); the C : N ratio is lower in the surface because of the higher nitrogen content of the plant debris (9 to 10). This grassland humus is less *acidifying* than the forest humus and it causes only a moderate decarbonation of the profile which is often richer in active carbonates, on comparable parent materials, than forested rendzinas (Duchaufour 1950).

## 2   Very humic calcimagnesian soils

In very humid, high-altitude climates, where the rate of organic matter decomposition slows down, it tends to accumulate in the profile and at the same time decarbonation of the fine earth occurs rapidly. The degree of incorporation of this organic matter with depth is very variable, depending on the local conditions of parent material and topography, and also on the altitudinal zone. In the upper, colder zones a particular kind of mor (**tangel** of Kubiena) tends to be formed.

The two essential constituents of the humic calcimagnesian soils of high altitude are organic matter that is generally only slightly transformed, and very coarse skeletal fragments of a white limestone. These soils have been given various names: organogenic rendzinas by Filipovski and Ciric (1969), or carbonated humic soils by the Swiss school. However, the use of the term **carbonate** would seem to imply the constant presence of $CaCO_3$ in the chemically active form in the soil, while most of the soils in this category are without such carbonate. Only some profiles developed on mobile talus slopes are exceptions to this point of view. Bottner has suggested the name humocalcic soils for the first group and humocalcareous for the second, terms that will be used in this book.

The most typical examples of humocalcareous and humocalcic soils have been described in France by Bartoli (1962, 1966), Duchaufour and Bartoli (1966) and Bottner (1972). Initially, they are formed on highly mobile talus slopes which are then gradually stabilised as a result of the establishment of forest vegetation; under its influence, active carbonate is rapidly eliminated, while less and less transformed organic matter tends to accumulate; these soils on talus slopes are much deeper (often exceeding 1 m) than the generally shallow karst infill soils, formed on horizontal rock platforms. The accumulation of organic matter in these soils on talus slopes results in a layering of different humus types, the deepest being the most calcareous (or the least acid) and the most transformed, the surface, in contrast, being generally base unsaturated and very slightly transformed. Within certain profiles the following sequence from bottom to top occurs:

(a) *Carbonate mull:* which is an altitudinal variety richer in organic matter than the carbonate mull of the plainlands and poorer in silicates. It forms large, irregular, black crumbs, 5 mm to 1 cm in size.

(b) *Calcic mull–moder:* this particular type of humus has been described by Kubiena (1953). It is made up of angular aggregates, very rich in humic compounds coming from the surface and precipitated by the calcium ion. Thus the humus is saturated but not calcareous.

(c) *Tangel (or calcic mor):* this is a kind of mor forming an A0 horizon (holo-organic, i.e. almost without silicate impurities), but less acid (it is often almost saturated with $Ca^{2+}$ and $Mg^{2+}$) and, because of this, biologically more active; it is more humified than the acid mor (double the exchange capacity, often 2 to 3 mEq/g of organic matter), very rich in fine coprogenic pellets, and having a *greasy* consistency, sticking to the fingers.

(d) *Acid mor:* very desaturated and less transformed; it has the usual mor characters: structure often fibrous or granular; it forms at the surface only when the very thick A0 horizon decomposes very slowly.

*Humocalcareous soils* (Duchaufour 1978: $V_3$) are characteristic of unstabilised talus slopes recently colonised by maple with ferns; the humus, a black carbonate mull, is very homogeneous throughout the profile with little or no decrease in the amount of organic carbon with depth. This reflects the importance of mechanical and biological mixing which is characteristic of this soil.

*Humocalcic soils* (Duchaufour 1978: $V_4$) on stabilised talus slopes generally as a result of an old and prolonged colonisation by coniferous forests, in contrast, are generally richer in organic carbon and entirely decarbonated (except sometimes at depth). Typically, they display a change in humus type from the surface downwards, which varies according to the age and degree of decarbonation: either mor, tangel, mull–moder; or tangel, mull–moder; or, in the least acid types, mull–moder, carbonate mull.

## 3   Brunified calcimagnesian soils

These soils differ from the preceding ones in that they are richer in fine-earth silicates, particularly clays, than the rendzinas; the active carbonate, in minor quantities compared to the clays, no longer has the possibility of forming the humocalcareous complexes characteristic of the rendzinas, all the more so as it tends to be eliminated very rapidly, in a temperate climate, at least in the surface horizons. In these circumstances a brown (B) horizon with a coarse polyhedral structure develops; the profile is of the A(B)C type, transitional to the brunified soils.

They differ from brunified soils by the presence in greater or lesser amounts of active carbonate in one horizon at least of the profile, which can even be abundant in certain warm temperate or dry subtropical climates, when the small amount of organic matter produced by scattered vegetation prevents the formation of humocalcareous complexes in appreciable quantities (Duchaufour 1978: $VI_4$). In a humid temperate climate, when the decarbonation occurs, two phases can be distinguished in soil development, according to whether the (B) horizon contains active calcareous material or not.

**Calcareous brown soil.** The humic horizon, most often without active $CaCO_3$, is a eutrophic mull, while the structural (B) horizon still effervesces with HCl. Note that, in a humid temperate climate, to maintain this profile there must be a cause of *rejuvenation* preventing the complete decarbonation of the (B) horizon. This could be erosion on marls, or colluviation of materials of very varied nature, so long as they contain carbonates (Duchaufour 1978: $VI_5$) or, finally, as a result of cultivation.

**Calcic brown soils.** The greater part of the (B) horizon with polyhedral structure does not effervesce with HCl; the presence of $CaCO_3$ in the fine earth can occur only below a depth of about 50 cm – that is to say, at the base of the (B) horizon, or within weathered C horizon material. It will be seen that this type of soil can have two very different origins, according to whether it is formed by the decarbonation of a marly limestone or marl (**monocyclic calcic brown soil**) or, on the contrary, as a result of a secondary recarbonation of a silicate-rich palaeosol of the terra fusca type (see Ch. 4); it is then a **polycyclic** or **polygenetic calcic brown soil**; in the first case, a Cca horizon of pulverescent nodules often occurs, as a result of the precipitation of carbonate transported from the top of the profile (Duchaufour 1978: $VI_6$), while it is absent in the second.

These two types of calcic brown soil, no matter whether monocyclic or polycyclic, have in common a high level of saturation of the absorbent complex in the (B) horizon, usually near 100%. This can be attributed to the calcium reserve at the base of the profile which, partially mobilised by animal activity (earthworms), allows the upper horizons to be saturated – at least the (B) in exchangeable calcium.

Note that on very indurated limestones, the monocyclic calcic brown soils very often pass into another type of soil which does not contain any active carbonate throughout the profile; the fine earth can even be partially base unsaturated (slight acidity), which is a reflection of the slowness of weathering of the skeletal limestone fragments. In this case, the profile belongs to the brunified class of soils despite the incontestable fact that it is related to polycyclic calcic brown soils (eutrophic or mesotrophic polycyclic brown soils; see Ch. 9).

## 4   Calcimagnesian soils and plant nutrition: agronomic properties

Rendzinas, despite their excellent aeration resulting from their very stable structure, have numerous faults which affect plant nutrition in terms of water supply and nitrogen and mineral nutrition. Being shallow and stony, they are often subject to dry conditions – water reserves are insufficient. In addition, the abundance of stones makes cultivation difficult. Fortunately, these unfavourable properties are decreased, or compensated for, on certain parent materials that are unindurated and are capable of storing capillary water in their weathered layer (marly limestone, chalk, deep colluvium).

As stated in Chapter 2, the nitrogen cycle is unfavourable: the level of mean annual mineralisation is slowed down by the active carbonate to less than 1% in an atlantic climate, which is very much less than in the case of a mesotrophic mull formed in the same conditions. In addition, losses of mineral nitrogen are considerable; these occur either in the gaseous form ($NH_3$) or, more commonly, as soluble nitrates that are carried downwards in humid periods, as nitrification is always very active in rendzinas. For most conifers that are unable to use the nitrate nitrogen when it is in excess, the deficiency in nitrogen is a frequent cause of chlorosis, which sometimes may be corrected by root mycorhiza (Le Tacon 1976). Finally, excess active carbonate sometimes raises the pH to about 8, which causes the insolubilisation of various necessary mineral elements: manganese, iron and minor elements such as boron. Phosphorus suffers a retrograde reaction to form the insoluble and only slightly mobile **apatite**; in addition, calcium tends to become the dominant element of the ionic exchange complex and soil solution, which can cause certain deficiencies in potassium and sometimes also in magnesium.

In natural plant formations, *califuge* species are not able to flourish in the presence of active carbonate, certain of them having a tendency to suffer from a lack of nitrogen, others of minor elements such as Mn or Fe (chlorosis). Experiments have shown that intensive agriculture can correct most of these problems except, in certain cases, physical defects such as lack of water; when these physical defects are reduced it is possible to obtain good harvests (particularly of cereals, that are not too water-demanding) on chalk soils by the frequent and regular addition of mineral fertilisers with a very definite composition, and by the maintenance of sufficient humus to guarantee a good structure and a sufficiently high exchange capacity; this is achieved in the chalky Champagne area of France.

These problems of rendzinas *sensu stricto* are considerably lessened in the brunified calcimagnesian soils (calcareous brown soils, calcic brown soils). In the first, the presence of a small quantity of active carbonate is itself a favourable factor, for it assures a better structure and aeration in the winter than for clay soils on marls, which are very deeply decarbonated. Monocyclic calcic brown soils on marly limestones are often excellent prairie soils, because of their good potential water reserve. In contrast, polycyclic calcic brown soils on indurated limestones are often too dry; their agronomic value is entirely dependent upon the thickness of the fine earth, preferably with few stones, that overlies the indurated rock; if this depth is sufficient, the soils are good for cereal cultivation (Berrichone Champagne area in France).

## III Formation of humus–carbonate complexes: biochemical and geochemical processes

It is known that the decomposition of fresh organic matter produces acid-soluble organic compounds and carbon dioxide which cause the solution and movement of active carbonate and hence its elimination from the upper part of the profile. Conversely, the active carbonate reacts with the organic matter and facilitates rapid humification by forming very stable, slightly transformed complexes. The final result of this interaction is very variable, depending upon the amount of these materials present. If acid organic material is produced in abundance, while the carbonate reserves of the soil are low, decarbonation – then decalcification of the absorbent complex – occurs rapidly. In the opposite situation, when the carbonate reserve is considerable compared to the acidifying soluble substances, it is the organic matter that is insolubilised and stabilised.

The humus of rendzinas is thus only slightly transformed, maintained at this youthful stage by the presence of active carbonate, in which it differs profoundly from chernozem humus that, in contrast, has been subject to a prolonged climatically determined transformation. *The stability of the humus–carbonate complexes of the rendzinas is evidently dependent on the maintenance of a sufficient quantity of active carbonate in the profile* and this precarious equilibrium is destroyed if decarbonation occurs. The humus then becomes a eutrophic or mesotrophic forest mull, which mineralises rapidly; it is the start of the process of brunification.

The biochemical equilibrium of rendzinas is thus dependent on two basic factors: on the one hand, a factor related to the mineral material which controls the production and the maintenance in the profile of a variable amount of active carbonate; on the other, a factor connected with the development of the organic matter itself and thus of the vegetation.

## 1   The freeing of active-carbonate by weathering of parent material

Depending on the hardness and degree of purity of the parent rock, the quantity of active calcium carbonate freed by weathering is very variable – a point, it should be remembered, that was briefly referred to in the first chapter. Two situations can be distinguished according to whether materials produced by weathering contain a great deal of active carbonate, or little to none at all.

*The formation of weathered materials, rich in active carbonate*, results from the rapid mechanical weathering of a marly limestone rich in silicate impurities and also of a very pure, friable white limestone, such as chalk. However, the high content of silicate impurities (particularly clays) in the first material compared to the low content of such impurities in the second, means that the possibilities of subsequent development, by decarbonation and brunification, are much greater for the first type of material than the second. This important difference will be discussed in Part IV.

*The formation of weathered material poor in active carbonate* occurs when silicate-poor, indurated limestones (sometimes crystalline, such as those of the Jurassic) are subject to surface weathering. Lamouroux (1971) showed that a surface skin was dissolved in rainy periods and, as the resulting bicarbonate solutions are of low concentration, the calcium thus dissolved is generally removed from the profile. Only the silicate clays remain behind after a certain degree of local movement or surface reworking, but as they are only present in small amounts it takes a very long time, in a cold temperate climate, to produce a layer sufficiently thick to form a real profile (terra fusca) at the limestone surface. In these circumstances the soils that develop have two characteristics: (i) they are very old and generally polycyclic; and (ii) commonly, they are very poor in active carbonate and thus a rendzina *sensu stricto* with an AC profile is not produced. If secondary recarbonation occurs, as a result of mechanical processes of transport or cryoturbation, it is of a limited kind and it leads only to the formation of transitional soils such as brunified rendzinas or calcic brown soils.

**Note** that in humid mountains, this surface weathering of limestone is much more rapid; if the parent material is heterogeneous, the fine-grained limestone particles are dissolved preferentially, leaving behind the coarser fragment, surrounded by a weathered layer containing active carbonate (Pochon 1978).

## 2 Production of organic matter by litter: humification

The organic matter supplied to the calcimagnesian soils by the vegetation shows considerable variation both in quantity and quality. As discussed in Chapter 3, this causes the speed of decarbonation and also the amount of humus stabilised in the humic A1 horizon to be very variable. As shown by Le Tacon (1976), *the speed of decarbonation of a marly limestone is almost proportional to the quantity of organic matter immobilised in the profile*. In this respect, forests have a greater effect than the grasses of prairies and meadows (Duchaufour 1950), which in turn are responsible for a decarbonation which is a little more rapid than that which is caused by cultivation of the soil. This explains why there are in general three types of rendzina, characterised by different types of vegetation and which correspond to three equilibrium levels of organic matter carbonates: the **forested black rendzina**, humic rich but poor in carbonate; **grey rendzinas** of meadows, poorer in humus but richer in carbonates; and **white rendzinas** on unindurated material that are exceptionally rich in carbonate (chalk), but poor in organic matter, often as a result of cultivation that has caused the humic horizon to be eroded.

The role of cultivation is in fact important in all cases; it breaks the original *humus–carbonate* equilibrium by decreasing, on the one hand, the amount of organic matter incorporated, particularly the most labile fractions – fresh organic matter, inherited humin, fulvic acids (Muller & Védy 1978) – and increasing the amount of active carbonate in the profile, on the other, by way of mechanical weathering. This particular equilibrium characteristic of the cultivated or anthropic rendzinas will be discussed again in Part VI.

The humification, blocked by carbonate at an early stage in the transformation of plant residues, was discussed in Chapter 2: the untransformed materials (fresh organic matter, remains of cell membranes) and the inherited humin resulting from *direct* humification (incomplete oxidation of lignin) accumulate in the A1 horizon where they form the greater part of the organic compounds. Another fairly important fraction is that resulting from the rapid insolubilisation by active carbonate of the phenolic precursors from the litter, which are mainly fulvic acids immobilised by calcium that remain extractable after decalcification; the humin of insolubilisation is of little importance, as it depends on the amount of free iron in the rendzina.

It was thought for a long time that they included a reasonable quantity of grey humic acids, with strongly polycondensed ring structure, similar to those occurring in isohumic soils. Actually, during paper electrophoresis a large proportion of these humic acids were immobilised in the position that grey humic acid would occupy, but Calvez (1970) and Dormaar *et al*. (1970) showed that this property of humic acids was not the result of polycondensation or decrease of charge, but rather of the presence of fine clayey impurities.

Figures 2.1 and 2.8 and Figure 7.1 deal with the composition of carbonate humus; it should be noted that the humin of Figure 7.1 is almost totally inherited.

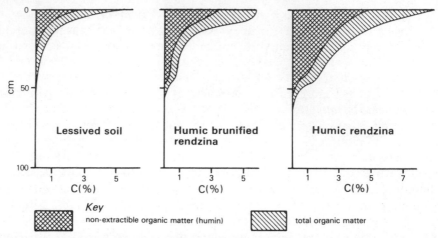

**Figure 7.1** Diagram of the distribution of organic matter as a function of depth (after Jacquin & Le Tacon 1970).

## 3 Biochemical mechanisms for the formation of the humocalcareous complex (Chouliaras *et al.* 1975, Brison 1978)

The effect of active calcium carbonate on the decomposition of fresh organic matter and the accumulation of *inherited* humin (young humin) that is still very close to fresh organic matter in composition, has been emphasised in Chapter 2. Calcium carbonate, dissolved and precipitated, covers the plant debris and protects it against microbial action so that biodegradation is slowed down. Nevertheless, oxidation does still occur causing the formation of carboxyl functional groups, characteristic of inherited humin, and the partial differentiation of this humin from fresh organic matter.

As has been said, the other major fraction of calcareous humus is made up of fulvic acids immobilised by calcium. The experiments of Brison (1978) showed that these fulvic acids resulted from an *immediate insolubilisation, of a purely chemical nature*, by active carbonate of the greater part of the soluble precursors which are transported from litters: these **calcium fulvates** are also particularly resistant to biodegradation (Linères 1977).

Brison was able to follow, by the use of lysimeters, the insolubilisation of soluble precursors from labelled maize litter within the A1 horizon of a rendzina (as compared to the A1 horizons of a brown soil and a podzol). In sterile conditions, 85% of the precursors are chemically insolubilised. In non-sterile conditions, all the soluble carbon disappears, with mineralisation complementing the insolubilisation (a part of the [14]C appears in the form of $Ca(HCO_3)_2$).

## IV Development of calcimagnesian soils in temperate plainlands

This development has three very different facets, depending on the nature of the calcareous material:

(a) On clayey–calcareous parent rocks (marly limestone), the rendzina occurs as a very transitory youthful phase which develops rapidly by brunification (except on slopes subject to erosion).

(b)  On unindurated chalk, with very high reserves of calcium carbonate, the rendzina represents a stable equilibrium.

(c)  Indurated limestones covered by layers of terra fusca and loam are generally unsuitable for rendzina formation, brunification occurs immediately, except when conditions particularly favour the process of secondary recarbonation.

### 1   Development of a marly limestone (Figs 7.2 and 7.3)

This development has two characteristics: (i) it is strictly dependent upon the vegetation type or the way the soil is utilised; and (ii) generally it is fairly rapid. Because of the abundance of silicate impurities in the parent rock, the dissolution of the carbonate is able to free a layer of loamy clay of reasonable thickness in a short time; the soils are recent, generally Postglacial, and are thus of the **monocyclic** type. Scheffer *et al.* (1960, 1962) have determined the speed of decarbonation on outcrops of Muschelkalk in central Germany. In 2000 years a brunified rendzina is formed, then a calcareous brown soil still with a little carbonate. A period of 8000 to 10 000 years is necessary for an almost entirely decarbonated calcic brown soil to form. Only at depth, under the typical polyhedral B horizon, is there a weathered Cca horizon which still effervesces. In certain cases, development goes as far as a lessived brown soil (Fig. 7.2):

lithosol → rendzina *sensu stricto* → brunified rendzina → calcareous brown soil → calcic brown soil → lessived brown soil

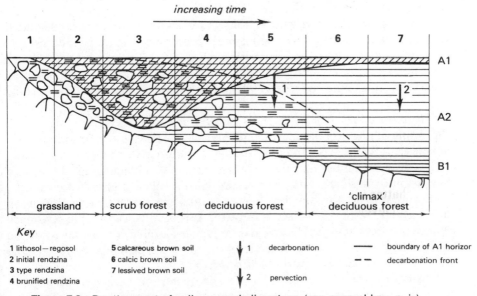

**Figure 7.2**  Development of soils on marly limestone (see general key, p. ix).

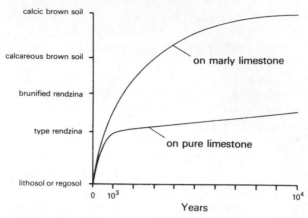

**Figure 7.3** Developmental phases of calcareous soils (after Scheffer et al. 1960).

But this development was studied under forest; that is to say, in conditions favourable to rapid decarbonation. Under grassland or when cultivated, the rate of these processes (as said previously) is slower; for erosion acts on the surface in a more intense manner, the soil being less protected, which tends to remove the surface layer at the same rate as it loses its carbonate; in these circumstances, the profile still contains a small amount of carbonate (all the greater as the climate is drier) in all of the horizons, even in the surface. In addition, as the biodegradation of organic matter is activated by cultivation, organic matter is not able to accumulate; the surface humus remains of little thickness. Development never gets beyond the calcareous brown soil phase.

It goes without saying that the two states of equilibrium thus described – calcic brown soil (forested climax soil) and calcareous brown soil (degraded soil as a result of cultivation or being under grassland) – are to be seen only on sites with gentle slopes; on less gentle slopes, development towards a brown soil is prevented by erosion; on upper slopes young rendzinas occur, while on lower slopes there are colluvial rendzinas or colluvial calcareous brown soils.

## 2 Development on friable, pure limestone (Fig. 7.3)

The result of the weathering of this type of material is the formation of a pulverescent white mass, initially without structure and thus not very permeable, composed almost entirely of calcium carbonate (Jamagne 1964); surface dissolution of this carbonate, if it occurs, leaves only a slight silicate residue. On the other hand, this powdery mass is particularly susceptible to erosion, so that *rejuvenation* of the surface is frequent. It is easy to see that the resistance of such a parent rock to decarbonation is considerable, no matter what the colonising vegetation. In these circumstances, no soil development is possible and the soil remains a rendzina *sensu stricto*.

This is the type intrazonal soil, often characterised by a *specialised plant association*, the whole forming a site climax. This appears to have been the case for most of the chalky Champagne of France, where the original vegetation appears to have been not a forest but a kind of **edaphic steppe** with scattered calcicolous bushes. In this situation the soil is low in humus and weakly coloured by organic matter (grey rendzina). Sometimes, if the weathered layer is sufficiently deep and if a fissured parent rock occurs where roots can penetrate, a deciduous forest can establish itself, and thus the soil can become more humic (forested black rendzina).

Such a forested rendzina formed *in situ* on unindurated calcareous materials has a number of characteristics that differentiate it from the colluvial rendzina. The incorporation of organic matter is less deep and less complete; the litter is frequently acidified at the surface and it develops towards a moder, or even to a calcic mor or tangel, with the formation of a thick A0 horizon, particularly under acidifying coniferous plantation species, which very frequently occur in such sites.

## 3 Development on indurated limestone

The soils formed on indurated limestones (poor in silicate impurities) have, as already stated, the maximum complexity. They are derived from a heterogeneous material – terra fusca – contaminated to a greater or lesser extent by aeolian loams and, having been subject to prolonged development, they are polycyclic. Generally, they possess inherited characters, either of hot climates (large amounts of iron oxides and sometimes even free alumina), or of cold climates (cryoturbation, solifluxion) or both, having been subject to these effects successively (Meriaux 1959, see Ch. 4).

To begin with, three statements can be made about the development of these soils. (i) The importance of the thickness of the surface unconsolidated layer is evident, for this layer is complex and very old and if it is eroded, it cannot be reformed. The factor of erosion thus plays a most important part, since it tends to eliminate the soil material itself. But erosion is closely related to the topography and local relief, which therefore take on a particular importance. (ii) In general, vegetation as a *direct* agent of soil development is unimportant, as it is unable to acidify or easily modify this strongly buffered and biologically stable material, but its *indirect* role is important for it is a more or less effective protection against erosion. In this respect, forest affords better protection than grassland and, all other things being equal, forest soils are deeper and more developed. Clearing is often the prelude to a rapid degradation by irreversible erosion of the surface layer. (iii) As the surface layer is decarbonated from the start, brunification is immediate; type rendzinas are rare on this kind of parent rock and are limited to lower-slope colluvium.

In these circumstances, soils that are characteristic of hard limestone platforms are either related to brunified soils (polycyclic eutrophic, or mesotrophic brown soils) or, where local topography and parent material allow secondary recarbonation of a part or the whole of the profile, transitional (brunified) calcimagnesian soils, i.e. **polycyclic calcic brown soils** or **brunified rendzinas**.

### Influence of topography (Gury & Duchaufour 1972, Baize 1971)

Three landscape sites are differentiated: horizontal structural platforms, solifluxion basins, dry valleys with steep slopes; for the last two, the degree of slope, weak and strong respectively, has played an important part in the reworking or the homogenisation of the unconsolidated materials (Fig. 7.4).

### Structural platforms

The unconsolidated layers overlying the limestone are almost in their original position and they have been truncated to a greater or lesser degree by erosion. The complete profile is to be seen only in areas distant from the edge of the plateau and well protected against erosion. Generally, three superposed horizons can be recognised: an upper loam (aeolian), a mixed middle horizon, and a lower clayey horizon (horizon-$\beta$ or terra fusca *sensu stricto*). The middle horizon, as can be seen from its microstructure, is complex: it contains both reddish **papules** of terra fusca, displaced and moved towards the surface by cryoturbation, and ochreous ferriargillans resulting from pervection from the upper loam. As a whole, this profile simulates that of a lessived brown soil and it should be recalled that this type of profile was taken as an example of a *complex soil* in Chapter 4 (see also Ch. 9 and Duchaufour 1978: $IX_6$). In less complete profiles, the upper pervected layer has been truncated or removed, so that they have the appearance of a brown soil

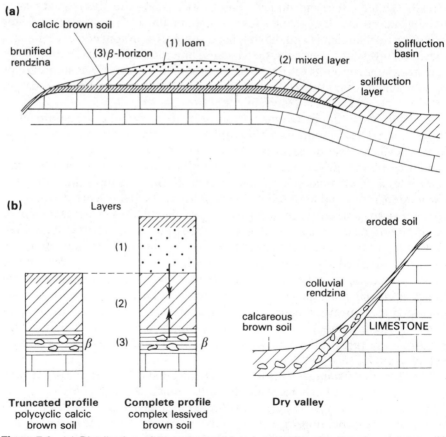

**Figure 7.4** (a) Distribution of the unconsolidated materials on a limestone plateau. (b) Polycyclic and complex soils on hard limestones.

with little or no pervection and more clay than the preceding one; only the lower, more pebbly horizons are effervescent. These are *polycyclic calcic brown soils*.

Finally, profiles at the plateau edge are of very little thickness; only the deep clayey layer remains, more or less mixed with pieces of limestone coming from the substratum, which causes a certain amount of secondary recarbonation and it is then a *polycyclic brunified rendzina*.

### Solifluxion basins

These are depressions surrounded by gentle slopes that are sometimes hardly perceptible. The clayey layer of decarbonation and the covering loams have been mixed and homogenised as a result of periglacial solifluxion: it is thus a homogeneous clay loam. When this layer is of little thickness (on slopes), the soil is a calcic (or eutrophic) brown soil; if it is thicker, as in the central parts of the basin, it forms a *lessived brown soil*, by pervection of ferruginous clays.

### Valleys with steep slopes

These are characterised by soil catenas where the contrast is still more marked; here it is no longer solifluxion alone which is responsible for movement in the unconsolidated layers, but an intense mechanical mixing, not only of the fine earth but also of limestone pebbles detached from the slope. Upper slopes are characterised by an erosional soil, lower slopes by colluvium where the addition of carbonate which is intimately mixed with the fine earth favours the formation of very humic rendzinas *sensu stricto* (colluvial rendzinas; see also Ch. 4, Fig. 4.8).

### Influence of the type of calcareous parent rock

Indurated limestones which are being considered here have generally been subject to fragmentation in the upper layer under the influence of ice. But their resistance to ice action can be very variable, depending on their induration and their crystallinity: the hardest, for example the Bajocian coral limestone of Lorraine, shows a maximum resistance to this process and it occurs as slightly fissured slabs impenetrable to roots. Thus in these circumstances the unconsolidated cover can never be enriched in carbonate, either mechanically or biologically; frequently it is slightly acidified, giving eutrophic or mesotrophic brown soils (see Ch. 9).

The less indurated limestones, the Bajocian oolitic limestone of Lorraine, are broken by ice action either into small slabs or into even smaller fragments that have been more or less mixed with the overlying clay by cryoturbation. In these circumstances, secondary recarbonation can occur at the base of the profile. Thus a particular kind of *calcic* horizon is formed, differing from those calcic horizons formed in monocyclic development. When secondary recarbonation affects the whole of the profile, which is then shallower, the soil is then a secondary brunified rendzina (polycyclic).

## V Development of calcimagnesian soils at high altitude

The two essential characteristics of humid mountain soils formed on limestones have already been given: (i) large-scale incorporation of organic matter with slow decomposition; (ii) tendency towards rapid decarbonation of the fine earth.

In spite of these general characters, the variability of soils formed on calcareous rocks in mountains is very great as a result of the particularly important role played by local factors – slope, marked relief, the nature of the parent material, local modifications of the hydrology, and great variations in the pedoclimate dependent on the exposure. The equilibria controlled by the *site* (site climax) are thus many and varied. Climatic sequences related to altitudinal vegetation zones are, however, an incontestable reality, but they are often masked by local factors so that evidence of them is only possible under certain conditions.

## 1   General bioclimatic factors: soil altitudinal zones (Bartoli 1966; Bottner 1972)

When bioclimatic development is not opposed by site factors, it leads to the formation of characteristic altitudinal soil zones related to increasing altitude. This zonation is seen in the particular development of certain profile characters: (i) acidification becomes more and more marked; (ii) the level to which decarbonation occurs becomes deeper on mixed calcareous and silicate parent materials; (iii) there is a weaker and weaker incorporation of organic matter with mineral material, which is accompanied by a tendency to form a surface holo-organic horizon (A0) of greater and greater thickness.

But this altitudinal zonation is often disturbed and prevented from developing by certain local factors, of which the following are the most important: (i) erosion or deposition preventing all acidification and causing secondary recarbonation; (ii) warm sites which, as shown by Ludi (1948) and Bartoli (1966), are caused by the increased intensity of radiation reached in the soil which compensates for the low air temperatures of high altitude; (iii) a lack of fine earth on the indurated limestone, preventing the formation of clay–humus aggregates (coarse-grained karst; Bottner 1972, Cabidoche 1979).

*Thus it is particularly at cold sites and on marly or sandy limestones with gentle slopes that the altitudinal sequences are best developed*: they can be defined in terms of the *altitudinal zonation of climatic humus types*, which allows the *analogous* soils of high altitude to be characterised on parent rocks as different as marly limestones, sandstones or crystalline rocks (Bottner & Paquet 1972).

The **montane altitudinal zone**, where the climatically determined vegetation is a mixed coniferous–deciduous forest, has a thick organic-rich mull, incorporated in the soil to depth (sometimes it is a mull-moder if there is a slight A0 development). The climatically determined soils on calcareous materials belong to the brunified class or the brunified calcimagnesian subclass. Their richness in organic matter and its depth of incorporation frequently causes them to be referred to as **melanised**.

The **sub-alpine zone**, with its predominantly coniferous forest with ericaceae, develops a humus characterised by an A0 horizon, moder first of all, then calcic mor (tangel) and finally acid mor. In this zone the soil, formed on materials not affected by erosion and sufficiently rich in silicate materials, shows a strong acidification with a tendency for podzolic development. This kind of development is more marked with altitude: podzols *sensu stricto* have been found even on terra fusca or on old calcareous moraines (Duchaufour 1978: $I_1$). Such altitudinal zonation is much less clear on coarse-grained karst, where this material controls the humification in a definite way.

In the **alpine zone**, the open alpine grassland replaces the forests. The increased effectiveness of solar radiation adds its effects to the raw plant material richer in nitrogen, to form a humus that is biologically more active than that of the lower forested zones. A mull–moder or even a moder is formed which is characteristic of shallow humic brown soils and lithocalcic soils (Duchaufour 1978: $V_5$).

## 2 Effect of local factors: parent material, topography

As shown by Bottner (1972), the altitudinal climatic zonation on outcrops of hard limestones is profoundly disturbed by local mechanical effects: erosion, lateral movement of material that separates the fine earth from coarser materials, and the frequency of secondary recarbonation that compensates for all processes of acidification. On *fine-grained karst* materials, a degree of climatic development occurs in certain cases (Duchaufour 1978: $I_2$), but this is never the case for *coarse-grained karst material* which, no matter what the altitudinal vegetation zone, is characterised by humocalcareous, humocalcic or humic–lithocalcic soils.

As shown by Cabidoche (1979), the parent material plays an essential role in this particular case, so that it can efface the control of the altitudinal climatic zone. As all the fine-grained and soluble residues of weathering are transported, the insolubilisation of humic precursors cannot occur; necessarily, there is formed a slightly transformed humus in which fresh organic matter and inherited humin are dominant.

With regard to the fine-grained materials, transported by erosion into the basins and badly drained depressions, they give rise to slightly developed stagnogleys in the snow hollows in the alpine zone while, in contrast, they are distinctly podzolised in peaty areas of the sub-alpine zone – spruce forest with *Sphagnum* (see Ch. 11).

Finally, the flaggy blocks of hard limestone, often stripped bare and sometimes fluted by erosion (**lapiaz**), are first of all colonised by a xerophilic pioneer vegetation (rock-dwelling plants) which, in very humid, cold sites, is subsequently replaced by a sub-alpine climatically determined vegetation (spruce, ericaciae, *Sphagnum*) which forms a thick layer of mor (humic lithocalcic soil with mor; Duchaufour 1978: $V_6$).

As an example, a detailed study will be made of soil development on a mobile talus slope gradually fixed by forest vegetation at cold sites in the montane altitudinal zone.

### Development on a mobile talus slope

Steep talus slopes, frequent in areas subject to avalanches, are characterised, at least initially, by a constant renewal of materials; the addition of calcareous material, of considerable size range, prevents the initiation of decarbonation. On warm slopes the pioneer vegetation (generally very scattered) persists, but on moist slopes a hygrophilic forest, favoured by the reserves of water stored in this very deep soil, is established fairly rapidly. This vegetation – maple with fern, in the pre-Alps – is adapted to the abundance of carbonate in the soil and, in particular, to the substratum which at this stage is unstable (Bartoli 1962).

The soil, very humic, pebbly and uniform, is in some ways like a rendzina, but there are some notable differences: it is deeper (often 1 m), much richer in organic matter (in general more than 16%), and of homogeneous composition throughout, as a result of the constant reworking to which it is subject. From this point of view it is typically **isohumic**; the humus is a calcareous mull, poor in fine-grained silicates, but rich in carbonates (*humocalcareous soil*; Bottner 1972).

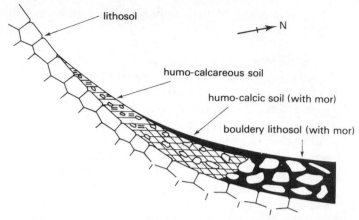

**Figure 7.5**  Soil catena on limestone talus slopes in mountains.

Development occurs when the talus slope is stabilised by plant colonisation of the upper slopes. The humus is then subject to a rapid transformation, related on the one hand to the general increase in organic matter production by vegetation and on the other to the prevention of addition of carbonate; the crumbs of the mull lose their carbonates at the same time as their organic matter increases. Humic compounds with small molecules, produced abundantly in the A0A1, migrate and precipitate in the peripheral zone of the crumbs (Bottner 1972). To begin with, the humic complexes remain saturated with calcium and are thus flocculated and the initial crumb structure is preserved. This particular type of forest humus, poor in mineral entities, particularly clay, was called **mull–moder** by Kubiena (1953) even though this name leads to confusion. Considering the whole of the profile, it is subject to a marked decarbonation, at least in the surface, but a *brunification* cannot be spoken of for the iron coloration is not apparent (*humocalcic soil*, Bottner 1972).

When acidification is sufficient at the surface, little by little an acidiphilic vegetation replaces the calcicolous vegetation and a layer of raw humus is not slow to form and to overlie the previous profile; it is a case of a weakly acid, raw humus that is rich in exchangeable calcium (calcic mor or tangel). In the lower part of the profile the extractable organic matter increases greatly. The pedoclimate is increasingly cold and humid as a result of the retention of water by this humic layer and, in these circumstances, the final stage of the succession is often represented by an association characteristic of the sub-alpine zone: spruce forest with *Sphagnum*, sometimes even pine with *Rhododendron*. Thus the profile has lost all its active carbonate; only the large blocks of slightly soluble limestone remain. It is a humocalcic soil with mor.

Before finishing this discussion, it should be said that the succession as described above in terms of time also occurs frequently in space, the mobile talus slopes occurring on upper slopes, the stabilised talus slopes and the soils with calcic mor, in contrast, occurring on the lower, more gentle slopes that have been colonised by plants for a considerable time (Fig. 7.5).

Table 7.1 gives some idea of the distribution of the different types of climax soils (climatic or site) in relation to the plant vegetation zones in mountains.

## VI   Classification of calcimagnesian soils: main types

This section goes back to the three subclasses that were differentiated at the beginning: (i) humic calcimagnesian soils – rendzinas; (ii) very humic

**Table 7.1** Soils developed on calcareous material (altitudinal sequence).*

| Altitudinal zone | Vegetation climatically determined humus | Parent material | Climatically determined soil development | Site-determined soil development |
|---|---|---|---|---|
| alpine | alpine grassland mull–moder moder | | acid brown soil or humic lithocalcic soil | (snow hollows) stagnogley |
| sub-alpine | coniferous forests with ericaceae mor moder | { coarse-grained karst material marly limestone or fine-grained karst material | humocalcic soil or lithocalcic soil with mor podzolised lessived soil or humic ochric soil | (depression) podzolic stagnogley (spruce forest with *Sphagnum*) |
| montane | mixed forest coniferous–deciduous thick mull | { karst material (often reworked) marly limestone and marls | calcareous brown soil or melanised calcic brown soil calcic brown soils or lessived brown soil | (talus slopes) humocalcareous soils humocalcic soils (depression) stagnogley |

* According to Bartoli (1966), Bottner (1972), Pochon (1978) and Cabidoche (1979).

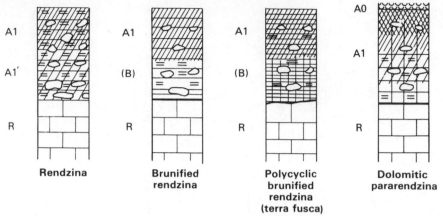

**Figure 7.6**　Humic calcimagnesian soils (AC profile) (see general key, p. ix).

calcimagnesian soils (under high altitude climates); (iii) brunified calci-magnesian soils.

## 1　Humic calcimagnesian soils: rendzinas (Fig. 7.6)

With the exception of the forested brunified rendzina, which has a slightly developed brown or ochreous (B) horizon compared with the A1, the soils of this group are characterised by having an AC profile.

*The A1 horizon is coloured by organic matter which more or less masks the brown colour of the iron.* Typical rendzinas are black or grey, depending on the quantity of organic matter incorporated, which is itself a function of the type of vegetation. Certain rendzinas on soft limestones or chalk are exceptionally rich in carbonates which completely mask the organic matter, which is generally not very abundant; these are immature rendzinas, still very close to **regosols** (white rendzinas).

However, there are rendzinas in which the humic A1 horizon has a brown colour, related to the abundance of free iron, although the profile is still of the AC type. As emphasised by Mückenhausen (1962), *it is necessary not to confuse the brown colour of the humic A horizon with true 'brunification', that is to say, the appearance of a (B) horizon with polyhedral structure below the humic A1 horizon.*

*The brown colour of the A1 horizon* is favoured by certain particular cir-cumstances; two examples will be given. The first concerns **anthropic rendzinas** (rendzinas of cultivation), formed on materials rich in iron (terra fusca). The cultivation of the soil has caused a decrease in the amount of organic matter accompanied by strong secondary recarbonation; the profile is less dark than the original. The second example is that of the **pararendzina**, described by German authors on dolomitic materials or sandy limestones, the weathering of which frees an important amount of sandy skeletal material, which decreases the proportion of humic complexes stabilised by carbonates in the profile. The colour is less dark, the crumb structures less well developed and less stable because of the sand not incorporated in the aggregates (Franz 1960, Mückenhausen 1962).

*The formation of a (B) horizon of little thickness* characterises the forested brunified rendzina, *sensu stricto*. The thick A1 horizon and its dark colour preserve the morphology and the main properties of the forested black rendzinas described at the beginning of the chapter. Only the quantity of active calcium carbonate is less, either as a result of the beginning of decarbonation (primary brunified rendzina), or because it is secondary brunified rendzina (or polycyclic) on hard limestones and terra fusca (Duchaufour 1978: $VI_1$).

**Dolomitic rendzinas** are poorly known, but two special varieties are to be noted, depending upon the pedoclimate and the nature of the dolomitic material: (i) **sandy rendzinas** (pararendzina type) often occur with moder (either xeromoder in a dry climate or acid moder in a humid climate); (ii) **rendzinas with a pseudosilt structure** are formed of magnesium humic complexes of insolubilisation with a very fine structure (Morocco, Middle Atlas).

**Rendzinas with gypsum** are characteristic of gypsocalcareous outcrops (central Alps). In a humid climate the differential dissolution of calcium sulphate causes a high concentration of active carbonate in the A1. In a dry climate the calcium sulphate remains in the profile: the soil then belongs to the salsodic class (see Ch. 13).

**Summary: rendzinas and pararendzinas**

(a) *Rendzinas with a thick A1 horizon and coloured by organic matter*: strong coarse crumb structure.
  (i) AC profile: *black or grey rendzinas*.
  (ii) A(B)C profile: slightly developed (B) horizon – *brunified rendzinas* (two varieties – primary and secondary brunified rendzinas).
(b) *Rendzinas with an A1 horizon coloured brown by iron* (or else with a partially degraded structure) – AC profile.
  (i) Cultivated rendzinas: *anthropic brown rendzinas*.
  (ii) *Pararendzinas*: importance of the sandy fraction (quartz or dolomite).
  (iii) *Dolomitic rendzinas with moder*.
(c) *Presence of gypsum: rendzina with gypsum* (AC profile).

## 2  *Very humic calcimagnesian soils* (Fig. 7.7)

The amount of organic matter is greater than in the type rendzinas. It is greater than 15% in typical A1 horizons with a crumb structure. When a (B) horizon with a polyhedral structure develops, it still contains organic matter in appreciable amounts and is often dark coloured.

Three groups can be differentiated and defined in summary form, as follows:

(a) Profile A0C (*or* A0R): soil characterised by the absence or the weak development of mixed organomineral A1 horizons – **humic lithocalcic soils**.
(b) Profile A1C (*or* A0A1C): A1 horizon with a well-developed crumb

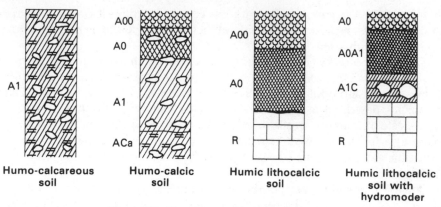

**Figure 7.7** Very humic calcimagnesian soils.

structure – **humocalcareous** and **humocalcic soils** (formerly carbonate humic soils).

(c)  Profile A1(B)C: presence of a (B) horizon rich in silicates and strongly structured (polyhedral) – **calcareous brown** or **humified (melanised) calcic brown soils**.

In fact, transitional forms are common, particularly between the first two groups. There are often humified lithocalcic soils that have, at the contact with the hard limestone, an A1 horizon several centimetres thick, but this horizon is always extremely rich in organic matter (more than 25%). Essentially, the profile is made up of a thick tangel or mor A0 horizon, containing more than 50% organic matter. Some of these humified calcimagnesian soils are formed on flaggy blocks of hard limestone which prevents any mixing of the mineral and organic materials; in others, in contrast, developed on very coarse talus materials, a certain amount of slightly developed organic matter gets down between the blocks; one such soil is transitional with the more humic, humocalcic soils (Duchaufour 1978: $I_3$).

**Humic lithocalcic soils.** These are characterised by their humus type: acid mor on slabs of very hard limestone (Duchaufour 1978: $V_6$); tangel on softer or finer-joined limestones; moder or hydromoder, with an organomineral A1 a little more developed under alpine meadows in humid sites (Duchaufour 1978: $V_5$): these are the *Pechrendzinas* of German authors (Kubiena 1953, Solar 1964).

**Humocalcareous and humocalcic soils.** These have a well developed mixed A1 horizon, of a black colour and coarse crumb structure, containing more than 15% organic matter in the fine earth. This A1 horizon penetrates more or less to great depths between the hard limestone blocks. Humocalcareous soils still contain active carbonate in the fine earth and their humus is a carbonate mull, relatively poor in silicates (Duchaufour 1978: $V_3$). Humocalcic soils, in contrast, are completely decarbonated in the fine earth and richer in organic matter than the preceding case, which is able to acidify in the surface and form an A0 horizon of calcic mor or tangel (Duchaufour 1978: $V_4$).

**Calcareous brown or melanised calcic brown soils.** This third group includes soils developed on fine-grained karst infillings (reworked terra fusca) or on outcrops of marly limestones affected by erosion. They are characteristic of the montane zone, while in the sub-alpine altitudinal zone they are replaced by more developed, more acid soils.

The dark A1 horizon with crumb structure contains more than 15% organic matter, and is slightly less thick than in the humocalcareous and humocalcic soils; it overlies a silicate-rich (B) horizon (derived particularly from inherited clays from the parent material) with a polyhedral structure; *often this horizon is still humic (more than 5% organic matter) and of a dark colour*, hence the qualification **melanised** given to these soils by several authors.

As in the soils of the plainlands, the calcareous brown soils are distinguished from the calcic brown soils by the presence of active carbonate in the fine earth of the first and its absence in the second; but even in the latter case there is often active carbonate in the weathering skin around the rock fragments (Pochon 1978).

## 3 Brunified calcimagnesian soils (Fig. 7.8)

These are soils of the plainlands which, as already stated, are formed from materials that are poorer in calcareous materials and richer in fine-grained silicates (particularly clays) than typical rendzinas. In these circumstances decarbonation occurs rapidly. The A1 horizon is no more than a eutrophic mull, poor in carbonates. The decomposition of organic matter occurs more rapidly; this horizon is poorer in humus and less thick than the A1 horizon of the type rendzina. The (B) horizon with polyhedral structure is generally well developed and often contains a certain amount of carbonates (calcareous

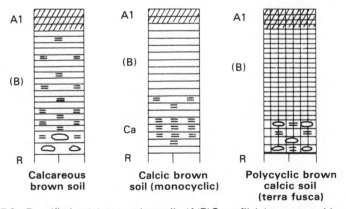

**Figure 7.8** Brunified calcimagnesian soils (A(B)C profile) (see general key, p. ix).

brown soil), but it is entirely decarbonated in the more developed calcic brown soil. Only the lower part of the profile (Cca horizon), often enriched by precipitation of the carbonates transported from the surface, still effervesces with HCl.

On certain materials, such as marls or soft marly limestones, the calcareous brown soils represent a phase of incomplete development compared with the calcic brown soil (or even in certain cases the lessived brown soil if the decarbonation has occurred to greater depths); the calcareous brown soil itself can have a very variable amount of organic matter and very different degrees of decarbonation, depending on its degree of development:

Slightly developed calcareous brown soil → developed calcareous brown soil → calcic brown soil → lessived brown soil

The subgroups of the calcareous brown soils are defined by their degree of development, by the kind of parent material (colluvial or *in situ*) and, finally, by the appearance of certain secondary characteristics (vertic or hydromorphic, i.e. mottled, with the appearance of diffuse rusty patches).

The calcic brown soils are of two types, depending on their origin and the nature of the parent material. Primary (or monocyclic) calcic brown soils on marly limestones are differentiated from secondary (or polycyclic) calcic brown soils on terra fusca and hard limestones.

**Summary**
(a) *Calcareous brown soils:* with an effervescent (B) horizon.
    *Subgroups:* slightly developed, vertic, colluvial, hydromorphic.
(b) *Calcic brown soils:* (B) horizon with no carbonate but generally base-saturated.
    *Subgroups:* monocyclic, polycyclic (on terra fusca).

# References

Baize, D. 1971. *Contribution à l'étude des sols les plateaux jurassiques de Bourgogne*. Spec. doct. thesis. Univ. Paris.

Bartoli, C. 1962. *Ann. ENEF* **XIX** (3), 329.

Bartoli, C. 1966. *Etudes écologiques sur les associations forestières de la Haute Maurienne*. State doct. thesis. Fac. Sci. Montpellier.

Bottner, P. 1972. *Evolution des sols en milieu carbonaté*. State doct. thesis, Fac. Sci. Montpellier; Mémoire no. 3, *Sciences géologiques*, Strasbourg.

Bottner, P. and H. Paquet 1972. *Science du Sol* 1, 63–78.

Brison, M. 1978. *Etude de l'humification par introduction* in situ *de mais marqué (*$^{14}$*C) dans trois types d'humus sous Hêtre*. Spec. doct. thesis. Univ. Nancy.

Cabidoche, Y.-M. 1979. *Contribution à l'étude des sols de Haute-Montagne*. Spec. doct. thesis, Univ. Montpellier (ENSA Montpellier): **I**, text; **II**, figures.

Calvez, C. 1970. *Contribution à l'étude des processus d'extraction et de characterisation des composés humiques*. Spec. thesis. Univ. Nancy I.

Chouliaras, N., J. C. Védy and F. Jacquin 1975. *Bull. ENSAIA* **XVII** (1), 65–74.

Dormaar, J. F., F. Jacquin and M. Metche 1970. *Soil Biol. Biochem.* 2 (4), 285–93.

Duchaufour, Ph. 1950. *Ann. ENEF* **XII** (1), 98–153.

Duchaufour, Ph. 1978. *Ecological atlas of soils of the world*. New York: Masson USA Inc.

Duchaufour, Ph. and .C. Bartoli 1966. *Science du Sol* **2**, 29–40.

Filipovski, G. and M. Ciric 1969. *Soils of Yugoslavia*. Belgrade: D. Jelenic.

Franz, H. 1960. *Feldbodenkunde*. Vienna: G. Fromme.

Gury, M. and Ph. Duchaufour 1972. *Science du Sol* **1**, 19–24.

Jacquin, F. and F. Le Tacon 1970. *Bull. ENSAIA* **XII** (1–2), 12–20.

Jamagne, M. 1964. *Pédologie*, Ghent **XIV** (2), 228–326.

Kubiena, W. 1953. *The soils of Europe*. London: Thomas Murby.

Lamouroux, M. 1971. *Etude des sols formés sur roches carbonatées. Pédogénèse fersiallitique au Liban*. State doct. thesis. Univ. Strasbourg.

Le Tacon, F. 1976. *La présence de calcaire dans le sol. Influence sur le compartement d l'Epicea commun.* (Picea excelsa *Link.*) *et du Pin noir d'Autriche* (Pinus nigra nigricans *Host.*). State doct. thesis. Univ. Nancy I.

Linères, M. 1977. *Contribution d l'ion calcium à la stabilisation biologique de la matière organique des sols*. Spec. doct. thesis. Univ. Bordeaux.

Ludi, W. 1948. *Die Pflanzengesellschaften der Schinige Platte bei Interlaken und ihre Beziehungen zur Umwelt*. Berne: Hans Huber.

Meriaux, S. 1959. *84th Congr. Soc. Savantes*, 387.

Mückenhausen, E. 1962. *Entstehung, Eigenschaften und Systematik der Böden der Bundesrepublik Deutschland*. Frankfurt: DLG.

Muller, J. C. and J. C. Védy 1978. *Science du Sol* **2**, 129–44.

Pochon, M. 1978. *Origine et évolution des sols du Haut-Jura suisse*. Doct. thesis, Neuchatel; *Memoire Soc. Helvetique Sci. Nat.* **XC**. Zurich: Fretz.

Scheffer, F., E. Welte and B. Meyer 1960 *Z. Pflanzener. Bodenk*. **90** (1–2), 18.

Scheffer, F., E. Welte and B. Meyer 1962. *Z. Pflanzener. Bodenk*. **98** (1), 1–17.

Solar, F. 1964. *Mitt. Öster Bodenkundl. Ges.*, Vienna **8**, 170.

# Chapter 8

# Soils with matured humus: isohumic soils and vertisols

The soils discussed in this chapter have a special kind of development, with regard to both organic matter (maturation) and the weathering complex (formation of swelling clays), resulting from large seasonal pedoclimatic variations in humidity and an abundance of the alkaline earth cations of calcium and magnesium. This twofold development leads to the formation of a very special clay–humus complex, which is the common characteristic of two large families of soils – the isohumic soils and the vertisols. However, differences between them justify their separation into two classes.

The climatically controlled maturation of organic matter has been discussed in detail in Chapter 2. Essentially, this involves the polycondensation of aromatic rings of humic compounds as a result of seasonal phases of marked desiccation. This same seasonal alternation of wetting and drying is responsible for the more or less important neoformation of swelling clays of the montmorillonite type, and also for the preservation in the profile of 2 : 1 clays inherited from the parent material. The clays build up a very intimate physicochemical bonding with a part of the organic matter; this stabilised organic matter is incorporated to considerable depth (often several decimetres) in the mineral profile and gives it the appearance of an A1C profile, with a very thick and dark-coloured A1 horizon. However, even though this fundamental process is common to both the isohumic and vertisolic classes of soils, they differ from one another both in the environmental conditions of their development and in several of their physicochemical characters.

**Environmental conditions.** Isohumic soils are *zonal soils*, the development of which is essentially bioclimatic: the effects of climate and vegetation are fundamental. In contrast, the vertisols, even though they are all to be found in those climatic zones with a dry season, are more closely dependent for their formation on site conditions of topography, parent material and drainage, which considerably increase the pedoclimatic seasonal contrasts, compared with those occurring in chernozems.

**Physicochemical characters.** The process of neoformation of swelling clays, which is limited in the chernozems, is of much greater importance in the vertisols. Only a small amount of organic matter is incorporated and stabilised in the vertisols and this has only a limited effect on structure. In contrast, the organic matter of chernozems, in addition to the dark-coloured stabilised fraction, has an important labile fraction subject to a rapid mineralisation–humification turnover, which plays a fundamental part in the development of the coarse crumb structure. Another difference to be noted is the manner in which the organic matter is incorporated in the mineral profile. In the vertisols this involves *vertical movements*, resulting from the alternating expansion and contraction of swelling clays, while these movements are of little importance in the isohumic soils, where the great thickness of the humic A1 horizon is mainly of biological origin.

Finally, the time factor is another difference, for, although chernozems develop relatively slowly, vertisols reach their equilibrium state much more rapidly (Parsons *et al*. 1973).

# A   ISOHUMIC SOILS

## I   General characters

The development of isohumic soils is mainly controlled by the general bio-climatic factors of climate and vegetation. The most typical of these, the chernozems and chestnut soils, are characteristic of the steppe zone with a climate that is too dry to allow tree growth and these were studied by the Russian school at the end of the last century. The overall development of these soils is affected by site factors only to a minor extent.

The climatically determined vegetation is thus not forest but is dominated by grasses that play a determinative role in the development of these soils and form either a more xerophilic **steppe**, or a more hygrophilic **prairie**. In hotter regions a bush vegetation, of the **maquis** or **garrigue** type, is added to the grassland vegetation.

Grasses produce a large amount of organic matter, not only at the surface but especially at depth by what can be called the **rhizosphere effect**, which involves the excretion of water-soluble compounds from roots and more particularly *the very considerable annual decomposition of roots 'in situ'*. The rapid transformation of these very nitrogen-rich compounds marks the beginning of a special biogeochemical cycle of nutrient elements and of an intense humification which, for the most part, occurs below the surface. A part of this humus develops by *slow maturation*, which gives the black, or at least very dark, colour characteristic of these soils. The intense animal activity (earthworms and burrowing animals) favours the deep incorporation of humus. The fact that there is little decrease of organic matter with depth is the reason for these soils being termed **isohumic**, even though they do not possess this property in the true sense of the word. In these conditions of intense biological activity, decarbonation of the parent material occurs rapidly as a result of the high level of $CO_2$ production; but because the clays and iron oxides are immobilised within aggregates and even within the coarse crumbs containing 2 : 1 clays and grey humic acids that are characteristic of these soils, pervection does not occur, except in transitional types.

Four subclasses are differentiated in terms of their climatic regimes and certain profile peculiarities which, in the case of two of them, indicates that they are transitional to other classes: one to brunified soils and the other to fersiallitic soils.

(a) *Saturated isohumic soils with an AC profile*. These are steppe soils developed in a cold, dry continental climate. They form a sequence of *chernozems, chestnut soils and brown steppe soils* as the climate becomes drier and the steppe vegetation less dense. *The occurrence of a diffuse calcic horizon at depth is typical of this subclass*.

(b) *Partially unsaturated isohumic soils with A(B)C or ABC profiles*. These are the prairie soils of a more humid, less cold climate; as will be seen, trees are not completely absent from these soils. The profile is transitional between those of the steppe – A1C – and the profiles of deciduous forest – A(B)C or ABC: the A1 horizon alone is more developed and of a darker colour (**mollic** horizon) than in the case of the brunified soils *sensu stricto*. *In this subclass the calcic horizon does not generally occur at depth*.

(c) *Fersiallitic isohumic soils*. These soils are characteristic of xerophilic plant formations (garrigues or maquis, and certain types of savanna), in markedly hotter subtropical or tropical climates. They are transitional to the so-called **rubified** fersiallitic soils, strongly coloured by iron oxides with a more or less reddish shade. The addition of the dark colour of the humus to that of the iron oxides gives a general dark red or cinnamonic colour. *In addition, the calcic horizon, which is always present, is generally strongly developed and sometimes indurated*.

(d) *Soils of arid regimes*. These include the **subdesertic grey soils** or **sierozems** that are relatively poor in organic matter and have mineral material that is only slightly weathered, which is responsible for the light colour of these soils. It is true that arid soils are not strictly isohumic so that placing them in this class leaves room for argument and it is a perfectly reasonable solution to place them in a separate class, as has been done in the American and FAO classifications. However, from the environmental and geographical points of view, the sierozems can be considered as being at the extreme end of the climatic sequence going from the chernozems to the driest brown steppe soils (**burozems**), the organic matter of which, moreover, has a certain degree of convergence with that of the sierozems.

## II  Isohumic soils with saturated complex (AC profile)

### 1  Environmental conditions and general characters

These are zonal isohumic soils, characteristic of the continental steppe with cold dry climates which are particularly well developed in the Ukraine, where they have been studied in detail by Russian authors, and are also found in certain regions with a similar climate in North America (Manitoba, Dakota); as will be further emphasised, the marked seasonal pedoclimatic contrast, both in terms of temperature and humidity, would appear to be absolutely essential in their formation.

The major zone of the Russian steppe is itself divided into several sub-zones, which form a succession from north to south of more and more xerophilic types of vegetation and profiles less and less rich in organic matter; forest still occurs on the northern boundary of the zone and then it disappears, giving way to less and less dense steppe vegetation. The fundamental role of *climatically determined drainage* ($\Sigma P - PET$ monthly) in the distribution of these plant formations and soils has been discussed in Chapter 3 and it will not be repeated here, except to say that the climatically determined drainage, which is not negligible in the northernmost zones where forests still occur, decreases and then ceases in the steppe zone. Thus the sequence from north to south is as follows:

(1) deciduous forest – *grey forest soil*;
(2) sparse forest – *lessived chernozem*;
(3) steppe forest – *thick humic chernozem (typic)*;
(4) dense steppe – *steppe chernozem (orthic)*;
(5) sparse steppe – *chestnut soil and brown steppe soil*.

The pervection of clay and sesquioxides is still to be seen under forest, the very thick mull having characters intermediate between the atlantic mulls and the chernozemic mulls. The genesis of these **grey forest soils** will be discussed again when the lessived soils are dealt with. Pervection decreases in intensity and practically stops in the steppe forest zone where the Bt horizon disappears. However, the decarbonation of the profile, which is almost total in the humid forest zone, becomes less and less so towards the south; the Bt horizon is replaced by a Cca horizon (pseudomycelial) which occurs nearer and nearer the surface (Fig. 8.1).

The greatest development of the humic horizon is to be seen in the middle zone – that of the steppe forest. It is in this zone that the annual addition of organic matter is at a maximum, both at the surface (forest phases) or at depth by the decomposition of roots (steppe phases); it is also here that climatic conditions are most favourable for the preservation of this humus. In the more humid forest zones the humification conditions are not so good as the humus is more labile; in more southerly drier regions the sparser steppe vegetation produces only small amounts of organic matter.

The principle of soil *zonality* can be applied particularly well in central and southern Russia, because the influence of local factors (such as the site) capable of disturbing the distribution of the climatic zones is reduced to a minimum, as there is relative uniformity in topography and parent material, for this is a great plain on which loess deposits are dominant, and loess is exactly the material, in terms of its average content of calcium carbonate and

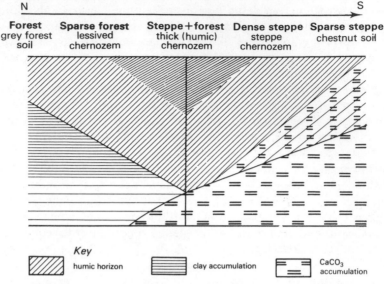

**Figure 8.1**   Climatically determined sequence on loess (Ukraine).

its moderate permeability, that is particularly favourable to chernozem genesis.

Nevertheless, it needs to be emphasised that, in other regions, it is possible for soils analogous to chernozems to occur, such as those discussed in Chapter 4, on materials other than loess; for example, on moderately acid, sandy materials, Franz (1960) differentiated the **parachernozems**, which have a less dark colour than the chernozems on loess. As the profile is less humic, the iron oxides are not completely masked by the organic matter and thus the parachernozems generally have a brown colour.

## 2   The chernozem profile: environment, morphology, biochemistry

Chernozems occur in the slightly arid continental zone with an annual rainfall of 400–500 mm. The seasonal contrasts of their pedoclimate are very marked. The very cold winters cause a deep and intense freezing of the profile; when thawing occurs, waterlogging and relatively anaerobic conditions result; heating and drying at the beginning of summer (when biological activity reaches its maximum) and marked desiccation in the second part of summer are the cause of humus maturation. The vegetation is either steppe forest, with patches of forest (particularly on slopes and small hills, the steppe vegetation being concentrated on the plateau), or steppe without trees in the more southerly and drier part. The **humic chernozem** of the steppe–forest zone **(typic)** is richer in organic matter (8–9%), more deeply decarbonated and still has a weakly developed (B) horizon with polyhedral structure (Duchaufour 1978: VII$_2$). The **steppe chernozem** (orthic) which corresponds to the modal type of A1–Cca profile will be described in greater detail (Duchaufour 1978: VII$_3$).

**Characteristics of the steppe chernozem.** In this steppe chernozem, the black A1 horizon is less thick (about 60 cm) and less rich in organic matter (5–6% at the surface) than that of the humic chernozem. The decrease in organic matter with depth is more rapid (Fig. 8.2). There is no (B) horizon with polyhedral structure; the Cca horizon, enriched in calcium carbonates by precipitation, directly underlies the A1. The A1 horizon is decarbonated, at least in its upper part, but frequently the base of the A1 overlaps with the Cca horizon (horizon A1Cca). Although in the typic humic chernozem the accumulation of calcium carbonate is only slight and in the form of a pseudomycelial whitish powder, in contrast, in the steppe chernozems the accumulation of calcium carbonate occurs as pulverescent patches and sometimes as localised and slightly indurated concretions. The structure is that of very stable, irregular crumbs of the size of a grain of wheat. In cultivated chernozems this is often destroyed in the Ap horizon, and it becomes cloddy or even massive. But generally the original porous crumb structure is preserved at the base of the A1, where the cereal roots are concentrated and spread out, in so far as it has been possible for them to penetrate the plough pan that often forms the base of the Ap horizon (Duchaufour 1978: VII$_2$).

**Crotovinas** are of frequent occurrence in chernozems. They are old burrows which occur particularly in the A1C and Cca horizons and are infilled with humus-rich materials from the A1 horizon.

The physicochemical and mineralogical characters can be summarised in a simple manner: uniform distribution of clay and free iron in the profile. The

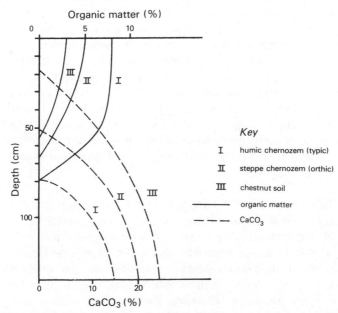

**Figure 8.2** The distribution in depth of organic matter and calcium carbonate in three types of isohumic soils.

free iron can be relatively abundant (sometimes 1% to 2%) but it occurs in a particular form that will be discussed in the following section. Only 2 : 1 clays are present, micaceous (illites) and montmorillonites in almost equal quantities. The exchange capacity is high (40–50 mEq/100 g) and normally saturated in basic cations, $Ca^{2+}$ and $Mg^{2+}$ being dominant. The great biological activity and the richness in nitrogen of the plant debris that gives rise to the humus are reflected in the low C : N ratio (about 10). This does not signify that the mineralising activity of the nitrogen is very great, it can be high at the surface at certain seasons but is always very low at depth where, as will be seen, stabilised forms of humus are dominant.

**Agronomic properties of chernozems.** Chernozems are noted for their great fertility: their physical and chemical properties are exceptionally favourable to plant growth. They are deep, with high porosity (65–70% in their natural state) which assures a good aeration, and their ability to retain water is adequate; the absorbent complex is saturated without the pH being too high or active carbonate being able to act unfavourably, as it does in the case of the rendzinas. The humus consists of two parts: a labile fraction with a very rapid mineralisation–humification turnover (which is mainly responsible for nitrogen nutrition), while the other fraction (which is stabilised after maturation) is responsible for the favourable structural properties and those of the absorbent complex.

However, it must be emphasised that these exceptionally favourable properties can be lost very rapidly if the soil is poorly cultivated; in fact, the best period for soil cultivation is short (between the period of spring-time saturation by water and summer desiccation). Intensive cultivation in an unfavourable season affects the structure, often in an irreversible way. In addition, the labile part of the organic matter is exposed to mineralisation at too rapid a rate, which leads to a wastage of nitrogen and a sharp fall in fertility; despite the low C : N ratio, of about 10, the annual amount of mineralisation becomes very low, often less than 1 part in 1000; the use of organic amendments or nitrogen fertilisers then appears to be necessary. In addition, often in the dry zones of the steppe, yields are adversely affected by a lack of water: over great areas of cultivation, wind erosion is extremely severe, all the more so as the soil structure has deteriorated; in the USSR tree shelter belts have been planted wherever possible.

## 3   *Humification and the formation of the clay–humic complex*

It has been known for a long time that the organic matter of chernozems is made up of two different fractions: one with a rapid turnover, particularly abundant at the surface (consisting mainly of fresh organic matter, young inherited humin, brown humic acids and fulvic acids); the other stable, with a slow turnover, which becomes the dominant fraction at the base of the A1. This has been confirmed by [14]C dating: the mean age of the organic matter of about 1000 years at the top of the A1 increases rapidly with depth where it exceeds 3000 years (Scharpenseel 1972, Ganzhara 1974), but the part of the

Photo: Toutain

Photo: Toutain

(a)

Photo: Bartoli

(b)

**Plate 1 Humification.** (a) Decomposition of beech litters: mull (on right) and moder (on left) (Bezange Forest, Meurthe-et-Moselle). Comparison of the rapid or slow decomposition of these two beech litters on Rhaetic sandstone in Lorraine. The only environmental difference between the two sites is in the amount of iron in the sandstone (twice as much at the mull site as at that with moder). (b) Calcic hydromor on limestone boulders, a sheltered site at an altitude of 1200 m (Creux du Van, Switzerland). At the surface there is a layer of *Sphagnum*; below the laminated L and F layers overlies the black H layer, which is constantly moist and 'silky' to the touch.

Photo: Lemoine

(a)

Photo: Lemoine

(b)

**Plate 2  Humification.** (a) Morphology of a chernozemic mull on loess (Oberkamersdorf, Austria). At a depth of 35 cm an irregular crumb structure; the white spots are a fine-grained precipitate of $CaCO_3$ ($\times 1.5$). (b) Morphology of a rendzina carbonate mull (Haye forest, Meurthe-et-Moselle). Similar crumb structure to the above, but of a smaller size; small limestone fragments present. Note that even though the morphology and the dark colour of these two types of humus are comparable, their development and biochemical compositions are totally different, which means that the rendzinas and chernozems must be differentiated from one another in classifications.

oto Duchaufour

(a)

(b)

**Plate 3  Movement of material within soils: illuvial horizons.** (a) Argillic horizon: complex lessived brown soil (loam and terra fusca) (Haye Forest, Meurthe-et-
Moselle). Polyhedral structure covered with clay skins and iron oxides, which give it a
glistening appearance. (b) Spodic horizon: podzol on Vosgesian sandstone (Biffontaine,
Vosges). Under the ashy horizon is the black horizon of humic accumulation, which in
places penetrates to greater depth; the mosaic of whitish patches is characteristic of very
humid spodic horizons.

(a)

Photo: Duchaufour

(b)

Photo: Duchaufour

**Plate 4  Climatically determined vegetation and analogous soils.** (a) Sub-alpine zone in the French Alps: pine forest with rhododendron on hard limestone (Moucherotte, Vercors; altitude 2000 m). *Pinus montana uncinata, Rhododendron* and *Vaccinium* on eroded limestone: humic lithocalcic soil. The convergence of the humification in (a) and (b) is to be noted in spite of the difference in the parent material. (b) Sub-alpine zone of the French Alps: pine forest with rhododendron on metamorphic rock (Belledonne massif, Champrousse; altitude 2000 m). *Pinus cembra* and *Picea excelsa* in a large clearing: the soil is a ranker or a mor podzol.

**(a)**

**(b)**

**Plate 5  Vegetation zones of tropical mountains.** (a) 'Paramo' alpine zone, central cordillera (Las Minas, Colombia), altitude 3700 m. Vegetation dominantly grass with shrubs (*Senecia* spp.); humic andic soil on volcanic ashes. (b) High altitude forest of the central cordillera of Colombia (altitude 2700 m). The very dense forest of bushy species is a climatic climax (hydromorphic andosol). The treeless depression is a *Sphagnum* bog, surrounded by an andic stagnogley which is a site climax.

Photo: Souchier

Photo: Duchaufour

**(b)**

Photo: Didier de Saint-Amand

**Plate 6   Zonality and palaeosols.** (a) Comparison of two podzols, a zonal Alpine podzol (Mt Corara, Romania; altitude 2100 m) and an intrazonal podzol with a water table from the Tropics (sandy coastal plain). The alpine podzol (on the left), as in the case of the boreal podzol, has a shallow profile: the ashy horizon is never greater than a few centimetres. In contrast, in the intrazonal tropical podzol (on the right) this horizon can be several metres thick. (b) Series of superposed palaeosols: Imberg terrace (Austria). The sequence from bottom to top: Rissian deposits capped by a hot climate ferruginous soil; a pebbly layer of Würm age, markedly affected by cryoturbation. At the surface a more or less brunified chernozem.

(a)

(b)

(c)

**Plate 7  Cryosols and rankers.** (a) Podzolic cryosol (Rondane, Norway). Deep horizon frozen; lichen vegetation; upper horizon with little humus and decolourised. (b) Alpine ranker, Laguna negra, Colombia (altitude 3800 m). Very sharp boundary between the humic horizon (40 cm) and the very hard quartzite substratum. (c) Landscape in which atlantic rankers occur on outcrops of crystalline rocks, in the region of Santiago de Compostela, Spain (altitude 1300 m). Low vegetation, mostly grasses and heather covering all of the plateau. Forest and brunified soils are restricted to the deep valleys where they are sheltered from the sea winds (in the foreground).

**(a)**

**(b**

**Plate 8   Isohumic soils.** (a) Characteristic vegetation of a tropical isohumic soil (Colombia, upper valley of the Magdalena; altitude 1000 m). Savanna with cactus and thorn bush (acacias), characteristic of the interior valleys sheltered from the humid winds with a rainfall of 700–800 mm and a long dry season. (b) Vegetation characteristic of fersiallitic isohumic soils (reddish chestnut soils) (sheltered coastal zone of the Peloponnesse, Greece). Bushy vegetation dominantly *Olea europoea, Pistacia lentiscus, Juniperus phoenicea*. Degradation by grazing of this vegetation-climax causes the patches of grass to increase in size.

Photo: Dommergues

(a)

Photo: Dommergues

(b)

Photo: Dommergues

(c)

**Plate 9   Vertisols.** (a) Vertisol landscape in the dry season (Dakar, Senegal). The herbaceous vegetation has withered; large contraction cracks have formed at the surface. (b) Vertisol landscape in the wet season (Addis Ababa, Ethiopia). The contraction cracks of the summer are filled with water and gradually close. (c) Mammelated microrelief called 'gilgai' caused by vertic movements (Uruguay).

Photo: Bottner

(a)

Photo: Mériaux

(b)

Photo: Robin

(c)

**Plate 10    Brunified soils.** (a) Acid lessived soil on marls (Vercors). Under mountain beech, in spite of the high calcium reserve at depth, the soil is more acid and more lessived than in the lowland forests: it is a case of climatic development. (b) Glossic lessived soil on Pliocene sandy loams (Longchamp forest, Cote-d'Or). Polycyclic soil; the glossic horizon is a partially rubified palaeosol; under the influence of the ameliorating litter of a hornbeam and oak forest, a recent surface brunification is to be noted. (c) Types of Bt horizon (argillic) on sandy material (reworked Fontainebleau sand, La Tillaie, Fontaine-bleau). A beta-horizon is formed above the limestone lenticle on the right; to the left where this chemical barrier does not occur, the Bt horizon forms superposed *lamellae*.

Photo: Carballas    Photo: Carballas    **(a)**

Photo: Robin    **(b)**

**Plate 11   Podzolised soils.** (a) Comparison of two analogous podzolised soils on different parent materials (Laurentian region, Canada, and Komi region, USSR). On the left, podzol with spodic horizon on a porous sandy moraine; on the right, hydromorphic podzolic soil with a *degraded* argillic horizon on an almost impermeable heavy loam. (b) Indurated humic podzol (Fontainebleau). The Bh horizon formed by calluna is black, indurated and very sinuous; lamellae or humic black bands are present. Note the super-position of several profiles connected with successive aeolian additions.

(a)

Photo: Becker

 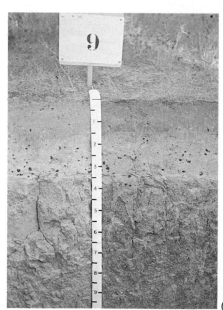

Photo: Carballas

Photo: Blume

(b

**Plate 12  Hydromorphic soils.** (a) Pseudogley landscape: clearing with Molinia in an oak forest (Charmes forest, Meurthe-et-Moselle). Within the clearing, the perched water table of winter comes to the surface and causes the surface segregation of iron by redox processes. (b) Comparison of a podzolic stagnogley (on the left, Siktivkar, USSR) and a planosol profile (on the right, Canhestros, Portugal). In the stagnogley the bleached A2 horizon is subject to the influence of an acid hydromor which causes the elimination of iron and the degradation of clays. In the planosol, which is only slightly humic and slightly acid, this horizon contains black concretions and overlies across an abrupt boundary a clay horizon with a prismatic structure.

(a)

Photo: Duchaufour

(b)

Photo: Duchaufour

(c)

Photo: Duchaufour

**Plate 13 Fersiallitic soils.** (a) Terra rossa on marble (Dyonysos, Greece). Old soil, formed by the surface weathering of a hard rock with few silicate impurities. (b) Acid, impoverished fersiallitic soil on shale colluvium (Kirki, Greece). There is a marked contrast between the decolourised A2 horizon, formed of fine-grained quartz and the argillic Bt horizon with its very red colour; the humus is a xeromoder. (c) Karst landscape on hard limestone (Torqual, Antequera, Spain); altitude 1300 m. This landscape is characteristic where hard limestone outcrops in a climate with markedly contrasted seasons: the residual red clay (terra rossa), transported by erosion, accumulates in depressions and is the parent material for more recent pedogenesis.

(a)

(b

(c)

**Plate 14 Ferruginous and ferrallitic soils.** (a) Ferruginous soil with plinthite (Australia). The plinthite is composed of red patches of haematite, yellow patches of goethite, decolourised areas and small concretions. (b) Ferrallitic soil of the humid Tropics (Zaire). The superposition of two levels on weathering is to be noted: the ochreous upper part is influenced by organic matter while the lower red part is not. (c) Appearance of dense hygrophilic forest (Buonaventura, Pacific coast, Colombia).

Photo: Duchaufour

(a)

Photo: Didier de Saint-Amand

(b)

**Plate 15  Ferruginous and ferrallitic cuirasses.** (a) Cuirasse landscape, formed by induration of plinthite (Meta Valley, Llanos, Colombia). An old surface has been dissected by the erosion and entrenchment of the rivers; the consequent lowering of the water table has caused the plinthite to indurate, which is responsible for the characteristic topography. (b) Old cuirasse in the Niger (Niamey). A very old and very thick plateau cuirasse, dissected by erosion.

(a)

Photo: Dommergues

Photo: Jacquin

(c)

Photo: Faivre

**Plate 16  Salsodic soils and sierozems.** (a) Saline soil with sulphate reduction (Niaya de Pikine, Uraguay). White saline efflorescences at the surface: the black parts of the profile correspond to reduced areas (iron sulphide), rusty patches are areas of sulphide oxidation (ferric hydrates). (b) Solodised solonetz (Czechoslovakia). The slightly acid, bleached horizon at the surface has been impoverished in sodic clays: the dark coverings around the columns are formed of humus and sodic clays. (c) Vegetation characteristic of a grey subdesertic soil (Guajira, Colombia). Tropical zone with low rainfall (less than 500 mm) and very high *PET*: sparse vegetation of xerophilic species: cactus and thorn bush. The salsodic soils cover more or less extensive areas in this climatic zone in which they are an *intrazonal facies*.

labile fraction, similar to that of the brown soils, remains comparatively minor even at the surface (small amount of brown AH, the ratio AF : AH less than 1).

Until fairly recently, the process of polycondensation of aromatic rings during maturation has been emphasised in most work, the fraction at the base of the A1 being mainly composed of grey-AH and transformed humin which forms with iron oxides an unextractable **adsorption complex**.

This strong polycondensation of rings has been demonstrated by several methods, particularly dating by $^{14}C$: the mostly aliphatic labile hydrolysable fraction (6N-HCl) has a low mean age (25 years) and the stabilised aromatic fraction has, in contrast, a very high mean age (Scharpenseel 1972, Campbell et al. 1967, Chakrabarty et al. 1974). Nguyen Kha (1973) has been able to reproduce this polycondensation experimentally (see Ch. 2) and has demonstrated the catalytic role of iron oxides.

But very recently another process of stabilisation in chernozems and vertisols has been discovered (Chernikov & Konchits 1978, Anderson 1979, Bruckert & Kilbertus 1981, Andreux & Correa 1981). This involves the finest-grained fraction of the swelling clays adsorbing compounds of small molecular size, almost totally (but not completely) aliphatic and rich in nitrogen. These compounds, normally biodegradable, are in this way stabilised and protected by the clays.

This type of stabilised organic matter becomes particularly important in vertisols, which are very rich in swelling clays.

**Influence of seasonal variations of pedoclimate on humification.** As has been shown by Ponomarcva (1974) and Ponomarcva and Plotnikova (1968), the important seasonal variations in the amount of carbon in the upper part of the A1 horizon are a good reflection of the fundamental effect of pedoclimatic alternations on the phases of humification; these fluctuations in the amount of carbon are evidently dependent on the labile fraction of the humus. According to these same authors, the production of water-soluble compounds by the rhizosphere of the steppe grasses is important; even in the presence of calcium, these compounds retain a certain degree of mobility and limited movement occurs. On the other hand, according to Aliyev (1966), the period of thawing which soaks the profile with water while it is still frozen at depth results in temporary anaerobic conditions which favour the accumulation and the preservation of these water-soluble compounds which are not subject to biodegradation at this low temperature (Dommergues & Duchaufour 1965).

In contrast, when the soil dries out and biological activity returns, these hydrosoluble compounds are subject to two possible processes. (i) The greater part is mineralised, producing a great amount of $CO_2$ and mineral nitrogen and it is at this time that the active decarbonation of chernozems occurs: *the calcareous material is dissolved in the biocarbonate form because of the high $CO_2$ pressure which occurs in the top of the profile; it reprecipitates within the Cca horizon as a result of a decrease in biological activity and thus of the $CO_2$*

*pressure*. (ii) A very small part is incorporated within the pre-existing humic compounds, generally after having moved only a short distance, *to be followed by aromatic ring condensation and stabilisation by clays in the following dry period*.

Thus the differences in the mean ages of humus determined at the top and the bottom of the profile are easily explained. The soluble labile fractions are stored, very temporarily, at the surface, where for the most part they are so rapidly biodegraded that they do not migrate at all. The small portion that does escape biodegradation moves towards the bottom of the profile where it is incorporated with very stable humic compounds.

The incorporation at depth of chernozem organic matter thus has an essentially biological explanation. The part played by the decomposition *in situ* of the roots of the steppe vegetation and the abundant production of water-soluble compounds which are distributed in depth are in fact complemented by the powerful mechanical action of the soil fauna, which mixes up the horizons and prevents all possibility of clay movement.

In most cases it would seem that these essentially biological processes are sufficient to explain the deep incorporation of organic matter in the chernozems. However, many authors have suggested other explanations which are certainly valid in some cases. In addition, it could be that several processes may operate simultaneously to produce this particular effect. For example, Gerassimov (1973) put forward the idea that a very slow aeolian addition of loess took place in the Holocene (a layer 1–2 cm thick in a century) which could have caused a gradual thickening of the humic A1 horizon. This theory is contested in part by Zolotun (1974) who believed in generally older, large-scale, discontinuous additions of loess. Kovda (1973) suggested a hydromorphic origin for the chernozem humus; in the presence of a relatively high phreatic water table, an anmoor type of humus would have formed; subsequent lowering of the water table and the resulting drying out of the profile would cause the humus to develop into a hydromull and then into a chernozemic mull. Such a process has probably occurred in the chernozems of the southeast of the USSR.

**Development of the mineral fraction: formation of the clay–humic complex.** The amount of clay in chernozems is only moderate (about 25% to 30%) and consists mostly of 2 : 1 clays (illites and montmorillonites) inherited from the loess which are well preserved in an environment rich in calcium and magnesium. However, there would seem to be no doubt that a certain amount of neoformed montmorillonite, that is difficult to determine, is produced as a result of seasonal contrasts in soil humidity (Valkov & Kryshchenkov 1973).

Iron is undoubtedly involved in the formation of the very stable clay–humic complex of the chernozems. However, its effect has not been satisfactorily explained, although it is certainly different from the way in which it acts in brunified soils (binding cation).

The *free* iron, extractable with dithionite-type reagents, is relatively abundant in the profile and can exceed 1%. A certain amount of free alumina can also be present. Thus, weathering is not a negligible phenomenon within the humic horizon, but these hydroxides appear to be of a particular type for they combine, on the one hand, with polymerised humic compounds as adsorption complexes, and, on the other, with neoformed clays that are often iron-rich montmorillonites: these processes are still more marked in the vertisols.

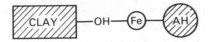

**Brown soil, clay–humic complex**

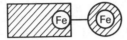

**Chernozem vertisol, clay–humic complex**

**Figure 8.3**  The role of iron in the formation of clay–humic complexes.

Many authors have noted the abundance of iron in the octahedral layers of neoformed mont-morillonites both those of vertisols and chernozems (e.g. Kornblyum 1967, Paquet 1969). In addition, Nguyen Kha (1973) noted the increasing importance of iron complexed by humic acids as their degree of condensation increases and as, simultaneously, the amount extracted decreased (gradual increase in the Fe : C ratio). Hess and Schoen (1964) attribute to this large amount of complexed iron the lack of effectiveness of pyrophosphate in extracting humus from chernozems and vertisols, the loss of functional groups also playing a not insignificant part.

These very polycondensed compounds are very closely bound to the clays. The bonds are stronger than those involved in the clay–humus aggregates of the brown soils; the ferric ions, even those incorporated with organic molecules and minerals, appear to strengthen the organomineral bonds (Scharpenseel 1968, Greenland 1971; Fig. 8.3). It is to be noted that in this kind of aggregate the humic compounds remain around the outside of the clays. However, recent research (Anderson 1979, Bruckert & Kilbertus 1980, Cloos 1981) showed that there is another type of bonding between very fine-grained swelling clays and certain organic compounds; this is so strong that it cannot be broken by any method of separation, either physical or chemical (Arshad & Lowe 1966, Dormaar 1973). This is **humin–clay bonding**, the stabilisation of which appears to result from the interlayer penetration of the clay platelets by small, dominantly aliphatic molecules, rich in nitrogen and COOH groups, which is those that are easily biodegradable, without the protection of clays (Anderson 1979).

## 4  *The main types of isohumic soils with AC profiles: classification*
The main types of soil that are closely related to the climate are given as a **climatic sequence**, which will emphasise the environmental conditions of their development. The problem of their classification will then be considered in which intrinsic physicochemical properties of the profiles will be preferentially used.

From the geographic point of view, the sequence from north to south in the USSR is both the most characteristic and the best known, as it has been studied for a long time; as previously stated, it clearly demonstrates the principle of zonality and it will be taken as a model. Other geographic areas where similar soils occur are not so well known (isohumic tropical soils) or are rather less favourable for their development, so that polycyclic or complex soils are more usually found (regions transitional between Central Europe and Atlantic areas).

**Zonal isohumic soils of the USSR.** The climatically determined sequence from north to south was discussed at the beginning of this chapter (see Fig. 8.1). The following succession of soils occurs in going from the more humid clear forest and steppe forest towards the less and less dense and more and more xerophilic steppe zone:

(a) *Lessived chernozem:* without a Cca horizon; argillic Bt horizon present (Duchaufour 1978: VII$_5$; is a danubian type from further south).

(b) *Humic chernozem (typic):* characteristic of the steppe forest zone, with a maximum accumulation of organic matter both in percentage and depth of penetration. Polyhedral structured (B) horizon is just beginning to form and the Cca horizon is deep and of the very diffuse pseudomycelial form (Duchaufour 1978: VII$_2$).

(c) *Steppe chernozem (orthic, i.e. modal):* a little less humic and A1 not so thick; (B) horizon absent; Cca horizon at shallower depth, with pulverescent concretions forming white patches (Duchaufour 1978: VII$_3$).

(d) *Southern chernozems* (or *chestnut chernozems):* less humic, less deeply decarbonated, transitional to the following type. The *danubian chernozems,* of a warmer climate, can be classed with this type in terms of its profile characteristics.

(e) *Chestnut soils (castanozem):* incompletely decarbonated, a little less rich in organic matter in the surface, which explains the lighter colour: 2.5% to 4% of organic matter in the surface (Duchaufour 1978: VIII$_2$).

(f) *Brown soil of the steppe (burozem):* more xerophilic type, transitional to the sierozem (2% to 2.5% organic matter), more rapid decrease in organic matter with depth; little or no decarbonation at the surface.

Figure 8.2 illustrates the variations in organic matter and carbonates with depth (for sketches of profiles see Fig. 8.7).

**'Intergrade' chernozems (with development partially controlled by the site).** The fundamental isohumic character is retained but certain *secondary* characters appear, which are the result of a reaction between the isohumic process and a physicochemical process typical of another class; only a few examples will be given.

(a) *Calcareous chernozem:* soil developed on materials rich in clay and carbonates, not very permeable and where the phreatic water table has been recently lowered; decarbonation has occurred much more slowly and it remains incomplete (Duchaufour 1978: VII$_4$).

(b) *Solonetzic chernozem:* the complex is partially saturated by $Mg^{2+}$ and $Na^+$, and the mobilisation of sodic clays has started.

(c) *Vertic chernozem:* exceptionally rich in swelling clays, characteristic of low-lying plains and alluvial depressions where the fine clays have been deposited by alluvial or colluvial processes; this is transitional to vertisols (Duchaufour 1978: III$_3$); in fact, it is often also an intergrade to an alluvial soil *sensu stricto* (see Ch. 6).

**Isohumic soils of hot climates (tropical isohumic).** These soils (which certainly exist) have not been studied to any extent so that, as yet, they are poorly known and that is why the name they have been given can only be regarded as being very provisional (Duchaufour 1978: VIII$_1$). The FAO classification relates them to the chestnut soils, although they are in fact poorer in organic matter (1–2%); in addition, this organic matter is more transformed by being richer in grey humic acids, which gives the profile a very dark colour. As in the case of the vertisols, the labile fraction of the organic matter disappears very quickly by mineralisation. In addition, the absorbent complex is often partially saturated by $Mg^{2+}$ and $Na^+$ ions, which indicates a tendency to alkalisation.

**Isohumic soils of central European climates with an atlantic tendency.** Although present-day climates of central Europe arc favourable to brunification and not to the formation of isohumic soils, in certain favourable sites in central and western Europe some partially brunified isohumic soils occur that are in fact polycyclic soils of biological origin and are entirely Postglacial, being characterised by two successive phases of humification: (i) an older phase of isohumic development; and (ii) a more recent phase of brunification (on this subject see Ch. 4). These soils are inherited from an earlier drier phase, with a more or less steppe vegetation; they occur particularly on loess or certain reworked calcareous loams; they have only been preserved in favourable climatic conditions – zones protected by a mountainous barrier (Limagne in France: Bornand *et al.* 1975), certain parts of Saxony, the Danubian plain in Austria, etc. (Duchaufour 1978: II$_2$). These soils contain two types of organic matter: one that is very old and inherited from the isohumic phase of development, almost inert and whose mean age, measured by $^{14}C$, is several thousands of years; and a newly incorporated type of organic matter with a rapid turnover (brunification). As a whole, the amount of humus is less than in chernozems *sensu stricto*.

Scheffer and Meyer (1962) showed the instability of these relic soils that, in a more humid climate, develop rapidly towards lessived brown soils and even hydromorphic lessived soils.

**Classification of isohumic soils with an AC profile.** The American classification uses pedoclimatic characters to differentiate the main subdivisions of the mollisols – boroll (cold pedoclimate), ustoll (warm pedoclimate), xeroll (dry pedoclimate) – but, as the discussion in Chapter 5 showed, this concept is very difficult to apply in practice. It appears that it would be better to use the morphological or biochemical characters of the profile in so far as they are a true reflection of the environmental conditions, particularly those of the pedoclimate.

The different kinds of profile with climatically determined development that have been distinguished can be easily classified, in terms of their colour, their content of organic matter, the degree of decarbonation, and the nature and depth of the calcic horizon.

The lessived chernozem with A1BtC profile transitional to a lessived profile should be grouped with the next subgroup and related to the brunizems.

As for the tropical isohumic soils, it should be possible when they are better known for them to be classified with the chestnut soils where they will form a special group.

As far as the intergrade isohumic soils are concerned (vertic or solonetzic chernozems, etc.), eventually they will be placed in a special group, provided that the classification is accompanied by developmental sequences that show to which other class each soil is related.

## III   Brunified isohumic soils; brunizems or phaeozems

### 1   General characters

This subclass can be considered as being transitional between the isohumic soils *sensu stricto* (with AC profiles) and the brunified soils where the humic A1 horizon is less developed compared to the (B) or B horizon.

This intermediate nature involves the environmental conditions of pedogenesis as well as profile characters. With regard to the environmental conditions, the **brunizems** of the old American nomenclature or the **phaeozems** of the FAO nomenclature are characteristic of the climatic zones that are more humid or less cold than that of the chernozems (because of this they are classed as **udolls** in the USA classification and not, as the chernozems are, as **borolls**). These soils are well developed under characteristic prairie vegetation in the area south of the Great Lakes in North America which, when cultivated, forms the cornbelt of the USA. Prairie vegetation is less xerophilic than that of the steppe, for there are scattered patches of trees. Originally it had the appearance of a meadow-woodland; but fire, which before cultivation occurred periodically, favoured the herbaceous vegetation at the expense of the trees. Over long time periods, fire has caused an alternation between prairie and forest vegetation.

Similar vegetation is characteristic of the brunizems of hot climates, such as in the pampas of Uruguay and Argentina.

### 2   Characters of brunizem profiles

As has already been said, this profile has mixed characteristics: the A1 horizon is of the **mollic** type, resembling in its dark colour and thickness the isohumic A1 horizon of the chernozem; the (B) and Bt horizon with marked polyhedral structure resembles the weathered (B) horizon or argillic Bt horizon in the lessived brown soils which are characteristic of the dense deciduous forest, with slightly more humid climates, such as occur in the St Lawrence region of northeastern America.

The A1 horizon is clearly different in several ways from the A1 of chernozems, even that of the transitional chernozems (for example, the lessived chernozem): it is less thick, and organic matter decreases more rapidly with depth. In addition, the absorbent complex is slightly unsaturated (pH slightly acid, 5.5–6). Finally, the amount of organic matter stabilised by maturation and bonding with montmorillonites, compared to that which has a rapid turnover, is less than that which occurs in the chernozems.

Generally, the clays are a mixture of illites and montmorillonites, and it is the finest of them (i.e. the montmorillonites) that are the most mobile and form argillans; they often carry with them mechanically a part of the humus to which they are bonded, which forms grey surfaces, very characteristic of the Bt horizon of brunizems. This process of pervection is similar to that which characterises the grey forest soils (see Ch. 9).

It should be remembered that, in contrast to the soils of the other subclasses, the calcic horizons are generally absent (or at a great depth); this is a result of the greater humidity. *As the climatically determined drainage increases, $Ca(HCO_3)_2$ is removed from the profile and not precipitated in the C horizon* (this has already been discussed in Ch. 3).

## 3 Main types

The **lessived brunizems**, with an argillic B horizon, are the most common and can be considered as the climatically determined type. Two groups can be differentiated: one characteristic of temperate climates (lessived brunizem of the American prairies, or **argiudoll**, in the USA classification) which corresponds to the type described above (Duchaufour 1978: $VII_1$); the other a tropical or subtropical brunizem, a soil of the pampas (the plains in the north of Argentina and Uruguay) which also occurs in certain dry valleys of Colombia. This **subtropical brunizem** is darker than the temperate type and is explicable in terms of a greater amount of humus maturation resulting from more marked seasonal climatic contrasts. The whole of the profile, even the Bt horizon with polyhedral structure, is black or dark grey (Lopez-Taborda 1967). The types that are richest in clay have a vertic character which makes them transitional to the vertisols.

**Non-lessived brunizems** have a simple weathered (B) horizon in which clay pervection at best is very slight, and they can be considered as youthful, incompletely developed soils. They occur on loess on gentle slopes where development has been slowed down, or on calcareous–clayey materials, rich in carbonates, where decarbonation occurs slowly. The C horizon, which effervesces with HCl, is generally at a shallower depth than in the lessived type (**hapludoll** of the USA classification).

## 4 Dynamics and agronomic utilisation

Brunizems are very fertile soils used in North America on a large scale for the cultivation of maize and soya beans (corn belt). From the chemical viewpoint, they are almost the same as the chernozems, but their more favourable hydrological regime and the possibilities of water storage in the Bt horizon make crops better able to resist summer droughts than those on chernozems, particularly the more southerly ones. In addition, they are indurated to a lesser degree in dry periods and are thus easier to work than vertisols. The tropical or subtropical brunizems are also used for the highly intensive cultivation of cereals (wheat and maize).

The brunizems of North America are frequently invaded by deciduous forest on their eastern and northern boundaries; then they develop rapidly, by increase in their acidity (lowering the amount of humus) and more intense pervection of clay, towards lessived brown soils (*grey–brown podzolic soil*).

## IV Fersiallitic isohumic soils

### 1 General characters

Here again it is a question of a transitional subclass between isohumic soils on the one hand and fersiallitic soils on the other. As for the preceding subclass, the **intergrade** (that is to say, intermediate) character is to be seen in all aspects: climatic, vegetational and profile morphology; climate is hot (tropical or subtropical) and semi-arid; *the rainfall is insufficient and the dry season too long to allow a dense forest to develop (even of the schlerophyll type), like the Mediterranean holly oak forests which are characteristic of the fersiallitic soils* sensu stricto.

The climatically determined vegetation is either thorn bush or cactus savanna, with xerophilic grasses (tropical zone) or a shrub formation of the maquis or garrigue type, as in the case of the **reddish chestnut soils** of the Mediterranean region. The frequent fires of the region have generally favoured the spread of xerophilic grasses at the expense of the shrubs, so that frequently a secondary steppe has replaced the primary vegetation.

It must be emphasised that these climatically determined shrub formations are not to be confused with secondary formations of the same type which have gradually established themselves in more humid areas after forest clearing by man. Then, it is a case of **degradation** which has led to the development of a degraded forest profile very different from that which is being dealt with here.

From the geographic point of view, these soils are particularly well represented on the African continent, where they occur in two climatic zones that are both semi-arid: one to the south of the Sahara (**brown and reddish brown sub-arid soils**, described by Maignien 1962); the other to the north, in the drier regions of North Africa (**reddish chestnut soils**). This second type, which has been studied in detail in North Africa (Boulaine 1957, Ruellan 1971), will be discussed. In addition, Gerassimov, who had described (under the name of **cinnamonic soils**) the reddish chestnut soils of the south of the USSR (dry regions of Georgia), was the first, in 1956, to point out the identity of these soils with those of North Africa; in both cases the profile is an isohumic–fersiallitic intergrade, and the vegetation, studied in detail in Georgia by Nakaidze (1965) and Zhukov (1975), is identical: it is a semi-woody or shrub formation interspersed by large patches of xerophilic grassland. The reddish chestnut soils of the Mediterranean region also have vegetation of this type (called **oleo-lenticetum**) in which the Phoenician juniper tree is important. Recently, Zonn and Kochubey (1978) have established the differences between fersiallitic and reddish chestnut soils in Libya.

**The profiles of reddish chestnut soils.** They have clearly intermediate characters: *the thick, dark-coloured humic horizon and the constant occurrence of a well developed calcic horizon at little depth are the isohumic characters; on the other hand, the abundance of free iron unincorporated in the humic complex, the rubification of the silicate materials, and the abundance of 2 : 1 clays are fersiallitic characters*. It is precisely because of the superposition of two colours – the dark grey of the organic matter and the red of the iron oxides – that they are called **marron** in French and **cinnamonic soils** in Russian (Gerassimov 1956). The calcic horizon is subject to phases of wetting and drying that are of greater intensity than occur in the more northerly isohumic soils, which results in it being indurated or even crystalline and called a **calcareous crust** or **petrocalcic** horizon, and is particularly important in the Mediterranean region.

Although the similarities between the reddish chestnut soils of Georgia and those of the Mediterranean periphery are considerable, there are, however, certain differences which, even though they are not too well understood, need to be considered. The reddish chestnut soils of North Africa are clearly older and have been subject to successive phases of climatic development so that they are generally *polycyclic*; in addition, they occupy particular sites in protected positions at the bases of eroded mountains. In these circumstances they are often developed on depositional material (alluvial or colluvial) which increases their complexity yet again.

## 2  Profile types of reddish chestnut soils: classification of calcareous crusts

In 1957, Boulaine described two basic types of reddish chestnut soils: one that was *little differentiated* with *diffuse* accumulation of carbonates, the other *differentiated* with *concentrated* carbonate accumulation (Duchaufour 1978: $VIII_3$ & $VIII_4$). A summary of their main characteristics will be given here, paying particular attention to the type of calcareous crust (Fig. 8.4).

**Morphology and physicochemical characters of the profiles.** Both profiles are, in fact, very similar. It is only in the decarbonation of the upper horizons and in the kind of calcareous accumulation that they differ. The A1 horizon, 30–45 cm thick, has a dark red colour which reflects the coexistence of organic matter, incorporated to depth (isohumism: 2–3% organic matter), and of a rubified silicate material. Immediately below there is either a horizon of calcareous accumulation with little structure ((B)Cca) or a (B) transition horizon, markedly red, very clayey and with a polyhedral structure.

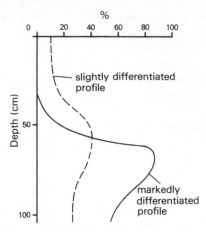

**Figure 8.4** Distribution in depth of calcium carbonate in two types of reddish chestnut soil (after Ruellan 1971).

The two types of profile are distinguished by the distribution of $CaCO_3$ in depth. In the first type, decarbonation is incomplete in the A horizon, and accumulation shows a gradual build-up in the Cca horizon; in contrast, in the second type decarbonation is complete in the A horizon, and the Cca horizon has a very high amount of carbonate across a very sharp boundary (from 0% to 70–90% carbonate). This is reflected morphologically in the presence of a discontinuous and incompletely indurated crust in the first case, and a much more massive and often indurated crust in the second (see Fig. 8.7).

Even in the type with a decarbonated A horizon, the profile remains saturated in $Ca^{2+}$ and $Mg^{2+}$ cations: the abundance of $Mg^{2+}$ ions freed by the precipitation of the calcium carbonate frequently causes the profile to be alkaline with a pH of up to 9 (Ruellan 1971).

**Types of calcareous crust.** Several authors (e.g. Durand 1959, Ruellan 1971) have suggested classifications of the calcareous crusts of the reddish chestnut soils of North Africa. These classifications are more or less the same in that they use as a basis two factors that are, as will be seen, a function of the age and the mode of formation of the crusts: (i) continuity or discontinuity of the accumulative materials within the profile; (ii) degree of crystallisation and induration of the carbonates; the very indurated types, often resulting from processes of resolution and reprecipitation in the crystalline form, being in crusts that were initially less indurated.

Thus the following are differentiated:

(a) *Pulverescent patches:* more or less discontinuous; pasty when moist, pulverescent when dry.
(b) *Nodules:* accumulation is discontinuous and forms spherical or elongated masses that are more or less indurated.

(c) *Nodular crusting:* massive formation resulting from the cementation of nodules by a chalky and generally less indurated material.

(d) *Layered crusting:* massive formation, homogeneous, chalky with horizontal more indurated *layers* resulting from the preferential circulation of water within the crust.

(e) *Zonal crust:* more or less layered, very indurated, highly crystalline at the surface; frequently resulting from processes of surface resolution and crystallisation of the upper part of the crust.

(f) *Flaggy calcareous material:* thick, indurated, massive; generally very old.

## 3   Development of reddish chestnut soils (Fig. 8.5)

Even though certain reddish chestnut soils (such as those of Georgia) are monocyclic, others (such as those of North Africa or generally around the Mediterranean) are polycyclic and have had a complicated history with multiple phases of development.

As already stated, these soils are generally located in relatively dry piedmont zones of strongly eroded limestone mountains which have a more humid climate. *The result of this is that not only has the rubified material been subject to several phases of development but it has been transported from where weathering started (mountains) to the place where it has been finally subject to pedological development.*

It has been shown clearly by Ruellan (1971) that, once the rubified silicate materials are deposited, two additional phases of pedogenesis occur successively.

(a) *Calcification* by lateral additions (Ruellan 1971): the greater part of the carbonates of the crust coming from the solution of upslope limestones by

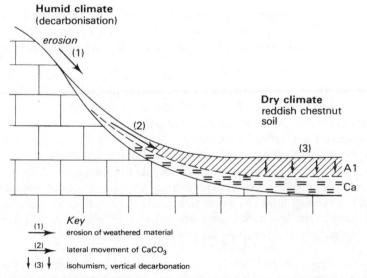

**Figure 8.5** Phases of development of reddish chestnut soils (piedmont zones).

running water and its precipitation down slope as a result of intense evaporation in zones with a locally dry climate.

(b) *Incorporation to depth of organic matter* by the vegetation (formation of isohumic profiles) accompanied, in some cases, by a redistribution of the carbonate previously accumulated by lateral transport, but this time in a vertical sense (Boulaine 1957).

The three phases will be examined in a more detailed manner.

**Formation and deposition of rubified material.** This material is generally inherited by colluviation from the red soils of mountains (terra rossa), formed in a more humid climate and transported by water into the depressions with a drier climate, where the reddish chestnut soils occur. Hubschmann (1967) and Ruellan (1971) showed by a detailed study of the heavy minerals, and also of the clays (particularly illites), that the rubified materials of the reddish chestnut soils were identical with those of the terra rossa, of relatively recent age (Tensiftian), occurring in the neighbouring limestone mountains. But terra rossa can only form by active decarbonation, in regions where the rainfall is clearly in excess of 1 m and thus much more humid than the zones in which reddish chestnut soils generally occur (Lamouroux 1965).

Nevertheless, it seems that during the formation of the embryonic (B) horizon with a polyhedral structure, which has occurred *in situ* in the course of formation of reddish chestnut soils, certain secondary processes have taken place. Thus there is a certain amount of *in-situ* neoformation of montmorillonites which, when added to the inherited illites, cause the polyhedral structural units to have shiny faces: these are **stress cutans** which mark the beginning of the development of vertic characters, a very likely phenomenon in such an environment (Boulaine 1957, Hubschmann 1967).

**Incorporation of carbonates by lateral transport.** This phenomenon has been investigated by several authors (Gile *et al*. 1966, Cointepas 1967, Ruellan 1971). Lateral transport of carbonates can, depending on circumstances, occur in two ways:

(a) By solution in the bicarbonate state, then reprecipitation in the dry zone by rapid evaporation of the solutions; first of all there is a formation of patches or pulverescent masses that indurate and develop into nodules, which are then welded together by the filling up of the spaces by diffuse carbonates, which gradually develops into a massive crust.

(b) By calcareous mudflows, on less indurated materials, as a result of their hydration during rainy periods; these flows stop at the bottom of the slope and consolidate as a crust when the mass congeals; the silicates that are moved at the same time are subject to a particular type of clay neoformation and attapulgites are formed (Paquet 1969).

Ruellan (1976) and Millot (1979) consider that very high carbonate concentrations can only be explained in terms of **epigenesis**, i.e. replacement of

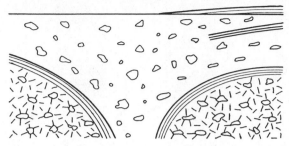

**Figure 8.6** An example of zonal and nodular crust formation in a steppe climate (after Gile *et al.* 1966).

pre-existing material (fine-grained quartz and certain silicates that are easily mobilised) by a carbonate; this explains how certain crusts are able to contain up to 90% $CaCO_3$.

In addition, the local phenomenon of resolution and recrystallisation, which occurs secondarily in the mass or more often at the surface of certain massive crusts, can lead to a more *intense induration* causing the formation of **zonal** or **lamellar crusts**, sometimes even of the very indurated **flaggy materials**. The development of total impermeability that results from this process occurs rapidly when the zone of induration is a few centimetres (or at most a few decimetres) thick. Muller (1954) attributes these surface-indurated crusts to man's destruction of the natural vegetation. First of all erosion occurred, then secondary crystallisation characteristic of the reddish chestnut soils of certain regions of Algeria. It is a case of true degradation.

Figure 8.6, taken from Gile *et al.* (1966), clearly shows the mode of formation of the two superposed nodular crusts of different age: a funnel of dissolution, covered by a zonal crust, cuts into the surface of an old crust which forms the general substratum. This funnel was then filled by more recent material which on drying caused the formation of a nodular crust, younger and less indurated than the preceding one. Indurated lamellar structures at the surface and the interior of this crust are caused by zones of preferential water circulation. Age studies by [14]C generally confirm that the massive crusting is increasingly younger from bottom to top (Buol 1965); while the deep nodular crusts have an apparent age of several tens of thousands of years, the less indurated, near surface crusts never exceed 2000 to 3000 years.

**Formation of isohumic horizon and vertical redistribution of carbonate** (Fig. 8.7). This is the terminal phase in the development of reddish chestnut soils; the formation of the isohumic horizon must be recent for it can occur only after the colonisation of the parent material by the vegetation from which it is formed. In fact, the isohumic appearance of the reddish chestnut soils is less characteristic than in the case of the soils of the preceding subclass, so that it is possible for some authors to argue about their isohumic nature. Even though incorporation of organic matter is deep, the decrease in its percentage is fairly rapid, so that the A1 horizon never exceeds 30–40 cm in thickness. As shown by Ruellan (1971), this appears to result from the mixed nature of the vegetation: xerophilic woody bushes alternating both in space and time (as a result

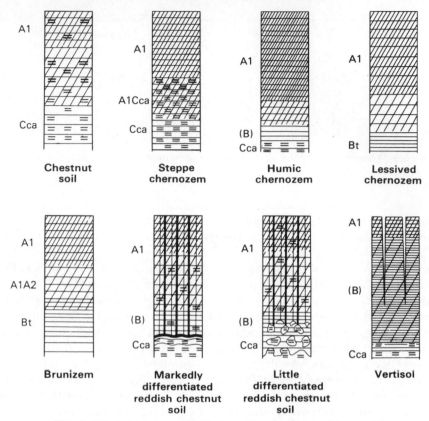

**Figure 8.7**　Isohumic and vertisol profiles (see general key, p. ix).

of fire) with expanses of grass; litter derived from the woody vegetation gives rise to a humic horizon at the surface which is generally rapidly eroded; it is the rhizosphere effect of the grasses that is responsible for the isohumic character which, given the particular conditions under which the A1 has been formed, is only slightly developed. In addition, except in a vertic environment, the organic matter of the reddish chestnut soils is less strongly **polycondensed** than that of the true isohumic soils, such as the chernozems (Ouezzani 1967). This appears to result from the fact that the humid phases do not cause such marked anaerobic conditions.

### 4　Main types of fersiallitic isohumic soils: utilisation

The incomplete state of knowledge at the present time makes it difficult to give a definitive classification of groups in this subclass. In these circumstances, only general guidelines will be attempted.

First of all, two major climatically determined units can be differentiated: (i) the sub-arid brown and reddish-brown soils of dry tropical regions; and (ii) the reddish-chestnut soils of subtropical regions or warm temperate climates.

Among the latter, it will probably be necessary to distinguish those with a monocyclic development from those with a polycyclic and complex development, such as the circum-mediterranean reddish chestnut soils that have been described.

Finally, as in the case of the other subclasses, it is necessary to differentiate intergrade groups. Among these, the most important are those that have a vertic development: these are very dark reddish-chestnut soils, formed in depressions, where the pedoclimatic contrasts are intensified by the topographic conditions and defective local drainage. Compared to the modal reddish-chestnut soils, the vertic reddish-chestnut soils are characterised on the one hand by strong neoformation of swelling clays and on the other by more intense processes of maturation and polycondensation of the humus, which gives the very dark colour (Duchaufour 1978: $VIII_4$).

As far as their utilisation is concerned, the reddish-chestnut soils are generally of low productivity, particularly as a result of the very dry climate which controls their formation (except where irrigation is possible) and their low water-storage capacity, resulting from their shallowness and the presence of the calcareous crust which is almost impenetrable to roots.

Vertic reddish-chestnut soils (often called 'tirs' in Morocco), when they are sufficiently deep and rich enough in organic matter, form the best of these soils for cultivation. In those areas with sufficient rainfall they are able to carry a good cereal crop.

## V   Soils of arid regions

As already stated, these soils formed in conditions which are almost constantly dry are not, strictly speaking, isohumic soils: organic matter is present in too small an amount. In addition, the climate is too constantly dry for the previously discussed process of maturation to occur. Most classifications differentiate them as a particular class: **aridisols** in the American classification, **xerosols** in the FAO classification. Provisionally, however, it would seem to be more logical to include this subclass within the isohumic class because of the environmental and geographical relationships. These soils replace, in the more arid steppe zones, either the brown steppe soils or the reddish-chestnut soils (Singer & Amiel 1974), between which there are several intermediate forms: as in the case of the brown soils, the little organic matter that there is (1–2%), is deeply incorporated. Finally, a calcic horizon is always present, often forming an undoubted indurated crust as in the reddish chestnut soils (the crust frequently contains gypsum in association with the carbonate).

**Sierozems** or **subdesertic grey soils** have little colour, as they contain little free iron and very small amounts of organic matter (generally no more than 1%), because of the slight amount of climatically determined weathering and the very weak development of vegetation.

Very frequently, profiles occur that are transitional to brown steppe soils when their organic matter is a little higher than normal (2–3%). Lobova

(1960) has referred to these soils as **sierozemic brown soils** (Duchaufour 1978: $XIX_6$). These subdesertic soils also have numerous intergrades with the salsodic soils. The dryness of the climate and the absence of climatically determined drainage frequently lead to the retention of the alkali ions that are more mobile than calcium in a humid climate, such as $Mg^{2+}$ and more particularly $Na^+$. This often causes a moderate degree of alkalisation of the profile.

**Improvement of sierozems.** Sierozems are fertile soils in so far as they can be irrigated, but generally they are used for extensive cattle rearing. However, it is necessary to say that in certain Mediterranean countries (e.g. Tunisia) xerosols are planted with commercial olive groves, even with an annual rainfall less than 200 mm, by the practice of dry farming, i.e. wide-spaced planting and surface cultivation to eliminate all competition by the steppe vegetation for the absorption of water.

# B  VERTISOLS

## I  General characters

These dark-coloured soils, rich in swelling clays, were differentiated for the first time, as a separate order, in the American classification of 1960; since then this idea has been adopted in most classifications.

## 1  Profile characteristics

Vertisols have one basic characteristic in common with the chernozems: the presence of clay–humic aggregates of a particular type, formed of a complex between *a swelling clay and a dark humic compound* in which the two constituents are so strongly bound to one another that it is difficult to dissociate them by normal reagents.

However, as emphasised at the beginning of this chapter, vertisols differ from chernozems in several important respects which amply justifies their being placed in a separate group. These characters can be summarised as follows:

(a) Abundance of neoformed or inherited swelling clays.
(b) Rapid turnover of the more labile fractions of the organic matter; the dark-coloured *stable* fraction, which in addition is not too abundant, being the only persistent factor in the profile.
(c) Mechanical mixing of all of the horizons by *vertic movements* resulting from seasonal variations of volume of the clay, causing an almost complete homogenisation of the profile to a depth of 60–80 cm.
(d) Generally very coarse structural elements, prisms separated in dry periods by large *contraction cracks*; abundance of *slickensides* that are evidence of vertic movement.

## 2  Environmental conditions

Vertisols are a typically *intrazonal* class of soils, in the sense intended by the first Russian classifiers. They occur in those climatic zones with strongly contrasted seasonal climates, one of which is *markedly dry* (Dudal 1967) and generally hotter than that which characterises the chernozems; for example, in Europe and in Africa they are well represented in the Mediterranean, tropical and Danubian regions (cf. the **smonitza** of the Balkans, Filipovski & Ciric 1969). But, for their full development they require special site conditions, particularly those of topography and parent material; *it is a case of sites with poor internal or external local drainage that, in pedoclimatic terms, intensify the seasonal contrasts of the general climate*. This generally occurs in depressions that have been choked up by a clayey material rich in calcium and magnesium, cations that are absolutely necessary (as was seen in Ch. 1) in the neoformation of swelling clays rich in silica (montmorillonites). In these particular circumstances, depending on the season, the profile rapidly changes from being markedly hydromorphic with complete saturation of the capillary pores to being very strongly desiccated in all horizons. All the morphological, physical, biochemical and mineralogical properties of the vertisols are explicable in terms of the seasonal contrasts of pedoclimate which are much more intense than for the chernozems.

## 3  Basis of classification

It would seem that the various classificatory schemes for vertisols are not completely satisfactory. The French classification differentiates one subclass of vertisols in which external drainage is possible from another in which there is no external drainage, the second being more typical than the first. In this book this idea will be retained; however, the basic criteria used will not be the more or less imperfect drainage conditions, but the profile characters that are a reflection of this state. The *vertisols sensu stricto*, with a very dark colour and exceptionally rich in swelling clay, are controlled by site conditions which increase the pedoclimatic contrasts to a maximum (drainage practically nil). The *vertic soils* that are *coloured* or **chromic** contain, besides montmorillonites, other types of clay and thus have *intergrade* characters towards other classes of soils; they occur at sites that allow moderate drainage, so decreasing somewhat the pedoclimatic contrasts.

## II  Vertisol profile: environment, morphology, properties

In addition to the climatic controls (occurrence of a dry season), those of parent material and topography, as already stated, are also very important.

(a) *Parent material rich in calcium and magnesium* either of a sedimentary nature (for example marls or fine-grained calcareous alluvium) or crystalline or eruptive basic rocks (such as basalt or dolerite) that free many $Ca^{2+}$ and $Mg^{2+}$ ions during weathering; in the first material the swelling clays (or semi-swelling if interstratified clays are involved) are inherited and preserved under the conditions of the environment; for the second type of material, in contrast, the swelling clays are neoformed as a result of complete and rapid chemical weathering, which is favoured in a generally hot climate with seasonal contrasts (Nguyen Kha 1973).

(b) *Topographic conditions* that accentuate the pedoclimatic contrasts by almost complete prevention of local drainage which occurs in basins, low-lying plains and closed depressions, that are generally choked up by clays resulting from weathering, or lateral additions from neighbouring slopes.

*Within soil catenas, vertisols are always situated on lower slopes, their characters becoming more and more marked and more typical as the zones that are less and less well drained are approached down slope.* An example of a typical catena in a tropical climate on crystalline rocks in Chad was given in Chapter 4 (Fig. 4.10, Bocquier 1973); Fig. 8.8 shows another example on marls in a mediterranean climate.

In the first example, the elements transported towards the lower part of the catena are soluble (silica, calcium and magnesium) and are involved in neoformation of clay, either as constituents or as catalysts. In the second example it is the finest clays, inherited from the material up slope, which accumulate and choke up the depression at the foot of the slope (Kovar *et al*. 1976).

## 1  Morphology and physicochemical characters (Duchaufour 1978: VII₆)

The most outstanding characteristic of the vertisol profile is its uniformity, resulting from the constant mixing of horizons by vertic movements; there is little differentiation to a depth of 80 cm to 1 m, with only slight variations in the generally very coarse structural elements being apparent. In these circumstances, the classical idea of horizons loses its significance; in particular, the (B) horizon with its well developed prismatic structure is purely structural and

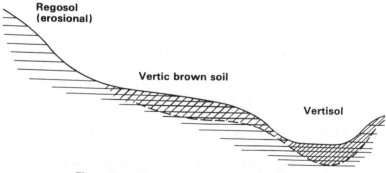

**Figure 8.8**  Vertisols, catena on marls (Spain).

only differs from the A1 in biochemical terms by the fact that its organic matter is sometimes slightly less.

In more favourable circumstances, when the slightly transformed organic matter still occurs in reasonable amounts in the surface, an A1 horizon is differentiated with a structure that is less coarse than the rest of the profile. In dry periods, angular crumbs are formed by contraction which form a *self-mulching* protection for the lower part of the profile against too much desiccation. Such an horizon, which is typical of vertisols often called **grumosols**, does not always occur and it disappears particularly where erosion is common.

In general, the typical structure of the vertisols is only to be seen in the dry season: the (B) horizon has fairly large, vertical contraction cracks (sometimes of 1–5 cm) delimiting large prisms; lower down, this horizon is replaced by a (B)C horizon of more irregular structure, where **slickensides** – those large oblique, polished surfaces caused by vertic movements – are dominant (Duchaufour 1978: $XX_3$). The frequent occurrence at the base of this horizon of carbonate concretions resulting from the precipitation of bicarbonates transported from the surface, or obliquely down slope, is to be noted. These calcareous concretions are formed both on crystalline and volcanic rocks as well as on sedimentary materials.

Despite the small amount of organic matter (about 2% distributed uniformly throughout the profile), the colour is generally dark, often black, which is related to the quality of this very transformed organic matter; the C : N ratio is of the order of 14 to 15 and thus higher than in the chernozems. The swelling clays (montmorillonites), often mixed with small amounts of illites and interstratified clays, account for some 40–60% of the total mass: they are responsible for giving the profile the greater part of its physical and chemical properties. In particular, their expansion and contraction cause vertic movements and the frequent occurrence of a mammelated surface relief known as **gilgai**. The very high cation exchange capacity of 40–80 mEq/100 g is mostly made up of *permanent* charges (differing from the andosols for example), resulting almost entirely from the abundance of swelling clays.

As far as iron oxides are concerned, although total iron is fairly abundant and it can reach 10% on certain very basic rocks (basalts), the ratio of free iron to total iron remains relatively low and the free iron itself can only be extracted by fairly strong reagents. As will be seen in the last section of this chapter, this index of weathering based on iron differs considerably, depending on the type of vertisol.

## 2  Agronomic properties: influence on plant nutrition

Vertisols are generally fertile and they count among the most fertile soils of tropical regions. They are frequently used for cotton growing and have been referred to as **black cotton soils**. However, even though they are highly fertile in chemical terms, the physical properties are sometimes unfavourable as a result of the large amount of swelling clays; the field capacity is certainly high, but as the wilting point is equally very high, the amount of available water as a reserve in the soil remains limited. In addition, vertic movements and deep cracking in dry periods have very unfavourable effects: breaking of absorbent roots, deep desiccation of the profile, compaction and increase in apparent density; the soil is difficult to work for long periods of time either as a result of waterlogging or to a marked induration in the dry season.

The self-mulching vertisols (i.e. with surface crumb structure (grumosols)) are from this point of view very much better than the surface-eroded and degraded vertisols; the self-mulching decreases the effect of contraction cracks and allows the maintenance of a certain amount of water in the structural (B) horizon. Certain artificial mulches, such as straw on the surface, are an effective supplement to the natural self-mulching, if this is considered insufficient.

# III  Development of vertisols

The fundamental characteristic of the profile is the presence of a dark-coloured clay–humic complex with special properties. After discussing humification, the formation of this complex will be briefly dealt with and also the way in which the profile is homogenised over such a great depth.

## 1  Humification

The humification of vertisols is similar in some ways, such as the high level of maturation of the humic compounds, to that of the chernozems, but it differs by the dominance, which is almost total, of dark-coloured humic compounds stabilised by iron oxides and clays. By classical methods of extraction (Duchaufour & Jacquin 1963, 1966), 80% of the humic acids extracted are of the *grey* type and the quantity of fulvic acids is very small; the greater part of the humic compounds remain non-extractable and form the **transformed humin** (Zonn 1967) which occurs in two forms, as has been seen with regard to the chernozems: one very polycondensed and bound to the iron, the other (aliphatic nitrogen compounds) stabilised by the fine-grained clays.

This indicates that the labile fraction of the fresh organic matter is subject to an even more rapid turnover than in the case of the chernozems, which seems to be related to the generally much hotter climate. It disappears almost entirely during the annual cycle, except in the surface of certain of the grumosolic vertisols. It is when the profile is drained again, after a period of saturation, that the slightly transformed organic matter is subject to an almost total mineralisation (high $CO_2$ and mineral N production), before biological activity is lowered again as a result of profile desiccation. Only a very small fraction of the phenolic precursors polymerise and stabilise. In Chapter 2, discussion of the experiments of Nguyen Kha (1973) showed the extraordinary stability of the humic acid and the humin of vertisols during seasonal alternations of wetting and drying. This stability contrasts with the very great tendency to biodegradation of fresh organic matter and the newly formed humic compounds.

## 2    The formation of the clay–humic complex

Even though the seasonal variations in profile moisture status only lead to the formation of a small quantity of stable humic compounds (2–3% – distributed to a considerable depth, it is true, which in part explains this low percentage), in contrast these variations favour clay formation on a much more considerable scale, for, as already stated, vertisols contain 40–70% of often very fine-grained swelling clays (montmorillonites). On sedimentary rocks (marls) they are inherited and are also formed by the physical subdivision of certain coarser clays, such as chlorites (Nguyan Kha 1973). On crystalline or volcanic rocks, *neoformation* is involved or *aggradation* by the addition of silica to the platelets of vermiculite (Seddoh & Pedro 1974, Kounetsron 1976). These two processes are favoured by the topographic position at the bottom of slopes; Paquet (1969) and Bocquier (1973) have demonstrated the upslope encroachment of neoformed montmorillonites around tropical inselbergs (see Ch. 4, p. 132).

While vertisols on sedimentary or crystalline rocks have similar properties, it is evident that the amount of weathering is very different in the two profiles; in these terms, the first can be considered as being much less developed than the second: the index of weathering of iron shows this difference very well. If these two kinds of profile, for example a vertisol on marls (Duchaufour 1978: VII$_6$) and a vertic eutrophic brown soil on crystalline schists (Duchaufour 1978: XVII$_2$), are compared, it can be seen that the ratio of free iron to total iron is just over 3% in the first and 50% for the second; even so, this second figure is low compared to that of the neighbouring climatically controlled ferruginous soils. A great part of the free iron is not extracted, even by powerful reagents of the dithionite type. *The iron freed by weathering is almost immediately reintegrated into the large molecules, both organic and mineral, that are rapidly formed in pedogenesis*: this process has been demonstrated by Hess and Schoen (1964) and Nguyen Kha (1973) as far as the iron complexed by the organic matter is concerned. Other authors have emphasised the *trapping* of iron within the octahedral layers of the neoformed clays (Kornblyum 1967, Paquet 1969). Zonn (1967) estimates the iron in the clays of the Indian **regur** to be 14%. As already stated with regard to the chernozems, this iron appears to play an undeniable part in the strong physico-chemical bonding between the humus and the clays (see Fig. 8.2).

## 3    Profile homogenisation by vertic movements

Montmorillonites can swell considerably when they pass from the dry to the wet state; the simple model of clay platelets hydrating is not sufficient to explain the great variations in volume of the soil mass which have been demonstrated. Tessier and Pedro (1976) and Tessier (1978) recently showed by stereoscan studies that the swelling clay molecules have a tendency to form a very flexible network, by edge to edge contact, instead of face to face as in the case of the illites: in the dry state, this network is squashed flat, while in the humid state, it swells like a sponge.

The homogenisation of the profile is the result of these large variations in volume of the clay mass. It has been proved particularly well by $^{14}$C dating (Scharpenseel 1972, Blackburn & Scharpenseel 1973) that the age of the organic matter was sensibly constant to a depth of 60–80 cm, depending on particular circumstances; below that, in contrast, the mean age increases very rapidly. The mechanism of homogenisation has been worked out by the application of various methods: field observations, study of the seasonal changes in the gilgai microrelief, and variations in the age of organic matter on mounds and depressions and at different depths (Blackburn & Scharpenseel 1973). Figure 8.9 shows the successive phases of profile homogenisation by developments from a very large fissure in the dry season, where the bottom is filled up by material coming from the sides of the fissure, everything then swelling in the humid period which causes a bulging of the surface that is then subject to erosion.

The consequences of profile homogenisation are many: the strong pressures that result help to reinforce the bonds between the clay and the humus; the swelling clay in this way catalyses the stabilisation of the humic compounds more rapidly than is the case with the chernozems (mean age not as great); *vertisol formation is rapid, which explains why certain very typical vertisols are in fact little developed and little weathered* (Parsons et al. 1973). The whole of the montmorillonite–humus complex has a black colour that has been reproduced experimentally by incubation under an artificial, strongly contrasted climate of a mixture of montmorillonites and polyphenols (Hess & Schoen 1964); but the quantity of organic matter is insufficient to give the soil a crumb structure, except sometimes in the surface horizon (self-mulching).

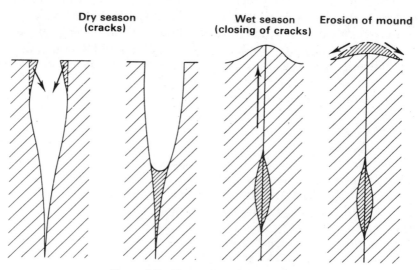

| Dry season (cracks) | Wet season (closing of cracks) | Erosion of mound |

**Figure 8.9** Phases of vertic movement.

# IV   Main types of vertisols: problems of classification

As previously stated, in most systems the classification of vertisols appears to have been insufficiently developed. For the most part, they take into account the more or less sudden variations in the pedoclimate and external and internal drainage conditions. For example, the American classification takes as a basis the importance, the nature and the persistence of contraction cracks in the dry season; the FAO classification only distinguishes two units based on the very dark (pellic vertisols) or less dark colour (chromic vertisols). This last concept appears to be the better in so far as it deals with an actual character of the profile that reflects the degree to which the influence of pedoclimate or of local drainage conditions are more or less favourable to vertisol formation; *in fact, it can be said that the dark colour expresses the degree of maturation of the organic matter*.

Taking this point of view and expanding it, the *vertisols sensu stricto* of dark colour can be contrasted with the *vertic soils* (coloured or chromic), which are generally intergrades to other classes and commonly have a brown, ochreous or reddish colouring. The use of the adjective *vertic* indicates a degree of vertisol formation that is not so great as when the noun is used.

However, although colour is apparently the most distinctive criterion, it is always associated with other no less important criteria: (i) *The kind of clays:* while clays of the montmorillonite family are generally dominant in the dark-coloured vertisols, there is a considerable amount of interstratified clays, illites and even kaolinites in vertic soils (Spaargaren 1979); (ii) *Profile homogenisation:* as shown by analysis, homogenisation is not so complete in vertic soils as it is in vertisols, for there are variations with depth in grain size and in the amount of organic matter; (iii) *Indices or structure:* in the dry season, the dimensions of fissures and slickenslides are less in the vertic intergrades than in the vertisols. It should be noted for soils that are transitional to those that are normally dark (chernozems, for example), these secondary criteria take over from colour, which loses all genetic significance (Duchaufour 1978: III$_3$).

## 1   Vertisols (dark)

These soils have a black or very dark grey colour, as their organic matter is very mature; homogenisation is almost perfect, the swelling clays being overwhelmingly dominant; finally, the structural indices are very marked. The *vertisols sensu stricto occur in climates and at sites that favour the most extreme contrasts of pedoclimate (almost total absence of local drainage)*. Two main groups are suggested:

(a) *Developed vertisols:* index of weathering based on ratio of free iron to total iron relatively high; clays of neoformation and aggradation dominant; parent material crystalline or volcanic.

(b) *Slightly developed vertisols:* i.e. with clays inherited from the parent material generally sedimentary (marls); the index of weathering is less; if the parent rock contains carbonates, the profile can contain an almost equivalent amount (Duchaufour 1978: VII$_6$).

## 2  Coloured (or chromic) vertic soils

The vertic development of these soils is not as strong as in vertisols *sensu stricto*, and they have a less dark colour, ochreous or reddish, determined by the iron oxides rather than the strongly condensed organic matter. As in the great majority of cases, the vertic soils are *intergrades* with other classes, a second adjective should show this relationship: for example, vertic fersiallitic soil, vertic reddish chestnut soil, etc. Nevertheless, for the **tropical vertic eutrophic brown soils**, which form an intergrade with the ferruginous class of soils, the original name is still used.

From the environmental point of view, vertic soils occur in pedoclimatic conditions in which the seasonal variations are not as great as in the vertisols *sensu stricto*. This decrease in seasonal contrasts is the result of two causes:

(a) Decrease in the length (or even complete disappearance) of the dry season in terms of the general climate: this is the case for soils in the most humid tropical zones and particularly for soils with vertic characters in the atlantic temperate zone; the absence of a dry season stops all recent maturation of organic matter; vertic soils of temperate climates inherit these characters either from parent material (*vertic pelosols* on marls with interstratified semi-swelling clays, Avery & Bullock 1977), or a previous development under a climate with greater seasonal contrasts (polycyclic vertic soils).

(b) The occurrence of a certain amount of external drainage, for example on a gentle slope (cf. Fig. 8.8), which decreases the seasonal contrast by preventing particularly the complete waterlogging of the profile in the wet season.

The *degree of development* is not indicated so well by the index of weathering (free iron : total iron) in vertic soils as in the case of vertisols, for the connection between this index and clay formation is not so close. This is because some of the iron no longer occurs either integrated within the humic compounds or within the crystal structure of the iron-rich montmorillonites. The result of this is that on equivalent materials the index of weathering is higher for a vertic soil than for a vertisol. In addition, it also explains why the chemical bonding of organic matter to montmorillonite is not so strong and also why the colour is brighter. Nevertheless, this index of weathering of vertic soils still varies over a considerable range; it can be low for certain vertic soils on slightly weathered sedimentary materials (vertic calcareous brown soils, vertic pelosols), while it is always relatively high (50%) in most tropical vertic eutrophic brown soils (Kaloga & Thomann 1971).

**Degradation of vertic soils.** Vertic soils can be subject to a secondary type of *degradation* (see Ch. 4) as a result of acidification or the formation of a water table at depth; as shown by Boulaine (1957), these hydromorphic vertic soils (with gley or pseudogley) are subject to two kinds of hydromorphism: a seasonal one in which the whole profile is waterlogged and another at depth resulting from the formation of a more or less permanent water table. Under the twofold influence of hydromorphism and acidification, iron-rich montmorillonites are weathered, freeing silicon and iron; the vertic characters of the profile decrease at the same time as the swelling ability of the clays; the iron freed – and hence the index of weathering – increases, at the same time as the ochreous or reddish colour of the profile becomes more marked; the vertic brown soil develops towards a soil of another class, a fersiallitic soil (mediterranean climate) or tropical ferruginous soil (*sol brun eutrophe ferruginise*: Perraud 1971).

As far as the hydromorphism is concerned, the iron is partially reduced, and rusty patches appear at the base of the profile (vertic soil with gley or pseudogley).

**Provisional classification: main types of vertic soils.** Lacking a definitive classification, the following provisional one is proposed: based particularly on environmental conditions, it is adapted from that of Filipovski and Ciric (1969). These authors classified the smonitzas of Yugoslavia as being *typic* and another class as being *transformed* – that is, they have been subject to the beginning of development that characterises another class.

(1) *Intergrade vertic soils.*
    (a) *Vertic characters only slightly developed* (temperate climate without a dry season):
        (i) polycyclic vertic brown soils;
        (ii) vertic pelosols.
    (b) *Marked vertic characters* (climate with dry season):
        (i) vertic alluvial soil;
        (ii) vertic calcareous brown soil;
        (iii) vertic reddish-chestnut and fersiallitic soils;
        (iv) vertic brunizem.
(2) *Vertic tropical eutrophic brown soils* (amount of weathering relatively high).
(3) *Vertic soils degraded by acidification or hydromorphism.*
    (a) Ferruginised vertic soils.
    (b) Vertic soils with gley.
    (c) Vertic soils with pseudogley.

Intergrade vertic soils will be studied in the chapters that deal with the soil groups concerned. Here, only the vertic tropical eutrophic brown soils will be dealt with, for they have special properties and form an important group.

*Vertic tropical eutrophic brown soils* (Duchaufour 1978: XVII$_2$). In Black Africa, they are characteristic of humid tropical climates that nevertheless have a short dry season; they occur on outcrops of basic crystalline or volcanic rocks, in conditions of poor drainage (Perraud 1971). Further north, in a dry tropical climate, they are replaced at equivalent sites by a vertisol *sensu stricto* with a dark colour. Kaloga and Thomann (1971) have clearly distinguished these two types of profile and emphasised the differences concerning the degree of maturation of the organic matter and the forms of iron; this study was within a particular climatic region (Haute-Volta, Senegal), and the more or less strong vertic development was dependent solely on the local drainage (see Fig. 12.3).

Apart from these two important differences, the other characters of the eutrophic brown soil profiles are rather similar to those of the vertisols: structure, high amount of clay with montmorillonite dominant, exchange capacity very high, saturation of the complex by $Ca^{2+}$ and $Mg^{2+}$; the authors admit, however, that the neoformation of montmorillonite is less important under the better drained conditions. In addition, the C : N ratio, clearly higher in the vertisol *sensu stricto* than the eutrophic brown soil, is also a reflection of differences in the humification.

As far as agronomic properties are concerned, the vertic eutrophic brown soils have the very favourable chemical properties of the vertisols, while the poor physical properties which have been emphasised decrease markedly.

# References

Aliyev, S. A. 1966. *Pochvovedeniye* 3, 71–80. (*Soviet Soil Sci.* 3, 306–14.)

Anderson, D. W. 1979. *J. Soil Sci.* 30 (1), 77–84.

Andreux, F. and A. Correa 1981. In Migrations organo-minérales dans les sols tempérés. *Coll. Intern., CNRS Nancy.* 1979, 329–40.

Arshad, M. A. and L. E. Lowe 1966. *Soil Sci. Soc. Am. Proc.* 30 (6), 731–5.

Avery, B. W. and P. Bullock 1977. *Mineralogy of clayey soils in relation to soil classification.* Soil Survey, Technical monograph no. 10. Harpenden: Soil Survey of England and Wales.

Blackburn, G. and H. W. Scharpenseel 1973. *INQUA Congress,* Chenilchmack.

Bocquier, G. 1973. *Genèse et évolution de deux topséquences de sols tropicaux du Tchad. Interprétation biogéodynamique.* State doct. thesis. Strasbourg.

Bornand, M., J. Dejou and J. Servant 1975. *C.R. Acad. Sci. Paris* 281D, 1689–92.

Boulaine, J. 1957. *Etude de sols des plaines du Chélif.* State doct. thesis. Algiers.

Bruckert, S. and G. Kilbertus 1981. *Plant and Soil* (in press).

Buol, S. V. 1965. *Soil Sci.* 99 (1), 45.

Campbell, C. A., E. A. Paul, D. A. Rennie and K. J. McCalum 1967. *Soil Sci.* 104 (3), 217–24.

Chakrabarty, S. K., H. O. Kretschmer and S. Cherwonka 1974. *Soil Sci.* 117 (6), 318–22.

Chernikov, V. A. and V. A. Konchits 1978. *Pochvovedeniye* 12, 84–8. (*Soviet Soil Sci.* 6, 685–90).

Cloos, P. 1981. In Migrations organo-minérales dans les sols tempérés. *Coll. Intern., CNRS Nancy,* 1979, 251–8.

Cointepas, J. A. 1967. *Bull. Bibl. Pédologie ORSTOM* XVI (2), 7–9.

Dommergues, Y. and Ph. Duchaufour 1965. *Science du Sol* 1, 43–59.

Dormaar, J. F. 1973. *Science du Sol* 2, 71–81.

Duchaufour, Ph. 1978. *Ecological atlas of soils of the world.* New York: Masson.

Duchaufour, Ph. and F. Jacquin 1963. *Ann. Agron.* 14 (6), 885–918.

Duchaufour, Ph. and F. Jacquin 1966. *Bull ENSA,* Nancy VII (1), 1–23.

Dudal, R. 1967. Soils argileux foncés des régions tropicales et subtropicales. *Coll. FAO,* No. 83, 172 pp.

Durand, J. H. 1959. *Les sols rouges et les croûtes en Algérie.* Algiers: Department of Irrigation.

Filipovski, G. and M. Ciric 1969. *The soils of Yugoslavia,* English edn. Washington: Nat. Sci. Foundation.

Franz, H. 1960. *Feldbodenkunde*. Vienna: Fromme.

Ganzhara, N. F. 1974. *Pochvovedeniye* **7**, 39–43. (*Soviet Soil Sci.* **6** (4), 403–7.)

Gerassimov, I. P. 1956. *6th Congr. ISSS*, Paris E (V.30), 189–93.

Gerassimov, I. P. 1973. *Soil Sci.* **116** (3), 202–10.

Gile, L. H., F. F. Peterson and R. B. Grossman 1966. *Soil Sci.* **101** (5), 347–60.

Greenland, D. J. 1971. *Soil Sci.* **111** (1), 34–41.

Hess, C. and U. Schoen 1964. *Al Awamia* **13**, 41.

Hubschmann, J. 1967. *Sols, pédogénèse et climats quaternaires dans la plaine des Triffa (Morocco)*. Spec. doct. thesis. Fac. Sci. Toulouse.

Kaloga, B. and C. Thomann 1971. *Cah. ORSTOM, Sér. Pédologie* **IX** (4), 461–505.

Kornblyum, E. A. 1967. *Pochvovedeniye* **11**, 107–21. (*Soviet Soil Sci.* **11**, 1527–39.)

Kounetsron, O. K. 1976. *Alteration des roches basiques du mont Agou au Togo*. Spec. doct. thesis. Univ. Paris VI. Versailles: INRA.

Kovar, J. A., C. L. Godfrey and G. W. Kunze 1976. *Soil Sci.* **122** (6), 339–49.

Kovda, V. A. 1973. *The principles of pedology*. Moscow: Nauka (2 vols).

Lamouroux, M. 1965. *Cah. ORSTOM, Sér. Pédologie* **III** (1), 21–42.

Lobova, E. V. 1960. *Bull. AFES* **5**, 269–82.

Lopez-Taborda, O. 1967. *Science du Sol* **2**, 97.

Maignien, R. 1962. *Sols Africains* **VI** (2 and 3).

Millot, G. 1979. In Alteration des roches crystallines en milieu superficial. INRA seminar. *Science du Sol* **2** and **3**, 259–61.

Muller, S. 1954. *Z. Pflanzener. Bodenk.* **65**, 107–17.

Nakaidze, E. K. 1965. *Pochvovedeniye* **11**, 31–41. (*Soviet Soil Sci.* **11**, 1288–97.)

Nguyen Kha 1973. *Recherches sur l'évolution des sols à texture argileuse en conditions tempérées et tropicales*. State doct. thesis. Univ. Nancy I.

Ouezzani, M. 1967. *Contribution à l'étude de la matière organique des sols vertiques de la plaine du Rharb*. Engng doct. thesis. Fac. Sci. Toulouse.

Paquet, H. 1969. *Evolution géochimique des minéraux argileux dans les altérations et les sols des climats méditerraneens et tropicaux*. State doct. thesis. Fac. Sci. Strasbourg.

Parsons, R. B., L. Moncharoan and E. G. Knox 1973. *Soil Sci. Soc. Am. Proc.* **37** (6), 924–7.

Perraud, A. 1971. *La matière organique des sols forestiers de la Côte-d'Ivoire*, State doct. thesis. Fac. Sci. Nancy.

Ponomareva, V. V. 1974. *Pochvovedeniye* **7**, 21–37. (*Soviet Soil Sci.* **6** (4), 393–402.)

Ponomareva, V. V. and T. A. Plotnikova 1968. *Pochvovedeniye* **11**, 104–18.

Ruellan, A. 1971. *Les sols a profil calcaire différencié des plaines de la Basse Mouluya*. State doct. thesis. Univ. Strasbourg: Mém. ORSTOM, no. 54.

Ruellan, A. 1976. *Bull. Soc. Géol. France* (7) **XVIII**, 1, 41–4.

Scharpenseel, H. W. 1968. *Proc. Symp. FAO/IAEA*, Vienna, 13–22.

Scharpenseel, H. W. 1972. *Z. Pflanzener. Bodenk.* **133** (3), 341–63.

Scheffer, F. and B. Meyer 1962. *Neue Ausgrabungen und Forschungen in Niedersachsen*. Hildesheim: Lax.

Seddoh, F. and G. Pedro 1974. *Bull. Groupe Fr. Argiles* **XXVI** (1), 107–25.

Singer, A. and A. J. Amiel 1974. *J. Soil Sci.* **25** (3), 310–19.

Spaargaren, O. C. 1979. *Weathering and soil formation in a limestone area near Pastena (Italy)*. Doct. thesis. Univ. Amsterdam; publ. no. 30.

Tessier, D. 1978. *Ann. Agron.* **29** (4), 319–55.

Tessier, D. and G. Pedro 1976. *Science du Sol* **2**, 85–100.

Valkov, V. F. and V. S. Kryshchenkov 1973. *Pochvovedeniye* **7**, 5–11. (*Soviet Soil Sci.* **5** (4), 385–91.)

Zhukov, A. I. 1975. *Pochvovedeniye* **5**, 20–27. (*Soviet Soil Sci.* **7** (3), 264–71.)

Zolotun, V. P. 1974. *Pochvovedeniye* **1**, 29–38. (*Soviet Soil Sci.* **6**, 1–12.)

Zonn, S. V. 1967. *Pochvovedeniye* **2**, 11–24. (*Soviet Soil Sci.* **2**, 156–70.)

Zonn, S. V. and M. I. Kochubey 1978. *Pochvovedeniye* **12**, 19–32. (*Soviet Soil Sci.* **6**, 637–49.)

# Chapter 9

# Brunified soils

## I  General characters

### 1  Environment and properties

Brunification is a climatically determined kind of pedogenesis, typical of regions with a temperate atlantic or semi-continental climate, where the natural vegetation (climax) is deciduous forest or a mixed evergreen–deciduous forest in the montane belt.

Recent research work, the details of which were given in Part I, showed that *it is the active iron, freed by weathering and bonded to the fine clays, which is central to the whole process of brunification*; it is responsible for the formation of the forest mull, for it causes rapid insolubilisation of humic precursors and it forms bridges between clay and humus molecules (**binding cations**) within clay–humic aggregates. In addition, it indirectly controls weathering, for even in acid conditions, the complexing agents are rapidly insolubilised in the A1 and made inactive, so that complexolysis (the weathering process characteristic of the podzols) is prevented. Weathering consists of a very gentle and gradual acid hydrolysis; the clays are either inherited (illites) or are moderately transformed (vermiculites); generally neoformation is very limited, as was emphasised in Chapter 1.

This discussion allows the environmental conditions necessary for brunification to be specified: *the amount of iron and fine clay is a limiting factor*; it should be recalled that there is a threshold value for the iron and clay content of a given material, below which podzolisation occurs, no matter what the vegetation (see Chs 2, 3 & 4). The state of *aeration* is no less important, for even in a partially reducing environment, although this occurs only seasonally, the iron is changed to the ferrous state and it can no longer play its esential part in the process of brunification; to a certain degree under acid conditions, the $Al^{3+}$ ion or $Al(OH)_n$ is able to take the place of iron but its effect is generally insufficient to prevent the degradation of the brown soil by hydromorphic processes.

In so far as the environment is well aerated and sufficiently rich in iron and clay (but not too much of the latter, so that all possibility of hydromorphic reduction is avoided), *vegetation* becomes the controlling factor. Generally, deciduous forests with a nitrogen-rich herbaceous substorey form an *ameliorating* litter with rapid decomposition, which is necessary for mull formation. Colonisation of the soil surface by an acidifying vegetation (*Calluna* for example) often marks the start of humus deterioration (moder) and of podzolic development; this again is a case of *degradation* of a brown soil.

The characters of the brunified soil profile can be readily explained in terms of the environmental conditions. The profile is of the A(B)C or ABtC type; the A1 humus is a mull which in the atlantic climate is not too thick or strongly coloured, while in contrast in a semi-continental climate it is thicker and more humic. The (B) or Bt horizon has a brown colour; this fairly dull colour is mainly the result of the poorly crystalline ferric oxide occurring as a very thin covering around the clay particles: *the close association between the clays and iron oxides within the mineral (B) or Bt horizons is one of the essential characteristics of brown soils*. Finally, it must be stressed that this (B) horizon is never calcareous; its pH and percentage base saturation are very variable depending on the soil type.

In summary, temperate brunification results from the interaction of various environmental factors: (i) a fairly humid temperate climate, of the atlantic or semi-continental type; (ii) an original vegetation of deciduous or mixed forest that forms an ameliorating litter; (iii) very varied parent materials that contain a sufficient quantity of iron oxides and fine clays, and are sufficiently permeable and well drained to be well aerated. Such parent materials can be formed by the weathering of both crystalline (granites, gabbros, mica-schists, etc.) and sedimentary materials (loams, sandstones, varied morainic deposits). Calcareous materials only give rise to brunified soils in as far as they have been subject to a prior decarbonation; thus the kind of materials involved are generally marls with large amounts of silicates. As to volcanic materials, if they form andosols in constantly humid climates, brunification occurs in hotter and drier climates (Hetier 1975); these brown soils on basalt have the peculiarity of being relatively rich in neoformed clays (montmorillonites or kaolinites).

## 2  Profile differentiation as a result of pervection

In many cases the brunified soil profile has a weathered (B) horizon of the **cambic** type, using the American terminology, where the index of clay pervection is not detectable. This is the case particularly when clay movement is slowed down by environmental factors, such as acidity or poor permeability, and also *when more clay is produced by weathering nearer the surface than at depth, such as occurs in granite regoliths, so that pervection is compensated for and it is not readily apparent* (Fig. 9.1).

In other cases, in contrast, particularly on sedimentary materials, where a large amount of clay exists prior to the brunification, weathering is not so strong and does not compensate for pervection, which becomes clearly evident, particularly if the environmental conditions are favourable: an **argillic** Bt horizon develops, with a particular microfabric in which clay and iron form argillans or ferro-argillans around the peds – thus the profile is of the ABtC type. Generally, the A horizon itself is then subdivided into two parts: a humic A1 (or Ah) and an A2 or Ae which is less humic and of a lighter colour as a result of the removal of some of the clay and the iron. The profile is then an A1A2BtC type.

*Thus, pervection appears to be a secondary process that is superposed in*

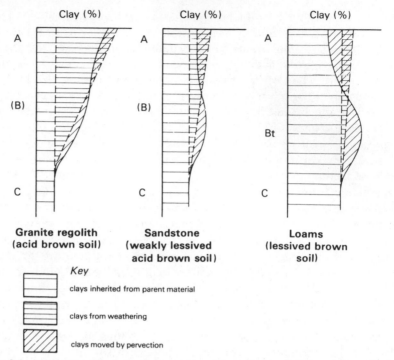

**Figure 9.1**  Influence of the type of parent material on clay development in brunified soils.

*certain circumstances (that are particularly related to the parent material) on the general process of brunification.*

## 3   Principles of classification

This last point would seem to justify the position taken in western European soil classifications, where lessived temperate soils (i.e. those with a Bt horizon) are considered as a subclass of the brunified soils and not as an entirely separate class. Thus the brunified soils can be divided into two subclasses: (i) those with an A(B)C profile and thus without pervection; and (ii) lessived brunified soils or more simply lessived soils, with an ABtC profile, in which pervected clay is apparent. It should be recalled that this approach is different from that of the American classification, where particular importance has been given to the morphological differentiation of the profile and, because of this, it has been thought necessary to place the non-lessived brown soils and the lessived soils in two different classes – **inceptisols** in the first case, **alfisols** in the second.

This also means that the American classification does not distinguish the process of temperate brunification, with its very particular kind of gradual weathering and humification controlled by iron, from the very different processes of weathering and also of humification characteristic of other climatic zones such as the Tropics and sub-Tropics. Thus, inceptisols are brown soils of all climatic zones and alfisols include all lessived soils. As emphasised in

Chapter 5, this appears to be seriously at odds with the environmental concepts developed in this book.

The third subclass of brunified soils that have been differentiated, the **boreal lessived soils**, raise many more difficulties which require discussion. These soils, which include the dernovo-podzolics and the grey forest soils of the north of the USSR, have several particular characters that differentiate them from the lessived soils and to an even greater extent from the temperate brown soils. In these soils the character of *pervection* itself (i.e. the movement of clay) is considered by many pedologists in the USSR to be secondary, with other processes, such as temporary hydromorphic conditions associated with winter freezing and associated degradation of clays (pseudopodzolisation), being more important. In fact, boreal lessived soils result from a very complex development which as yet has not been completely elucidated, so that they are difficult to include in western classifications. The FAO has solved this problem by creating two special classes: the **podzoluvisols** corresponding to the dernovo-podzolics, and the **greyzems** to the grey forest soils. Rather than accept this easy solution, it is preferable to consider them as intergrade soils with complex development involving pervection, hydromorphism and podzolisation: *for, as will be seen, similar intergrade soils also occur within the subclass of temperate lessived soils as podzolic lessived soils, hydromorphic lessived soils, degraded lessived soils*, etc., where generally the humus is no longer a mull and the partially reduced iron is only able to play its catalytic role in brunification very imperfectly. Often, these soils, in terms of their morphology and physicochemical processes, are very similar to the boreal lessived soils. Thus, until more information is available, it would seem to be logical to place all of these various groups of intergrade lessived soils with complex pedogenesis, which occur in both western and eastern Europe, within one subclass, and this is the reason for the adoption of this provisional solution.

In summary, three subclasses of brunified soils are differentiated, the first two being closely related (except for certain intergrades), the third, in contrast, having rather different characters:

(a) temperate brown soils with A(B)C profile;
(b) temperate lessived soils with ABtC profile;
(c) boreal or continental lessived soils.

## II Temperate brown soils

Temperate brown soils are brunified soils without visible pervection of clay and iron. In certain cases, nevertheless, slight pervection occurs, but this does not cause the development of a lighter A2 and an argillic Bt with cutans. Thus the profile is of the A1(B)C type, the weathered (B) horizon corresponding in general to the American cambic horizon. The A1 horizon is a forest mull, moderately acid, of little thickness or poor in organic matter in an atlantic climate, but becoming thicker and more humic in a continental climate where a certain maturation of organic matter occurs. The brown (B) horizon has a polyhedral structure which is, however, less well developed on sandy materials.

## 1  The brown soil profile: environment, morphology, geochemical and biochemical characters

**Environment.** The environmental conditions controlling the formation of temperate brown soils are considered here only from the point of view of the differentiation of the non-lessived (A(B)C profile) and the lessived (ABtC profile) subclasses. Therefore, in this section emphasis will be given to those factors of the environment that slow down or prevent pervection. Elsewhere, when considering intergrade brown soils, environmental factors responsible for transitions to other classes will be considered in terms of the physicochemical processes involved.

*The degree of pervection is determined to a large extent by the composition and nature of the parent material*, but climate, and particularly rainfall distribution, is not without effect on parent materials that are themselves favourable to pervection. Thus brown soils are particularly widespread in somewhat continental temperate climates where summer rainfall is dominant, for (as stated in Ch. 3) this means its almost immediate evaporation so that pervection is not favoured, which is different from the situation in the atlantic region with its uniform distribution of rainfall. Because of this, less and less pervection occurs in favourable materials from west to east in Europe. However, it should be remembered that when continentality becomes very marked (USSR plains), another kind of pervection occurs which is characteristic of grey forest soils. This point will be returned to when boreal lessived soils are discussed.

As far as *parent materials* that prevent pervection and hence favour the development of brown soils with A(B)C profiles are concerned, they can be considered under the following headings:

*Crystalline materials.* For these materials there is *compensation* between weathering and pervection (which is generally limited, particularly in acid conditions). Clay formation by microdivision or transformation is more important at the surface, particularly within the humic horizons: even if a moderate degree of pervection occurs, it is more than compensated for by this clay formation, so that percentage clay tends to increase from bottom to top of the profile (Souchier 1971, Fig. 9.1).

*Sedimentary materials.* On this type of material, in general relatively poor in weatherable minerals, clay formation remains limited. Thus it is not able to compensate for the pervection of pre-existing clays, particularly if conditions are favourable to such pervection (on this subject, see Ch. 3). *In an atlantic climate, pervection normally occurs on reasonably permeable, non-calcareous, loessial loams, with little acidity and containing easily dispersed fine clays (montmorillonites)*. Sedimentary materials that do not allow the formation of argillic horizons are generally of two different types:

(a) *Very acid materials, poor in calcium and magnesium*. These materials generally contain coarser clays which are less susceptible to movement (often a considerable amount of kaolinite); in addition, aluminium ions in different forms, $Al^{3+}$, $Al(OH)^{2+}$, $Al(OH)_2^+$, neutralise the negative charges of the dominant clays, generally vermiculites, and they have developed by the formation of interlayer **islets** towards Al-vermiculites (Guillet *et al*. 1981). Everything is **flocculated** and resists movement. In these circumstances, pervection is a slow process which has not been able to affect certain recently deposited acid loams (acid solifluction deposits of late Würm age in the Ardennes). Only certain older and more porous materials (Triassic sandstone) show evidence of a limited pervection, but this only produces a variety of an acid brown soil that is sometimes referred to as *weakly lessived*, where the B horizon is not argillic, being only slightly richer in clay than the A horizon, and this is not discernible to the naked eye (Fig. 9.1).

(b) *Materials rich in clay, free iron and calcium*, such as the decarbonated clays of the terra fusca palaeosols (see Ch. 7), where the permeability is less, the very abundant insoluble iron gives aggregates very resistant to water dispersion and $Ca^{2+}$ ions in solution favour clay flocculation. In addition, the very high level of biological activity in this environment (earthworms) mixes all horizons and prevent all visible signs of accumulation.

Calcic brown soils on terra fusca are generally not subject to pervection and have been discussed in Chapter 7. Eutrophic or mesotrophic brown soils formed in the same conditions on indurated flaggy limestones, and where no horizon effervesces with HCl (apart sometimes from very coarse pebbles), generally belong to the brunified class of soils, although in terms of their development they are very near to calcic brown soils (Duchaufour 1978: $IX_4$). However, it should be remembered that if the terra fusca is diluted by the mixing or incorporation of a certain amount of aeolian loam, pervection is again favoured (complex lessived brown soils).

**Main profile characteristics.** Profile morphology as a whole shows little differentiation and is very similar no matter whether it is an acid brown soil (very base unsaturated on granite or sandstone) or a eutrophic brown soil (with a higher degree of saturation on basic materials). Mull with crumb structure on no great thickness is poorly differentiated from the (B) horizon, into which it gradually merges (Duchaufour 1978: $IX_1$).

It is the same for the basic geochemical and biochemical characters which are identical in both types of profiles, acid or eutrophic: only secondary characters are different. The humification, totally controlled by iron, has been discussed in detail in Chapter 2: here, it need only be remembered that there is a dominance of slightly condensed humic compounds, very susceptible to microbial biodegradation (rapid turnover), such as fulvic acids (abundant), brown humic acids, and insolubilised humin, which is quantitatively the most important. The *stable* fraction is not completely absent, but is in a very small amount compared to that of the chernozems (less than 20%, O'Brien & Stout 1978).

The hydroxides (Fe and Al), always amorphous and cryptocrystalline, fix the organic matter to the clays in the form of a salt-complex in which the two constituents, mineral and organic, remain extractable by pyrophosphate (pH 10). With regard to the insolubilised humin, it forms **adsorption complexes** with the iron hydrates in which the two constituents are resistant to extraction reagents.

These two types of complex are the cement of the aggregates of the mull, whose formation is favoured by earthworms. It is *the almost complete absence of true mobile complexes that allows the brown soils* (**cambic horizon**) *to be differentiated from soils with a podzolic tendency* (**spodic horizon**, see Ch. 10).

In brown soils, the dominant clays are inherited (illites) or transformed (vermiculites), and there is little neoformed clay; in acid conditions the appearance of free aluminium ions favours the formation of **aluminous intergrades** by a fairly moderate degradation of the clay platelets (Al-vermiculite).

**Comparison of acid and eutrophic brown soils.** The various acid or eutrophic monocyclic and, to an even greater extent, the eutrophic polycyclic, brown soils are differentiated in terms of some of their secondary characters, particularly the amount of exchangeable bases and free iron, which are always much higher in eutrophic soils.

A comparison will be made between an acid brown soil on sandstone (or granite) and a eutrophic brown soil on basalt. It should be recalled that Hetier (1975) showed that brown soils replace the andosols on basalt in the Massif Central below 700 m, in protected valleys with relatively dry climate and a high degree of insolation.

As far as base saturation is concerned, it should be noted that although the (B) horizons differ considerably from one another according to the type of material, the A1 horizons differ much less: S : T varies slightly, about 40–70%, pH always being moderately acid, 5 to 6 in lowlands (these figures are distinctly lower in humid mountains); it is the same for the C : N ratio which is always between 12 and 15, reflecting *the climatically determined convergence of humus development under the influence of the biochemical cycles* (analogous soils of Pallmann; see Ch. 4 and Fig. 9.2). With regard to the amount of free iron, it can be four to five times higher on basalts than on sandstones and twice as high as on granite.

Clay development, and hence the exchange capacity, differs. In the eutrophic types there is, besides the micaceous clays, a considerable amount of kaolinite and montmorillonite: exchange capacity is relatively high and always greater than 0.5 mEq/g of clay, sometimes reaching 0.8 or 1 mEq/g of

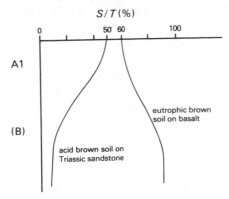

**Figure 9.2** Comparison of the degree of base saturation of acid brown soils and eutrophic brown soils under forest.

clay. In the acid brown soils, the acidity is responsible for aluminium being freed from the beginning of weathering so that the clays, that are almost wholly of the micaceous type, are transformed to vermiculites or even Al-vermiculites, which lowers the exchange capacity (Jamagne 1973) to about 0.3 mEq/g or even 0.2 mEq/g if the material contains inherited kaolinite. As is well known, in the A1 horizons the much higher values of the exchange capacity are mostly the result of the organic matter and they converge in both soil types (10–20 mEq/100 g soil).

The polycyclic eutrophic brown soils on terra fusca (on outcrops of indurated limestone) have specific properties that are for the most part inherited from the parent material (Duchaufour 1978: IX$_4$). This profile is rich in clay (more than 40%) and iron (4–5% of the element; free aluminium can reach almost 1%). On the oldest terra fusca a considerable amount of kaolinite can be mixed with the micaceous minerals.

**Agronomic properties and utilisation.** Brown soils as a whole are favourable to plant growth. The rapid cycling of nutritive elements (nitrogen, phosphorus and bases) resulting from the presence of a biologically active mull and the stable structure assuring that aeration is good are generally factors contributing to high fertility, provided that there is sufficient depth of soil and it does not contain too many pebbles or still-unweathered rock fragments. However, it is necessary to distinguish between the *acid* and the *eutrophic* types, for they behave very differently when cultivated.

Acid brown soils have a whole collection of physical and chemical properties that are clearly unfavourable; the more or less degraded clays only have a weak exchange capacity and the exchangeable cations themselves are markedly out of equilibrium, particularly in mountainous areas where the bases (calcium, magnesium and potassium) are easily leached and replaced by toxic ions such as $Mn^{2+}$ (also present in the *easily reducible* $Mn^{3+}$ form) and more particularly $Al^{3+}$. The biological return of bases to the surface is less effective in a humid climate and, when the pH falls below 5, $Al^{3+}$ becomes the dominant ion of the complex, where it can amount to 80% of the total exchange capacity (Juste 1965, Espiau 1978). It is known that these two cations have a toxic effect on roots, conifer seedlings in forests being particularly sensitive to the effect of manganese. Cultivated plants are not resistant to either of these ions. Liming often remedies this state of affairs by insolubilising both these toxic ions and the soils that result are **resaturated acid brown soils**, which can be mistaken because of their high level of saturation for eutrophic brown soils, but from which they are clearly different in the poorer *quality* of the clays. In addition, their deeper horizons often remain acid and retain their unfavourable properties.

In mountainous areas, the acid brown soils are generally under forest or grassland. As these soils are shallow and often too dry, grassland for the most part occurs on cool slopes where a favourable humus is maintained as a result of earthworm activity. Forests (beech or fir) are often degraded and sparse and occur on warm slopes, where colonisation by ericaceae (*Calluna* or *Vaccinium*) is frequent and causes a transformation of the mull humus to moder, together with the start of a podzolic degradation (ochric brown soil).

As far as **eutrophic brown soils** are concerned, because of their pH, their aerated structure and the nature of their clays (high exchange capacity), they have optimal conditions for the growth of cultivated plants: even the eutrophic brown soils on terra fusca can be very fertile provided that they are of sufficient depth and without hard pebbles in the cultivated layer.

## 2   Development of brown forest soils

Drawing on the work of Souchier (1971), Lelong and Souchier (1972) and de Coninck *et al*. (1979), the processes of formation of an acid brown soil on old indurated parent materials will be discussed. A brief comparison will then be made with brunification on basic materials (diorite or even basalt, from the work of Hetier 1975). As far as polycyclic eutrophic brown soils on terra fusca are concerned, their formation was studied at the same time as the calcic brown soils in Chapter 7 (calcimagnesian soils), to which the reader is referred.

### Development of acid brown soils on crystalline or metamorphic rocks

*Environmental conditions and pedogenesis*. The amount of weatherable minerals (which is reflected particularly in the amount of iron) of the Vosgesian granites studied by Souchier (1971) is the fundamental factor controlling pedogenesis on this material.

The relation between pedogenesis and the amount of iron can be represented as follows:

|  | | *Soil* | *Humus* |
|---|---|---|---|
| $Fe_2O_3$ greater than | 4% | brown soil | mull |
| $Fe_2O_3$ | 2.5–4% | ochric soil | mull–moder |
| $Fe_2O_3$ | 1.5–2.5% | brown podzolic soil | moder |
| $Fe_2O_3$ less than | 1.5% | podzol | mor |

The humus type mull, mull–moder, moder or mor is also a reflection of the vegetation:

(a) fir woods with fescue grass characterise the mull;
(b) fir woods with *Luzula albida* and *Deschampsia fluxuosa* characterise the mull–moder;
(c) fir woods with *Vaccinium* and mosses (sometimes *Calluna*) characterise the moder and the mor.

There is no doubt that vegetation has a determinative effect on humus formation (see Part I for the role played by ameliorating species, such as fescue grass, compared to acidifying species such as *Vaccinium* or *Calluna*). But vegetation itself is only a secondary cause acting as an intermediary between parent rock and humus, for *vegetation of the ameliorating type occurs only on granites that are rich enough in iron and weatherable minerals*. Thus the effects of both causes, the chemistry of the parent rock and that of the vegetation, are additive; but the former is dominant as it is involved at the same time, both directly and indirectly (see Ch. 2), in humus formation.

*Process of weathering.* In a temperate climate, weathering of the biochemical type is incomplete and gradual (as compared to that of the Tropics); the percentage weathering as defined by Souchier (A1 of the weathering complex/total A1 in C horizon) is at a maximum in the surface (i.e. the A1), where it is about 45–50%; the *weathering complex* consists of clays, *amorphous or cryptocrystalline free oxides* (particularly Fe and Al) and minerals that have already lost ions and become porous (they can be determined by repeated extraction with sodium hydroxide at 90°C). The amount of material lost in solution can be determined by using the method of balances based on standard quartz.

In general, clay increases in an upward direction apart from a slight decrease in the A, because of a minor amount of pervection. However, this is not sufficient to appear on the diagram (Fig. 9.3a) and the index of clay movement is not significant. The free oxides are fairly uniformly distributed, for their movement is compensated for by a larger production in the A1 and their preservation in the clay–humic complex; *they are present in much smaller quantities than the clays*. Clay formation is thus the characteristic process of brunification (compared to podzolisation).

The sequence of clays produced by the transformation of micaceous minerals (biotites or primary chlorites) or of inherited pre-existing clays (in the case of deep regoliths) is:

mica ⟶ illite ⟶ vermiculite ⟶

Al-vermiculite
(sometimes Al-chlorite in impeded conditions)

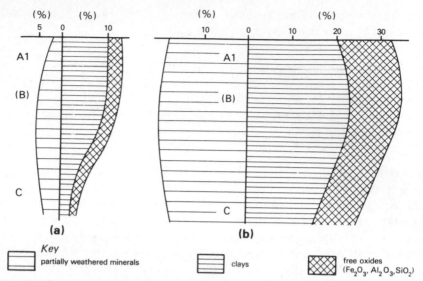

**Figure 9.3** Weathering complex of an acid brown soil in granite (a) and a mesotrophic brown soil on diorite (b) (after Souchier 1971).

In addition the vermiculites can be subject to the beginning of degradation and the aluminium freed in this way is either removed by drainage or added to the free oxide fraction. With regard to the feldspars, they do not form clays by neoformation, as in the Tropics, but they weather by releasing ions (some of which are transported out of the system, while others are absorbed by the clay–humic complex) and form part of the free oxide fraction after polymerisation.

**Development of the eutrophic (or mesotrophic) brown soils on diorites, labradorites or basalts** (Dejou 1967, Lelong & Souchier 1972, Hetier 1975, Dejou *et al.* 1979). As already stated, brown soils richer in bases occur on basic rocks. Where these rocks are well crystallised (diorites) with relatively slow weathering in a humid montane climate, a **mesotrophic** type results (S : T moderate in the (B)), but where the rock is a rapidly weathering microcrystalline one (basalt) and the climate is a sufficiently dry lowland type, a eutrophic soil occurs (S : T 80–90%).

Comparing the weathering complex on diorite with that on granite (Fig. 9.3), it is apparent that *the weathering complex is much more important in the first case in all three components: partially weathered minerals, clays and free oxides*; and in addition the amount of clay decreases with depth while the other two components, and more particularly the free oxides ($Fe_2O_3$, $Al_2O_3$, $SiO_2$), remain comparatively uniform.

The more detailed clay mineral investigation on the same soil by Lelong and Souchier (1972) showed that, compared to the acid brown soils, there is a much greater variety of clays. Besides the micaceous clays derived by inheritance and transformation, a considerable amount of kaolinite and also montmorillonite occurs, the latter apparently being formed by the soluble silica coming from the surface and penetrating into the tetrahedral layers of vermiculite. It is a special kind of *transformation* favoured by the presence of the $Ca^{2+}$ ion in the environment. With regard to kaolinite, it is the result of neoformation within the regolith in a prepedologic stage, as emphasised by Dejou (1967) and Hetier (1975) – on this subject see Chapter 6 and also Chapter 1.

The high amount of bases has another important consequence concerning the aluminium cycle and it also partly explains the relative abundance of amorphous materials in this soil: exchangeable $Al^{3+}$ assumes minor importance or disappears entirely from the absorbent complex (25–30% of the S value – sum of $Ca^{2+}$, $Mg^{2+}$, $K^+$, $Na^+$ – in the mesotrophic brown soil at pH 5, and is not present at all in the (B) horizons of eutrophic brown soils), so that, as the aluminium is freed by weathering of primary minerals, it is insolubilised and polymerised all the more rapidly as the amount of $Ca^{2+}$ and $Mg^{2+}$ in the environment is greater. This can be represented as follows:

$$\xrightarrow{\text{increasing amounts of } Ca^{2+} \text{ and } Mg^{2+}}$$

$$Al^{3+} \qquad Al(OH)^{2+} \qquad Al(OH)_2^+ \qquad Al(OH)_3$$

The alumina thus insolubilised retains and adsorbs a considerable amount of silica when the acidity is very slight (see Ch. 1). As iron hydroxide is also very abundant (often more than 8%), there is in total a large amount of amorphous (or poorly crystalline) material, which gives to this soil an **andic** character (andic brown soil, see Fig. 6.5). Thus brown soils on basalt have certain special characters: (i) abundance of *amorphous* or poorly crystalline oxides; (ii) large amounts of neoformed clays (or clays of aggradation) such as montmorillonite and kaolinite, part of which results from the partial crystallisation of the amorphous materials: these two factors are responsible for raising the exchange capacity of the clays; (iii) a slight pervection of clay to be seen in the B horizon (**slightly lessived brown soil**).

## 3   Main types of brown soils: classification (Fig. 9.4)

**Basis of the classification.** The pedogenic study of brown soils shows the importance of the base content of the soil material (related to that of weatherable minerals) which allows two great groups of temperate brown soils to be differentiated – **acid brown soils** and **eutrophic brown soils**. Mesotrophic brown soils, which are slightly acid in the surface (pH 5), occupy an intermediate position, but from the properties of the (B) horizon they are rather nearer to the eutrophic type. Note that polycyclic brown soils in terra fusca, despite the differences owing to the nature of the original material (in particular the higher amount of clay), have characters that are very like those of monocyclic eutrophic brown soils and it is thus easy to consider them as a subgroup or at least as a *related group*.

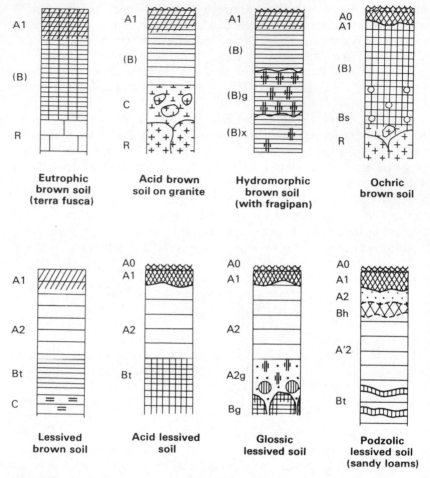

**Figure 9.4** Profiles of brown and lessived soils (see general key, p. ix).

The acid brown soils *resaturated*, as a result of cultivation (sometimes called anthropic), form a subgroup that is not to be confused with the eutrophic brown soils *sensu stricto*.

Both acid and eutrophic brown soils can be divided into several subgroups of which two are common to both of them. These are the **very humic brown soils** (sometimes called **melanised**) with very thick mull-type humus that occurs in mountains and the **slightly lessived brown soils** without an argillic horizon which, as stated previously, occur more frequently on basic materials relatively favourable to this process. However, they also occur on acid material provided that it is very sandy and extremely porous (Triassic sandstone in the Vosges, see Fig. 9.1).

A particular variety of acid brown soil associated with gently sloping sites on old massifs has been described which involves loamy material, generally of mixed origin and more or less reworked. This has a particular type of horizon, a **fragipan**, which is developed at depth, very compacted and often partially indurated. Acid brown soils with fragipan require separate consideration.

Finally, the intergrade soil types can be considered as a distinct brown soil unit: **ochric brown soils** (podzolic intergrades), **andic brown soils** (andosol intergrades, see Ch. 6), **hydromorphic brown soils** (hydromorphic intergrades). The ochric brown soils are very important in granitic mountains and will be studied in detail. The hydromorphic intergrades are of two types: (i) with fragipan (with pseudogley), a horizon of little permeability and waterlogged at certain seasons; and (ii) **gleyed brown soils** with a reducing water table at depth, leaving the surface horizons aerated (see Ch. 11).

**Proposed classification**

(1) *Acid brown soils:*
    (a) modal;
    (b) humic;
    (c) weakly lessived;
    (d) with fragipan;
    (e) anthropic (resaturated).
(2) *Eutrophic or mesotrophic brown soils:*
    (a) modal;
    (b) humic (monocyclic);
    (c) weakly lessived;
    (d) modal (plainlands)   polycyclic
    (e) humic (mountains)   (terra fusca)
(3) *Intergrade brown soils:*
    (a) modal ochric brown soil (and montane humic ochric brown soil);
    (b) andic brown soil;
    (c) hydromorphic brown soil (with pseudogley or gley).

**Special study of acid brown soils with fragipan.** This involves acid brown soils formed on loams that have not been subject to pervection for one of the following reasons. (i) The material is a decarbonated, acidified old loam, but one which has been reworked, at least in the surface, as a result of the Würm glaciation; its acidity and low permeability have made pervection so slow that it has not had time to develop. (ii) More recent loessial loams mixed by cryoturbation with weathered material from an older bed rock (for example, schists or primary quartzites of the Ardennes); the reserve of carbonates, strongly diluted in a material of acid weathering, has been eliminated rapidly, but pervection has not been able to occur.

Very frequently in these circumstances the argillic horizon has been replaced by a special horizon (at a depth of 50–80 cm) called a fragipan (American classification), which results not from illuviation but from compaction that has reduced the non-capillary porosity to almost nothing. The apparent density is considerable (1.8–2.0). This very compact horizon remains brittle and can be easily broken, hence its name. It often has secondary hydromorphic characters as a result of the decrease in its permeability. It has a structure in which roughly hexagonal prisms are separated from one another by vertical white veins (filled with deferrified clays) which have a rusty border (Duchaufour 1978: $XX_5$). In addition, there sometimes occurs in the interior of the prisms discontinuous areas of clay cutans that are generally being degraded.

In the case being considered, the fragipan is situated immediately below the weathered (B) horizon which is better aerated and occupied by the root system which generally does not penetrate into the fragipan, except in certain cases, when roots are found in the white zone bordering the prisms. Fragipans also occur below argillic horizons in certain acid lessived soils (particularly the **glossic** ones) and these will be considered in the next section.

There is great argument as to how fragipans have been formed. Pedologists are in agreement only about the fact that acid water preferentially circulates in the system of vertical cracks surrounding the prisms. The Americans (Soil Taxonomy 1973, Hannah et al. 1975, Harlan & Franzmeier 1977) differ from European pedologists regarding both the age and the mode of formation of fragipans. For most American research workers it is a relatively recent, postglacial, pedogenic process, the alternate wetting and drying of an acid material having a tendency to cause sufficient pressures to destroy the structure and result in an overall compaction. The contraction cracks are formed in dry periods and they serve as avenues of penetration of infiltrating acid water, which causes a deferrification by local reduction and eventually leads to the penetration of a small amount of fine clays into the gaps between the prisms; hence the formation of the white network characteristic of the fragipans.

For European pedologists (Lozet 1969, Jamagne 1973) it is an older process of geomorphic rather than pedologic origin. The fragipan horizon is considered to be a periglacial relic, resulting from a freeze–thaw compaction. The white cracks can then be considered as the prolonging in depth of an old glossic network, caused by ice. According to Kerpen (1960) and Mückenhausen (1973), this kind of very old horizon is related, at least in certain cases, to a plastosol, i.e. a very weathered palaeosol, which is then covered by a more recent weathered material that is less compacted and more aerated.

Thus there would seem to be at least two types of fragipan: one, monocyclic, of recent pedologic origin, resulting from the structural degradation of an acid loam, which occurs most commonly in a climate with a continental tendency; the other much older, where it is undoubtedly a palaeosol feature, inherited from periglacial compaction, which is common in western Europe and is characteristic of acid brown soils that have both a polycyclic and complex development.

**Ochric brown soils** (Manil 1959, Duchaufour 1978: $IX_2$). This type of soil is characterised by the beginning of a podzolisation which is as yet very slight. It forms an intergrade between brunified and podzolised soils. It differs morphologically from an acid brown soil in that the humus is degraded to a mull–moder and sometimes a moder, with a partial breakdown of structure and a darker colour; the (B) horizon also has a more intense ochreous colour and the structure is characteristically flocculated (fluffy). However, generally the profile remains little differentiated. In biochemical terms the following distinctive characters are to be noted: while iron is distributed uniformly throughout the profile, free aluminium has begun to migrate, fulvic acids are more abundant in the (B) horizon than in the acid brown soils. Bruckert's test (Bruckert & Metche 1972) shows a *marked increase in mobilisable complexed iron compared to the iron bound to clays* (10%, or more of the total free iron occurs in the first form), which gives the profile a more intense colour; but in these very aerated conditions, the amount of complexing agents is still insufficient to cause a marked movement of iron. Note that, at high altitudes, in mountains, **humic ochric soils** are dominant (Duchaufour 1978: $IX_3$): it is a climatic soil *sensu stricto* which has a general distribution in the upper montane zone, even on materials relatively rich in weatherable minerals (calc-alkaline granites: see *Soil map of France*, Saint Die sheet 1977).

In contrast, at lower altitudes, as shown by Souchier (1971), the ochric brown soil tends to occur more on granites (or sandstones) that are not as rich in weatherable minerals and iron as those characteristic of the acid brown soils. The vegetation on this type of material is an acidophilic forest (for example beech forest with *Vaccinium myrtillus* and *Deschampsia flexuosa* in the low montane zone); however, the presence of a certain amount of grasses decreases the podzolising effect, which remains limited.

The change from an acid brown soil to an ochric soil can also result from a process of degradation. The development of scattered forest cover owing to excessive clearing causes a vegetational change by increasing the number of acidifying species (ericaceae) which in turn leads to a secondary modification of the humification (moder, producing large amounts of fulvic acids), even on material initially favourable to the formation of an acid mull. Such an ochric soil represents the first stage in the gradual transformation of a *cambic* horizon into a *spodic* horizon (*direct podzolisation*; see Ch. 10).

**Note** that **andic ochric soil**, of the particular type described by Hetier (1975), Loveland and Bullock (1976), Avery *et al.* (1977) and Boudot and Bruckert (1978), is extremely rich in amorphous materials and is developed on acid volcanic rocks or sandstones containing volcanic materials (**greywackes**). There is little or no redistribution of oxides and the fluffy structure is, in contrast, strongly developed.

## III   Temperate lessived soils

### 1   General characters and environment

The lessived soils have in common with the brown soils the characters resulting from the process of brunification (formation of a forest mull and a particular type of weathering). But here the (B) horizon is divided into two parts by the occurrence of a process of clay movement that is essentially mechanical: the light-coloured A2 horizon impoverished in clay, iron and exchangeable bases and the Bt argillic horizon enriched in these same entities. In the French classification, it is necessary for the index of clay movement (clays in A2 : clays in Bt) to be at least 1 : 1.4 on homogeneous materials for it to be considered as a lessived soil. In addition, the presence of argillan-type coverings is necessary. If these criteria are not satisfied, such as when clay pervection is very weak, the soil is a **weakly lessived brown soil** which, as previously stated, is classed with the brown soils.

The pervection is accompanied by an acidification of the upper part of the profile which, when it is marked, opens the way for secondary processes such as hydromorphism as a result of the degradation of structure of the Bt horizon or, nearer the surface, the deterioration of humus with the start of podzolisation; in both cases, intergrade soils are involved, the properties of which deviate more and more from that of the original brown soils of the non-lessived type.

**Environment.** Lessived soils, like the forested brown soils, are characteristic of the deciduous forest (or mixed forest in the montane zone) of central and atlantic Europe. It should be recalled that on similar materials the intensity of pervection tends to increase as the climate becomes more of the atlantic type and less continental in going from east to west across Europe.

As already stated, the nature and the composition of the parent material is very important. Deep unindurated sedimentary materials, sufficiently, but not excessively, porous and non-calcareous, at least in the surface, are very favourable to pervection. The most typical of these are the aeolian loams of different ages where the intensity of pervection and the differentiation of the profile appear to be related to the age of the loam, as will be seen in section on *development*. Lessived soils also occur on old alluvium, glacial deposits, moraines and certain sedimentary marls (after decarbonation). Sandy materials are subject either to an immediate podzolisation or alternatively to a special kind of pervection, according to whether the critical threshold value of free iron and clay (discussed previously) is attained. These lessived soils on sandy materials have certain specific morphological characters. In addition, the climatic equilibrium with the forest is unstable; usually the vegetation is more or less degraded (sparse forest or heath), characterised by the start of a secondary podzolisation. The soil is thus either an intergrade or a secondary podzolic soil.

Terra fusca on hard limestone is not generally subject to pervection except when contaminated by a sufficient quantity of loam to form a thick enough layer, poorer in calcium and in iron, overlying the hard limestone (complex lessived brown soil).

## 2    Type profiles: morphology, properties (Fig. 9.4)

All lessived soils have a certain number of morphological or physicochemical processes in common. The profile is of the A1A2BtC type where the A1 is a mull of little thickness, moderately acid, similar to that of the brown soils, or a mull–moder in developed types (exceptionally a moder). The Bt horizon generally has a polyhedral structure characterised by the presence of coverings of clay and iron (**cutans** or, more precisely, **argillans**). The organic matter decomposes quickly and is concentrated in the surface (A1 horizon); the decrease in organic matter with depth is rapid, the Bt horizon generally contains little (less than 1% coming from the *in-situ* decomposition of roots). There is no movement of fulvic acids from the surface downwards. The C : N ratio is always low in the B, generally less than 10, which is the result of ammoniacal nitrogen fixation by the clays. Exchangeable bases show two maxima, one at the surface resulting from the biogeochemical cycle, the other at depth resulting from the retention within the argillic horizon of bases moved from the surface. The degree of base saturation is thus minimal in the A2 and it should be noted that in very transformed and very old profiles the degree of base saturation is considerably lowered in all horizons, particularly in the Bt (and even more particularly if it is a *relic* horizon in a polycyclic lessived soil).

Three basic types of lessived soils classed in terms of their degree of acidification and increasing pervection will be compared: (i) lessived brown soil; (ii) acid lessived soil; and (iii) glossic lessived soil. All other kinds of lessived soils are, in fact, variants of these three or intergrades towards other classes. The *index of movement* of iron and clay is generally used to differentiate the first two types; nevertheless, even though pervection is rather more marked in the second (index 1 : 2 to 1 : 3) than in the first (index between 1 : 1.4 and 1 : 2), in fact this criterion is of very doubtful significance owing to the frequent heterogeneity of the soil material, so that *the means whereby the three types of profile can be differentiated must be based on the whole of the criteria considered together*.

The described profiles will be interpreted according to the work of Dudal (1953), Duchaufour and Souchier (1966), Duchaufour *et al*. (1973), Begon and Jamagne (1973a,b) and Jamagne (1973, 1978).

**Lessived brown soil** (Duchaufour 1978: $IX_5$). *Morphology:* A1, very active mull, very well aerated (C : N 12 to 15) with very gradual change to the A2. The A2 and Bt horizons are only slightly differentiated in terms of colour (A2 light brown, Bt a little darker). The contrast in structure and texture is more marked (Bt polyhedral, with argillan coverings of a brown colour). Porosity is high throughout, particularly in the A1 and A2; it is still 40–45% in the Bt, which because of this is penetrated by numerous roots, and it has no sign of hydromorphism (and thus there is no segregation of iron oxides).

(a) *The dominant clays* are micaceous (illites), sometimes with an open structure (vermiculites) but without a great amount of aluminium-rich types. *The finest clays (montmorillonites), inherited from the parent material, are preferentially pervected and they form most of the cutans in the Bt*; the exchange capacity of the clays is high and greater than 0.5 mEq/g (Jamagne 1973).

(b) *The degree of base saturation* shows a marked maximum in the A1 and Bt, which has been discussed previously. This degree of saturation is always greater than 50% in A1 and 25% in the A2. There is practically no exchangeable $Al^{3+}$ ions, except sometimes in small amounts in the A2, in the more acid types.

**Acid lessived soil** (Duchaufour 1978: $X_1$). *Morphology:* as a whole the profile is more differentiated. The humus is a darker mull–moder, less well structured with a clear change to the A2 (higher C : N of about 20). A2 is lighter, less porous, often with a lamellar structure. The A2/Bt boundary is often sharp with the Bt having a more intense colour than the A2 and with ochreous and rusty patches alternating with more or less decolourised argillans. These weakly hydromorphic characters become stronger in the **degraded lessived soil**, distinguished by Jamagne and which is characterised by whitish pulverescent patches on the A2/Bt boundary. The porosity of Bt is lowered considerably and can decrease to 30%. In these degraded and hydromorphic types rusty patches and concretions can occur at the A2/Bt boundary.

(a) *Micromorphology* shows there is a tendency for the iron and clay to migrate separately, but at comparable speeds. Thus clays form partially decolourised argillans of secondary illuviation (de Coninck & Herbillon 1969, Federoff 1973) next to, but separate from, rusty patches of iron concentration.

(b) *The clays* show the beginning of degradation by **aluminisation** (Jamagne 1973). The Al-vermiculites and the Al-chlorites are dominant; montmorillonites *sensu stricto* tend to disappear from the upper part of the profile and even from the argillans of the Bt horizon. Degradation of micaceous clays is reflected in an increase of the ratio free Al : clay, which changes from 0.6% in the lessived brown soil to between 1% and 2%.

(c) *The acidity* of the whole profile is much more marked than for the lessived brown soils; the base saturation is less than 25% in the A2 where the amount of exchangeable $Al^{3+}$ compared to the total exchange capacity can reach 35%. The exchange capacity of clays throughout is always a little lower than in lessived brown soils (0.40 mEq/g), particularly at the top of the profile.

**Glossic lessived soil** (Duchaufour 1978: $X_3$). *The morphology* of the upper horizons shows little change from the previous type, but there is a very marked change in texture, structure and colour across the very sharp A2/Bt boundary. Besides which, at least in those soils which have been affected only slightly by surface disturbance, the A2 can be seen penetrating deeply down into the Bt horizon as glossic structures, that sometimes extend to great depth as a fine network filled with decolourised clays of the fragipan type. The material filling these glossic structures is loamy and pulverescent at the top and becomes more clayey with depth (Duchaufour 1978: $XX_6$).

Clay *degradation*, already to be seen in the acid lessived soils, increases considerably, accompanied by a very strong acidification: $Al^{3+}$ is present in abundance at all levels and it becomes the dominant cation, over the other exchangeable bases, at least to the bottom of the A2. The clays are always aluminous intergrades, but there appears at the bottom of the glossic structures a new type, the **montmorillonites of degradation**, which result from a removal of the aluminium from certain vermiculites and which are very different from the primary montmorillonites that are transported in the lessived brown soils.

The ratio free Al : clay exceeds 2% and in the *powdery* zones, where degradation reaches its maximum (at the top of the glossic structures), amorphous **allophane**-type materials occur (Bullock *et al*. 1974). Finally, the exchange capacity of the clays decreases considerably in the A2 and often reaches 0.3 mEq/g and sometimes even less.

The contrast between the upper and lower part of the profile, already considerable in the glossic lessived soils of the *monocyclic* type, becomes even greater again in *polycyclic* and *complex* profiles that characteristically have two layers, the deeper, older layer being a palaeosol. The amount of free iron in this deep palaeosol is thus very high because of the almost complete weathering of the primary minerals (ratio free iron : total iron, 80–100%). Its colour is ochreous, and sometimes slightly rubified. In these circumstances it is obvious that the index of movement of the free iron is without significance and it diverges clearly from that of the clay (ratio of free iron : clay is higher in the Bt than in the A: see Azvadurov & Constantinescu 1970). Another characteristic of these polycyclic soils is the great abundance of kaolinite, inherited from the older pedogenesis and brought to the surface by periglacial processes. Thus clay resistant to movement tends to accumulate at the surface, so that in the surface horizons, the exchange capacity is lowered to as little as 0.2 mEq/g of clay.

**Agronomic properties of lessived soils.** While lessived brown soils on loams are ideally suited to intensive cultivation (particularly cereals and sugar beet in the north of France and Belgium), the acid lessived soils with more or less hydromorphic conditions in the B, and to an even greater extent the glossic lessived soils are, in contrast, much less fertile and pose serious problems of utilization.

*The lessived brown soils* on loams are very favourable, first of all because of their depth and their structure. The Bt horizon is still well aerated and contains many roots, which find in it reserves of bases and water in the dry season. The nitrogen cycle and the state of the absorbent complex are very satisfactory. However, the excess of fine particles compared to the sands is responsible for some degree of structural instability, particularly if the soil is cultivated in unfavourable conditions, so that a certain degree of impermeability and the formation of a plough pan can result.

*The acid lessived soils*, particularly the degraded and hydromorphic facies, are of mediocre value. The B horizon is more or less without oxygen and has few roots. The formation of temporary water tables in the degraded horizons (A2B) decreases even more the depth of soil that can be used. The appearance of aluminous ions, which accompanies the acidification, is particularly harmful (causing poisoning of the absorbent roots) and needs to be counteracted by liming; in addition, the decrease in the quality of the clay can be seen in the low exchange capacity. These soils would generally require drainage before cultivation, which would necessitate considerable capital investment. For this reason they are often left under forest and poor quality grasslands. As far as polycyclic glossic lessived soils are concerned, in which kaolinite is dominant from the surface, they should be left under forest.

In addition, as the growth of deciduous species gives little return on these soils, they are generally planted to little-demanding coniferous species.

## 3   The stages of pervection: study of the processes

It can be said that the three profiles of the lessived soils that have been dealt with above are three stages in the same developmental process. At the beginning the process is simple, since it only involves the movement of fine clays. It becomes more and more complex in the more advanced stages, because it involves other processes, such as those connected with hydromorphism and podzolisation. Finally, when the glossic lessived soil is polycyclic – which seems to be generally the case – certain characters are inherited from an old phase of ferruginous development in a hot climate, and the profile then reaches its maximum complexity.

**Lessived brown soil stage: decarbonation and mechanical movement of fine clays.** When the parent materials (in the case of loessial loams) contain a limited amount of calcium carbonate, its movement as soluble bicarbonates is favourable to pervection which occurs very rapidly in decarbonated conditions (Blume & Schlichting 1964, Burnham 1964, Schwertmann 1965). The solution of carbonates not only releases entrapped fine clays (montmorillonites) but also causes the formation of large pores whereby the clay can move. Such movement is limited, however, for it is purely mechanical and does not affect the coarser micaceous clays.

It is certain that in these conditions of good aeration the ferric iron remains bound to the clay and they move together. The ratio of free iron : fine clay is, in fact, constant throughout the profile (Gebhardt 1964, Jamagne 1973). The ochreous-coloured argillans are, according to de Coninck and Herbillon (1969) and Federoff (1973), ferri-argillans of primary illuviation, and their ratio of iron to clay is always high, about 5% (Jamagne 1973). All of which proves the simultaneity of the movement of these two entities, which remain bound to each other. As was explained in Chapter 2, organic matter is not involved at all in this movement. It should be recalled *that the iron bound to the coarser and less mobile micaceous clays insolubilises the soluble humic precursors from the A1 horizon.*

Several authors (e.g. Zottl & Küssmaul 1967, Schwertmann 1968) have determined clay balances with the aim of proving that the gains of the B horizon accord reasonably with losses from the A horizon. By using this method on various materials it has even become possible, in certain cases, to determine the rate of the process. It is evident that such balances can only be determined on parent materials that are absolutely homogeneous, which is the case for certain loess deposits. In addition, it is necessary to demonstrate this homogeneity by the constancy throughout the profile, of the ratios of certain grain-size fractions that are not mobile and do not contain weatherable minerals. The graphical representations of balances by Kundler (1961) are incomplete, for they do not take account of certain factors such as density variations and clay formation by weathering. However, an approximate idea of the importance of pervection is given in Figure 9.5 by the area representing the losses from the A and the gains of the B horizon.

In addition, Zöttl and Küssmaul (1967) and Schwertmann (1968) have determined more exact balances by using soil columns and making the necessary corrections for compaction (varia-

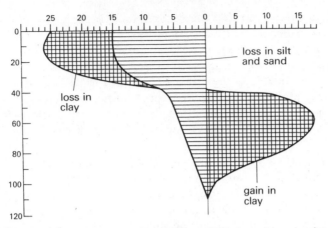

**Figure 9.5** Graphs of the losses and gains of clay, silt and sand in a lessived brown soil on glacial deposits (after Kundler 1961). Note that the losses of silt and sand are in the form of $CaCO_3$.

tions of density) and of clay genesis resulting from weathering. This clay formation by physical breakdown, according to Zöttl and Küssmaul, amounts to about 100 kg/m² which corresponds closely with the decrease in the amount of fine silt. These corrections being made, it can be said that the losses from the A2 exactly balance the gains in the B. The quantity of clay moved in the loess under forest was determined by Zöttl and Küssmaul as being 75 kg/m². Schwertmann for his part estimated that, on a calcareous moraine, the Bt horizon lost 64 kg/m² of calcareous material by decarbonation, but gained 60 kg/m² of clay by pervection. The two values are thus of the same order of magnitude.

**Pervection under acid, poorly aerated conditions.** This stage follows the preceding one with the disappearance of the calcic reserves at depth and is characteristic of the acid lessived soils. Three environmental changes are involved in modifying the process: (i) slight deterioration of the humus; (ii) acidification; and (iii) compaction and decrease in the aeration. In these circumstances there is *a partial reduction of the free iron and eventually a very transitory complex formation at certain seasons*. This causes the clay–humus aggregates to break down, accompanied by the freeing of a certain amount of clay and iron. The movement of these two entities, favoured in the case of the clay by the dispersing effect of the soluble organic matter, is able to occur, but at a very slow rate. In fact, the most mobile of the fine clays has been almost entirely eliminated from the A horizon during the preceding stage. The aluminium ion in its different forms decreases the mobility of clays which are this time the coarser aluminium-rich vermiculites. In addition, *the iron and clay move independently of one another*. The form of accumulation of these two entities in the Bt horizon illustrates this separate migration well. The iron forms localised rusty patches, or even small concretions, the clay forms decolourised argillans (secondary argillans of illuviation: de Coninck & Herbillon 1969) in which the ratio of iron : clay is near to 1% instead of 5%, as it is in the ferri-argillans (Jamagne 1973). As a whole, the index of clay movement increases only moderately compared to the preceding stage.

It is generally believed that the water-soluble organic matter coming from the A1 horizon plays a certain part in this stage of pervection, both as an agent of clay dispersion (Souchier & Duchaufour 1969, see also Ch. 3) and as a reducing or even complexing agent of free iron. However, this differs from the situation in podzols as the effect is only very short lived, for this organic matter is labile and from the time that the soil dries out and is aerated again it is rapidly biodegraded. The small quantity of organic matter with very low C : N ratio which occurs in the B horizon is proof of this.

**Process of degradation: formation of 'glossic' horizons** (Begon & Jamagne 1973b, Jamagne 1973, Bullock *et al*. 1974, de Coninck *et al*. 1976). It should be remembered that the word degradation is used here in the physicochemical and mineralogical senses. Clays are particularly affected by this process and are increasingly weathered, releasing aluminium (and also a small amount of iron which occurs in the octahedral layers). This degradation was already taking place at the end of the preceding stage at the boundary between the A2 and the slightly permeable Bt horizons, where the acid water has a tendency to stagnate to form very temporary perched water tables. This acid water moves down the vertical fissures that bound the prisms of the B horizon and form zones of preferential drainage. The process of degradation affects the material that fills these fissures and then extends downwards causing the formation of glossic structures.

As was seen with regard to fragipans, the origin of the fissures is still a point of argument. To some they are the result of wetting and drying of the B horizons and are thus of relatively recent pedologic origin. This is probably the situation for the monocyclic glossic lessived soils which seem to be uncommon in western Europe. In the case of glossic palaeosols, the fissures would seem to have a periglacial origin, as was explained in the fragipan section.

At the bottom of the A2 horizon, and also in the upper part of the glossic forms (where they widen out at the top of the Bt), the amount of clay decreases, partly as a result of its movement towards the bottom of the glossic fissures and partly as a result of its hydrolysis and destruction (primary chlorite). Only a small amount of clay persists and this is fairly coarse and relatively inert, with low exchange capacity (inherited kaolinite and Al-chlorite). Silt-size quartz, freed from aggregates, forms whitish powdery patches. Thus all movement is rapidly stopped within the A2 horizon while, on the contrary, as noted by Bullock *et al*. (1974), *redistribution of clays occurs locally within the glossic structures*, which leaves in their upper parts only a skeletal quartzose powder (Duchaufour 1978: $XX_6$). In addition, the clays moved to the bottom of the glossic structures are subject, under the influence of acid waters, to a process of marked *degradation*, for X-rays have often shown a decrease in the degree of crystallinity. The previously deposited finest clays tend to disappear (decrease in the ratio of fine clay : coarse clay, in the clay skins that border the glossic structures: Jamagne 1973). The aluminium of some of the Al-vermiculites is expelled and several authors (Magniant *et al*. 1973, Ducloux & Ranger 1975, de Coninck *et al*. 1976) have noted the formation of very open types of *montmorillonite of degradation* at the bottom of the glossic structures. At the same time, the amount of amorphous material of the allophane type increases by the polymerisation of the freed aluminous ions (Bullock *et al*. 1974).

At this stage the process of pervection is complicated by a *process of podzolisation of the localised hydromorphic type*, similar to that which will be described with regard to the formation of an A2g horizon of a podzolic pseudogley (see Ch. 11).

In the extreme cases of degradation described by Begon and Jamagne (1973b) in the old alluvial soils of the Aquitaine basin, the upper part of the glossic structures becomes less and less discernible and the contact between the A2g and Btg is more and more abrupt. These authors have referred to this process as being a planosolisation, by comparing it with the tropical planosols which have a very abrupt textural and structural boundary between the A2 and the Bt horizons (planosolic lessived soil).

**The special case of complex and polycyclic glossic lessived soils.** These soils which occur on the oldest loam deposits (pre-Würmian) are very common in western Europe, and the general process of formation has been described in Chapter 4 (see Fig. 4.13) and in Great Britain by Bullock and Murphy (1979), where they arc referred to as **palaeo-argillic brown soils**.

It should be remembered that the profile of these soils is complex, being made up of two parts. The upper part was reworked by cryoturbation, which microfabric investigations show involved the top of the glossic structures, and has since been affected by recent pedogenesis. Below this the *in-situ* palaeosol persists and consists of the lower vertically oriented parts of the glossic structures, although frequently the polycyclic fragipan below the glossic horizon is the only thing that is recognisable (Duchaufour 1978: $X_4$: *hydromorphic lessived soils with fragipan*).

These complex and polycyclic lessived soils have certain distinctive characters inherited from the older pedogenesis, such as the particular abundance of kaolinite, very strong weathering giving a particularly high ratio of free iron : total iron and a general ochreous and sometimes a reddish colour, which are the characters of the **plastosols** described by Kubiena (1953) and Kerpen (1960).

The formation of glossic structures, here of periglacial origin, is probably of Würmian age and thus subsequent to that of the plastosol *sensu stricto*.

## 4 Development of lessived soils: influence of environmental factors

If the three fundamental environmental factors of time, parent material and the effect of man, are compared, the most important, according to Jamagne (1973), is without a doubt that of time. In principle, *in the sequence that has just been discussed, the older the material, the greater the development*. However, the other two factors, parent material and the effect of man, have affected the general process by increasing or decreasing its speed. In particular, when the material has a coarser surface texture (sandy loam), acidification is more rapid. In addition, the more pronounced the pervection and the acidification, the more the upper part of the profile can be affected by the complementary process of *podzolisation*, which gives rise to a new type of soil of mixed character – a **podzolic lessived soil** – which opens the way for a more marked podzolisation which then affects the whole of the profile.

**The time factor: age of parent material** (Jamagne 1969, 1973). It will be recalled that in Chapter 4 the effect of the age of parent materials was discussed in terms of short-cycle (Postglacial) soils on Würm III loams, *old* monocyclic soils on Würm I materials and polycyclic soils on pre-Würmian materials. These three aspects of the influence of parent material age will now be discussed again.

On recent loess, the soil resulting from short-cycle pedogenesis rarely gets further than the *lessived brown* stage. Generally, the C horizon still contains calcareous reserves, and acidification is moderate, for it is compensated by calcium being brought to the surface by biological means. The equilibrium between oak or beech forest with mull flora and the soil is relatively stable.

On old loams (Würm I), the acid lessived soils that are considered as belonging to the *old monocyclic type* generally have an acidophilic forest with associated *Pteridium aquilinum* and *Deschampsia flexuosa*. This soil–vegetation equilibrium is less stable than the preceding one and more likely to be subject to hydromorphic effects as a result of the formation of a relatively thin and not very persistent layer of stagnant water at the junction of the A2 and Bt horizons. This pervected soil is of the *degraded* type, and in extreme cases it can even have a glossic appearance.

Finally, on the oldest of materials (Rissian loams or moraines), the soil is complicated by the superposition of successive stages of pedogenesis. These are the complex and polycyclic glossic lessived soils, the surface of which has often been disturbed by Würmian cryoturbation.

These observations can be summarised as follows:

(a) *loessial loam* (Würm III): lessived brown soil (recent monocyclic);
(b) *old loam* (Würm I): acid lessived soil or degraded lessived soil (old monocyclic);
(c) *Rissian loam:* glossic lessived soil (complex and polycyclic).

**Composition of parent material.** The presence of active carbonate, even in moderate amounts in the C horizon, undoubtedly prevents acidification, so that, as long as this reserve occurs at no great depth, the lessived brown stage is practically never exceeded. However, if the unconsolidated layer overlies hard limestone with slow weathering, and if in addition it has a sandy porous texture, a certain degree of acidification and even a surface podzolisation can occur. The Bt horizon formed by the precipitation of the fine clays transported from the surface, on contact with the calcareous materials, has a special morphology and structure resulting from the mixture of the transported clays and those which come from decarbonation. This is the **beta** horizon described in Chapters 3 and 4 (Duchaufour 1978: $II_6$ & $X_2$).

When there is no reserve of calcium carbonate within the root zone or that reached by earthworms, acidification occurs all the more quickly as the unconsolidated surface layer is richer in quartzose sand, or it is not so thick, for in these conditions the reserve of exchangeable calcium of the profile is rapidly exhausted; $Ca^{2+}$ is then replaced by $Al^{3+}$, and the degradation of the clay occurs rapidly.

On very sandy loams or certain flinty clays with a sandy cover, the process accelerates to such a point that the first stage (mechanical movement of fine clays with acidification) is very transient and often in practice does not occur. The local climax on such materials is an acidophilic forest which, moreover, is in unstable equilibrium with an acid lessived soil. When an impermeable palaeosol occurs at depth (clay-with-flints), the soil that characterises this type of forest is a hydromorphic degraded lessived soil. In contrast, if the porous material is very thick, the Bt horizon takes on a particular appearance which consists of markedly parallel bands or **lamellae** enriched in clay and iron distributed over a considerable thickness (Scheffer & Meyer 1962, Van-damme & De Leenheer 1968, see Figs 9.4 & 6).

In both cases, it is a question of a very unstable equilibrium, particularly susceptible to degradation under human or biotic influence. *All increase in the*

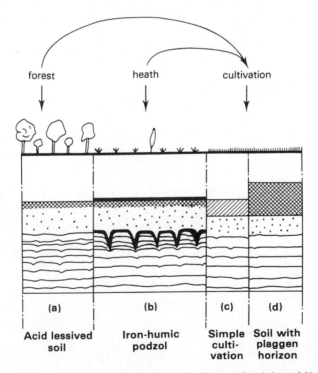

**Figure 9.6** Development of soils and vegetation in the sandy plains of North Germany (after Scheffer & Meyer 1962). Note that the litter and raw humus collected in (a) and (b) are used in the simple cultivation (c) or plaggen (d), i.e. by massive additions that raise the level of the soil.

*speed of pervection of clays and iron favours the initiation of podzolisation by the formation of a surface mor*. This is an *indirect* secondary podzolisation, i.e. one that follows on from a previous phase of pervection.

**Effect of man: environmental degradation** (Figs 9.6 and 7). Environmental degradation of soils under human influence was discussed in Chapter 4; it must not be confused with the physicochemical degradation of deep horizons that has just been discussed. What is involved is a modification by man of the natural deciduous forest vegetation (climax) and its replacement by a secondary or degraded heath-type vegetation (with *Molinia coerulea* when moist and *Calluna vulgaris* when dry, depending on the nature of the soil and the parent material). This soil in turn is subject to a modification of pedogenesis in a direction which gradually diverges from the climax.

This degradation is the result of very old practices which have been at the expense of the forest throughout history: excessive clearing, grazing within forests, prevention of regeneration, fire, collection of litter, etc; litter and even the forest humus thus collected was used, after treatment, as a fertiliser in cultivated areas (Fig. 9.6). These practices, frequently repeated, had as their main effect the breaking of the biogeochemical cycle of the nutritive elements, in particular those of the bases and nitrogen. This resulted in an acidification and an impoverishment of the surface horizons, a decrease in the biological activity and thus a deterioration of the humus. The ameliorating flora (grasses and mull plants) was replaced by a less demanding, but acidify-

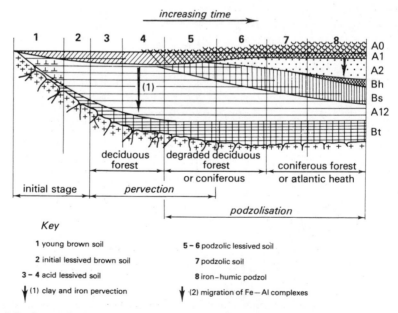

**Figure 9.7** Stages in the development and degradation of brunified soils on sandy loams (see general key, p. ix).

ing, flora (*Calluna*) and a humus of moder or mor type replaced that of the mull.

In fact, the stability of the soil–forest equilibrium which is characteristic of the climax varies considerably depending upon its initial state. In certain cases it is very stable as it is well buffered against degradation, in others it is very sensitive to change. In addition, the new equilibrium is strictly dependent on the initial properties of the profile and the kind of materials involved, particularly the texture of the surface horizons, the occurrence of calcareous reserves at depth and finally the presence or absence of a deep hydromorphic horizon, which control in a determinative manner the secondary pedogenesis induced by the change of vegetation. Three cases will be considered.

*Würmian loess: Lessived brown soils*. Because of the calcium reserves at depth, Würmian loess is resistant to all processes of degradation. From observations that have been made in southwestern Germany (Werner 1964, Glatzel & Jahn 1967), it was shown that the intensive cultivation of spruce over several centuries, in spite of its very acidifying character, has not caused great changes in the deeper soil horizons but only the formation of a surface mor, accompanied by a slight hydromorphic development for a few centimetres in the underlying horizons (presence of rusty patches).

*Old loams: Acid lessived soil and glossic lessived soil with hydromorphic conditions at depth*. The initial equilibrium is, in contrast, much less stable in this case, *degradation is marked by a rise in the level of hydromorphic processes towards the surface*. The A2 horizon is decolourised; at the same time as the iron is concentrated at certain points in the horizon, forming rusty patches or concretions, the humus is degraded to hydromoder. This is a pseudogley characteristic of humid heathlands with *Molinia coerulea* (see Ch. 11):

acid lessived soil ⟶ hydromorphic degraded lessived
soil ⟶ pseudogley

*Sandy loams:* Initially, this material contains little clay and iron at the surface and pervection removes what there is, so that the environment rapidly becomes suitable for podzolisation: *it is indirect secondary podzolisation* (in contrast to direct podzolisation which causes the transformation of the cambic (B) horizon of a brown soil to a spodic horizon and which has been discussed with regard to the ochric brown soil). The initial equilibrium is, as already stated, unstable and degradation occurs very quickly in such an environment. This can occur as a result of clearings being colonised by heath vegetation as well as by the introduction of acidifying forest species (Norway pine and maritime pine), very often the effects of both pines and ericaceous species occur together. Two cases can be differentiated according to whether an almost impermeable layer occurs at depth or not.

(a) *First case.* The initial profile has an almost impermeable Bg horizon (often a palaeosol) at depth. This is the case of certain clays with flint which are sandy at the surface, with a clay loam at depth; *two processes then occur simultaneously, one starting at the surface and moving downwards – podzolisation – the other starting from the impermeable layer and rising – hydromorphic degradation.* The new soil is thus a podzolic lessived soil, or even a podzolic soil with hydromorphic conditions at depth, also called a podzolic soil with pseudogley, which is the final stage of development and will be discussed in the next chapter:

acid lessived soil ⟶ hydromorphic podzolic lessived soil ⟶ podzolic soil with pseudogley ⟶ podzol with pseudogley

(b) *Second case.* No impermeable hydromorphic horizon occurs at depth. Podzolisation progresses to depth without hindrance: following the *lessived podzolic* and *podzolic* stages an iron–humic podzol develops. Here again, the superposition of the B horizons allows the history of the profile to be worked out. The newly formed spodic horizon (of Bh and Bs type) occurs above the old horizons of the lessived profile of the original forest, and is either of the beta type where indurated limestone occurs at depth, or of the banded or lamellae type where a very deep sandy material is involved (cf. Figs 9.4 & 9.6):

acid lessived soil ⟶ podzolised lessived soil ⟶ podzolised soil ⟶ iron-humic podzol

## 5   Main types of lessived soils: classification

**Basis of classification.** The three stages of the developmental sequence that have been discussed correspond very naturally with the three fundamental groups that are characteristic of the lessived subclass of soils. Related groups or subgroups that will be distinguished differ from the type profile either by their complex or polycyclic development or in a minor character, for example a greater or lesser degree of hydromorphism in the B horizon. The intergrade soils with the podzol class form a separate group: podzolic lessived soils.

The proposed classificatory scheme is as follows: there are four basic groups divided into subgroups in terms of: (i) whether the deeper horizons have hydromorphic characters; (ii) whether pedogenesis is monocyclic or, in contrast, polycyclic (and complex); and (iii) whether particular horizons, such as a fragipan, are present.

(1) *Lessived brown soil:*
    (a) modal: subgroups non-hydromorphic and hydromorphic;
    (b) complex: subgroups non-hydromorphic and hydromorphic.
(2) *Acid lessived soil:*
    (a) modal;
    (b) degraded (hydromorphic);
    (c) with fragipan.

(3) *Glossic lessived soil:*
  (a) modal (monocyclic);
  (b) polycyclic and complex.
(4) *Podzolic lessived soil:*
  (a) modal subgroup;
  (b) hydromorphic subgroup (or with pseudogley).

**Note** that all such soils as the degraded and hydromorphic acid lessived soils and also those with a fragipan, as well as the various types of glossic lessived soils, are often given the collective name *lessived soil with pseudogley*, which suggests that the temporarily hydromorphic condition at depth which is characteristic of these soils is of the same kind as the pseudogley. Nowadays the more exact terms tend to be preferred.

(a) *Planosolic lessived soils and complex lessived soils with fragipan* (Duchaufour 1978: $X_4$) are in fact glossic lessived soils in which the Btg horizon has been transformed, either by very strong hydromorphic degradation at the top of the glossic structures or by complete reworking of the upper part of the horizon; they can be considered as being closely related to the glossic lessived soils (Begon & Jamagne 1973b).

(b) *The fragipans* characteristic of group (2) are very probably not as old as those of group (3). They can also differ in their mode of formation, recalling the two types of fragipan, monocyclic and polycyclic, which were distinguished in the case of acid brown soils and which also occur in the lessived soils. No matter how they are formed, the fragipan horizon of lessived soils is situated below the *argillic* horizon, whether reworked or not, that has the maximum amount of clay.

This section will conclude by giving certain additional details of the complex lessived brown soils and podzolic lessived soils.

**Complex lessived brown soils** (Duchaufour 1978: $IX_6$). These profiles are formed from two geologically superposed layers: the surface one is generally an aeolian loam, of no great thickness; the deeper one with a finer texture is either a palaeosol (terra fusca, horizon beta on hard limestone) or a decarbonated marl. The first type is generally not hydromorphic at depth, while the second is in the wet season, leading to a segregation of iron, and mottling in the deeper layer.

The genesis of these complex profiles has been dealt with previously (Chs 4 & 7).

On the question of whether these soils are better classified as lessived or brown soils, it can be argued that the initially more clayey deeper layer acts in a similar way to an argillic horizon, without being one in the true sense of the word: however, microfabric studies show that generally the upper part of this layer is definitely enriched in clays coming from the surface loam (Duchaufour 1978: $XX_8$ & $XX_9$; see also Fig. 7.4b). Thus, very often on hard limestones the Bt horizon is a mixture of loams and terra fusca as a result of cryoturbation, and microfabric studies allow a very clear differentiation to be made between argillans of illuviation and the more or less disoriented papules of the terra fusca which have been mechanically incorporated with the loams.

**Podzolic lessived soils** (Duchaufour 1978: $X_2$, Robin 1968). Podzolic lessived soils are the first stage of a sequence of degradation of lessived soils towards podzols. This degradation begins on generally porous materials (sandy loams) when eluviation has caused a sufficient impoverishment of the top of the profile in bases clay and iron and so that, as a result of a change in humus, a true podzolisation can start. Thus, it is an *indirect podzolisation, the way for which is prepared by the eluviation*. The ashy horizon, characteristic of podzols, as well as the spodic BhBs horizon are developed at the top of the previously impoverished A2 horizon. Thus there are two superposed profiles, a podzolic profile at the top, which is only very slightly developed as yet (25–30 cm), and the older lessived profile, much more developed and which keeps its original characters.

The podzolic profile of the upper part has a black moder, an ashy horizon of a few centimetres and the beginnings of a spodic horizon as a band or a series of roughly horizontally aligned brown patches. The *old lessived profile* consists of a yellowish brown A2 horizon and a Bt horizon characteristic of the parent material, which is either a beta horizon on hard limestone or a banded or lamellae structured one on deep sand, or again a Btg hydromorphic horizon, if there is an impermeable layer at depth that was initially present or one that has resulted from pervection. In this last case, it is a **hydromorphic podzolic lessived soil** (or a **podzolic lessived soil with pseudogley**).

# IV  Boreal lessived soils

The zone of the boreal lessived soils, in the north of the Russian plain, marks the transition between the podzol zone (boreal coniferous forest) and that of the steppe forest (lessived chernozem). The characteristic vegetation in the north is mixed forest (spruce and birch) then, more to the south, deciduous forest (oak and lime). In both cases the ground vegetation is herbaceous, characteristic of mull or mull–moder and is not ericaceous, as in the case of the podzols.

The **dernovo-podzolic soil**, which in European terms can be compared to the glossic lessived soil, is characteristic of the first type of vegetation and the **grey forest soil** of the second.

The process of pervection that has been described (particularly that which is characteristic of the most developed stages of the western regions) certainly plays a determinative part in the formation of these soils, but it is often complicated by a rather special process of *hydromorphic podzolisation* favoured by the boreal climate. The temporary hydromorphic conditions, connected with the melting of snow and the formation of a surface water table overlying an as yet frozen subsoil, give these soils an appearance different from the lessived soils of western Europe. Certain Russian authors (e.g. Zonn 1973a,b) have referred to this mixed process as **pseudopodzolisation**. It has been well described in the work of Zaboyeva *et al*. (1977), and Sokolova and Targulian (1977). In a word, the boreal lessived soils and in particular the grey forest soils have unique characters that justifies the creation of a particular class – a position, moreover, which has been taken by the FAO (**greyzems**).

## 1 Main types

**Dernovo-podzolic soils** (Duchaufour 1978: $X_5$). The profile of the A1A2gBt type is very similar to the glossic lessived soil described previously, but the degree of hydromorphic degradation of clays is much more intense than in the temperate soil type, even though the humus is a fairly active biological mull (instead of a mull–moder), because of the vegetation. The A2g horizon is almost entirely decolourised at its base and is poor in clay (inherited kaolinite is dominant as a result of relative accumulation) with a pulverescent structure. The Btg horizon, more clayey, compact with an ochreous colour, is deeply embayed by glossic structures that narrow towards the base. Degradation is at a maximum at the top of the glossic structures which are filled with a whitish pulverescent material, studded with small rusty concretions: the boundaries of the glossic structures are covered by oriented clay skins, and also with a small amount of organic matter, iron oxides and aluminium. As in most lessived soils, base saturation is at a maximum in the A1 and in the Btg and minimal in the A2 where the pH is acid (about 5).

The clays are subject not only to a marked aluminisation (presence of Al-vermiculite and Al-chlorite) but also an acid hydrolysis leading to a partial destruction of the clays themselves.

The balance sheet determinations of Targulian *et al*. (1974), Glazovskaia (1974) and Sokolova and Targulian (1977), in which the silt fraction of the C horizon was used as a standard, indicate an important loss of clay from the profile. The presence of iron–humus and aluminium–humus complexes within the clay skins bordering the glossic structures in the Btg is another proof of degradation and even partial amorphisation of the clays.

**Grey forest soils** (Duchaufour 1978: $X_6$). The profile of the grey forest soils differs from that of the dernovo-podzolic soils in several important ways. The darker A1 horizon is very thick and consists of a chernozemic mull. The pervected A2 horizon is reduced to a thickness of a few centimetres and is characterised by a white powdery covering of the more or less rounded polyhedra. The Bt horizon is less hydromorphic than the dernovo-podzolic soil and it does not have glossic structures. The clay skins are humus rich and coloured brown; they also contain iron and free aluminium.

In general, the soil is less acid than the preceding one; while the A1 is practically saturated in $Ca^{2+}$ and $Mg^{2+}$, the A2 and Bt are moderately acid.

An important character of the grey forest soil is the special way in which the humic compounds are distributed. While the grey humic acids are dominant in the A1, the mobile compounds, particularly the fulvic acids, are dominant in the Bt. According to Ponomareva and Plotnikova (1975), the composition of humus of the grey soils is intermediate between that of chernozems and brown forest soils.

Two varieties of grey soils can be differentiated: the dark grey soils (such as the profile described) and light grey soils in which the A1 horizon is not so thick or so dark and where, in contrast, the A2 horizon of white pulverescent material is more strongly developed. This second type is transitional to the dernovo-podzolic soils.

It should also be noted that certain boreal lessived soils occur in northwestern Canada, where they are referred to as **grey wooded soils** and also **grey luvisols**. They are characteristic of calcareous morainic deposits. In spite of their name, they are related to the dernovo-podzolic soils of the USSR, but the humus, although slightly acid, is generally a not very thick moder or mor.

## 2   Development

Pervection is complicated here by a process of hydromorphic podzolisation (*pseudopodzolisation* of Zonn 1973a,b), very similar to that which occurs locally (particularly at the level of the glossic structures) in temperate glossic lessived soils, but with two marked differences. *It is a general phenomenon that affects the whole of the A2; in addition, it occurs very rapidly even on recent parent materials that are still calcareous at depth.*

The process of pervection, i.e. the mechanical movement of clays, which was denied for a long time by pedologists of the USSR, is now admitted by most of them (Karavaeva 1972, Glazovskaia 1974, Targulian *et al*. 1974). The micromorphological study of Bt argillans shows the similarity of their structures with those of the atlantic lessived soils (Yarilova & Rubilina 1975). However, certain authors (Rode 1966, Zaidelman 1974) denied the occurrence of a pervection in terms of the mechanical movement of fine clay. According to them, clays are destroyed in the A, the more or less soluble mobile constituents alone migrate and reform clays by neoformation in the Bt. This would seem to be invalidated by clay mineral investigations which show that the clays in the Bt horizon are the same type as those of the A horizon, being interstratified clays or more or less transformed micaceous minerals and not neoformed clays (on this subject, see Ch. 1).

Even though the phenomenon of pervection is no longer in dispute, that of hydromorphic podzolisation (pseudopodzolisation) is and this time certain western pedologists are opposed to it, while in contrast Russian workers are almost unanimously in favour and draw attention to several lines of evidence. Thus the balance sheet method, using the C horizon as a standard, shows that there is an important loss of clay from the whole of the profile. This clay undergoes destruction by acid hydrolysis or even complexolysis (see Ch. 1), as seems to be proved by the presence of organometal complexes of iron and aluminium in the clay skins of the Bt horizon (Glazovskaia 1974, Suvorov 1974, Sokolova & Targulian 1977). The coexistence of a slightly acid biologically active mull with these podzolic characters is surprising to western pedologists, for it is completely outside their ordinary experience, but it is explicable in terms of a pedoclimate that is very different from that of an atlantic lessived soil. The hydromorphic and even anaerobic phase, subsequent to the surface thawing, is responsible for the persistence in the profile of soluble organic compounds that are only slightly biodegradable at low temperatures. It is in this phase that the complexing of the iron in the ferrous state and the degradation of clays occur. The phase of increasing biological activity in the spring and that of summer desiccation are responsible for the good humification of the surface horizon. In fact, the accumulation in the B horizon of iron and free aluminium is only partial: balances show that a great amount is carried out of the profile by lateral flow within the water table or by vertical drainage (see lysimetry experiments of Ponomareva & Sotnikova 1972, Belousova 1974).

*The grey forest soils* have a pedogenesis that is even more complex than that of the dernovo-podzolic soils: (i) pseudopodzolisation is less marked; (ii) humification is characterised by a maturation similar to that of the chernozems; (iii) finally pervection, which involves the more or less swelling fine clays, transports at the same time a part of the non-extractible organic matter which they have fixed (**organoclay humin**, see Ch. 8). This humin coexists in the Bt with the mobile complexes of iron and aluminium (fulvic acids).

# References

Avery, B. W., B. Clayden and J. M. Ragg 1977. *Soil Sci.* **123** (5), 306–18.
Azvadurov, H. and M. Constantinescu 1970. In *In memoriam*, N. Cernescu and M. Popovat (eds). Pedologie, Seria C, no. 18, 155–71. Bucharest.
Begon, J. C. and M. Jamagne 1973a. *Science du Sol* **4**, 223–39.
Begon, J. C. and M. Jamagne 1973b. In *Pseudogley and gley. Trans Comms V and VI, ISSS*, E. Schlichting and U. Schwertmann (eds), 307–18. Weinheim: Chemie.
Belousova, N. I. 1974. *Pochvovedeniye* **12**, 55–69. (*Soviet Soil Sci.* **6**, 694–708.)
Blume, H. P. and E. Schlichting 1964. In *Experimental pedology*, E. G. Hallsworth and D. V. Crawford (eds), 340–52. London: Butterworth.
Boudot, J. P. and S. Bruckert 1978. *Science du Sol* **1**, 31–40.
Bruckert, S. and M. Metche 1972. *Bull. ENSAIA* **XIV** (2), 263–75.
Bullock, P. and C. P. Murphy 1979. *Geoderma* **22**, 225–52.
Bullock, P., M. H. Milford and M. G. Cline 1974. *Soil Sci. Soc. Am. Proc.* **38** (4), 621–8.
Burnham, C. P. 1964. *8th Congr. ISSS*, Bucharest **III** (VII, 25), 1303–10.
Coninck, F. de and A. J. Herbillon 1969. *Pédologie*, Ghent **XIX** (2), 159–272.
Coninck, F. de, J. C. Faurot, R. Tavernier and M. Jamagne 1976. *Pédologie*, Ghent **XXVI** (2), 105–51.
Coninck, F. de, E. Van Ranst, M. E. Springer, R. Tavernier and P. Pahaut 1979. *Pédologie*. Ghent **XXIX** (1), 25–69.
Dejou, J. 1967. *C.R. Acad. Sci. Paris* **264**, 37–41.
Dejou, J., J. Guyot and J. Trichet 1979. *Ann. Agron.* **30** (1), 63–88.
Duchaufour, Ph. 1978. *Ecological atlas of soils of the world*. New York: Masson.
Duchaufour, Ph. and B. Souchier 1966. *Conf. Sols. Medit.*, 401–6. Madrid: ISSS.
Duchaufour, Ph., M. Becker, J. M. Hetier and F. Le Tacon 1973. In *Pseudogley and gley. Trans Comms V and VI, ISSS*, E. Schlichting and U. Schwertmann (eds), 287–93. Weinheim: Chemie.
Ducloux, J. and J. Ranger 1975. *Ann. Soc. Nat. Charente-Maritime* **VI** (2), 116–32.
Dudal, R. 1953. *Agriculture* **1**, 119–63.
Espiau, P. 1978. *Science du Sol* **3**, 167–83.

Federoff, N. 1973. In Pseudogley and gley. Trans. Comms V and VI, ISSS, E. Schlichting and U. Schwertmann (eds), 295–305. Weinheim: Chemie.

Gebhardt, H. 1964. Bilanzanalytische Untersuchungen zur Silikatverwitterung und zum Stofftransport in feuchten und nassen Holozenboden. Doct. thesis. Univ. Gottingen.

Glatzel, K. and R. Jahn 1967. Sudwestdeutsche Waldboden im Farbbild Schriftenreiche der Landesforstverwaltung, no. 23, Stuttgart.

Glazovskaia, M. A. 1974. 10th Congr. ISSS, Moscow VI, 102–10.

Guillet, B., J. Rouiller and J. C. Védy 1981. In Migrations organo-minérales dans les sols tempérés, 49–56. Coll. Intern. CNRS, Nancy 1979.

Hannah, W. E., L. A. Daugerthy and R. W. Arnold 1975. Soil Sci. Soc. Am. Proc. 39 (4), 716–22.

Harlan, P. W. and D. P. Franzmeier 1977. Soil Sci. Soc. Am. J. 41 (1), 93–8, 99–103.

Hetier, J. M. 1975. Formation et évolution des andosols en climate tempéré. State doct. thesis. Univ. Nancy I.

Jamagne, M. 1969. In Commun. VIIth INQUA Congr., Paris, 359–72.

Jamagne, M. 1973. Contribution à l'étude pédogénétique des formations loessiques du Nord de la France. Doct. thesis. Univ. Gembloux.

Jamagne, M. 1978. C.R. Acad. Sci. Paris 286D, 25–7.

Juste, C. 1965. Contribution à l'étude de la dynamique de l'aluminium dans les sols acides du Sud-Ouest atlantique; application à leur mise en valeur. Engng doct. thesis. Fac. Sci. Nancy.

Karavaeva, N. A. 1972. Pochvovedeniye 1, 126–9. (Soviet Soil Sci. 1973. 5 (1) 11–13.)

Kerpen, W. 1960. Die Boden der Versuchgutes Rengen. Wissenchaft. Bericht Univ. Bonn, 5.

Kubiena, W. 1953. Soils of Europe. London: Thomas Murby.

Kundler, P. 1961. Z. Pflanzener. 95 (2), 98–110.

Lelong, F. and B. Souchier 1972. Sciences de la Terre XVII (4), 353–79.

Loveland, P. J. and P. Bullock 1976. J. Soil Sci. 27 (4), 523–40.

Lozet, J. 1969. Les sols à fragipan du Condroz. Aspects physiques, minéralogiques et chimiques. Sci. agron. doct. thesis. Louvain.

Magniant, D., P. Dutil and M. Jamagne 1973. Ann. Agron. 24 (2), 219–40.

Manil, G. 1959. Pédologie, Ghent IX, 214–26.

Mückenhausen, E. 1973. Anal. Edafol. y Agrobiol. XXXII (1–2), 1–20.

O'Brien, B. J. and J. D. Stout 1978. Soil Biol. Biochem. 10 (4), 309–17.

Ponomareva, V. V. and T. A. Plotnikova 1975. Pochvovedeniye 9, 63–73. (Soviet Soil Sci. 7 (5), 564–73.)

Ponomareva, V. V. and N. S. Sotnikova 1972. The biogeochemical process in podzolic soils. Leningrad.

Robin, A. 1968. Contribution à l'étude des processus de podzolisation sous forê de feuillus. Spec. doct. thesis. Fac. Sci. Paris.

Rode, A. A. 1966. Pochvovedeniye 4, 76–86. (Soviet Soil Sci. 4, 437–44.)

Scheffer, F. and B. Meyer 1962. Neue Ausgrabungen und Forschungen in Niedersachsen. Hildesheim: August Lax.

Schwertmann, U. 1965. Mitt. Deutsch. Bodenk. Ges. 4, 129–30.

Schwertmann, U. 1968. Sitzungsber. Ges. Naturforsch. Freunde, Berlin 8 (1), 16.

Soil Taxonomy 1973. Soil Survey Staff, USDA, Washington DC.

Sokolova, T. A. and V. O. Targulian 1977. In Problems of Soil Science, 479–92. Moscow: Nauka.

Souchier, B. 1971. Evolution des sols sur roches cristallines à l'étage montagnard (Vosges). State doct. thesis. Univ. Nancy I.

Souchier, B. and Ph. Duchaufour 1969. C.R. Acad. Sci. Paris 268D, 1849–53.

Suvorov, A. K. 1974. Pochvovedeniye 2, 3–10. (Soviet Soil Sci. 6 (1), 18–25.)

Targulian, V. O., N. Karavaeva and I. Sokolov 1974. 10th Congr. ISSS, Moscow VI, 93–101.

Vandamme, J. and L. De Leenheer 1968. Pédologie Ghent XVIII (3), 374–406.

Werner, J. 1964. Standort Wald und Waldwirtschaft, 55–68. Stuttgart.

Yarilova, Y. A. and N. Y. Rubilina 1975. Pochvovedeniye 6, 12–22. (Soviet Soil Sci. 7 (3), 281–90.)

Zaboyeva, I. V., G. V. Rusanova, A. V. Sloboda and E. G. Kuznetsova 1977. In *Problems in soil science*, 431–55. Moscow: Nauka.

Zaidelman, F. R. 1974. *Podzolisation and gleyzation*. Moscow: Nauka.

Zonn, S. V. 1973a. *Soil Sci.* **116** (3), 211–17.

Zonn, S. V. 1973b. In *Pseudogley and gley. Trans Comms V and VI, ISSS*, E. Schlichting and U. Schwertmann (eds), 221–7.

Zöttl, H. W. and H. Küssmaul 1967. *Anal. Edafol. y Agrobiol.* **XXVI** (1–4), 381–94.

# Chapter 10

# Podzolised soils

## I  General characters

The podzol, or **ashy soil**, was defined at the end of the last century by the Russian school, under the leadership of Dokuchaev, as a *zonal soil* characteristic of the boreal zone of the **taiga**, i.e. of the coniferous forests with ericaceae. The profile consists of three horizons very different in their colour, morphology and other properties: the A0 horizon is a brown or black mor or raw humus, the A2 horizon is ashy and without structure, and the B horizon is strongly coloured by the accumulation of amorphous organic and mineral compounds (particularly hydroxides of aluminium and iron); this is the *spodic* horizon of the American classification. The way in which this soil develops can be summarised quite simply: the very active organic compounds produced by the litter and the mor cause the silicates, including certain clay minerals, to weather biochemically by complexolysis (Razzaghe-Karimi 1974, Razzaghe-Karimi & Robert 1975, see Ch. 1); all the complexed materials migrate leaving a residue of fine quartz below the mor (the ashy A2 horizon) and accumulate in the B, forming the spodic horizon (or more often, horizons) when both a dark humus-rich Bh and a rusty coloured, sesquioxide-rich Bs horizon are formed.

However, podzols do not occur only in the boreal zone (or at high altitudes in the alpine and sub-alpine zones) but also in the temperate zone, particularly where there is a humid atlantic climate, although they are less widespread, occurring only at particular sites and under particular vegetation. They occur very occasionally in tropical regions, where their formation is dependent upon very particular site conditions. *Thus, while podzols are generally distributed in very cold boreal climates that favour the formation of mor, towards the south their distribution becomes more restricted and dependent upon increasingly special conditions of site and vegetation.*

### 1  Environment
Environmental conditions controlling the formation of podzols will be discussed in order of decreasing importance: climate, parent material, vegetation.

**Climate.** All other things being equal, a low mean temperature causes a slowing down of litter decomposition and a massive production of active water-soluble compounds; this factor is absolutely fundamental since the podzols are generally distributed in cold climates, boreal or alpine (and sub-alpine). Under these climates, practically all parent materials are podzolised, except certain limestones and some parent rocks that are exceptionally rich in iron which buffers the process.

The atlantic climate, particularly in its most humid form, is relatively favourable in so far as the weak insolation slows down litter decomposition and humification. But under its influence podzolisation is slower and profile differentiation is generally less marked, only reaching its maximum if other environmental conditions, such as parent material and vegetation, can add their effect to that of climate.

**Parent material.** It follows that in an atlantic climate podzolisation is closely related to the type of parent material; *from this point of view, the conditions of podzolisation are complementary to those responsible for the formation of other classes of soil, particularly those considered as being climatic, such as the brunified soils* (and also, of course, the calcimagnesian soils and the andosols). In this respect, the basic part played by the *environmental threshold* should be remembered, which is dependent on the amount of fine clay and free iron (which in the case of crystalline rock means the amount of weatherable minerals) and, as was emphasised in Part I and in Chapter 9, the development of a podzolic profile requires the presence of no more than a very small amount of them.

Of course, other properties of the parent material are implied by this first condition, such as strong acidity, low amounts of calcium and reasonable permeability in its upper part. The presence of an *acid water table* capable of lateral movement favours the rapid formation of a particular group of podzols: the hydromorphic podzols, which will be considered separately.

*In summary, podzols that are the best developed and most differentiated occur on coarse-textured quartzose material, at least at the surface, on very gentle slopes and poor in weatherable minerals.*

**Vegetation.** It has often been thought, and written, that podzols are associated with a particular kind of vegetation which is acidifying and mor producing (conifers and ericaceae) and this is true in most cases. However, when parent material conditions are very favourable, particularly in an atlantic climate, podzolisation can occur no matter what the vegetation, even under a deciduous forest of oak or beech which generally forms a mull (Dimbleby 1962, Munaut 1967). However, in these circumstances, horizon differentiation is not so marked (*podzolic* soil). It is only under acidifying vegetation (coniferous forest and ericaceae) that the podzol profile is fully developed (iron–humic or humic podzols). It should be remembered that this is precisely the vegetation that is generally distributed throughout the boreal zone, while in the atlantic climate it is preferentially restricted to more favourable parent materials such as quartzose sands and sandstones. As in the case of the brown soils, there is a close relationship in a humid temperate climate between the parent material and its vegetation.

**The environmental conditions of podzolisation: conclusions.** If the environmental conditions controlling the soil–vegetation relationships as given in Chapter 4 are considered, it can be seen that podzols have a very different environmental significance depending on the zones in which they occur.

As previously stated, the soil–vegetation equilibrium is zonal in the boreal zone, and it is thus a **climatic climax** and the same holds for the high-altitude podzols. In the atlantic zone, two kinds of podzol can be distinguished. First, those of **site climaxes**, where the podzol is not fully developed (podzolic soils) and its formation is controlled by local site conditions, particularly parent material. They can even occur under deciduous forest, which generally shows little tendency to cause podzolisation. Secondly, there are, in contrast, the *podzols of degradation*, generally with well developed characters and highly differentiated profiles (iron–humic podzols), that are the result of forest destruction by man which causes a change in the original vegetation to one of *Calluna* heathland.

Finally, tropical podzols, as developed in lowland sandy areas with a phreatic water table, clearly provide another and one of the best examples of a site climax.

## 2    *The basis of the classification*

Two main subclasses of podzol are differentiated: one in which drainage is good, at least in the surface; while the other involves hydromorphic conditions, that is to say *an acid water table is present which can reach the surface in certain seasons*. The presence of such a water table has a twofold effect on soil development. First, it reduces the iron and so makes it more soluble, which considerably increases the speed and makes easier the process of podzolisation, while it prevents any kind of brunification. Secondly, if (as is generally the case) lateral water flow occurs, it causes lateral movement of iron and its local accumulation by reoxidation, as very concretionary horizons (ironpans or placic horizons).

This causes a classificatory problem that is difficult to resolve, for in the next chapter a class of hydromorphic soils is defined also in terms of a water table that in certain seasons comes right to the surface. The placing of an exact boundary between these two classes is difficult. *However, the soil is considered to be a podzol if the hydromorphic conditions are such that they do not prevent the formation of horizons characteristic of podzols, such as a grey or ashy A2 and a B of the spodic type or concretionary ironpan.* Of course, there are many intergrades between podzolised and hydromorphic soils, in particular certain stagnogleys (see Ch. 11).

The definition of the groups needs even greater care. It would seem to be better to avoid the use of purely climatic criteria (boreal or atlantic podzol) as far as possible, for reasons given in Chapter 1, and use instead the terms **primary podzol** and **secondary podzol** (that is to say, of degradation) which are a better reflection of the relationship between the soil and vegetation and thus of the environmental conditions of development. In addition, by making use of the degree of profile development, which is a fundamental environmental factor, three stages can be distinguished of increasing profile differentiation and maturity:

*brown podzolic soil* ⟶ *podzolic soil* ⟶ *podzol*

In a simplified form, this has been done in the FAO classification, and in that of Great Britain (Avery *et al*. 1977), but the Soil Taxonomy of the USA puts all of them in one group (haplorthod).

It is also to be regretted that in this last classification the spodosol order is defined in terms of the presence of *one diagnostic horizon only – the spodic horizon*. This causes the exclusion from this order of all those podzols that for one reason or another have no identifiable spodic horizon. However, certain of them undoubtedly, as will be seen, are subject to a process of weathering, and even to a certain extent of movement, related to *complexolysis* which has been defined as the fundamental process of podzolisation.

## II   Podzolised soils with little or no hydromorphism

This subclass includes most podzols, that is those that are *well drained in their surface horizons*, even though the iron that moves in the profile does so as a ferrous complex. However, this does not exclude the occurrence of hydromorphic conditions at depth. Thus, certain degraded or glossic lessived soils with sandy surface horizons and Btg horizons at depth, formed by pervection, can change into a podzol with pseudogley, in which temporary hydromorphic conditions only occur at depth and do not affect the surface horizons, which remain subject to the same processes of weathering and migration as podzols that do not have this pseudogley horizon. *Thus, podzolised soils that are well drained throughout and podzolised soils with pseudogley should be placed in the same subclass*.

First of all, the meaning of three terms commonly used in this chapter – podzolised soils, podzolic soils and podzols – should be remembered. While the expression *podzolised soils* is very general and can be applied to this class of soils as a whole, the term *podzolic soils* refers to incompletely developed and slightly differentiated profiles which distinguishes them from the fully developed podzols. However, in an atlantic climate, it needs to be emphasised that the use of these last two terms implies considerable environmental differences. Thus, podzolic soils are incompletely developed because they are in equilibrium with a near-climax forest (sometimes sparse and partially degraded), which is either deciduous in the lowlands or mixed deciduous–coniferous in the mountains. In contrast, the iron-humic or humic podzols with well developed profiles are generally secondary *podzols of degradation*, that result from the clearing and gradual destruction of the original forest by man and its replacement by a heath with ericaceae, particularly *Calluna*.

One of these secondary atlantic iron–humic podzols will be described as the type profile and at the same time criteria will be given for the differentiation of *podzols* from *podzolic soils*.

## 1   The type profile: atlantic iron–humic podzol

**Morphology of the iron–humic podzols** (Duchaufour 1978: $XII_1$ & $XII_2$). The profile is of the A0A1A2BhBsC type.

*A0 (or 0)* is a wholly organic horizon 0–10 cm thick, formed of incompletely decomposed organic matter, the structure of which is fibrous or laminate. It is often made up of three parts: the little layer (L), the incompletely decomposed fermentation layer (F), in which plant structure can still be seen, and thirdly the finely granular black humified layer (H) with small rounded coprogenic aggregates of lower arthropods.

*A1 (or Ah)*, 2–5 cm thick, is a grey to black mineral horizon, rich in organic matter and with single-grain structure.

*A2 (or Ae)*, of variable thickness, is a whitish or ashy horizon with grey streaks containing humus. It has a friable single-grain structure which is called *ashy* if the quartzose material is very fine.

*Bh*, a black layer, a few centimetres thick, of amorphous organic matter accumulation which in places penetrates to considerable depth along old root channels or stems of heath plants. It has a friable, intergranular aggregate structure (pellets), often called *flocculated*.

*Bs (or Bfe)*, 5–15 cm thick, of a rusty colour as a result of the accumulation of amorphous hydrates of iron and aluminium, together with fairly large amounts of lightly coloured fulvic acids. The hydrates form a skin-like covering around the quartz grains which are either cemented together as incompletely indurated ironpan or give rise to a flocculated structure, similar to that of the Bh horizon, which is thus not cemented.

*Bt (or Btg):* the presence or absence of this old horizon at depth is dependent on whether podzolisation has been direct or indirect. *Its presence indicates that a phase of pervection preceded the indirect podzolisation* and is also evidence of the former presence of primary oak forest. It can have a varied form depending on the nature of the original material, for example as a $\beta$ horizon over hard limestone, or as a banded or laminated type on sandy materials (see Ch. 9).

Micromorphological study confirms the macroscopic observations. In the A1 and A2 horizons, the quartz grains have been stripped of all coverings and are translucent; small coprogenic aggregates (pellets) and a certain amount of plant debris, still showing signs of some organisation, occur in between the quartz grains, but there is no bonding between them. In contrast, the amorphous compounds in the BhBs horizons *form coverings similar in appearance to the crazed pattern that develops on china glazes* (Altmuller 1962, Righi 1975). If they are very rich in organic matter, these coverings become detached from the quartz grains and form isolated pellets, some of which have been transformed by micro-arthropods and mixed with dead root residues in coprogenic aggregates (Righi 1975, Robin 1979). In contrast, if these amorphous compounds are very rich in iron and poor in organic matter, as they are in the Bs horizon, they act as a cement and indurate by the crystallisation of the iron oxides (goethite). This is followed by general concretionary development, forming what may be called ironpans or orstein.

## Biochemical and geochemical characters

*Humification*. There are two horizons rich in organic matter, the surface A0 horizon and the Bh horizon; but while the A0 horizon is almost exclusively organic, the Bh horizon, enriched by the migration of water-soluble precursors that are insolubilised or polymerised *in situ*, only contains 4–10% organic matter. While the surface mor is only slightly developed (dominance of slightly transformed, fresh organic matter, and young inherited humin), the humus of the Bh or Bs is, in contrast, for the most part formed of humic compounds of insolubilisation: humic acids are dominant in the Bh, the fulvic acids with little colour in Bs. The C : N ratio in particular is very high, more than 30, both in the A0 and in the Bh; *the high amount in the B horizons of organic compounds of insolubilisation (more than 1%), of which most remains extractable (at least 60%) and with C : N ratio greater than 20, is a criterion that distinguishes the podzolised soils (even in the slightly developed podzolic phases) from the lessived soils.*

*Absorbent complex*. The podzol is very acid in all its horizons (pH less than 4). The exchange capacity is high in the A0 (more than 100 mEq/100 g), moderate in the B (about 25–30 mEq/100 g) and is minimal or almost non-existent in the A2. Consequently, base saturation is only of significance in the A0 and B; it is very low in all the mineral horizons (less than 10%) but is sometimes a little higher in the A0, which reflects a certain accumulation of bases, connected with the biogeochemical cycles.

*The clays*. The clays have been almost entirely eliminated from the A2 horizon, either by movement or destruction resulting from biochemical weathering. Those which remain are extremely resistant to weathering: either kaolinite (inherited from the parent material and, as it is not mobile, it accumulates relatively) or *montmorillonite of degradation* with layers that are very open but impoverished in charges which make it relatively inert (Guillet *et al.* 1975, Vicente & Robert 1975). In contrast, the clays that have migrated to the B horizon are generally micaceous clays more or less transformed to vermiculite and Al-chlorite during the gradual acidification of the profile.

*The amorphous hydroxides of iron and aluminium* are completely removed from the A2 (as well as the soluble or amorphous silica) and they accumulate in the Bh or Bs where they form, together with the organic matter of insolubilisation, the cement of the structural units: *the index of movement of iron (A2 : Bh or A2 : Bs) exceeds 1 : 10 in the atlantic iron–humic podzols*. It is a very good criterion whereby podzols can be distinguished from podzolic soils. With regard to alumina, it generally exceeds 2% and forms with the colloidal silica, which can also be abundant, a compound of the allophane type (Bartoli *et al.* 1981). Alumina is more mobile than iron, at least in sandy more aerated conditions, and the index of movement can reach a maximum of 1 : 50 or even 1 : 100 and this generally at a depth greater than that of maximum iron accumulation (Bs horizon).

**Distinctive characters of podzols and podzolic soils.** The less developed podzolic soils are distinguished from the iron–humic podzols by less distinctive characters and not as clearly differentiated horizons. Generally, the humus is a *moder* with a fairly thin A0 horizon, while the A2 is not completely ashy and still contains free iron. The accumulation of organic matter in the B, which never exceeds 3%, is not too obvious, so that it is often difficult to make the distinction between the dark Bh and the iron-rich Bs, which is so characteristic of the iron–humic podzols, where it is attributable to heath vegetation, particularly *Calluna*.

The index of iron movement in podzolic soils is about 1 : 5 and indurated ironpans never occur. This is very different from the iron–humic podzols where the high index results from the effect of *Calluna* and specifically from the formation of the black Bh horizon.

Certain authors differentiate *iron podzols* from podzolic soils by using one criterion only and that is in terms of the A2 being ashy, while in podzolic soils it is not completely so, but only decolourised. This criterion would seem to be too vague to be used as it is almost impossible to define the boundary between being decolourised and being ashy. For this reason it is preferred not to use the iron podzol as a valid classificatory unit, at least until it is possible to define it in a more exact way.

**Identification of spodic horizons.** The problem of the distinction between cambic (B) horizons (brown soil) and spodic B horizons (podzols) is difficult to resolve, particularly for the *intergrade ochric soils*, so that a chemical criterion is absolutely necessary. Certain ochric soils are very like brown soils; others have a podzolic tendency (cryptopodzolic); others again are andic (Avery *et al.* 1977, Boudot & Bruckert 1978, see Ch. 9).

Most classifications use as a chemical criterion the amount of Fe and Al extracted by 0.1 M Na-pyrophosphate (Soil Taxonomy, Avery *et al.* 1977, Baril & Tran 1977), either as an absolute value or as a Fe + Al : clay ratio. However, this test is not completely satisfactory, for the reagent used is not capable of selective solution of the true mobile complexes of Fe and Al, for it also extracts salt complexes bound to clays (characteristic of the brown soils) and some adsorption complexes immobilised in the surface if free iron and aluminium are very abundant (andic character). Bruckert and Metche (1972) and then Bruckert and Souchier (1975) suggested a very selective reagent with regard to the true mobile complexes: Na-tetraborate at a pH of 9.7. This enables the aluminium and iron of these complexes to be determined ($Al_{Na}$

and $Fe_{Na}$). The $Fe_{Na}$ is then compared to the free iron ($Fe_d$) which is determined by oxalate–dithionite extraction and if:

$$Fe_{Na}/Fe_d \begin{cases} \text{is less than 0.10, it is a cambic (or andic) horizon} \\ \text{is greater than 0.20, it is a spodic horizon} \end{cases}$$

It must be emphasised, however, that this test is not applicable to differentiated podzols with a B horizon enriched in iron, for by then the mobile complexes have developed into adsorption complexes; but in such a case morphological criteria are sufficient to distinguish these soils.

## 2 Podzols and vegetation: utilisation

While the boreal and sub-alpine podzols are climatic climax soils and are not necessarily unproductive and sometimes carry good forest, this is not the case for the iron–humic podzols of the atlantic heaths which have, in all their horizons, properties very unfavourable to plant growth. Without considering the barren heathlands, only non-demanding acidophilic species can grow (pines) on such soils, in the degradation of which they also help; deciduous species cannot survive except for birch and American red oak in certain circumstances. Unfavourable properties differ from one horizon to another; unfortunately, far from cancelling one another out their effects are additive.

Horizon A0 adsorbs a great deal of water and in the wet season becomes waterlogged and without oxygen, and it prevents the build-up of water reserves in the deeper mineral horizons for dry season use. In addition, the mor is inactive, with a C : N ratio greater than 30. Annual nitrogen mineralisation is non-existent or very low. This horizon can store a certain amount of bases ($Ca^{2+}$, $Mg^{2+}$) as complexes, or even in the exchangeable form, but the acidity prevents their adsorption by roots (antagonism of $H^+$ ions). Phosphorus is transported and immobilised in the B by excess alumina.

The mineral horizons have unfavourable physical and chemical properties: A2 is without structure, having had all the fine-grained materials removed from it; B, in contrast, is often indurated and impenetrable to roots. From the chemical point of view, they are both extremely poor in exchangeable cations: A2 particularly behaves almost as a sterile medium. Manganese and free aluminium toxicity occurring in the B horizon prevents the growth of neutrophile species.

It is also necessary to emphasise that the biogeochemical cycle of nutritive elements is different in podzols from that of the lessived brown soils. In the latter, it is responsible for maintaining a sufficient reserve of exchangeable bases, which are thus assimilable, in the A1 and Bt; in contrast, in the podzol the biogeochemical cycle causes a gradual immobilisation, not only of the bases but also of nitrogen and phosphorus in a form that is only slightly assimilable within the surface A0 horizon.

For this reason, it is the humic horizons – A0 and A1 – that are preferentially colonised by roots of conifers: the barren A2 horizon is an absolute barrier to root penetration (only certain species with tap roots can reach the B

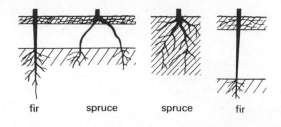

        fir              spruce          spruce            fir

**Podzolic soil        Brown soil   Deep podzol**

**Figure 10.1**   Comparison of root development of coniferous seedlings on various soils.

horizon of deep podzols; see Fig. 10.1). *The biogeochemical cycle of nutritive elements has a tendency to be concentrated to a greater and greater extent in the surface humic horizons by the effect of direct exchange between humus and absorbent roots.*

Podzols are used particularly in re-afforestation, which involves considerable soil cultivation, destruction of the vegetation and mixing of the surface humic and mineral horizons. In this process it is essential neither to remove the raw humus nor to bury it too deeply, but to activate it by addition of fertilisers and calcic amendments (Thomas slags), which neutralises the toxic effect of Mn and Al. For the same reason, it is not recommended that ironpan be brought to the surface by cultivation, but it must be broken up *in situ* by subsoiling if it is an obstacle to root development. To maintain a sufficient microbial activity, the introduction of deciduous species (birch, red oaks) in clumps or in lines within the mass of conifers has been recommended.

## 3   Biochemical mechanism of podzolisation

**General outline.** Podzolisation results from a markedly acidified humus (mor) producing large amounts of soluble or pseudosoluble organic compounds which migrate to depth; this movement begins with simple organic anions which are dominant in the brown podzolic and podzolic phases, then more polymerised humic compounds accumulate in the Bh horizon in the iron–humic podzol phase (Bruckert 1970). These soluble or dispersed compounds not only reduce and remove all of the free iron from the A1 and A2, but also cause the weathering by complexolysis of the mineral part of the absorbent complex and the removal of the aluminium and the silica so produced. The iron and aluminium oxides form complexes with the microbially resistant soluble organic compounds and they migrate in this state, while silica does so in the soluble form.

Finally, in the A2 of type podzols only a small amount of clay minerals remain: everything else, including the finest-grained material, is detrital quartz (Duchaufour *et al.* 1950). Silica is precipitated in the B horizon at the same time as the iron and aluminium hydroxides: the alumina fixes the silica and gives rise to a mixed gel comparable to that of the allophane of andosols (Bartoli *et al.* 1981).

**Formation of complexing organic compounds.** In Chapters 1 and 3 a detailed account was given of weathering by complexolysis and movement by way of chelates. Here, an account will be given of how these processes operate in podzols.

Bruckert (1970) and Razzaghe-Karimi and Robert (1975) showed the part played by soluble organic compounds coming from the litter and the A0 horizon of the mor in the formation of complexes with Al and Fe; certain aliphatic acids (citric and oxalic for example) appear to play a major part, particularly in the weathering phase, but certain phenolic polymers, the molecular weights of which do not generally exceed 2500 to 3000 (precursors of fulvic acids), but which can reach and pass 5000 in very developed *Calluna* podzols (Boudou 1977), are also effective complexing agents. This was confirmed by Berthelin (1976), who has also shown the part played by certain micro-organisms (*Pseudomonas*) in the development of these complex-forming polymers. According to this author, the acidifying litters are involved more particularly in an indirect manner by favouring a partial sterilisation of the soil: *the antimicrobial compounds produced by this litter modify the microbial metabolism and stimulate the production of strongly complexing compounds (oxalic acid and polymers)*.

As a general rule, complex formation occurs at a very low pH (4) and in periods of heavy rainfall, which thus coincides with the short-term waterlogging of the soil profile, the Eh then being at a minimum at the A0/A2 boundary, *the iron is generally complexed in the ferrous state*. Such true complexes are not only mobile but relatively stable; they are resistant to a rise in the pH and Eh which can occur at the bottom of the A2. Under conditions of better aeration, in contrast, salt complexes are formed that are less mobile and less stable; precipitation occurs more rapidly (brown podzolic or podzolic soil, with little if any A2 development).

**Weathering of silicates by complexolysis** (Souchier 1971, Belousova *et al.* 1973, Fig. 10.2). Contrary to what is thought, *on average* the amount of weathering of complex silicates in a podzol is not as great as that which occurs in a brown soil: this is essentially because of the small amount of weatherable minerals in the podzol parent material. Minerals difficult to weather, such as muscovite and orthoclase, are generally resistant to the effect of these active soluble organic compounds. However, the amount of weathering is very variable, depending on the horizon; thus it is at a maximum in the A0A1 and minimal in the B because the covering of the minerals in this horizon by the surface-derived amorphous materials has a protective effect.

The production of amorphous materials by polymerisation of soluble materials ($SiO_2$) or complexes (iron and aluminium) after fairly considerable movement is much more important than clay formation. In the extreme leaching environment of a true podzol, the individual layers of the micaceous clays open up and an Al-vermiculite is formed, subsequent removal of the inter-layer Al produces a poorly crystalline *montmorillonite of degradation*, which

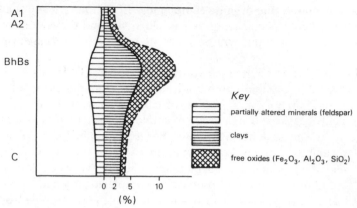

**Figure 10.2** Weathering complex of a humic podzolic soil on granite (after Souchier 1971). Compare with the weathering complex of an acid brown soil and a mesotrophic brown soil (Fig. 9.3).

in extreme cases is entirely destroyed (Fig. 10.2). The Al-ions that migrate react with previously moved illite–vermiculite to form Al-chlorites in the B horizon.

De Coninck *et al.* (1968), Souchier (1971) and Guillet *et al.* (1975) showed, either by the method of balances or experimentally, that a considerable amount of clay destruction occurs in the A2 horizon. Sawney and Voigt (1969) and Razzaghc-Karimi and Robert (1975) demonstrated that citric and other complexing acids destroyed the octahedral layer of clay minerals leading to structural disorganisation and the formation of amorphous products. Berthelin (1976) and Schnitzer (1981) obtained the same results with the reaction of phenolic polymers of microbial origin on biotite. Vermiculisation and partial destruction occurred leading to the formation of a white pulverescent residue. Owing to their rapid mineralisation, the effectiveness of the complexing aliphatic acids as weathering agents is restricted to the surface horizons. At lower levels they are replaced by phenolic polymers that act mainly as *carriers* of Fe and Al during their migration (Robert *et al.* 1979).

**Migration and precipitation of complexes.** Under conditions that are permeable and poor in free oxides, such as in quartz sands, the complexes formed have very low cation : anion ratios, which keeps them in the soluble state and assures their mobility. *These complexes are insolubilised in the B horizon by the sudden rise in the cation : anion ratio* – or, if it is preferred, the decrease in the anion : cation ratio (Petersen 1976, Fig. 10.3). The biological activity which is not negligible at the top of the B is involved in biodegrading part of the complexing anions, which raises the cation : anion ratio and frees a part of the Al and Fe. In addition, the previously freed hydrates of Fe and Al in the B immobilise by adsorption the newly transported complexes (adsorption complexes).

The selective extraction of complexes and the determination of free iron for a type podzol (Bh horizon on Vosgian sandstone) give the following results (Bruckert 1970, Table 10.1).

It is known that despite the *true complexes* being present in such minor amounts, migration mostly occurs in this form.

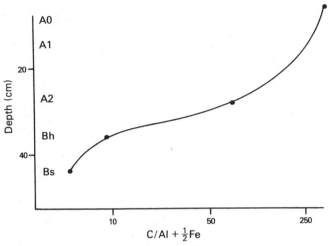

**Figure 10.3** Development of the $C : Al + \frac{1}{2}Fe$ ratio in the mobile complexes extracted from a Danish podzol (after Petersen 1976).

**Table 10.1** Selective extraction of complexes.

|  | Iron % | Aluminium % |
|---|---|---|
| true complexes (mobile) | 6 | 5 |
| salt complexes | 24 | 18 |
| adsorption complexes and free oxides | 70 | 77 |

Bruckert showed, in fact, that biological decarboxylation occurring in the B horizon is one of the main causes of initial precipitation of a part of the complexes; the complexing ability thus being decreased, a part of the complexed iron and aluminium is liberated and immobilised, then polymerised; the hydrates of free iron and aluminium thus formed cause in their turn an insolubilisation by adsorption of soluble silica and new complexes carried by subsequent movement. A general biodegradation of complexing organic anions then occurs gradually, so that the amounts of free alumina and iron oxides increase yet again (Berthelin 1976, Fig. 10.4).

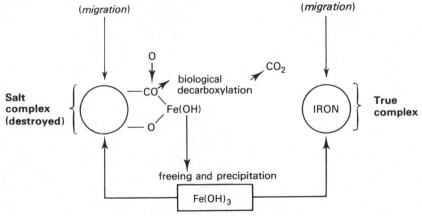

**Figure 10.4** Biochemical processes of insolubilisation of iron complexes within a spodic horizon.

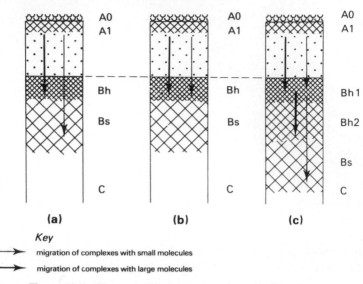



Key

→ migration of complexes with small molecules

➤ migration of complexes with large molecules

**Figure 10.5** Phases in the formation of spodic B horizons.

The formation of the two superposed spodic horizons remains to be explained (Fig. 10.5), one with large humic acid molecules (Bh horizon), the other with small fulvic acid molecules (Bs horizon). It seems that spodic horizons pass through two or sometimes even three stages in their formation. (i) In *phase one*, both horizons are formed at the same time by the movement of small molecules (simple aromatic acids) to a greater depth in the profile than the larger molecules (polymers); after precipitation each of these types of molecule very gradually polymerises, as a result of seasonal climatic changes and the catalytic effect of the free iron in the B horizon, to form the humic acids, dominant in the Bh horizon, and the fulvic acids, dominant in the Bs. It should be remembered also that *aluminium is more mobile than iron and that it accumulates to a greater extent in the Bs horizon*. (ii) The *second phase* occurs in well developed iron–humic podzols The amount of free iron hydrates, which are responsible for insolubilisation, increases with the age of the podzol, and this happens more particularly in the Bh horizon, so that eventually all complexes, no matter what the size of their molecules, are insolubilised in the Bh horizon which thus forms an absolute chemical barrier that cannot be crossed by mobile complexes. The $^{14}C$ determinations by Guillet (1972) confirmed this by showing that the mean age of the Bs was always greater than that of the Bh. (iii) The *third phase* occurs in very old, very well developed podzols and involves a thickening of the B horizon with no change in the position of its upper boundary (Franzmeier & Whiteside 1963, Harris 1970) by a remobilisation of organic matter, as well as sesquioxides, in its upper part and its reprecipitation lower down the profile (Bruckert 1970, Figs 10.5 & 6).

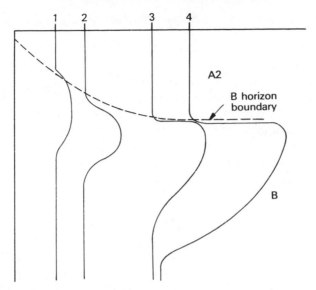

**Figure 10.6** Phases in the development of a podzol, showing the sequential distribution of amorphous organic and mineral compounds (after Franzmeier & Whiteside 1963, Franzmeier 1965, Harris 1970).

## 4   Influence of environmental factors on podzolisation

As stated in the introduction to this chapter, the podzols have a different environmental significance depending on the climatic zones in which they occur. In the boreal zones, podzolisation is the generalised process of the coniferous forests or taiga and occurs over a considerable range of different materials (climatic climax). In an atlantic climate, the podzols are more restricted: they can be either **forested podzolic soils** on sandy or quartzose materials (site climax) or **podzols of degradation** (or secondary podzols) the development of which has been caused by human action in which forest has been replaced by heathland.

**Podzolisation in a boreal or alpine climate** (Duchaufour 1978: $XI_6$). In the boreal zone as well as in the high mountains, podzolisation is a zonal phenomenon, i.e. a generalised process that takes place on a range of materials. Only calcareous materials and certain very basic parent rocks (basalts and gabbros) appear to prevent it, at least partially, and development is then related in this last case to **cryptopodzolisation** (see Ch. 6). These profiles with little differentiation are called **podbur** by the Russians (Targulian 1971, Sokolov & Targulian 1976).

Even though biochemical weathering is intense, the processes of migration are reduced so that in general, compared to the atlantic podzols, the profiles of boreal podzols are much reduced in thickness. Thus, while the mor horizon is very thick (A0 20 to 30 cm), the ashy A2 horizon with powdery structure is not more than 8 to 10 cm thick, the B horizon can vary considerably in thickness, as it increases in thickness with age, while the A2 does not (Fig. 10.6).

The generally reduced thickness is explicable in terms of the climate, for podzolisation can only occur during the short period of surface thawing. As the subsoil remains almost permanently frozen and hence impermeable, the spodic horizon must form near the surface. In fact, boreal podzols with a well characterised spodic horizon (referred to by Russian authors as podzols with humus and iron illuviation) only occur on the most sandy of materials (reworked regoliths, sandy moraines). On less permeable loamy materials, the spodic horizons are almost completely lacking; only the upper horizons of a thick A0 mor, and a white and powdery A2, are identical with those of podzols formed on coarser materials. At depth, a Btg horizon occurs which is richer in clay and iron and brown in colour, often glossic, with white powdery streaks fringed by concretions. However, in general the process is always one of podzolisation, as has been established by Russian workers using the method of clay balances (Targulian 1971) in which it is estimated that, when compared to the C horizon, the A2 has lost almost 50% of its clay by acid hydrolysis.

At present, no complete answer can be given for the absence of a spodic horizon on loamy material. However, a great deal can be explained in terms of the strongly hydromorphic conditions of the whole profile, for in these circumstances biological activity remains low and the processes of decarboxylation and polymerisation, which both result from microbial oxidation (phenoloxydases), cannot occur. In these circumstances, ferrous iron can no longer be re-oxidised and the complexes remain soluble and are removed either laterally via the water table or by drainage to depth (pseudopodzolisation, Zonn 1973).

Climatically determined soils formed on loams can be called **boreal hydro-morphic podzolic soils** and are a perfect example of an *analogous soil*, when compared to the podzols with a spodic horizon, formed on sandy materials, in the sense intended by Pallmann (see Ch. 4).

**Podzolisation on plainlands in an atlantic climate.** Numerous authors have commented on the differences between the *primary* forested podzols (or rather podzolic soils that are less developed than podzols) which are true *site climaxes*, characteristic of certain very sandy parent materials, and the secondary iron–humic or humic podzols that are the result of *degradation*, under the influence of heath plants and *Calluna* (see Ch. 4 on this subject).

*The forest podzolic soils.* On very quartzose sands, impoverished in clays and weatherable minerals, podzolisation occurs very rapidly, even under deciduous forest that palynology shows to have been originally oak (Dimbleby 1962, Munaut 1967), by the migration of soluble organic compounds before they can be biodegraded. Podzolisation goes back to the end of the atlantic period (4000 years ago), but the fairly labile organic matter is subject to a rapid biodegradation in the B horizon and does not accumulate beyond 1–3%, with a MRT of only 200–300 years measured by $^{14}$C: figures that give no idea of when exactly podzolisation started. The cycle of iron and aluminium is another special characteristic of these podzolic soils, for as the amount of free iron and weatherable minerals in the parent materials is very small, the biological mobilisation of iron and aluminium (see Ch. 3) becomes more important than chemical mobilisation (Juste 1965). The major part of the iron and aluminium migrates into the B horizon in the form of complexes, formed in the litter by way of the biogeochemical cycle. In these circumstances the amount of iron that accumulates in the B is not very great and the distinction between Bh and Bs horizons becomes very vague.

The possible presence at depth of an argillic horizon, under the spodic horizon, is proof of a pervection prior to this podzolisation (*indirect podzolisation*). Depending on the parent material, this argillic horizon can be very varied (banded Bt, hydromorphic Btg, or a $\beta$-horizon on limestone). In fact, podzolisation frequently does not occur without there having been a preliminary pervection of clay and iron, for it cannot start until the amount of clay and iron falls below a threshold value. This situation occurs all the more quickly as a result of certain practices that increase the acidity, such as the collection of litter as is done in the forests west of the Paris basin.

*The podzols of degradation (iron–humic and humic)*. The causes and the stages of degradation have been discussed in Chapter 4, where it was shown that degradation is initiated by the colonisation of the forest soil by heath plants and *Calluna*. For this to occur, forest destruction must be over a considerable area, for the heath is light-demanding (Duchaufour 1948, Dimbleby 1952, Galoux 1954, Munaut 1967, Anderson 1979, Robin 1979). It has been shown (Guillet 1972) that the phenolic precursors coming from the *Calluna* litter are particularly involved, for as a result of their resistance to biodegradation they are able to migrate, taking along the small amount of iron there is in the A2, and accumulate as a characteristically black Bh horizon that is often indurated (humic pan). As well as the increase of organic matter in the Bh horizon, the index of iron movement increases and the horizon contrast becomes more marked. Generally, as this degradation is the result of indirect podzolisation, the Bh horizon is more recent than the Bt horizon of the previous forest stage, and therefore overlies it (see Figs 9.6 and 9.7).

Note that on sands very poor in clays and iron-rich minerals (Stampian sands of the Paris basin), *Calluna* has been established for some 3000 to 5000 years. In these circumstances the profile has neither a Bt nor Bs horizon, but only a thick, organic-rich, iron-poor indurated Bh horizon (humic pan) and the soil is an **indurated humic podzol** (Robin 1979).

**Podzolisation in mountains in an atlantic climate.** Podzols of this type have been studied in the Vosges by Souchier (1971), Guillet (1972) and Guillet *et al.* (1975), both on the Triassic sandstones of the low Vosges and the granites of the high Vosges. This involves a special case of **direct podzolisation** (Chirita *et al.* 1970), in which both the fairly small amount of clay and the iron and aluminium complexes are mobilised at the same time, i.e. pervection and podzolisation occur simultaneously, to produce a mixed BtBs horizon.

Under the mixed forest of beech and fir, on both sandstones and granite regolith, the occurrence of the four soil types – acid brown soil, ochric brown soil, brown podzolic and podzolised soil – is dependent on the amount of free iron and clay that these parent materials contain, while iron–humic podzols under *Calluna* are restricted to the sandstones of the low Vosges.

*The low Vosges sandstones.* Under the mixed beech–fir forest cover, the soils never go past the slightly humic podzolic stage. The iron–humic podzols of degradation are restricted to areas that were cleared or cultivated about 1000–2000 years ago where the slopes have a southern exposure. Since then, a forest of Norway pine has been established by man, which has caused some of the *Calluna* to be replaced by less light-demanding acidophilic plants (*Vaccinium myrtillus, Pteridium aquilinum*). An account of this work (Guillet 1972) was given in Chapter 4 (Fig. 4.11).

The detailed study of the morphological and biochemical development of an iron–humic podzol profile on sandstone, together with a balance sheet of the clay minerals, enabled Guillet *et al.* (1975) to evaluate the stages of weathering and movement within the profile during its development, first of all under forest and then under *Calluna*.

During the forest stage, the organic compounds coming from the litter only cause a moderate degree of weathering by complexolysis but act as energetic dispersing agents of micaceous clays, which causes their migration. Only the micaceous clays are pervected (the inherited kaolins remain in the A2) to produce very characteristic argillans in the B horizon, with which are associated amorphous materials from the weathering of primary minerals.

During the *Calluna* stage (degradation), clay pervection appears to stop; in contrast, some of the micaceous clays are degraded and become completely amorphous, which is reflected in both clay balance determinations and the visible degradation of B horizon argillans. Complexolysis reaches its maximum intensity; but a great part of the complexes formed are removed from the profile; those that accumulate in the Bh are rich in humic compounds that are very resistant to biodegradation (black Bh of *Calluna*).

*High Vosges granites.* Souchier showed that the forest podzolic soils are concentrated on very quartzose granites that are very poor in weatherable minerals (Brézouard and Kagenfels granites). An important difference from the podzolisation on sandstones at low altitudes should be emphasised; the fir tree forests of the high Vosges are at an altitude of 800 to 1100 m, corresponding to the middle or upper montane zone. Under the influence of the resulting moderately low temperatures, the complex-forming organic compounds mineralise slowly and accumulate in all horizons, particularly the B, as irregular patches. However, they remain in the fulvic acid state and have a reddish-brown colour and not black as in the iron–humic podzols. The result is a **humic podzolic soil** (Duchaufour 1978: $XI_3$).

As a result of the difficulties connected with their steep slopes, the forested high Vosges was not settled in the Middle Ages, as was the low Vosges: hence there is no sign of degraded soils.

## 5  *Main types of well drained podzols* (Fig. 10.7)

**Basis of classification.** As stated in the introduction, the classification of podzols is essentially dependent on their degree of development and thus the

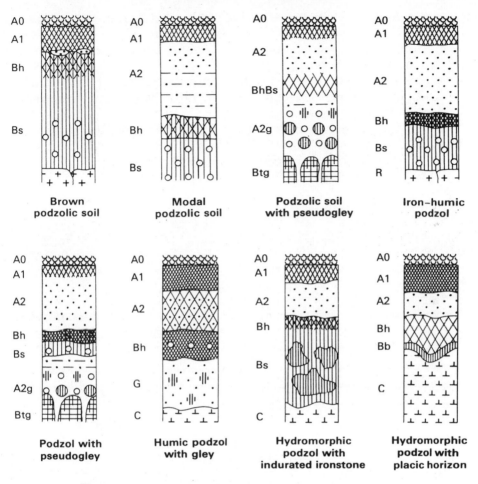

**Figure 10.7** Podzolised soil profiles (see general key, p. ix).

amount of profile differentiation is taken as the basis of the developmental sequence:

ochric brown soil ——→ brown podzolic soil ——→ podzolic soil ——→ podzol

These four phases of overall development of podzolised profiles can be differentiated by using well known criteria.

*Ochric brown soils.* Some are very near to acid brown soils; for others the ochric (B) horizon corresponds to that of a spodic horizon (Bruckert & Souchier 1975); but at this stage of development its recognition is difficult and it is for this reason that all of the ochric brown soils have been placed in the *brunified* class (see Ch. 9).

*Brown podzolic soils*. Profile A0A1BhBsC; horizon B clearly spodic (quartz grains are covered), situated below an A1 (quartz grains transluscent). Horizon A2 is not present or it occurs as discontinuous patches, sometimes forming an ashy parting.

*Podzolic soil*. Profile A0A1A2BhBsC; well developed A2 but not ashy; spodic horizon only slightly humic (Bh hardly to be seen) and sometimes it does not occur in the hydromorphic type.

*Podzols*. They have a maximum of profile differentiation, the decolourised ashy A2 horizon sharply overlies the black (or dark brown) Bh horizon and the strongly ochreous or rusty coloured Bs horizon; in general they are termed iron–humic (or humo-ferric, Avery 1980).

Two types are differentiated:

(a) *primary podzol (boreal or alpine)* with a profile that is not too deep;
(b) *secondary podzol* (atlantic) with a deeper profile that, as previously stated, results from the degradation of a forest profile.

Depending on whether the podzolisation by degradation of forest soils occurs at a more or less rapid rate (direct or indirect podzolisation), these secondary iron–humic podzols may or may not have an argillic horizon at depth, i.e. below the spodic horizon. This enables several groups of atlantic podzols to be differentiated:

(a) *humic podzols*, often with a pan (A0A1A2Bh);
(b) *simple iron–humic podzols* (A0A1A2BhBs);
(c) *iron–humic podzols with argillic horizon* (A0A1A2BhBsBt);
(d) *iron–humic podzol with β-horizon* on indurated limestone (A0A1A2BhBsβ);
(e) *iron–humic podzol with pseudogley*, the deep Btg horizon being choked with perverted clay (A0A1A2BhBsBtg).

Thus, this last group of podzols has temporary hydromorphic conditions, but only at depth, and it should not be confused with hydromorphic podzols where the whole of the profile is subject to poor drainage. As iron–humic podzols have been discussed as the type profile, some additional details of the other groups only (brown podzolic soils, podzolic soils and humic podzols) will be given at this point.

**Brown podzolic soils** (Duchaufour 1978: $XI_1$). The profile is of the A0A1BhBs type, the A2 horizon being absent or occurring only as white patches or as an ashy parting at the A1Bh boundary. As in all podzolic soils, the Bh horizon is brownish and diffuse, grading into the Bs horizon, which is much thicker (20–30 cm) and has a characteristically strong ochreous colour. It is mellow, with a flocculated (fluffy) structure (spherical intergranular aggregates), generally of chemical origin (Clayden 1970) but sometimes it is coprogenic (Righi 1975). Ironpan formation never occurs.

This soil is characterised by moderate podzolisation (weak index of movement, no A2) and by a biological activity that is much more intense than that which characterises the more developed soils of the series, and there are many roots in the B horizon. On the sandstones of the low Vosges, the brown podzolic soils are very good forest soils with a productivity that is two to three times higher than that of the iron–humic podzols and almost the same as that of the acid brown soils.

*Environment of the brown podzolic soils.* The environmental conditions are intermediate between those characteristic of the acid brown soils or ochric soils on the one hand and podzolic soils on the other. For instance, this is so for the iron and weatherable mineral content of well drained brown podzolic soils (Souchier 1971) and in the case of the vegetation, which gives rise to a fairly active litter owing to the presence of certain ameliorating species, such as when there is a covering of grasses (*Molinia*) or old degraded oak coppices on warm slopes on Vosges sandstone.

The steepness of slopes also plays an important part, for it can prevent the formation of an A2 horizon, when other conditions are favourable for podzolisation, by causing the lateral movement of organometal complexes. At the top of slopes, brown podzolic soils are relatively thin, whereas in contrast, at the bottom in the well aerated colluvial zone, they are very deep with an ochric Bs horizon up to 30 cm thick which is also explicable in terms of lateral movement.

At high altitudes, the brown podzolic soils become very rich in humus and the dark-coloured A1 and Bh horizons can be several tens of centimetres thick. Nys (1975) described such a very humic brown podzolic soil (which he indeed called a humic podzol) at 950 m altitudes, on the plateau of Millevaches, on a very acid two-mica granite where all horizons are extremely rich in organic matter (14% in the Bh). This type of podzolised soil, frequent in the very humid mountains, is also related to the cryptopodzolic rankers, but the movement of aluminium and iron is much more marked because of the low amount of weatherable minerals in the parent material (see Fig. 4.6).

**Podzolic soils.** The podzolic soils differ from the brown podzolic soils in having a well developed A2. Two types can be differentiated: (i) non-hydromorphic podzolic soils with easily recognisable spodic horizons, although less clearly differentiated than in the case of the podzols; and (ii) podzolic soils with hydromorphic conditions at depth. These hydromorphic conditions, even though they are only temporary (pseudogley), slow down the formation of the spodic horizon by preventing the processes of oxidation necessary for the precipitation of the complexes and the formation of the organic or mineral insoluble amorphous compounds that form the active part of the spodic horizon. In these circumstances, this horizon is completely absent (boreal podzolic soil) or reduced to diffuse brown patches.

The first category consists of the well drained podzolic soils on sandstone or granitic regolith:

(a) *Modal podzolic soils* (Duchaufour 1978: $XI_2$).
(b) *Humic podzolic soil of high altitude* (Duchaufour 1978: $XI_3$).

The second category consists of the hydromorphic podzolic soils:

(a) *Podzolic soils with pseudogley* (atlantic: Duchaufour 1978: $XI_4$). This is a degraded soil where, if the temporary water table formed seasonally above the pseudogley horizon occurs near the surface of the soil, development never gets beyond an early stage. However, when this is not the case a complete iron–humic podzol is developed in the material above the A2g horizon.

(b) *Boreal hydromorphic podzolic soil*. This is the analogue of the boreal podzol, but formed on loamy material, which has been described in Section 4 above.

**Humic podzol** (Robin 1979). This is a podzol with a strongly differentiated profile formed on quartz sands with a low content of iron and clay. A very black and very thick Bh horizon is formed and when this is indurated (**humic pan**), which is often the case, it belongs to a separate group, *the indurated humic podzols*.

## III  Hydromorphic podzolised soils (acid water table)

This subclass consists of those podzols whose development is controlled by a permanent, or almost permanent, acid water table, frequently subject to slow lateral flow (even if the slope is very slight); the acid water table causes a reduction and a mobilisation of ferric oxide which can then be transported either vertically or laterally over fairly considerable distances.

These soils are typically intrazonal and are site climaxes, related to special site conditions which allow the water table to form; certain of them result from a *hydromorphic degradation* owing to a modification of the vegetation. Evidently there are numerous intergrades, for example with the preceding subclass (particularly with the boreal hydromorphic podzolic soils) and also with the subclass of stagnogleys (of the hydromorphic class) in very humid mountains (see Ch. 11).

There are fundamental differences between the development, and consequently the character, of these hydromorphic podzols and the podzols of the previous subclass, which can be summarised as follows.

The presence of an acid water table becomes the main environmental factor, with the others losing their importance. This is particularly the case with parent material where even one that is rich in ferric iron can be podzolised owing to the solution and removal of the iron in the ferrous state. It is the same in the case of vegetation which no longer needs to be particularly acidifying: in hot climates, this kind of podzolisation is no longer incompatible with a fairly active mull.

The order of mobility of iron and aluminium is reversed. *While aluminium is more mobile than iron in the podzols of the previous subclass, the reverse is the case with hydromorphic podzols* (Juste 1965). Iron is often eliminated from the profile, being transported by the lateral flow of the water table for great distances, and it re-precipitates in zones where re-oxidation is possible. However, aluminium, often of biological origin, moves vertically and its precipitation in the Bh horizon occurs rapidly.

The iron, being mobilised mainly in the form of salts (ferrous bicarbonate), precipitates at first as an amorphous oxide (ferric hydrates) and rapidly becomes crystalline. In the Bh horizon (poor in iron) the abundant organic matter prevents all crystallisation of iron while, in contrast, in the iron-rich horizons there is insufficient organic matter to prevent crystallisation (Schwertmann 1966, Guillet et al. 1975). Thus goethite is formed which makes up the cement of the indurated iron pan.

The site conditions that allow the formation of the water table and give it a fair degree of mobility are very important, particularly that of the local topography. Hydromorphic podzols always form part of a topographic catena, and the profile has considerable local variations, both for those occurring in the sandy lowlands and those on high-level platforms or mountain slopes. Of course, these variations are more gradual and less abrupt in the first case than the second. In addition, the water tables are formed in different ways, for the podzols of the sandy lowlands are associated with a phreatic water table with slow lateral flow and iron movement over great distances (**groundwater podzols**); the podzols of the very humid mountain areas result from the formation of a perched water table, as a result of rainfall, often after a secondary modification of the vegetation, which means that reduction and precipitation of iron occurs over much shorter distances.

## 1 The type profiles: podzols of sandy lowlands with a phreatic water table

As an example, the hydromorphic podzols of the Landes of Gasgony of southwestern France will be considered (Juste 1965, Jambu & Righi 1973, Righi 1975).

In this extensive sandy plain with very permeable materials, the acid phreatic water table flows slowly and transports iron. This movement is away from profiles that are relatively high in the landscape and it results in the almost complete removal from them of the reduced ions of $Fe^{2+}$ and $Mn^{2+}$ and their subsequent concentration and deposition in lower areas, often as accumulations of great thickness (see Ch. 4, p. 132 et seq. & Fig. 10.8). Thus it is possible to differentiate the following kinds of profile:

(a)  upslope podzols poor in iron: **hydromorphic humic podzols;**
(b)  intermediate, often mixed podzols: **humic podzols with concretions;**
(c)  zone of large-scale iron deposition: **podzol with an indurated concretionary iron pan.**

**Hydromorphic humic podzol** (Duchaufour 1978: $XII_3$, Righi 1977). The profile type is A0A1A2Bh because of the absence of ferric iron accumulation, but sometimes a gley horizon can occur at the base of the profile, which is

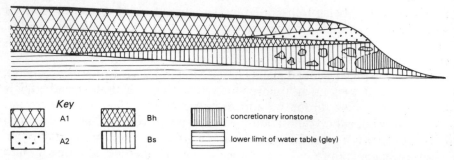

Key

A1            Bh            concretionary ironstone

A2            Bs            lower limit of water table (gley)

**Figure 10.8**  Distribution of Bh and Bs horizons in a catena of podzols with phreatic water table (Landes of Gasgony).

either white in acid sandy conditions or greenish with rusty patches because of the iron in more clayey conditions; the succession of horizons is then A0A1A2BhG (humic podzol with gley).

The humus is a hydromor or hydromoder that impregnates the A horizon with organic matter so that the A2 is less distinct and is often simply a rather lighter band. The quartz grains in the 30–40 cm thick A horizon are translucent, while in the Bh horizon they are covered and cemented together, by the humic compounds of insolubilisation from the surface, so that it is black at the top and coffee brown at its base. There are also small rounded aggregates within this horizon (pellets) whose origin, according to de Coninck *et al.* (1973), is thought to be coprogenic. The C : N ratio is lower than in well drained podzols with ericaceae and is often lower than 20. If the organic matter is greater than 3%, the Bh horizon becomes more compact and forms a humic pan which, however, is less indurated than the iron pan of the third type of profile, for even though few roots penetrate it, it can be easily broken. In addition it should be noted that even though the pan does not contain iron, it does contain significant amounts of complexed aluminium. This humic pan is similar to that of the well drained humic podzols, but in this case it is the hydromorphic conditions that have eliminated the iron by reduction. The characteristic vegetation of this soil is a heath with *Molinia coerulea* and *Erica tetralix*.

**Humic podzol with concretions: intermediate type.** This type differs from the preceding one in that iron has begun to accumulate; it can form a ferric Bs horizon as a continuous band, but generally the precipitation of ferric oxides occurs locally in patches, in the better aerated zones: thus large irregular concretions are formed, replacing old dead roots.

**Podzol with indurated concretionary ironstone** (Duchaufour 1978: XII₄). This type of profile is characteristic of the zones where the water table reaches the surface, at breaks of the slope low in the landscape or around the edges of valleys. The iron and the manganese oxidise and precipitate to form a mass of very indurated ironstone which can be several decimetres thick; the A2 horizon, better drained, is less humus rich, more decolourised, than in the preceding profiles; in zones with a grass vegetation where the roots are more superficial the iron oxidises near the surface in the form of more or less platy indurated discs (Schlichting 1965: *Raseneisenstein*).

**Utilisation of hydromorphic podzols.** The heath soils were improved up to the middle of the 19th century by drainage and re-afforestation, generally with maritime pines whose only effect has been to accelerate the process of podzolisation. More recently, diversification into agriculture has occurred (maize). However, not all of the hydromorphic podzols can be cultivated, such as those where there is a surface deposition of ironstone. Only the fairly undeveloped humic podzols (i.e. up to the podzolic stage with a slightly compact humic pan) can be cultivated if the water table does not reach the surface in winter and falls to a depth of 1–1.5 m in the summer. Calcic amendments are essential to remove aluminium toxicity; deficiencies are frequent and must be dealt with by the addition of minor elements (for example copper).

## 2 Development of hydromorphic podzols with a phreatic water table (Fig. 10.9)

The acid organic matter favours the reduction of iron by lowering the pH and the Eh of the water table. In these circumstances the iron is totally reduced and made soluble, but only a small part of this forms organic complexes as most occurs as a very mobile ferrous salt (Juste 1965, chemical mobilisation)

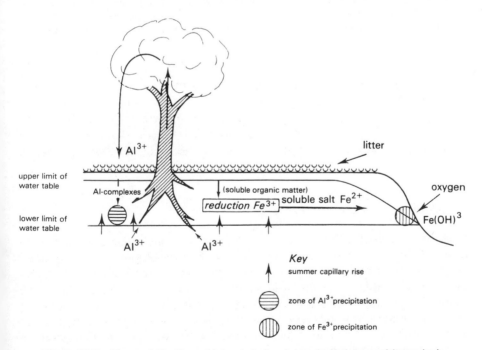

**Figure 10.9** The mobilisation of iron and aluminium in hydromorphic podzols.

which is transported by the water table often over great distances (Daniels *et al*. 1975). In contrast, the aluminium which is concentrated, often in considerable amounts in the Bh horizon, has a biological origin, as shown by Juste (1965). It accumulates in pine needles and hence in the litter, whence it is transported as organic complexes by infiltrating water, in which form it is less mobile than the iron salt: it moves vertically in the profile, rather than laterally, and accumulates in the Bh (Fig. 10.9). *This explains why the accumulations of aluminium on the one hand and iron and manganese on the other are separated in space and why the very indurated iron-rich deposits contain little aluminium and organic matter*. In the humic podzols, aluminium replaces the iron, thus it is the sole agent involved in the precipitation of the organometal complexes. As is well known, physical causes are also involved to a major extent in the precipitation of these complexes, for their movement occurs in summer when the water table is at its lowest level and the Bh horizons form at the upper limit of the capillary rise from the water table. Thus another difference between the behaviour of iron and aluminium becomes apparent for *it is the water table that moves the iron laterally and it is rainwater that moves the aluminium complexes vertically*.

Several research workers have reported this special segregation of iron and aluminium that is so characteristic of hydromorphic podzols. Schlichting (1965) has done this for hydromorphic podzols on the sands of northern Germany; Jacquin *et al*. (1965) have reported the presence of strongly polycondensed humic acids (grey AH) containing complexed aluminium in the humic pan in the Landes of Gasgony; in similar conditions in the USSR, Chekalova (1965) has reported the aluminium–humus complexes developed in place, while the iron is completely dissolved and removed from the profile. Finally, Menut (1974) has made similar observations with regard to the acid peats. They behave as humic podzols, for while iron is removed via the water table, the biological aluminium moves as a complex in the mass of the peat and accumulates at its base (see Ch. 11: Peats).

As already stated, the physicochemical conditions of this precipitation favour the rapid crystallisation of ferric and manganic oxides. This crystallisation is the reason for the very great degree of induration of such deposits, which can either consist of large irregular concretions within cavities or old roots, or a massive deposit in the zone where the water table emerges at the surface.

## 3  Main types: classification

**Basis of the classification.** Hydromorphic podzols can be classified either in terms of the climatic conditions controlling their development, which allows temperate and tropical hydromorphic podzols to be differentiated, or in terms of the water table – the phreatic water table of the lowland plains or the temporary perched water table resulting from rain. Because development in a tropical climate gives soils special characteristics, such as great profile depth and a more marked release of the alumina, it is felt that classification based on climate is more fundamental.

**Temperate hydromorphic podzols**
(1) *Phreatic water table of the sandy lowlands:*
    (a) humic subgroup;
    (b) humic subgroup with concretions;
    (c) subgroup with indurated concretionary iron pan.
(2) *Temporary perched water table* (resulting from rainfall):
    (a) podzol with a thin iron pan (placic horizon).

**Tropical hydromorphic podzols**
(1) *Lowlands of quartzose alluvium:*
    (a) humic subgroup.
(2) *Lowlands of ferrallitic alluvium:*
    (a) subgroup with secondary gibbsite.

Two very special kinds of hydromorphic podzols will be considered in more detail: the hydromorphic podzol with a placic horizon (humid temperate climate) and the tropical hydromorphic podzol with secondary gibbsite.

**Hydromorphic podzol with placic horizon** (Duchaufour 1978: $XII_5$). This type of hydromorphic podzol is characteristic of mountainous areas with a very humid climate, where heavy rainfall and low evapotranspiration favour the formation of a surface water table in sands or deep regoliths overlying indurated acid rocks; on plateau sites these acid water tables flow very slowly, while, in contrast, on slopes movement is more rapid. Under the influence of the water table a *stagnogley* is formed first of all, with accumulation of a peaty type of organic matter (anmoor or hydromor) and mobilisation of the iron in the ferrous state (see Ch. 11). Then the stagnogley develops towards a podzol by the formation of **placic horizon** which has a very indurated sinuous band of ironstone, formed by the precipitation of iron and manganese transported in a zone of marked contrast in grain size which makes it more aerated, less humic and less acid. It is this abrupt change in the grain size that would seem to cause the lateral flow of the water table and thus favour the formation of this horizon (**Bändchenpodzol** of the German classification, **lacaquod** of the US classification and the **stagnopodzol** of Avery *et al*. 1977).

This type of hydromorphic podzol can be considered as one of the final phases of the development of peaty stagnogley:

$$\text{stagnogley} \longrightarrow \text{humic stagnogley with placic}$$
$$\text{horizon} \longrightarrow \text{podzol with placic horizon}$$

The podzol phase is distinguished by the appearance of a light grey or ashy A2 immediately below the peaty humic horizon.

In the formation of this type of podzol and the origin of the placic horizon, there are still several problems that have not been solved. Certain authors (Crampton 1963, Glatzel *et al*. 1967) suggested that it involved the degradation of a forest ochric or podzolic soil by a decrease in the *PET*, which increased the degree of hydromorphism and the formation at the very surface of an acid water table. This in turn causes the well drained moder to become a peaty hydromor which immediately mobilises the iron and manganese, concomitantly with the hydromorphic degradation of the clays. This process is often the result, as is particularly the case in the Black Forest of West Germany, of man's destruction of the forest and its replacement by a humid heath with less evaporative ability.

Crampton (1963) studied a similar development, in the humid mountains of Scotland, of a forest podzolic soil on a loam with little permeability and a prismatic structure. When hydromorphic podzolisation started, so did reduction in the centre of the prisms. The placic horizon was formed initially as patches around the more aerated periphery of the prisms, which subsequently joined up with one another.

**Tropical hydromorphic podzols** (Duchaufour 1978: $XII_6$). They are characteristic of the sandy coastal plains with phreatic water table of very humid equatorial regions. Thus they are typically intrazonal soils owing their formation to very special site conditions, which counteract the climatically determined soil development (ferrallitic soils) that takes place in well drained areas.

Two conditions control their formation: (i) the presence of a permanent water table; and (ii) a parent material impoverished in weatherable minerals and with a sandy texture, which allows the circulation of the water table and favours a very rapid acidification. However, it can also affect transported ferrallitic material containing quartz, kaolinite and a large amount of free iron, because *this iron is no longer an obstacle to podzolisation in as far as it is easily reduced and eliminated via the water table*.

Vegetation does not play any special part. Turenne (1975) showed that this hydromorphic podzolisation can occur under a reasonably active mull–moder, formed by a grass savanna. The alternation of relative drying and wetting of the humic horizon favours the phases of polymerisation–depolymerisation of the organic matter (see Ch. 2 on this subject). In periods of water saturation, water-soluble phenolic polymers with a molecular weight of about 3000 are formed, which cause the hydrolytic degradation of kaolinites and the formation of considerable amounts of free alumina. Part of this develops by incomplete crystallisation to form secondary gibbsite (Turenne 1975).

The profile is much more developed than that of temperate hydromorphic podzols, but is also clearly less humic, for the organic matter still biodegrades rapidly in periods of relatively good aeration, on the one hand because it is relatively labile and on the other because of the high temperatures. The brown or black Bh horizon is built up very slowly (very high MRT) by the gradual accumulation of the most resistant fractions. The ashy A2 horizon (or white with rusty patches) is very thick (sometimes more than 1 m); the B horizon is particularly enriched in organic matter on quartzose parent material, poor in kaolinite. In contrast, on parent material that contains kaolinite, a large amount of free alumina can be detected (amorphous and also as secondary gibbsite), this free alumina coming from the partial degradation of kaolinite. This allows two subgroups of tropical hydromorphic podzols to be differentiated depending on whether or not they contain secondary gibbsite.

# References

Altmuller, H. J. 1962. *Z. Pflanzener. Bodenk.* **98** (3), 247–57.

Anderson, S. T. 1979. Brown earth and podzol: soil genesis illuminated by microfossil analysis. *Boreas*, Oslo **8**, 59–73.

Avery, B. W. 1980. *Soil classification for England and Wales*. Soil Survey Technical Monograph **14**. Harpenden: Soil Survey of England and Wales.

Avery, B. W., B. Clayden and J. M. Ragg 1977. *Soil Sci.* **123** (5), 306–18.

Baril, R. and T. S. Tran 1977. *Can. J. Soil Sci.* **57** (3), 233–45.

Bartoli, F., E. Jeanroy and J. C. Védy 1981. In Migrations organo-minérales dans les sols témperés, 281–90. *Coll Intern., CNRS*, Nancy, 1979.

Belousova, N. I., T. A. Sokolova and N. A. Tyapkina 1973. *Pochvovedeniye* **11**, 116–32. (*Soviet Soil Sci.* **5** (6), 692–708.)

Berthelin, J. 1976. *Etude expérimentale de mécanismes d'altération par des microorganismes hétérotrophes*. State doct. thesis. Univ. Nancy.

Boudot, J. P. and S. Bruckert 1978. *Science du Sol* **1**, 31–40.

Boudou, J. P. 1977. *Etude comparée des constituants organiques de l'eau de gravite de deux écosystémes forestiers des basses Vosges greseuses*. Spec. doct. thesis. Nancy.

Bruckert, S. 1970. *Influence des composés organiques solubles sur la pédogénèse en milieu acide*. State doct. thesis. Univ. Nancy; *Ann. Agron.* **21** (4), 421–52 and **21** (6), 725–57.

Bruckert, S. and M. Metche 1972. *Bull. ENSAIA*, Nancy **XIV** (2), 263–75.

Bruckert, S. and B. Souchier 1975. *C.R. Acad. Sci. Paris* **280D**, 1361–4.

Chekalova, M. I. 1965. *Pochvovedeniye* **2**, 25–35. (*Soviet Soil Sci.* **2**, 133–9.)

Chirita, C., A. Vasu and C. Rapaport 1970. Studii technice si economice, pedologie. In *In memoriam*, M. Cernescu and M. Popovat (eds). Pedologie, Serie C, no. 18, 193–203. Bucharest.

Clayden, B. 1970. Soil survey, micromorphological techniques. *Technical Monographs* **2**, 53–68.

Coninck, F. de, A. J. Herbillon, R. Tavernier and J. J. Fripiat 1968. *9th Congr. ISSS*, Adelaide **IV**, 353–65.

Coninck, F. de, D. Righi, J. Maucorps and A.-M. Robin 1973. *4th International Working Meeting on Soil Micromorphology*, Kingston, Canada, 1–23.

Crampton, C. B. 1963. *Welsh Soils Discussion Group*.

Daniels, R. B., E. E. Gamble and C. S. Holzhey 1975. *Soil Sci. Soc. Am. Proc.* **39** (6), 1177–81.

Dimbleby, G. W. 1952. *J. Ecol.* **4** (2), 332–41.

Dimbleby, G. W. 1962. The development of British heathlands and their soils. *Oxford Forestry Memoirs* **23**.

Duchaufour, Ph. 1948. *Recherches écologiques sur la chênaie atlantique française*. State doct. thesis. Montpellier; *Ann. ENEF.* **XI** (1).

Duchaufour, Ph. 1978. *Ecological atlas of soils of the world*. New York: Masson USA Inc.

Duchaufour, Ph., R. Michaud and G. Millot 1950. *Ann. Géol. Appl. et Prospect Minière* **III**, 31–62.

Franzmeier, D. P. 1965. *Soil Sci. Soc. Am. Proc.* **29** (6), 737–41.

Franzmeier, D. P. and E. P. Whiteside 1963. *Mich. Agric. Exptl St. Q. Bull.* **46**, art. 1, 2, 3.

Galoux, A. 1954. *La chênaie sessiliflore de la Haute Campine. Essai de bisociologie*. Groenendaal: Edit. Stat. Rech.

Glatzel, K., R. Jahn and S. Müller 1967. *Sudwestdeutsche Waldboden im Farbbild Schriftreihe der Landesforstverwaltung Boden Württemberg*. Stuttgart: Selbstverlag der Landesforstverwaltung.

Guillet, B. 1972. *Relations entre l'histoire de la végétation et la podzolisation dans les Vosges*. State doct. thesis. Univ. Nancy.

Guillet, B., J. Rouiller and B. Souchier 1975. *Geoderma* **14** (3), 223–45.

Harris, S. A. 1970. In *Paleopedology. Origin, nature and dating of paleosols*, D. H. Yaalon (ed.), 191–8. Amsterdam: Israel University Press.

Jacquin, F., C. Juste and P. Dureau 1965. *C.R. Acad. Agron. Fr.* **51**, 1190–97.

Jambu, P. and D. Righi 1973. *Science du Sol* **3**, 207–219.

Juste, C. 1965. *Contribution à l'étude de la dynamique de l'aluminium dans les sols acides du Sud-Ouest atlantique*. Engng doct. thesis. Fac. Sci. Nancy.

Menut, G. 1974. *Recherches écologiques sur l'évolution de la matière organique des sols tourbeux*. Spec. thesis. Univ. Nancy.

Munaut, A. V. 1967. *Recherches paléoécologiques en basse et moyenne Belgique*. Doct. thesis. Univ. Louvain.

Nys, C. 1975. *Science du Sol* 3, 207–11.

Petersen, L. 1976. *Podzols and podzolisation*. Copenhagen: DSR.

Razzaghe-Karimi, M. 1974. *Evolution géochemique et minéralogique de micas et phyllosilicates en présence d'acides organiques*. Spec. doct. thesis. Univ. Paris VI.

Razzaghe-Karimi, M. and M. Robert 1975. *C.R. Acad. Sci. Paris* **280D**, 2645–8.

Righi, D. 1975. *Science du Sol* 4, 315–21.

Righi, D. 1977. *Génèse et évolution des podzols hydromorphes des Landes du Medoc*. State doct. thesis. Poitiers.

Robert, M., M. Razzaghe-Karimi, M. A. Vincente and G. Veneau 1979. In Alterations des roches cristallines en milieu superficiel. INRA seminar. *Science du Sol* 2–3, 153–87.

Robin, A.-M. 1979. *Génèse et évolution des sols podzolises sur affleurements sableux du bassin parisien*. State doct. thesis. Univ. Nancy.

Sawney, B. L. and G. K. Voigt 1969. *Soil Sci. Soc. Am. Proc.* 33 (4), 625–9.

Schlichting, E. 1965. *Chemie der Erde* 24 (1), 14–29.

Schnitzer, M. 1981. In *Migrations organo-minérales dans les sols témperés*, 229–34. Coll. intern. CNRS, Nancy, 1979.

Schwertmann, U. 1966. *Nature* 212 (5062), 645–54.

Sokolov, L. A. and V. O. Targulian 1976. *Soviet Soil Sci.* 4, 405–16.

Souchier, B. 1971. *Evolution des sols sur roches cristallines à l'etage montagnard (Vosges)*. State doct. thesis. Univ. Nancy.

Targulian, V. O. 1971. *Soil formation and weathering in cold humid regions*. Moscow: Nauka.

Turenne, J. F. 1975. *Modes d'humification et différenciation podzolique dans deux toposéquences guyanaises*. State doct. thesis. Univ. Nancy.

Vicente, M. A. and M. Robert 1975. *C.R. Acad. Sci. Paris* **281D**, 523–6.

Zonn, S. V. 1973. *Soil Sci.* 116 (3), 211–17.

# Chapter 11

# Hydromorphic soils

## I  General characters

Hydromorphic soils are characterised by the reduction or localised segregation of iron, owing to the temporary or permanent waterlogging of the soil
pores which causes a lack of oxygen over a long period. Depending on the
circumstances, the ferrous iron either accumulates in the profile and gives it a
greenish (or sometimes bluish) grey colour, or it is mobilised and moves very
locally, forming rusty patches or concretions of ferric iron within mineral
horizons.

In terms of the environment, hydromorphic soils occur in all climates and
form, together with their characteristic specialised plant associations, *site
climaxes* related to special features of the parent material or to poor drainage.
This is the situation for the gleys, most of the stagnogleys, the pelosols and the
planosols. Others, such as the pseudogleys and some stagnogleys, are the
result of processes of degradation owing to the build up of a surface water
table following the partial or complete destruction of the forest cover.

Thus hydromorphic soils are characterised by having a special environment
in which water controls the developmental processes and gives the profile
particular characters, which are different from those developed in comparable
but well drained conditions. To begin with, it is necessary to differentiate two
great groups of hydromorphic soils:

(a) *Hydromorphic soils sensu stricto* have a reducing water table within
    which the effect of redox processes on iron are plainly visible: pseudo-
    gley, stagnogley, gley and peats.
(b) *Related hydromorphic soils* are transitional to other classes. Although
    such soils are waterlogged at certain seasons, this is the result of a com-
    plex process in which absorption by capillary pores is important.

This type of hydromorphism is favoured by the swelling or semi-swelling clays
that these soils generally contain. Generally, redox processes in these soils are
of relatively little importance and they are complicated by additional processes which in certain cases can become dominant, such as the impoverishment of surface horizons in clays (pelosols and planosols).

Conversely, certain soils with a true phreatic water table are not considered
as hydromorphic soils, in as far as these water tables are not reducing, which
has been seen to be the case in alluvial soils where the water table varies
considerably in height and is relatively rich in dissolved oxygen owing to the

rapid rate of water renewal. Certain kinds of alluvial soil in which water-table fluctuations are somewhat less (1 to 2 m), or that have a less rapid water circulation, are transitional to hydromorphic soils (alluvial soils with gley at depth, semi-gley; see Ch. 6).

As far as true hydromorphic soils are concerned, there are two consequences arising from the lack of oxygen within the water table: (i) the reduction and mobilisation of iron oxides which differ considerably with variations in the Eh and pH of the water; and (ii) it may affect the speed of decomposition of fresh organic matter and the processes of humification. Thus it is necessary to discuss the influence that the various kinds of water table have on these two aspects of pedogenesis in hydromorphic soils.

## 1 Water tables at the surface and at depth: influence of Eh and pH on iron

Considering the $Fe^{2+}-Fe^{3+}$ equilibrium as a function of Eh and pH in the soil solution, as was done in Chapter 3 (see Fig. 3.4), two basic facts can be emphasised: (i) the reduction of ferric iron occurs at a lower Eh in neutral (and thus rich in calcium) than in acid conditions; (ii) in the absence of complexing organic anions, the solubility of ferrous iron is very low and is practically nil above a pH of 6.5. Thus, at neutrality the $Fe^{2+}$ has little mobility and it tends to accumulate in an insoluble form in the profile, to which it gives a greenish-grey colour.

These circumstances explain the contrast in hydromorphic processes that occurs in perched water tables formed from rainfall and within phreatic water tables at depth, in comparable climatic conditions.

Perched water tables, supplied by rainfall with dissolved oxygen, often have a relatively high Eh. In these circumstances a strong acidity and the presence of complexing organic matter are absolutely necessary for the reduction and mobilisation of iron. As these water tables are only temporary, when they disappear the complexing organic anions are biodegraded and the iron re-oxidised to the insoluble ferric state, forming patches or concretions (*pseudogley*).

In phreatic water tables at depth, the Eh is very much lower (see Fig. 11.5) and it can fall below zero in hot periods, particularly if the water table is not very mobile. In these circumstances iron reduction can occur even in neutral or slightly alkaline conditions (rich in calcium), but then the ferrous iron occurs as ferrous carbonate or insoluble complex salts and it persists in this form at least in the lower horizons (*gley*). It is only under more acid conditions that the mobilisation of iron starts, often leading to a decolouration of the gley horizons (*white gley*).

Thus, there is a close relationship between the kind of water table on the one hand and of redox processes and iron mobilisation on the other, which will allow these two types of hydromorphism to be placed in two different subclasses. However, it is to be noted that in a cold climate, processes occurring in very acid perched water tables to a certain degree are like those occurring in acid water tables at depth (*stagnogley*).

## 2   Hydromorphism and humification

It is known that all marked deficits of oxygen, accompanied by a lowering of the Eh, slow down humification and cause a peaty development of the organic matter, this effect being strengthened by the acidity (see Ch. 2). But, as previously stated, the nature and the properties of water tables are very variable, in terms of their seasonal duration, their degree of reductive power or acidity, and whether they affect the surface humic horizon over long periods or not at all. Thus the effect that water tables have on humification differs considerably according to circumstances: being very slight in the case of a pseudogley with a temporary perched water table with slight powers of reduction, and at a maximum in the peaty gleys and more particularly the organic hydromorphic soils (peats), for here the water table is permanent, highly reducing and varying little in level. The slowing down of organic matter decomposition and humification is thus very great, even though it is far from being entirely prevented, particularly in calcic conditions.

In these circumstances, at first sight, it would seem that hydromorphic soils can be classified in terms of their organic matter content, which increases with the degree of hydromorphism. An attempt along these lines was made in the French classification of 1967 in which three subclasses were differentiated: (i) mineral hydromorphic soils, (ii) moderately organic hydromorphic soils, and (iii) organic hydromorphic soils (peats).

However, with the exception of the third subclass (peats) which is special, a more detailed study shows that taking humification into account at the sub-class level is not satisfactory if, as it is generally admitted, it is the dynamics of iron in reducing conditions which should be the main base of the classifica-tion. For one such system, based on this threefold grouping, puts in the same subclass soils in which iron reactions are very different. Thus low humic gleys are classed with pseudogleys simply because the reductive water table does not reach the surface horizon, which remains aerated, and organic matter development is hardly affected by hydromorphism. But, as will be seen, these two environments are not at all comparable, either in terms of pedoclimate, or of pH and Eh, or even in profile morphology.

Thus the nature of humification, or in fact the degree of accumulation of slightly transformed organic matter, is only taken into account at a lower classificatory level, below that of the subclasses, with the well known excep-tion of the completely organic hydromorphic soils, the peats. This will cause the stagnogleys to be differentiated from the pseudogleys within the subclass of soils with perched temporary water tables, as in general the first is much more humus rich than the second (here again, as will be seen, there are exceptions). In the same way, within the subclass of soils with permanent phreatic water tables the mineral gleys are differentiated from the humic gleys.

## 3   Classification of hydromorphic soils

**Ranking of the criteria used.** From the previous discussion it would appear possible to rank the criteria used in the classification of hydromorphic soils.

*Importance of the redox processes of iron.* This allows the differentiation of soils that have a true water table (at the surface or at depth), in which these processes are important, from those in which the hydromorphism is a complex process related to the swelling of clays, in which the redox processes of iron are not important except in some transitional forms.

*Nature and time of persistence of water table.* Obviously this criterion is only used to differentiate within the hydromorphic soils *sensu stricto*, belonging to the first category mentioned above. Thus are differentiated:

(a)  perched water table of *pluvial* origin, pseudogley, stagnogley;
(b)  permanent phreatic water table, gley;
(c)  strongly reducing water table with little or no variation in height, causing a considerable accumulation of organic matter, peats.

*Development of organic matter and humification.* Obviously this criterion is the basis for the differentiation of the peats of the third type of hydromorphism mentioned above. In other cases it allows the differentiation, within each of the subclasses of the divisions or groups, according to the degree of organic matter accumulation and its humification.

As far as the soils with temporary perched water tables are concerned, the slightly humic pseudogleys are differentiated from the more humic stagnogleys, in which, however, the organic matter is less transformed.

In those soils with a permanent water table, the humic gleys are differentiated from the mineral (or low humic) gleys.

Some planosols have a chernozemic or vertic humus (these are the **mollic** planosols).

*Acidity and the degree of reactivity of the water tables.* The high content of very acid organic compounds of certain water tables not only increases their ability to reduce and complex with regard to iron, but it can also cause clay degradation by acid hydrolysis (see Ch. 1). This process is comparable to pseudopodzolisation which has been discussed with regard to the boreal soils. When it occurs, such hydromorphic soils can be called *podzolic*. This allows the differentiation within certain subclasses of different phases of development: (i) slightly transformed types, i.e. slightly acid with a moderate degree of iron mobilisation; (ii) modal types with a marked amount of mobilisation or reduction of iron; (iii) podzolic types with a hydromorphic degradation of the clay minerals.

### Proposed classification

*Soils with marked redox processes: water table present*
(1)  *Perched water table of pluvial origin.*
    (a)  Temporary water table, moderately reducing: **pseudogley**. The iron does not remain in the ferrous state but is reoxidised in the dry season to form patches or concretions: slightly humic profiles.

(b) Permanent stagnant, strongly reducing water table: **stagnogley**. The iron is partly in the ferrous form and is often removed from the profile. There is a tendency for slightly transformed organic matter to accumulate.

(2) *Permanent phreatic water table with variable acidity: gley*. Redox potential often very low with an accumulation of reduced iron at the bottom of the profile, which can be partially eliminated in acid conditions.

(a) Height of water table very variable: **low humic gley**.

(b) Variations in height of water table moderate: **humic gley**.

(3) *Permanently waterlogged organic hydromorphic soils: peats*.

*Related soils: hydromorphism resulting from capillary absorption by a clayey material, with surface impoverishment in clay* (generally redox processes are reduced)

(1) *Capillary absorption dominant*; impoverishment limited: **pelosols**

(2) *Formation of surface water table ephemeral* as a result of a marked impoverishment: **planosols**.

**Discussion: comparison with other general classifications.** The difficulties of defining and classifying exactly the various kinds of hydromorphic soils is the reason for the great variety of classifications that have been used. This makes it difficult to compare them with each other. However, it is possible to distinguish three main trends:

(a) *The American classification* considers hydromorphism as a secondary process superposed on the main pedogenetic processes, which alone are used to characterise the order. Thus there is no such thing as a hydromorphic order (apart from that of the histosols, peats). A hydromorphic sub-order is, however, differentiated in each of the main orders and is indicated by the prefix *aqu* (*aqu*alf, *aqu*ent, *aqu*ept, *aqu*od, *aqu*ult).

(b) *The German classification* (Mückenhausen 1975) takes the opposite viewpoint and from the start distinguishes three main divisions: terrestrial, semi-terrestrial (which corresponds to the hydromorphic soils of this book) and the sub-aquatic soils.

(c) *The FAO and English classification* (Avery 1973) adopts a viewpoint similar to that of this book by differentiating a special class (or two, to say nothing of peats) with hydromorphic processes of development such as have been described above. Thus the FAO classification clearly differentiates the *gleysols*.

However, even in this third group the definition of hydromorphic soils used is very much more restrictive than that suggested in this book. It is limited to hydromorphic soils *sensu stricto*, i.e. those that have a fairly reducing water table, and consequently excluding those hydromorphic soils referred to as *related* which, as stated previously, form intergrades to other classes: the

**pelosols** and **planosols**. The first belong to a separate class in the English classification, and the second to a separate class in the FAO classification.

However, it would seem to be preferable to take a rather broader approach and consider the pelosols and the planosols as profiles that are related to the true hydromorphic soils for the following three reasons:

(a) Even though usually of minor importance, redox processes do occur and in certain cases are able to be of considerable importance. This is so for the pseudogley–pelosols with rusty patches and certain planosols affected by ferrolysis (see Ch. 1).

(b) It would seem to be better to avoid increasing the number of classes, even if this means that their classificatory position creates problems that are difficult to resolve, by defining subclasses qualified as being *related*, forming transitions to other classes.

(c) The subclass in question is characterised by the process of impoverishment resulting from the special properties of the clays, which will be discussed in greater detail later. Although this is not a redox phenomenon, but more particularly a mechanical one, it is, however, closely related with a certain type of hydromorphism which is sufficient to justify the inclusion of pelosols and planosols in the hydromorphic class.

## II  Soils with perched water table of pluvial origin (Fig. 11.1)

The stagnogleys and pseudogleys, soils with a temporary perched water table, have been differentiated for a long time, in both German and English classifications, from the gleys with permanent phreatic water tables (Avery 1973, Schlichting 1973). Avery in particular distinguishes surface water gleys from ground water gleys.

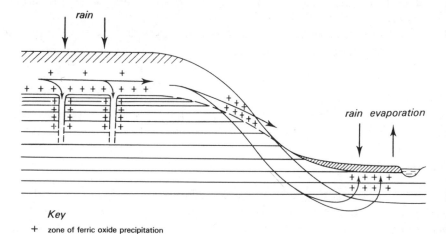

*Key*

+   zone of ferric oxide precipitation

**Figure 11.1**  The movement of water and the formation of water tables in (a) pseudogley and (b) gley soils: the closer the horizontal lines, the more impermeable the material.

*The pseudogleys differ from the gleys in terms of their pF gradient which increases from top to bottom of the profile, while the opposite is the case for the gleys.* They occur in climates that are sufficiently humid, at sites where topographic conditions and parent materials slow down or prevent internal or external drainage; for example, depressions in which an almost impermeable layer occurs at a shallow depth which prevents the complete removal of rainwater at certain seasons. Thus they are *site climaxes* or soils resulting from a *degradation*, when vegetational modification by man has caused the formation of the surface water table.

Thus pseudogleys and stagnogleys have a comparable development; however, they are distinguished from one another by the time of persistence of the water table, which is determined essentially by the climate. The first are characteristic of the atlantic plains, where high summer *PET* (potential evapotranspiration) causes the complete disappearance of the water table during this period. The second occurs in cold humid regions, where the low *PET* (compared to the high rainfall) allows the water table to persist practically throughout the year. Consequently, the processes of reduction and of mobilisation of iron, and also the accumulation of organic matter, are less marked in the pseudogleys than the stagnogleys.

**Pseudogley.** Although iron is mobilised in the ferrous state during periods when the water table exists, it does not persist in this form. After having moved a short distance, it precipitates after re-oxidation, with the disappearance of the water table, forming rusty patches or concretions which give the surface mineral horizons a mottled appearance. In addition, conditions of aeration within the humic horizons are generally sufficient to maintain a reasonably rapid *turnover* of organic matter so that the humus remains a mull–moder or, in the most unfavourable conditions, a hydromorphic moder; peat does not develop.

**Stagnogley.** The iron mobilised in the reduced and complexed state is not re-oxidised. Most remains in the ferrous state and is thus very mobile, conditions being very acidic; either it remains in the profile in this form, or it is removed from the profile, so that the mineral horizons are more or less completely decolourised. In addition, except in the initial stage, the organic matter is affected by hydromorphism and it decomposes slowly to form a hydromor or even, in extreme cases, an acid peat. It will be seen that an acid peat, in most cases, can be considered as an extreme case of the formation of a peaty stagnogley accompanied by a very large accumulation of incompletely decomposed organic matter.

In certain special cases when there is a sharp change in the pH and Eh at a certain level in the profile, a precipitation of the reduced iron can occur by oxidation. It forms a rusty indurated Bs horizon that can occur either as isolated concretionary masses or as a *placic* horizon, a thin very sinuous iron pan, already described with regard to hydromorphic podzols. These are the **molkenpodzol** of Kubiena (1953) or stagnogley with placic horizon.

## 1  Study of type profiles: environment and characters

**The formation of the surface water table.** From the environmental point of view, apart from general climatic conditions (atlantic humid temperate climate), the fundamental condition for the development of pseudogleys and stagnogleys lies in a lack of drainage, both internal and external. It is particularly necessary that the topography prevents the too-rapid lateral flow of the water table; slow movement, however, can occur and it is even frequent in both types of soil. *It is particularly the sudden decrease in the non-capillary porosity at a relatively shallow depth (30 cm to 40 cm at a maximum) which causes the formation of the perched superficial water table.* Thomasson and Bullock (1975) showed that the coarser porosity, which is relatively high in the zone where the water table circulates, is lowered by at least 5% in the deeper little-permeable layer, sometimes called the **floor** of the water table. While this term suggests a much sharper boundary than is actually the case, as will be seen, this almost impermeable floor can be formed in various ways, such as by the occurrence of two materials of different grain sizes, the accumulation of clay in a Bt horizon by pervection, or by compaction of deeper horizons, either recently or in a former period, to give rise to a fragipan.

Although the reasons for the formation of water tables are the same for both pseudogleys and stagnogleys, the generally more humid and colder climate of the latter causes them to have a very different pedoclimate *where the water table persists throughout, or almost throughout, the year, which is not the case for the pseudogleys.* The cold combining with the humidity causes the organic matter to decompose slowly and form a thick A0 horizon; the reducing and active water-soluble compounds biodegrade only slowly and cause a massive reduction and mobilisation of iron, generally accompanied by a hydrolytic degradation of clays. These processes are of lesser importance in the pseudogley, except in certain exceptional circumstances.

### Profile characteristics: morphology, biochemistry, geochemistry

*Pseudogley* (Duchaufour 1978: XIII$_2$). A pseudogley of degradation will be taken as an example. This occurs under a partially degraded acidophilic oak forest (clearings with *Molinia coerulea*) on loams on an old terrace, which is very frequent in the forests of the Paris basin.

*The type of humus* is a hydromoder, sometimes a hydromor (in the most humid climates with a montane tendency). The poorly structured brownish-black A1 horizon is always thicker than the moder on well drained sites.

*The A2g horizon*, light grey with scattered rusty patches and massive structure, is occupied by the water table in the wet season. Black iron–manganese concretions are formed at the Bg boundary.

*The Bg horizon* forms the floor of the water table, being compact with a finer texture and having a prismatic structure in the dry season. There are zones of preferential drainage down the contraction cracks between the prisms, or vertically oriented old root channels, which are marked by white bands bordered by rusty streaks or sometimes concretions. This particular feature is explicable either in terms of recent hydromorphism or the inheritance of a previously developed glossic structure or a type of fragipan.

The decomposition of organic matter is slowed down by the strong acidity (pH less than 5) and temporary anaerobic conditions, although the C : N ratio never exceeds 20 under a grass cover. Iron mobilisation occurs since iron is concentrated in patches and more particularly in concretions in the upper part of the Bg horizon, but the A2 horizon is never as completely lacking in iron as it is in the stagnogley (the amount of free iron, as the element, rarely falls lower than 0.2%). In the Bg horizon the vertical white bands of material between the prisms are lacking in iron, and their rust-coloured margins often contain 15 to 20 times more free iron, in the ferric form. In addition, even though the organomineral complexes play a large part in iron segregation, they do not persist, for the complexing anions are mineralised in the summer and there is no organic accumulation in the Bg.

The processes of iron mobilisation are more marked in very acid conditions (pH equal to or less than 4) and particularly when they are complicated by an acid hydrolysis of clays. This is the situation in the **podzolic pseudogleys** (Duchaufour 1978: $XIII_3$) in which the conditions approach those in the most developed type of stagnogley. In some cases a small amount of organic matter of insolubilisation, with a high C : N ratio, and amorphous alumina have been detected in the upper part of the Bg horizon. But, generally, the accumulation of both mineral and organic amorphous compounds, if it occurs at all, is low compared to that of the well drained podzols.

*Stagnogley* (Duchaufour 1978: $XIII_5$). A developed stagnogley of a mountainous area is taken as an example. It is formed on loam in an undrained depression on a high level platform under an acidifying vegetation of spruce and *Sphagnum* (the natural or artificial '*Sphagnopicetum*' of the ecologists).

This stagnogley almost always shows signs of a *podzolic* development as a result of the strong acidification (pH less than 4 in the surface) and to the persistence of active organic compounds, so that it is possible to call them podzolic stagnogleys. In comparison with the pseudogley, its main characters are: (i) considerable development of an A0 horizon which is a hydromor; (ii) complete decolourisation and almost total elimination of the iron from the A2g horizon; (iii) persistence of a small amount of organic matter of insolubilisation in the upper part of the B horizon. The Bg horizon which forms the impermeable floor is mottled with rusty, ochreous or greenish (owing to ferrous iron) patches, without the formation of white streaks of material. The free iron of this horizon results from weathering *in situ*, in either the ferric or ferrous form depending on the local redox conditions. *Generally, there is no sign of reprecipitation of the mobilised iron, in either the A2g or Bg horizons, which is an important difference from the pseudogleys*. (This precipitation occurs only in special circumstances to form particular Bb horizons such as the placic horizon (Schlichting 1973, Duchaufour 1978: $XIII_6$).)

In addition, clay balances (Schweikle 1971) show a considerable loss of the initial clay content, partly owing to lateral transport and partly also to acid hydrolytic degradation. Here again the Al and Si that are freed do not accumulate in the profile.

**Utilisation of the pseudogley and stagnogley.** The pseudogleys are among the most unfavourable soils for plant life. Asphyxiating in wet periods, they behave as dry soils in the summer owing to a lack of root penetration into the Bg horizon which is itself dried out. The bleached podzolic pseudogleys, sometimes impermeable from the surface, are almost totally useless. Their vegetation is either a very poor forest or a wet heath with *Molinia*, made up of species resistant to phases of asphyxiation and desiccation. Porosity can fall to 35% in the A and 30% in the Bg. From the biochemical point of view these soils are equally poor; root activity is weakened by anaerobic conditions in the spring and absorption of ions is limited. New nitrogen compounds are formed, rather than ammonification, which disturbs the nitrogen cycle; toxicity by the soluble noxious ions, $Al^{3+}$, $Fe^{2+}$, $Mn^{2+}$, is very frequent for these ions cannot be removed from the A horizon, as in the case of the podzols (Périgaud 1963).

These soils cannot be used for agriculture. The wet heathlands and degraded forests should be reforested (maritime pine and red oak in the west, on the acid palaeosols of the heaths of Poitou). The drainage is often fairly poor but it may be sufficient to eliminate the temporary water table in the case of the less well developed pseudogleys. In Poitou, re-afforestation is done after using a heavy disc plough to eliminate the *Molinia*; the subsoiling effect is only very temporary but it allows a good initial root establishment.

Natural grasslands are always very poor and they revert very quickly to heath. Agronomists advocate the introduction of temporary grass covers with deep roots by the use of fertilisers and suitable amendments, as the best means of regenerating the structure and improving the natural drainage.

The stagnogleys are even more unfavourable than the pseudogleys. Only certain very undemanding forest species that can resist asphyxiation by a surface concentration of roots (spruce) can prosper. The roots are so totally concentrated in the surface humic-rich horizons that the biogeochemical cycle of nutritive elements tend to be confined to these horizons and growth becomes very slow. The utilisation of these soils poses problems that are identical with the utilisation of acid peats.

## 2  Development of pseudogley and stagnogley: biochemical processes

In spite of the great differences that exist in the biochemical development of these two groups of soil, it is possible to consider them together to distinguish in a general way three major phases in their development.

(a) *Initial phase:* water table less strongly reducing (or existing over less time); slightly acid conditions with the humus remaining as a mull; in these circumstances the mobilisation and segregation of iron is slight (pseudogley) or its reduction is incomplete (stagnogley). The differentiation of A2g and Bg horizons is not very marked.

(b) *The modal phase* corresponds to the profiles described in the previous section. The acidification is more marked and the humus starts to deteriorate. In the pseudogley, iron segregation is considerable and in the stagnogley its elimination by reduction is complete or almost complete in the A2g horizon. The differentiation of the A2g and Bg is very marked.

(c) *Phase of hydromorphic podzolisation.* This is a biochemical process similar to that of podzolisation in better drained conditions, for it involves clay degradation by acid hydrolysis and complexolysis with the freeing of aluminium and silica. This phase, which is relatively infrequent in the pseudogleys of the lowlands, is, in contrast, general in the stagnogleys and it happens so quickly that in most of them the intermediate modal phase is short-circuited. In addition, the stagnogleys have certain special developments that are specific to them: peaty development, accumulation of iron oxides in more or less indurated horizontal or sinuous bands on parent materials that have abrupt variations in Eh or pH at depth.

**Development of the pseudogley.** *In the initial phase* the water table is not very persistent and is only weakly reducing. The humus remains a mull with high biological activity. The iron is very incompletely mobilised and it forms rusty patches on an incompletely decolourised yellowish-brown background. In most cases the water table remains at depth and does not reach the upper part of the profile (the A1 and upper part of the A2 horizons are normally well drained). Thus the soil should be classed as a brunified soil like the hydromorphic lessived soil (or with pseudogley). It is only when the A2 horizon shows signs of hydromorphism (rusty patches) that the soil should be considered as belonging to the hydromorphic class. It is then a **slightly developed,** or **initial, pseudogley** (Duchaufour 1978: $XIII_1$).

*The second phase* corresponds, as previously stated, with a much more marked segregation of iron, in more acid and temporarily more reducing conditions, causing a marked profile differentiation.

In a humid climate, the temporary water table is present throughout the cold period, during which, however, it is subject to marked variations in level. In Lorraine, according to Becker (1971), it is very near the surface from December to May, of course with considerable differences depending on whether the years are wet or dry. As the rainwater brings with it dissolved oxygen, the water table is not very reducing. The Eh never falls below 400–500 mV (Gury 1976). The part played by acidity is fundamental in mobilising the iron, first of all because the complexing organic compounds are more abundant in acid conditions, and also because for such high Eh values it is the pH that finally determines the reduction of iron (see Ch. 3).

In these circumstances, as shown by Schwertmann and Fischer (1973) as well as by Kaurichev and Shisshova (1967), *iron mobilisation must occur by the complexing of iron in the ferrous state*. The iron separates from the clay, which is not mobilised and is, as yet, at this stage only weakly degraded. The part played by the organic matter and the microbial enzymatic activity in this process have been emphasised by several authors (e.g. Berthelin 1976, Munch & Ottow 1977).

The iron–organic matter complexes thus formed have a very short existence. They migrate locally and concentrate in certain zones of the profile, such as around the absorbing roots in the A2g horizon and around the prisms in the Bg horizon. When the water table disappears as a result of an increase in both the *PET* and the mean temperature, processes of oxidation occur, the complexing anions are biodegraded, and the iron precipitates in the ferric state: initially in the amorphous state and, as it dries out, with an increasing degree of crystallinity, in the sequence, ferrihydrite, lepidocrocite, goethite (Schwertmann & Taylor 1973, Thomasson & Bullock, 1975, Guillet *et al.* 1976). According to Blume (1968a,b), concretion formation is favoured by the rapid crystallisation of iron oxides in a loamy sand which dries out rapidly. This is generally the case in the A2g horizons where the water table is concentrated.

In contrast, in the prismatic Bg horizon which generally has a clay–loam texture, the processes are different: the iron is mobilised and transported in zones of preferential drainage, particularly along contraction cracks which are decolourised as the iron is removed. Part of this iron is eliminated by drainage to depth, but another part migrates to the interior of the prisms, and there it is gradually re-oxidised when the profile dries out. In these conditions, the crystallisation being very gradual, it forms patches or rusty streaks (Blume 1968a,b; Vaneman *et al.* 1976). Exceptionally in more sandy conditions, concretions are also able to form around the prisms in vertical zones.

**Note** that in pseudogleys with a circulating water table, a part of the complexed iron is transported laterally and precipitated as large irregular concretions (**grepp**) at a point where there is a break in slope. These concretionary pseudogleys are transitional to the hydromorphic podzols with *indurated ironstone* described in Chapter 10 (Begon & Jamagne 1973, Gury 1976, Begon 1979).

The *third phase* is *hydromorphic podzolisation* (pseudopodzolisation). In very acid conditions, in the presence of a hydromoder humus, or even hydromor, the influence of water-soluble organic compounds becomes increasingly strong. To the reduction and mobilisation of iron is added the process of freeing of silica and aluminium by weathering and acid hydrolysis of clays. Although it differs in certain ways, this process is related to that of podzolisation under conditions of better drainage. It is about the same as the pseudopodzolisation of Russian pedologists (Zonn 1973, Akhtyrtsev 1974, Zaidelman 1974), used to explain certain aspects of boreal pedogenesis in hydromorphic conditions (see Ch. 9). In western Europe, Gury (1976) studied biochemical and pedological aspects and Berthelin (1976) the microbiological aspects of this process. By using lysimetric methods and comparing well drained acid lessived profiles with adjacent hydromorphic profiles, the first author showed that there was a considerable loss of aluminium from the clays, owing to the massive production of soluble organic compounds (10 times more abundant than in well drained conditions). First of all, this aluminium is found as a complex in the A1 horizon, then when the water table rises to the level of the humus-rich horizons, the complexes are diluted and then biodegraded, the $Al^{3+}$ being transported laterally in the free state. While the mobilised iron and manganese are partially precipitated within the concretions formed in the upper part of the Bg horizon, or around the decolourised vertical bands in the Bg, it is not generally so for aluminium, of which the greater part is removed from the profile.

This peculiarity of hydromorphic pseudopodzolic soils is explicable by the very great mobility of the $Al^{3+}$ in very acid conditions, even when it is no longer in the complexed state. It is not like the $Fe^{2+}$ ion, which is susceptible to sudden increases of the Eh that cause it to become insoluble in the ferric state. The process of hydromorphic degradation of clays has been reproduced experimentally by Kanivets (1973), and this confirmed the field results of Gury.

Under the influence of complexing organic agents, the Al-chlorites lose their interlayer aluminous ions and are changed into montmorillonites of degradation, but other clays are degraded as in the podzols by direct weathering of the octahedral layer: Berthelin (1976) has reproduced these processes in the laboratory by using anaerobic microflora on certain clays such as the primary chlorites, vermiculites and biotites. A partial destruction of these minerals occurred together with the formation of a whitish amorphous aluminosiliceous residue.

**Development of the stagnogleys** (Fig. 11.2). As previously stated, the initial phase with limited reduction occurs in slightly acid stagnogleys with high biological activity, but when the acidity increases the *podzolic* phase is quickly reached, for the intermediate phase is so transient that it is rarely seen. On the contrary, some stagnogleys have a peaty development and a special type of iron accumulation.

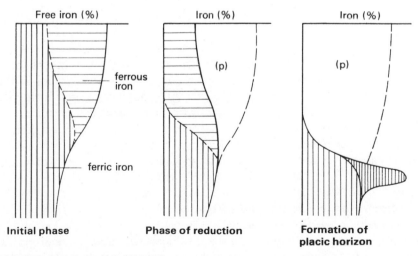

Free iron (%)     Iron (%)     Iron (%)

ferrous iron

ferric iron

(p)     (p)

**Initial phase**     **Phase of reduction**     **Formation of placic horizon**

(p) = losses by vertical and lateral transport

**Figure 11.2** Phases in the development of a stagnogley: reduction and mobilisation of free iron.

*Initial phase:* The reduction of iron is incomplete and it still occurs in the ferric state in the profile, which has an olive–beige colour. In fact, ferrous and ferric iron co-exist and at this relatively high pH the ferrous iron is only slightly mobile. However, some is able to migrate and accumulate at the bottom of the profile, either as concretions or as a rusty parting at the point of contact with the better aerated zone (Duchaufour 1978: $XIII_4$).

*Developed and podzolic phase:* In this case the environment is much more acid (pH less than 4) and a hydromor forms on the surface. Under its influence, the Eh decreases much more than in the case of the pseudogley and varies between 0 and 200 mV at the surface (see Fig. 11.5). In these circumstances the reduction of iron becomes total, but simultaneously the ferrous iron is mobilised and removed from the profile which is thus rapidly impoverished in iron. The A2 horizon is completely decolourised. Clay degradation by acid hydrolysis occurs at almost the same time. These processes are almost identical with those involved in the formation of the podzolic pseudogley but in this case they take place at a much faster rate and the results are much more noticeable.

Schweikle (1971) studied in a very detailed manner the development of stagnogleys on Triassic sandstones of the Black Forest. Clay balances are strongly negative and show that clay is being destroyed by hydrolysis. All the soluble entities, organic matter, aluminium and iron, are removed from the profile, particularly by lateral movement, for in high-altitude sites the water table is generally mobile. In certain cases, however, all these transported entities accumulate in better aerated zones at a lower level. A similar example of a catena was given in Chapter 4 (see Fig. 4.9).

*Phase of peaty development; formation of placic horizon* (Duchaufour 1978: $XIII_6$). When anaerobic processes become even more developed, the hydromor is partially transformed into a *peat*. The water table is then absorbed by the organic horizons like a sponge. In these humus-rich stagnogleys, frequently occurring around true peat bogs, the peaty horizon can be 20–30 cm thick. The water table flows towards the lower lying peat bog. Complex formation and the downslope transport of iron reaches a maximum. If there occurs within the soil material a grain-size discontinuity causing a local rise of Eh, the transported iron is precipitated to form a Bb horizon of more or less aligned concretions, or even a *placic horizon* (sinuous thin iron pan) comparable to that occurring in certain hydromorphic podzols (see Ch. 10). In extreme cases, a hydromorphic compact indurated ironstone is formed under the peat layer.

## 3   Influence of environmental factors on the development of pseudogleys and stagnogleys

It is not necessary to discuss again the local site conditions (topography and parent material) which are seen from the preceding sections to be necessary prior factors in the formation of surface water tables. Particularly with regard to the parent material, the considerable change in the non-capillary porosity between the almost impermeable Bg horizon and the Ag horizon normally occupied by the water table can be the result of various causes which, as will be seen in the following section, are used to differentiate subgroups. For example, in the pseudogleys it is possible to distinguish **primary pseudogleys** (with two layers, in which the lower one is more compact or of a finer texture), **secondary pseudogleys** (in which the impermeable Bt has been formed by blocking of the pores by pervected clay) and **complex pseudogleys** (with an almost impermeable palaeosol at depth).

It would appear preferable to emphasise the bioclimatic conditions that, for a given material and topography, favour the formation of water tables and influence, generally through the type of humus, the degree to which they are active. For this purpose the lowland atlantic climates giving rise to pseudogleys are differentiated from humid mountain climate in which stagnogleys develop. (The boreal zone (taiga) has boreal stagnogleys, similar to the sub-alpine stagnogleys.)

**Environmental conditions for the formation of a lowland pseudogley.** As shown by Becker (1971), the factor of potential evapotranspiration (*PET*) plays a fundamental role in the formation of perched water tables. In terms of the general climate, first of all, the atlantic type of climate has relatively gradual seasonal variations in *PET* that favour pseudogley development, while the continental climate, with sharper seasonal variations of *PET*, favours the formation of **planosols**. Then, on the local level, the formation of clearings in the forest cover leads to the development of wet heathlands (*Carex brizoides*, *Molinia coerulea*), which in turn leads to a lowering of the *PET* and thus a rise of water table towards the surface (Becker 1971, Aussenac 1975). The forest soils characteristic of the near-climax dense forests but on poorly drained sites, as previously defined, are *soil analogues* of those which occur at better drained sites. Thus the humus remains a moderately acid and well aerated mull and, depending on the depth of the water table and the nature of the almost impermeable horizon at depth, the soil can either be lessived soil with hydromorphism at depth or an acid brown soil overlying a fragipan or, in the most unfavourable case, an as yet slightly developed pseudogley (water table nearer the surface but not very active). In such circumstances, the forest is always in an unstable equilibrium and silviculture requires care, for it is necessary to avoid sudden clearances which locally cause a lowering of the *PET* followed by an upward movement of the water table. The mull is then replaced by a hydromoder (or even a hydromor), biological activity decreases and soil compacts, which increases still more the process of hydromorphism. This degradation by *accentuating the hydromorphism* causes the formation of a **modal pseudogley** or even a **podzolic pseudogley** in extreme circumstances.

Becker (1971) studied the stages of degradation in the Charmes forest in Lorraine, which is a beech forest on old loams covering an alluvial terrace. The initial soil of the dense forest is an acid brown soil overlying an almost impermeable palaeosol of the fragipan type. The herbaceous flora is basically *Poa chaixii* on acid mull, or *Luzula albida* on humus that is a little more acid of the

**Table 11.1** Humus and water table depth.

| Type of vegetation | Type of humus | Depth of water table (cm) | Colour of the A2g | Index of iron B/ iron A2 |
|---|---|---|---|---|
| (1) *Poa chaixii* | mull | 40 | light brown 10YR6/3 | 2.54* |
| (2) *Luzula albida* | mull–moder | 30 | light brown 10YR6/3 | 2.25* |
| (3) *Carex brizoides* | hydromoder | 11 | grey 10YR7/2 | 7.01 |
| (4) *Molinia coerulea* | hydromor | 1 | light grey 10YR7/1 | 8.8 |

*Note that this index for types (1) and (2) in fact represents a difference in initial composition between the surface acid brown soil and the reworked deep fragipan; it varies somewhat from one point to another.

mull-moder type. In clearings colonised by *Carex brizoides*, the soil is already a modal pseudogley. Finally, the appearance of *Molinia coerulea* as a dense cover is an indication of the formation of a true hydromor, accompanied by the beginning of a pseudopodzolic development. The general outline of the degradation is as follows:

acid brown soil ⟶ slightly developed pseudogley ⟶ modal
pseudogley ⟶ pseudogley with a podzolic tendency

The author showed that a perfect correlation existed between the depth of the water table in winter, the kind of humus, the degree of decolouration of the A2g, and the segregation index of the iron (iron B/iron A1), which is shown in Table 11.1.

**Environmental conditions for the development of stagnogleys in mountainous areas.** While the formation of lowland pseudogleys is generally the result of degradation, which controls the different stages of hydromorphic development, the formation of stagnogleys in mountainous areas differs in that the various stages of their development are related to the altitudinal vegetation zones.

(a) Alpine zone (alpine meadow): initial or slightly developed stagnogley (site climax).
(b) Sub-alpine or upper montane zone: podzolic stagnogley (site climax); the boreal stagnogleys belong to this type.
(c) Middle montane zone: here the degree of development is more related to a process of degradation.

One example from each of these zones will be discussed.

*Alpine zone.* Stagnogleys occur particularly in depressions where snow accumulates in the so-called 'snow-hollows'. The soil is waterlogged throughout most of the year, but the water table can disappear for a short period in summer, as a result of the very strong insolation that occurs in this zone. In addition, the vegetation is an alpine meadow dominated by grasses that form a humus which preserves a certain degree of biological activity (mull–moder). In these circumstances, the environment is only slightly acid, no matter what the parent material, and the water table does not have sufficient reducing powers to reduce all of the free iron present. Thus an **initial** (or **slightly developed**) **stagnogley** is formed (Duchaufour 1978: XIII$_4$).

*Sub-alpine and upper montane zones*. Here the depressions are occupied by a sub-alpine forest, dominantly spruce with *Sphagnum*, lycopods, etc., which form a slowly decomposing acidifying litter (hydromor). Under the forest canopy, the water table persists for a longer time than under the alpine meadow and in addition it is much more acid and has greater powers of reduction. Here the stagnogley is generally of the *podzolic* type (Richard 1961, Duchaufour 1978: XIII$_5$).

*Middle montane zone* (Glatzel *et al*. 1967). Some very typical examples of the degradation of slightly hydromorphic acid brown soils (or initial stagnogleys) which correspond to a climax forest (beech–fir) have been described in the Black Forest of West Germany. A considerable increase in the degree of hydromorphism has been caused by a modification of the natural vegetation which generally has occurred in two stages: (i) a very old general deforestation, resulting in colonisation by *Molinia coerulea*; (ii) recent re-afforestation by a dense, pure stand of spruce. Under the influence of a humid heath, the water table has risen towards the surface; then the acidifying litter of the spruce has increased its acidity and reducing powers, which has resulted in a pseudopodzolic type of development complicated very often by the formation of a peaty humus (secondary development of *Sphagnum*) and of a placic, or thin ironpan, horizon: a part of the iron mobilised in the surface horizons precipitates on contact with the deeper mineral horizons that remain better aerated and less acid. The peaty stagnogley and the podzols with a placic horizon (scc Ch. 10) thus have a comparable origin, the first forming at sites where water table circulates more slowly. Both kinds of profile are often associated in the field.

## 4   Summary: main types of pseudogley and stagnogley

This will be restricted to enumerating the *groups* that have been differentiated previously by taking account of two basic factors: (i) *the nature and the duration of the hydromorphism* causing different profile characters to develop, particularly related to the mobilisation of iron and the transformation of organic matter (distinction between pseudogley and stagnogley); (ii) *degree of development* as related to the presence of a water table that is more and more active and acid.

The subgroups are differentiated in each category in terms of the nature and origin of the almost impermeable floor of the water table.

**Groups** (Duchaufour 1978: Pl. 13, Figs 11.3 & 4).

*Pseudogley*
*Initial pseudogley* (slightly developed). The humus remains as a mull; segregation of iron incomplete: rusty patches on a yellowish-brown background in the A2g and Bg. Little contrast between

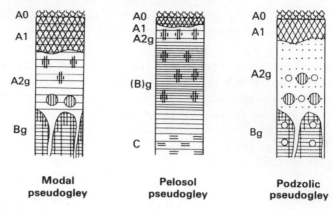

**Figure 11.3** Pseudogley profiles (see general key, p. ix). Note that the pelosol pseudogley is classed with the pelosols.

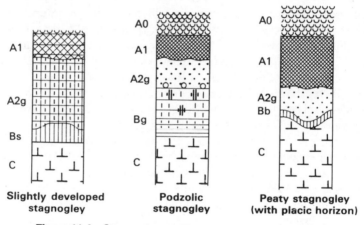

**Figure 11.4** Stagnogley profiles (see general key, p. ix).

the horizons A2g and Bg; the Bg horizon does not have a prismatic structure in which the prisms are separated from one another by vertical white lines of material.

*Modal pseudogley:* moder humus (or hydromoder). Segregation of iron is more complete in the partially decolourised A2g. Concretions occur at the base of the A2g. Prismatic structure in the Bg in which the prisms are separated from each other by vertical white bands of material (which are sometimes glossic if it is a palaeosol).

*Podzolic pseudogley:* acid hydromoder (or hydromor). Degradation of structure in the completely decolourised A2g; infiltration of organic matter with high C : N ratio; and weathering of clays and production of free-Al, which is sometimes difficult to detect in the profile.

### Stagnogley

*Initial stagnogley* (slightly developed). A still not too thick mull–moder humus; co-existence of ferrous and ferric iron in the A2g; formation of rusty patches or a band at the base of the profile.

*Modal stagnogley.* Hydromor; complete reduction of free iron of which most is removed from the profile; A2g completely decolourised. Generally no horizon of iron accumulation, or of concretions.

*Podzolic stagnogley.* Hydromor; complete reduction and removal of iron; degradation of clays with production of free aluminium as in the case of acid pseudogleys.

*Humic stagnogley with a placic horizon (or hydromorphic ironpan)*. Considerable development of humic horizons that become peaty; frequent occurrence of a placic horizon or one with concretionary ironstone, but which is not absolutely necessary.

**Subgroups**
*Primary subgroups:* difference in porosity between A2 and B caused by a difference in structure or by the occurrence of two geological layers with different texture.
   *Secondary subgroups:* thickening and filling up of the pores of a near-surface Bt horizon by prior pervection of clay.
   *Complex or polycyclic subgroup.* The almost impermeable deeper horizon is formed of a *glossic* or *fragipan* type palaeosol in which the structure has been destroyed and the porosity is low.

**Note** that these subgroups occur in both pseudogleys and stagnogleys.

# III   Soils with a deep phreatic water table: gleys

Gley soils formed in depressions or alluvial plains that have a water table supplied from a subterranean source (see Fig. 11.1). Very frequently such water tables differ from perched water tables in having little acidity and being rich in calcium; but they are particularly poor in dissolved oxygen and much more powerfully reducing than the water tables formed by rainwater (Eh is often negative; Blume 1968a,b, Callame & Dupuis 1972). This explains why *iron reduction occurs even at neutrality, and even in the presence of active carbonate*. But the iron thus reduced is only slightly mobile at this pH and it accumulates at the base of the profile to which it gives a greenish or bluish colour (horizon Gr). Another difference from those soils with a perched water table is in the Eh gradient, for gleys are more reducing at the bottom than in the middle of the profile (where a partial re-oxidation in the form of rusty patches often occurs: horizon G0), while in the pseudogley the situation is reversed and even more so in the stagnogley with a placic horizon, where the iron is re-oxidised in the lower part of the profile (Kaurichev & Shisshova 1967, Fig. 11.5).
   Gleys are typically very localised *intrazonal* soils that form, with their very special kind of vegetation (alder clumps, meadows with tall sedge and rushes), site climax equilibria. Their frequent occurrence in the lower-lying parts of alluvial plains explains why they are so frequently associated, on the one hand, with alluvial soils in areas where the water table is subject to considerable variations in height and is not reducing, and on the other, with eutrophic peats in areas where the water table, in contrast, is constantly very near the surface and strongly reducing. In addition there are many intermediate soils such as the **alluvial soils with gley** (or **semi-gley**) in which the iron is incompletely reduced and occurs only in the deeper horizons and in contrast the **peaty gleys**, or gleys with **anmoor**, which are transitional to the peats.
   The factors of the water table that favour the formation of gleys can be summarised as follows: (i) subterranean source of water table supply, but with very slow lateral circulation; (ii) variations in height of water table not as

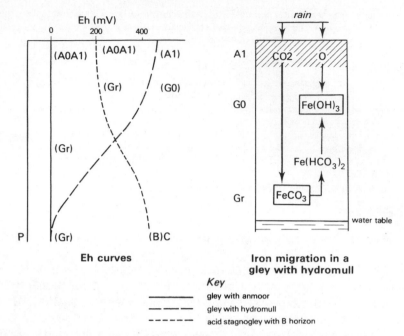

**Figure 11.5** Gley and stagnogley: redox processes and mobilisation of iron.

great (1 m at the most) as those of the water tables of alluvial soils; (iii) a more or less strong lowering of the Eh owing to the presence of soluble organic matter.

## 1 The morphology and biochemistry of the profile

A eutrophic humic gley in which the height variation in the water table is moderate and which appears to have the best developed characters will be taken as the type example (Duchaufour 1978: XIV$_2$). There are three basic horizons:

(a) *The A1 horizon*, organomineral, dark brown or black, thick (20–30 cm), containing more than 8% organic matter: at the surface it is a **hydromull** with a well aerated crumb structure (this is the part of the horizon that is outside the zone of water table oscillation), becoming an **anmoor** towards the base (this is the zone which is reached by the water table when at its highest), with a massive structure and a plastic consistency. Often at the base of the A1 horizon, carbonate concretions occur if the parent material contains carbonates. The biological activity is considerable in the aerated part (the hydromull), it decreases markedly in the more or less anaerobic anmoor. However, as stated in Chapter 2, certain dark anmoors which have had a long period of development have a very high degree of humification with a certain degree of maturation of the humic compounds.

(b) *The G0 horizon* (oxidised gley) corresponds with the zone of water-table oscillation and is characterised by rusty patches, sometimes weakly concretionary. The structure is polyhedral or prismatic. The iron hydrates form skin-like coverings around the peds, which is the complete opposite of that which occurs in the Bg horizon of the pseudogleys (Blume 1968a,b).

(c) *The Gr horizon* (reduced gley) is uniformly greenish or bluish and has permanently hydromorphic conditions. Its upper limit is almost coincident with the lowest level to which the reducing water table falls. The structure is massive; the free iron is mostly in the ferrous state, either as ferrous carbonate or as a complex salt (vivianite). Note that, generally, the amount of free iron decreases from the G0 to the Gr, which is an indication of its *upward movement*. The maximum amount of free iron occurs in the G0 horizon.

In the type profile, all the horizons are reasonably saturated in exchangeable bases.

**Variations in the gley profile and their causes.** The morphological and physicochemical characters of the profile described can be modified as a result of two basic environmental factors: (i) differences in the highest and lowest level reached by the water table; and (ii) variations in the chemical composition or, in other words, the acidity of the water table.

*Variations in the level of the water table* (Fig. 11.6, Mückenhausen 1975). If the water table varies in level more than in the type example (between the limits of 0.3 m and 1.0 m), generally the profile is less reducing and the

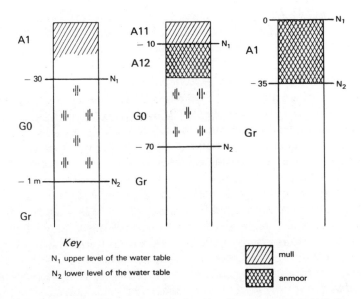

Key

$N_1$ upper level of the water table

$N_2$ lower level of the water table

mull

anmoor

**Figure 11.6** Influence of the water table level on the development of gley profiles.

horizons are less clearly differentiated. The humus is a mull that is not so thick, while, in contrast, the G0 horizon is thicker and the lower Gr horizon is less completely reduced (olive tinge). This is a **low humic gley**, often qualified as being alluvial, for it is generally formed in recent alluvium and is consequently associated with true alluvial soils. It has been stated previously that several transitional types of soil occur (Duchaufour 1978: $XIV_1$, which is a good example of an intergrade profile that can be qualified as being an **amphigley** or **pseudogley–gley** profile, in which the two profiles are superposed on one another).

If, in contrast, the variation in height of the water table is very slight (0–0.4 m and often less), the humus is subjected to strong temporary anaerobic conditions and is no longer a hydromull but a coherent and plastic dark brown or black anmoor, 0–30 cm thick, with a very low level of biological activity. Even though it is very rich in organic matter, anmoor is not completely organic as is a peat (less than 30% organic matter). The G0 horizon is lacking, the profile being uniformly greenish (sometimes bluish) and thus of the Gr type. The ferrous iron is very much more dominant than ferric iron throughout the profile (Nguyen Kha & Duchaufour 1969). The profile is thus a **reduced gley with anmoor** (Duchaufour 1978: $XIV_3$).

*Acidity of the water table*. The acidity of the water table controls the *mobility of the ferrous iron* which can be transported laterally by water-table movement in acid conditions. In extreme cases a podzolic development is favoured similar to that discussed for the podzolic pseudogleys and stagnogleys. In the **podzolic gleys** (Duchaufour 1978: $XIV_4$), the ferrous iron is gradually removed from the profile so that the Gr tends to become lighter in colour (*white* gley). The rusty patches are less numerous and more weakly developed. In contrast, analysis detects a vertical migration of vulvic acids carrying aluminium, of essentially biological origin, which is similar to the developments in a hydromorphic humic podzol (see Ch. 10).

## 2   Biochemical processes of gley development

The developmental processes of acid water tables (podzolic gleys), which are similar to those of the hydromorphic humic podzols, will not be discussed again. The only difference between them is that the water table being higher in the gleys, the polymerisation of the aluminium–organic complexes does not occur and the organic matter does not get past the weakly coloured fulvic acid stage. A Bh horizon is not formed.

Instead, discussion will concentrate on reactions between organic matter and iron, in gley soils developed in a non-acid or slightly acid environment, which is a common occurrence. It has been previously stated that, in such circumstances, ferrous iron is immobilised in the Gr horizon as slightly soluble $FeCO_3$. Schaefer (1967) showed that *iron mobilisation is the result of high $CO_2$ production by the biological activity in the humus-rich horizons, and that such mobilisation is only effective over short distances, generally in an upward direction*. Rain water transports the $CO_2$ downwards when the water table is at its lowest level (see Fig. 11.5) and it reacts with the $FeCO_3$ to form the soluble $Fe(HCO_3)_2$ which moves upward by capillarity and re-oxidises in the middle of the profile, owing to the oxygen content of the descending rain-water, to form rusty $Fe(OH)_3$ patches characteristic of the G0 horizon.

$$FeCO_3 + CO_2 + H_2O \longrightarrow Fe(HCO_3)_2$$
$$2Fe(HCO_3)_2 + O + H_2O \longrightarrow 2Fe(OH)_3 + 4CO_2$$

Berthelin (1976) and Munch et al. (1978) confirmed the considerable importance of micro-organisms in the different phases of these redox processes involving iron. For this to occur, two conditions must be fulfilled: (i) the humus needs to be a hydromull with strong biological activity and thus producing a great deal of $CO_2$; the upper level of the water table must be such as to leave the upper part of the A1 well aerated; and (ii) the water table must fall low enough in the dry season to allow the re-oxidation and precipitation of the iron in the middle of the profile (formation of G0 horizon). However, when the water table is higher, the conditions are more anaerobic and more reducing throughout the profile, the humus being an anmoor with little biological activity which causes the production of $CO_2$ to be insufficient to mobilise the iron. Finally, even if this mobilisation occurs, the iron cannot be re-oxidised and form rusty patches characteristic of the G0. This is a **reduced gley with anmoor**.

**Dynamics of gley development by a lowering of the water table.** The succession of profiles seen as a water table is naturally or artificially lowered in given conditions allows the stages of gley development to be specified: (i) permanent anaerobic or semi-anaerobic conditions: no oxidised gley, the profile is of the A1Gr type (gley with anmoor); (ii) water table subject to temporary lowering; the upper part of the anmoor is aerated and becomes a hydromull, the G0 horizon appears; profile of the A1G0Gr type; (iii) water table is considerably and permanently lowered; all the anmoor becomes active and acquires a crumb structure; the humus is a very active mull which develops by desiccation and polymerisation into a thick mull reminiscent of chernozems (black soils of valleys). In the mineral horizons, the ferrous hydrate reoxidises and binds onto the clays, which is a process of brunification. These soils could possibly be called **alluvial isohumic** soils, with a variable degree of brunification according to the circumstances.

## 3   The properties of gleys and their utilisation

This dynamic study of gley soils will be used as a basis for evaluating their agronomic properties. Gleys never behave as dry soils, as do the pseudogleys, *for conditions are always asphyxiating below the level of the permanently reducing water table which limits the utilisable depth of soil.* Thus, from the point of view of fertility there is a very great difference between the reduced gley with anmoor, which is very unfavourable for plant growth, and the gley with hydromull and a G0 horizon, which is much more favourable. Thus the aim of agronomists must be to produce an oxidised gley with hydromull from a reduced gley, by drainage; to lower the water table from its highest position without affecting its lowest position too much, which would endanger the water supply to plants by capillary rise in dry periods.

Gleys with hydromull, no matter whether they are humic or low humic (alluvial gleys), have intense microbiological activity concentrated in their surface horizons and strong powers of nitrification. This makes them suitable for use as pastures, or for market gardening, or the growth of poplars.

The acid gleys, and even more so the podzolic gleys, are, in contrast, unfavourable to plant growth and they carry peaty meadows with large sedges which can be improved by re-afforestation with non-demanding coniferous species (maritime pines in the south-west of France).

### 4 Conclusion: main types of gley (Fig. 11.7)

Around the four fundamental types – low humic gley, humic oxidised gley, reduced gley with anmoor, and podzolic gley – which have been discussed in detail (Duchaufour 1978: Pl. 14), there are a great number of intermediate types with more or less complex development, which will be considered briefly.

The **semi-gleys**, or alluvial soils with gley at depth, are transitional to alluvial soils *sensu stricto*, which do not show any sign of redox phenomena. Ferrous and ferric iron co-exist, giving the profile an olive to yellowish-brown colour. Rusty patches are concentrated at depth and often are poorly developed.

The **amphigley** or **pseudogley–gley** show both types of development: an alluvial gley of the preceding type occurs at a depth of greater than 1 m; this is overlain by a compact almost impermeable horizon that contains black concretions similar to the pseudogley, in which a true perched water table of pluvial origin is formed, with the phreatic water table occurring at the lower level (Duchaufour 1978: $XIV_1$).

**Slightly developed gleys with acid hydromull** have been described by Jambu and Righi (1973) in the south-west of France and these are transitional

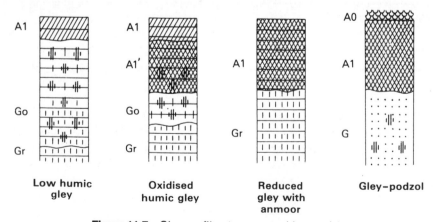

**Figure 11.7**  Gley profiles (see general key, p. ix).

between soil with a neutral (or calcareous) hydromull and the gley podzolics (these last with hydromor or hydromoder). The humus is acid (pH 5–5.5), but there is still a great amount of biological activity. The C : N ratio stays at less than 20 because of the dominant grass vegetation; in these circumstances, the water table is still very weakly reducing and the iron is incompletely reduced and little is removed from the mineral horizons.

## IV Organic hydromorphic soils: peats

In permanently waterlogged conditions, organic hydromorphic soils or peats develop which are frequently several metres thick. With the exception of mineral material resulting from lateral flow, which is frequent in the case of low moor peats (eutrophic peats), the material is wholly organic, the only mineral elements being those contained originally in the vegetation and thus of biological origin. Owing to the strongly anaerobic conditions which slow down the biological activity considerably, the decomposition of fresh organic matter is very slow and incomplete, so that annual additions exceed the losses by mineralisation. The soil gradually thickens until the equilibrium point is reached, the environment becoming gradually more aerated in the surface, which increases the speed of decomposition.

Two main types of peat can be distinguished in terms of site, the origin of the water table, the biochemical properties and finally the vegetation composition, which plays the part of the parent material by the accumulation of its debris: the **eutrophic peats**, also called low moor peats (that is to say, of valleys) or **infra-aquatics**; and the acid, high moor peats, sometimes also called **supra-aquatics**. The first type forms in alluvial valleys under the influence of a very high phreatic water table that does not fluctuate in level to any great extent. They can be genetically related to the humic gleys with anmoor, of which they represent the most extreme state of the developmental sequence. The vegetation is composed of varied hygrophilic plants, reeds, large sedges and mosses. The second type is generally formed in very humid mountainous areas as a result of the accumulation of rainfall in undrained depressions on acid materials. On these grounds they can be considered as being related to the developmental sequence of the peaty stagnogleys, with which they are frequently associated, as stated previously. In both cases, it is the spongey mosses, of the *Sphagnum* genus, which form the peaty material.

In fact, many of the acid peats have a slightly different origin: they have a water table supplied from subterranean sources, as in the case of the eutrophic peats, but in this case the water is acid and therefore poor in calcium. These peats, situated at a lower altitude in a less humid climate, are connected genetically with the humic podzolic gleys which often surround them. The vegetation is no longer made up solely of *Sphagnum*, but other plants are present (grasses and sedges). In the Vosges, for example, both types occur: the high moor peats *sensu stricto* with *Sphagnum*, supported by rain at high altitudes (above 1000 m), and the acid peats of low altitude, supported by the water of lakes (Gérardmer).

*1   Morphology and biochemistry of peat profiles* (Duchaufour 1978: XIV$_5$ & XIV$_6$)

Peats are formed by the accumulation of great amounts of incompletely decomposed organic matter, and thus generally have only a low degree of humification. Nevertheless, this humification can vary a great deal depending on the local conditions of peat development. To characterise peat, it would seem to be essential to take account of this degree of humification, in terms of either its morphology or biochemical characters.

The American classification uses morphological characters based on the amount of *fibres* (in fact, plant debris with organised structures). **Fibrists** (slightly developed peats) have more than two thirds of their mass formed of fibres; **lenists** or **hemists** have a third to two thirds of their mass composed of fibres. Finally, **saprists** (or developed and humified peats), also called **muck**, have less than one third of their mass composed of fibres.

Menut (1974) used biochemical methods to characterise the degree of humification. The amount extracted allows an evaluation to be made of fulvic and humic acids. Unfortunately, there is no method whereby humins can be separated from fresh organic matter. However, measurement of the exchange capacity allows an indirect measurement of the degree of transformation of this fresh organic matter, in particular by oxidation of lignins (direct humification: *inherited humin*).

The fundamental difference between acid and eutrophic peats is, on the one hand, because of differences in development and humification (which will be discussed in the next section) and, on the other, owing to the composition of the plant material (in part resulting from the amount of calcium in water). The richness in nitrogen and exchangeable bases (particularly calcium) is much higher in the eutrophic peat (pH 6–7), than in the acid peat (pH ⩽ 4). The C : N ratio is greater than 30 in the undeveloped acid peats and less than 20 in the eutrophic peats. These peats are also distinguished by their *ash* content, which never exceeds 5% in the acid peats, but in contrast exceeds 10% in the eutrophic peats. If peats have not been contaminated by mineral materials, these differences reflect differences in the biological accumulation of bases, heavy cations ($Al^{3+}$ and $Fe^{3+}$) and silica, which is much more important in the plant debris of eutrophic peats than in acid peats.

**Morphology.** *Eutrophic peats* are made up of two horizons: a *saprist* or *muck* surface, well aerated, humified, dark coloured, with a fine crumb structure overlying an incompletely decomposed, brown fibrous *hemist*.

The *acid peat* generally has four horizons:

(a)  a horizon of the *saprist* type which is, in fact, a hydromor or even a mor, owing to a stronger humification of the peaty material as a result of increased surface aeration;
(b)  a very thick middle horizon of dark brown fibrous peat in which there are successive hemist and fibrist layers;
(c)  at depth there is another blacker horizon of the saprist type which is

clearly more humified (the origin of this horizon will be explained later and has been referred to by Menut (1974) as A0Bh by analogy with the podzols);

(d) at the absolute base there is a black mixed organomineral horizon, with a massive structure, referred to as Bh by Menut.

## 2   Peat development: biochemical processes

Peats are far from being inert materials. A temporary drying out of the surface horizons is common and this is generally speeded up by a modification of the vegetation, such as a colonisation by woody plants (willows and birch) or, for acid peats, by mor plants (*Calluna*, *Molinia*) which cause the strictly hydrophilic species to regress (particularly the *Sphagnum* of the acid peats). This desiccation causes an increase in the humification, leading to the formation of a muck (or saprist) in neutral peats and of a hydromor or an acid mor in the case of an acid peat. This development of surface horizons has an effect on changes in the whole of the profile, particularly in the acid peats which have a development that is almost the same as that which has been described for the hydromorphic humic podzols (Ch. 10). Menut's work (1974) shows that the peat behaves as a profile in which the various horizons react with one another. It also provides important information on the processes of humification and the part played by the biogeochemical cycle in these particular environments; *in the acid peats, both avenues of humification (direct and indirect), which have been discussed in Chapter 2, can be clearly recognised because in this particular case they are spatially distinct from one another.*

**Development of eutrophic peats.** Under the influence of surface aeration, humification occurs very quickly in the surface, for it is favoured by the high pH and the high nitrogen content. Compared to the underlying peat, this development is reflected in a great increase in the exchange capacity (which can be double that of the underlying peat and exceed 200 mEq/100 g organic matter), and also in the amount of material extracted which indicates the neoformation of fulvic and humic acids.

The two types of humification occur simultaneously: *direct* (by oxidation of the lignin that remains insoluble) and *indirect* (by the insolubilisation of the phenolic precursors). In the present case, owing to the abundance of cations of biological origin ($Al^{3+}$, $Fe^{2+}$ or $Fe^{3+}$, $Ca^{2+}$, etc.) occurring in a complexed state, the insolubilisation of the precursors takes place immediately (i.e. in the surface) as in the case of a mull, and the two processes cannot be separated in space. Obviously this makes an evaluation of the processes involved more difficult than in the acid peats.

**Development of acid peats.** The drying out of the surface means that a humification similar to that of the eutrophic peats occurs, but much more slowly and to only a moderate extent owing to the strong acidity and the low nitrogen content. Here, there are two fundamental differences: (i) iron in the ferrous state ($Fe^{2+}$) is mobilised and entirely removed from the profile; and (ii) the cations supplied by the biogeochemical cycle (the only ones involved are $Al^{3+}$ and the alkaline earths) are not sufficient to insolubilise the humic precursors at the top of the profile. Mobile organometal complexes are formed which are transported downwards and are precipitated on contact with the mineral horizons (owing to the clay) to form an accumulative horizon comparable to the Bh of podzols. This development is clearly reflected in the increasing amounts of fulvic and humic acid that are extracted in going from the top to the bottom of the profile (about 10% in the surface A0 and more than 50% in the A0Bh and Bh), and also by an increase in the amount of aluminium transported (two to three times more in the Bh than in the A0). Indeed, the biological aluminium acts here as a *tracer*.

Thus in the acid peats the two avenues of humification are separated in space. Direct humification is dominant at the surface (this is reflected in an increase in the exchange capacity which, however, remains less than that of the eutrophic muck: less than 200 mEq/100 g organic matter). Indirect humification by insolubilisation of precursors is different in that it occurs at depth after migration. Thus it can be seen that the two types of *saprist* of the acid peats – that of the surface and that at depth – have a very different origin. This has also been verified by microfabric studies.

## 3   Utilisation of peats

Peats, no matter what their type, have asphyxiating and reducing conditions unfavourable for plants. The better aerated surface horizon is, in fact, generally not deep enough to allow root development of species other than the hydrophilic ones well adapted to this environment. Thus lowering the level of the water table is necessary to improve the peats, for this causes an increase in biological activity and thus favours humification and also increases the depth of soil that can be used. But the results thus obtained are very different, depending on the type of peat: while the acid peat, even when aerated, retains the character of a mor and remains infertile, the calcic peat, owing to its high base saturation (80–100%) and its low C : N ratio, can be fertile and valuable for certain demanding market garden crops, as is the case for the horticulture of the Somme Valley.

## V   Related hydromorphic soils: capillary absorption and impoverishment of materials rich in fine clays

The *pelosols* are slightly developed soils in humid temperate climates, and the *planosols* are much more strongly differentiated soils of hot (or continental) climates with marked seasonal contrasts, and up to now they have never been associated together in a classification. However, recent work in various countries has shown that there is undoubtedly a certain degree of genetic relationship between these two great soil families: *pelosols can be considered to be the initial phase of planosol development, with planosols being the developed phase, the differences in the degree of development being mainly the result of the climatic factor.* In the same way where the climate has strong seasonal contrasts, the formation of vertisols is often a phase preliminary to planosol formation (Brinkman 1977). But, for the reasons given in Chapter 8, vertisols are a separate group. Pelosols and vertisols, however, have certain properties in common:

(a) The parent materials are clay rich, either inherited (pelosols) or neoformed (vertisols).
(b) The initiation of planosol development is, in both cases, the result of an impoverishment by lateral removal of fine clays from the surface horizon. Hydromorphic development by redox processes only occurs subsequently and thus is secondary and generally of no more than moderate intensity.

Thus it is possible to define the two groups in the following way:

(a) *Pelosols* are only slightly developed owing to the clay-rich nature of the parent rock, which makes it susceptible to erosion. The profile shows little differentiation and hydromorphism results from temporary capillary absorption by the mass of clay. The fairly moderate degree of impoverishment of the top of the profile has been demonstrated by Nguyen Kha (1973) by the use of the method of balances, using quartz as the standard of comparison, and confirmed by drainage basin lysimeter experiments (Nguyen Kha *et al*. 1976). Redox processes are restricted and the environment remains practically saturated with $Ca^{2+}$ and $Mg^{2+}$.
(b) *Planosols:* the marked impoverishment in everything other than quartz (clays, oxides of iron and alumina, organic matter etc.) leads to a loss of colour in the surface horizon and the formation of an *albic* A2 horizon (Baize 1981). Then, very ephemeral surface water tables are formed which in certain circumstances favour redox processes in the residual iron.

## 1 Pelosols

Pelosols are slightly developed soils, formed on very clay-rich sedimentary materials, in a humid temperate climate (of the atlantic type), without a marked dry season (which is an important factor). Hydromorphism results from absorption by the capillary-sized pores, owing to the swelling of clays in rainy periods. It is a very temporary phenomenon and it causes only very limited reduction, except in more acid transitional types (pseudogley–pelosol). Generally, up to this point, lateral impoverishment as demonstrated by Nguyen Kha *et al*. (1976) is limited, and only modifies slightly the morphology and the physicochemical composition of the profile.

**Type profile** (Duchaufour 1978: $XV_1$). A **vertic pelosol** will be taken as an example, which is more representative of the group that occurs in Lorraine, on gentle slopes on outcrops of Keuper Triassic clays (Nguyen Kha 1973). There is little profile differentiation. Note that the mull is more humus rich and thicker than the normal forest mull (15–20 cm thick, with organic matter content more than 10% and C : N ratios abnormally high, 18–20, which is attributed to the fixation by the clay of young inherited humin, poor in nitrogen, which is reasonably abundant).

The (B)C horizon is olive grey or sometimes brownish grey and decarbonated (but saturated in $Ca^{2+}$ and $Mg^{2+}$, this last cation often being very abundant). The structure is typically coarse polyhedral or prismatic. The slickensides, the major contraction cracks of the dry season, are similar to those of the vertisols (hence the qualification *vertic*). The very abundant clays (50% or more) are the semi-swelling interstratified type and iron- and magnesium-rich chlorites inherited from the parent material. The free iron is of little importance and does not play any part in the grey colouration of the clays (Kornblyum 1967).

A decrease in the amount of clay occurs both in going towards the A, owing to surface impoverishment, and the C horizon; Nguyen Kha showed the importance of the processes of physical microdivision of the coarser clays which occurs in wetting and drying cycles after decarbonation of the C horizon.

**Agronomic properties of the pelosols.** These soils are exceptionally rich from a chemical viewpoint, but excessive clay is responsible for poor physical properties. Apart from the humus-rich horizons, whose structure and aeration is the result of the organic matter, this soil is either plastic and asphyxiating in wet periods or as hard as a brick in dry periods. The deep cracking resulting from the contraction of the clays contributes to the drying out. Improvement by drainage is not very effective as the excess water is retained by the capillary pores. Crop growth is hardly practical owing to the difficulties of cultivation. Two uses are possible: (i) *forest* – pedunculate oak and hornbeam, owing to the strength of their root penetration, can get into deep horizons; (ii) *grassland* creates a thick A1 horizon with crumb structure and well aerated in which roots are concentrated. It is, however, very much affected by summer droughts.

**Developmental processes** (Nguyen Kha 1973, Nguyen Kha *et al.* 1976). Even though, as previously stated, hydromorphism is essentially the result of the saturation of the capillary pores of the (B) horizon, free water is still able to circulate at certain seasons in the A1 and the upper part of the (B) horizon, because the irregular crumb structure and the contraction cracks, in fact, create in the top of the profile a network of coarse pores. It is this circulation of free water, particularly in its lateral movement, that is responsible for the removal of the fine clay and the impoverishment of the surface horizon to which reference has already been made. However, this impoverishment is always limited and never gives rise to an *albic* horizon as in the planosols.

As a whole, development is moderate, as is shown by the small amount of weathering (the ratio of free iron : total iron stays less than 30%, which is clearly less than the ratio that is characteristic of temperate brown soils). This weathering increases markedly when the environment becomes acid and rusty patches appear (pseudogley–pelosol). Chlorites are particularly affected and they lose small amounts of both iron and magnesium at about the same rate. Free $Mg^{2+}$ is retained in part by the absorbent complex. As for the free iron, it is mobilised in the ferrous state and partially removed from the profile. *In the vertic pelosols, the redox processes of iron remain very restricted and do not cause the formation of rusty patches by re-oxidation of the free ferrous iron.* It is different in the pseudogley–pelosols where all the processes – weathering of the iron-rich clays, mobilisation and localised re-oxidation of the free iron oxides – are much more marked.

**Main types of pelosols.** In addition to the vertic pelosol, it is possible to distinguish a brunified pelosol, a pseudogley–pelosol and finally a pelosol–gley.

*Brunified pelosol* (Duchaufour 1978: $XV_2$). At level sites under deciduous forest, a brunification of surface horizons occurs under moderately acidifying forest humus. Weathering is more important, and the more resistant illite clays become relatively more abundant. Clay–iron–humus complexes are formed that are characteristic of the brunified soils and the colour becomes clearly browner in the surface. This is a soil *analogue* formed by the convergent development of humus, owing to the climatically determined atlantic oak forest.

*Pseudogley–pelosol* (Duchaufour 1978: $XV_3$, see Fig. 11.3). Pelosols can develop towards a pseudogley of a special type, the pseudogley–pelosol, which does not have a true surface water table like that of the pseudogley described in Section II of this chapter. The greater degree of weathering of the clays is caused by a stronger acidification, either primary (when the parent material does not have such a great reserve of bases), or secondary as a result of a moderate degradation of the humus (for example, under conifer plantations). Under the influence of acid organic compounds, the iron-rich clays free increased amounts of iron, first of all in the ferrous state when the profile is waterlogged, which changes to ferric iron when the profile dries out and is precipitated along the better aerated contraction cracks.

Note that there is another type which is very like the pseudogley with perched water table in which iron segregation, in the form of rusty patches, occurs in the surface few centimetres as a result of the very temporary and weakly reducing surface water table formed in the wet period in the impoverished layer. In this case, the (B)C horizon retains its original appearance. Frequently, however, iron segregation occurs simultaneously both at the surface and at depth.

*Pelosol-gleys.* These are formed at the bottom of slopes as a result of ferrous iron moving laterally. This ferrous iron, rather than the clay, is responsible for the greenish colour of the (B)G horizon, which is more intense than in vertic pelosols.

## 2  Planosols

**General characters.** The formation of planosols can be considered as starting with the same basic developmental process as the pelosols, but their development is much stronger and, differing from pelosols, causes a considerable degree of profile differentiation. The upper part of the profile, impoverished in mobile entities, is decolourised and loses its original structure (albic horizon), while the lower part only develops to a slight extent and keeps its original clay-rich character. *The essential character of planosols is the very abrupt boundary between the albic A2 or A2g horizon and the argillic Bt horizon, in terms of colour, texture and structure.* Seasonal pedoclimatic variations, particularly those associated with water content, are also very marked, the profile going very rapidly from being completely waterlogged to being very desiccated.

*Environment.* The special water relationships of these profiles are the result of planosols being located in hot climates, or at least in continental climates, with marked seasonal variations; while, in contrast, pelosols are typical of humid temperate climate with no great seasonal variations. The planosols, according to Dudal (1973), are preferentially located in prairie or steppe zones and are generally associated with brunizems or vertisols. Thus they are particularly well developed on the plains of Argentina and Uruguay where the brunizems are the zonal soil. In North America the planosols are associated with the brunizems of the temperate prairie. In this climatic zone, two types of hydromorphic soil occur owing to site conditions: **humic gleys** with hydromorphism at depth and **planosols** with surface hydromorphism. In the Soil Taxonomy the first are classed as **aquolls**, the second as **albolls** (argialboll, Kleiss 1973).

These differences between the climatic zones where pelosols and planosols occur explain certain of the special characters of these two groups of soils:

(a) The clays are generally inherited from the parent materials in the pelosols, while to a greater extent they can be ascribed to climatically controlled pedogenesis in the planosols. This last group is generally polycyclic soils having previously been subject to a vertic or isohumic development.

(b) The water relationships are very different in the two types of soil and there are much greater contrasts in the planosols. Torrential rain alternates with phases of strong desiccation of the profile. The permeability of the deeper horizons is such as not to allow the removal of the excess water by vertical drainage; thus very ephemeral surface water tables are formed with a lateral flow which causes selective erosion of the profile and impoverishment of the upper horizons (on this subject, see also Ch. 12).

**The type profile: morphology and geochemical characters.** The type profile can be considered as being a normal planosol (Duchaufour 1978: $XV_5$) of the A1A2Bt type, the A2 horizon being albic. In contrast, the Bt horizon is very clay rich, most of this clay predating the process of planosol formation. A small amount only comes from vertical pervection, which gives this horizon an **argillic** character and allows the symbol Bt to be used.

The humic A1 horizon is generally a base-rich (eutrophic) thick mull, with a crumb structure and thus of the **mollic** type according to the American classification; but in many cases this humic horizon is almost entirely missing, with an **albic** (decolourised) horizon occurring at the surface.

The A2 or A2g albic horizon is without structure and has a sandy loam texture in which quartz is dominant. In more developed types, small black iron–manganese concretions occur.

The Bt horizon is a clay-rich horizon (more than 30% and generally 40% to 60%) generally of a swelling or semi-swelling 2 : 1 type clay; however, kaolinites sometimes occur, particularly on sedimentary clays or reworked ferrallitic materials. The structure and morphology reflect prior developments; the peds are either angular polyhedra, vertic prisms or rounded columns (old salsodic soils which have lost their $Na^+$ ions). Colour is very variable – black, grey or red (if fersiallitic or ferruginous material is involved) – and in its upper part the horizon has irregular hydromorphic patches. In addition, at its junction with the A2 there is a characteristic zone of corrosion which rounds off the top of the prisms and the polyhedra and covers them in a white, residual, silica-rich powder, clearly visible in Plate 15 (Duchaufour 1978).

Generally, this horizon is slightly enriched in clays by vertical pervection, but this remains so limited that often it cannot be seen. In contrast, in certain cases, illuviated cutans are very thick and easily seen, for they contain not only clay but also brown or even black organic matter, which appears to have been transported at the same time as the clay: *planosols with clay–humus illuviation occur frequently in tropical mountains* (Duchaufour 1978: $XV_4$); they have also been reported on Formosa (Wang 1972).

**The physicochemical process of planosol formation.** The processes of planosol formation, which are multiple and complex, are as yet not well known. However, by considering the partial investigations of this problem that have been made in various countries (Cardoso & Bessa 1973; Conea *et al*. 1973; Baize 1974, 1981; Faivre 1977) and particularly the excellent synthesis of Dudal (1973), it is possible to make a provisional statement about the major features of the successive phases of planosol formation.

The initial phase is practically the same as that described for the pelosols, but it is much more rapid. It should be recalled that it is a question of surface impoverishment which removes laterally a part of the more dispersible clays (Baize 1981). *This impoverishment prepares the way for planosol formation just as pervection prepares the way for podzolisation on sandy, parent materials in the temperate zone*. In fact, it favours the formation of water tables in the

extreme surface, often temporary after storm rains, which in turn is responsible for increased impoverishment. Note that in many cases this impoverishment is associated with a vertical movement of humic clays that are deposited
on the surface of prisms in the Bt horizon (brown or black cutans).

The second phase is the result of the appearance (often not essential) of
redox processes, but these are generally limited, for the environment is still
only very slightly acid and rich in calcium, the water tables that are relatively
rich in dissolved oxygen being very temporary. At this stage small concretions
can form and it is at this time that processes of acidification of mineral origin
related to **ferrolysis** can occur, as discussed in the first chapter. It can cause
the beginning of degradation of the clays remaining in the A2 and also of
those in the top of the B (Bartelli 1973, Brinkman 1977).

The third phase is, in fact, rare in the planosol climatic zone. It involves a
marked degree of acidification that is rarely primary (a very acid parent
material), but more often secondary, as a result of the removal of exchangeable cations, transported at the same time as the entities of the absorbent
complex (clay, organic matter). Under the influence of this acidification, the
surface humus degrades, the production and preservation of water-soluble
organic compounds increase, which allows complex formation with $Al^{3+}$ and
$Fe^{3+}$, and certain clays can be hydrolysed. This is reminiscent of the processes
described with respect to the formation of acid or even podzolic pseudogleys
(Baize 1981). It needs to be emphasised that this ultimate process occurs only
exceptionally in hot climates, which are unfavourable to the formation and
preservation of water-soluble complexing compounds. However, this can
occur locally in transitional areas with a warm temperate climate or those with
a continental tendency. This process has been described in the south-west of
France by Begon and Jamagne (1973), Begon (1979) and in central France
by Baize (1974, 1981).

Depending on the local climatic conditions, as to whether the *acid pseudogley* (with moder or hydromoder) or the *planosol* character (sharp A2/Bt
boundary) will be dominant, this type of intergrade soil can be referred to as a
**pseudogley–planosol** or a **planosolic pseudogley**.

*Summary* (Fig. 11.8, Table 11.2)
The mechanism for the formation of the surface albic horizon has not yet
been completely worked out. The process of impoverishment (which, strictly
speaking, refers only to the mechanical transport of dispersed clays together
with the oxides of iron and aluminium to which they are bonded, is common
to all soils of hot climates and will be discussed in the next chapter) *is rather
special in the case of most planosols in that clay and organic matter are affected
simultaneously (together with a certain amount of vertical pervection)*. The
contrast between the A2 horizons, sometimes totally lacking in humus, and
the Bt horizons with humus-rich cutans, easily visible in the photographs of
Plate XV of Duchaufour (1978), is complete proof that in the *pelosol* phase
the clays, dispersed in the A1 by rainfall, transport with them a certain

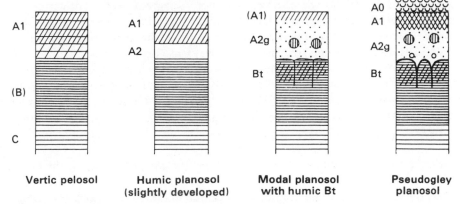

| Vertic pelosol | Humic planosol (slightly developed) | Modal planosol with humic Bt | Pseudogley planosol |

**Figure 11.8** Stages in the development of a pelosol to a planosol and a pseudogley with increasing impoverishment and acidification (see general key, p. ix).

**Table 11.2** Planosolisation.

| Soil | Pelosol ⟶ | Planosol ⟶ | Planosol–pseudogley |
|---|---|---|---|
| process | moderate impoverishment | strong impoverishment | strong impoverishment |
| | no acidification | weak acidification | strong acidification |
| | no redox processes | limited redox processes | strong redox processes |
| | | no effect of organic matter | organic matter involved |

amount of organic matter (Nguyen Kha *et al*. 1976) and that in the planosol phase this process is more marked (Faivre 1977).

The question can be asked as to what form the organic matter is in when transported? Is it as *soluble precursors* coming from the rhizosphere which are responsible for the dispersion of the clay? Or is it as parts of the developed clay–humus complex of a chernozemic or vertic type, whose structure was destroyed by the violent rains and the material dispersed to form suspensions within which the fine clays and the humified organic matter remains bonded? It is this second hypothesis which is now favoured. The black organic matter of the Bt horizon of planosols has, in fact, the properties of *organoclay humins* associated with the very fine, and hence dispersive, clays, which have been described with respect to the chernozems. This simultaneous migration of fine clays and associated humic compounds (which is similar to that of the grey forest soils, see Ch. 9) has been demonstrated by Faivre *et al*. (1975) and Andreux and Correa (1981).

**Types of planosol** (Fig. 11.8). As yet, there has been no definitive classification of planosols, both because it is only in recent times that their investigation has been initiated and also because there is a great number of varieties in many different environments. Even though no valid classification has been made, it is possible, from what is known of their characteristic development processes, to give certain indications of what may form the possible bases of a classification. There are two possible methods of classifying planosols: the first based on recent developments of the upper part of the profile and the degree of surface planosol development; the second based on the origin and nature of the underlying clay horizon, most planosols being, as stated previously, of a polycyclic type.

*Classification based on the recent development of the upper part of the profile* (degree and appearance of planosol development)

*Humic (or mollic) planosol:* soils with organic matter similar to that of neighbouring isohumic profiles. The A1 is well developed with a dark colour and a saturated, absorbent complex. Apart from the decolouration of the albic horizon, the signs of hydromorphism are limited (see *mollic planosol*, FAO 1971).

*Modal planosol:* a marked disappearance of organic matter in the upper part of the profile. Beginning of acidification; hydromorphic characters more marked; presence of small black concretions within the albic horizon.

*Pseudogley–planosol:* presence of an acid organic horizon (moder); very marked redox processes; formation of organomineral complexes (fulvic acids) in the A2; beginning of clay degradation with freeing of aluminium.

*Planosol with clay–humus illuviation:* formation of dark cutans (brown or black) around the Bt prisms (Duchaufour 1978: $XV_4$).

*Classification based on prior development* (nature of the B horizon)

*Vertic planosol:* B horizon has vertic characters: dark colour, abundance of swelling clays; slickensided *vertic* structure.

*Solodic planosol* (Duchaufour 1978: $XV_5$): near to the previous type but the abundance of the exchangeable sodium ions in the B is evidence of a prior development dominated by the sodium ion (solonetz and soloth with a natric horizon). The planosol is formed by the elimination of the sodium ion from the B and by an increase in the hydromorphic characters at the surface (degradation by corrosion of the top of the columns of the old *natric* horizon); on this subject see Chapter 13 (p. 430).

*Fersiallitic (ferruginous) planosol* (Duchaufour 1978: $XV_6$): the B horizon is rubified and its character is evidence of a prior fersiallitic or ferruginous development.

**Agronomic properties of planosols.** It is difficult to give a brief overall view of the agronomic properties of planosols because of their great variety. The properties of planosols are, in fact, dependent on both the degree of hydromorphism of the upper horizons and the kind of clays present in the B which can be inherited from a prior pedogenesis or simply from parent material. Certain planosols are submerged by water for a part of the year; others, in contrast, havé very restricted hydromorphic characters. The most humid can be used, because of their high mineral fertility, for rice cultivation, but then it is necessary to irrigate in the dry season. When irrigation is not possible, the extreme hardness of the B horizon makes dry season cultivation impossible. They are commonly under grass (the pampas of Argentina), but they suffer from drought because of an insufficient depth of root penetration, owing to its prevention by the compactness of the B horizon.

# References

Andreux, E. and A. Correa 1981. In *Migrations organo-minerales dans les sols tempérés*, 329–40. Coll. Intern., CNRS, Nancy, 1979.

Akhtyrtsev, B. P. 1974. *Pochvovedeniye* 9, 14–26. (*Soviet Soil Sci.* 6 (5), 507–18.)

Aussenac, G. 1975. *Couverts forestiers et facteurs du climat: leurs interactions, conséquences écophysiologiques chez quelques résineux*. State doct. thesis. Univ. Nancy.

Avery, B. W. 1973. *J. Soil Sci.* 24 (3), 324–38.

Baize, D. 1974. *Science du Sol* 1, 5–22.

Baize, D. 1981. In *Migrations organo-minerales dans les sols tempérés*, 365–76. Coll. Intern., CNRS, Nancy, 1979.

Bartelli, L. J. 1973. *Soil Sci.* 115 (3), 254–60.

Becker, M. 1971. *Etude des relations sol-végétation en conditions d'hydromorphie dans une forêt de la plaine lorraine*. State doct. thesis. Nancy.

Begon, J. C. 1979. *C.R. Acad. Sci. Paris* 288D, 481–4.

Begon, J. C. and M. Jamagne 1973. In *Pseudogley and gley. Trans Comms V and VI ISSS*, E. Schlichting and U. Schwertmann (eds), 307–8. Weinheim: Chemie.

Berthelin, J. 1976. *Etude expérimentale de méchanismes d'altération par des microorganismes hétérotrophes*. State doct. thesis. Univ. Nancy.

Blume, H. P. 1968a. *Z. Pflanzener. Bodenk.* 119 (2), 124–34.

Blume, H. P. 1968b. *9th Congr. ISSS*, Adelaide IV, 441–9.

Brinkman, R. 1977. *Geoderma* 17 (2), 111–44.

Callame, B. and J. Depuis 1972. *Science du Sol* 2, 33–60.

Cardoso, J. C. and M. T. Bessa 1973. In *Pseudogley and gley. Trans Comms V and VI ISSS*, E. Schlichting and U. Schwertmann (eds), 335–40. Weinheim: Chemie.

Conea, A., C. Oancea, A. Popovat, C. Rapaport and I. Vintila 1973. In *Pseudogley and gley. Trans Comms V and VI ISSS*, E. Schlichting and U. Schwertmann (eds), 221–7. Weinheim: Chemie.

Duchaufour, Ph. 1978. *Ecological atlas of soils of the world*. New York: Masson.

Dudal, R. 1973. In *Pseudogley and gley. Trans Comms V and VI ISSS*, E. Schlichting and U. Schwertmann (eds), 275–85. Weinheim: Chemie.

Faivre, P., F. Andreux and Ph. Duchaufour 1975. *C.R. Acad. Sci. Paris* 281D, 981–4.

Faivre, P. 1977. *Science du Sol* 2, 95–110.

FAO 1971. *Soil map of the world IV*. South America. Paris: UNESCO.

Glatzel, K., R. Jahn and S. Müller 1967. *Sudwestdeutsche Waldböden im Farbbild. Schrift. Landesforstverwaltung*. Baden – Württemberg, Stuttgart: Selbstverlag der Landesforstverwaltung.

Guillet, B., J. Rouiller and B. Souchier 1976. *Bull. Soc. Géol. Fr.* XVIII (1), 55–8.

Gury, M. 1976. *Evolution des sols en milieu acide et hydromorphe sur terrasses alluviales de la Meurthe*. Spec. doct. thesis. Univ. Nancy.

Jambu, P. and D. Righi 1973. *Science du Sol* 3, 207–19.

Kanivets, V. I. 1973. *Pochvovedeniye* 7, 51–60. (*Soviet Soil Sci.* 5 (4), 433–42.)

Kaurichev, I. S. and V. A. Shisshova 1967. *Pochvovedeniye* 5, 66–78. (*Soviet Soil Sci.* 5, 636–46.)

Kleiss, H. J. 1973. *Soil Sci.* **115** (3), 194–8.

Kornblyum, V. A. 1967. *Pochvovedeniye* **11**, 107–21. (*Soviet Soil Sci.* **11**, 1527–39.)

Kubiena, W. 1953. *The soils of Europe.* London: Thomas Murby.

Menut, G. 1974. *Recherches écologiques sur l'évolution de la matière organique des sols tourbeux.* Spec. doct. thesis. Univ. Nancy.

Mückenhausen, E. 1975. *Die Bodenkunde.* Frankfurt-am-Main: DLG.

Munch, J. C. and J. C. Ottow 1977. *Z. Pflanzener. Bodenk.* **140**, 549–62.

Munch. J. C., Th. Hillebrand and J. C. Ottow 1978. *Can. J. Soil Sci.* **58**, 475–86.

Nguyen Kha 1973. *Recherches sur l'évolution des sols a'texture argileuse en conditions tempérées et tropicales.* State doct. thesis. Univ. Nancy.

Nguyen Kha and Ph. Duchaufour 1969. *Science du Sol* **1**, 97–100.

Nguyen Kha, J. Rouiller and B. Souchier 1976. *Science du Sol* **4**, 259–67.

Périgaud, S. 1963. *Ann. Agron.* **14**(3), 261.

Richard, J. L. 1961. *Les forêts acidophiles du Jura.* Doct. thesis. Univ. Neuchâtel. Berne: Huber.

Schaefer, R. 1967. *Caractères et évolution des activités microbiennes dans une chaîne de sols de la plaine d'Alsace.* State doct. thesis. Fac. Sci. Orsay, Summary.

Schlichting, E. 1973. In *Pseudogley and gley. Trans Comms V and VI ISSS*, E. Schlichting and U. Schwertmann (eds), 1–6. Weinheim: Chemie.

Schweikle, V. 1971. *Die Stellung der Stagnogley in der Bodengesellschaft der Schwartz-waldhochfläche auf S.O. Sandstein.* Stuttgart: Sulz Kreis Horb.

Schwertmann, U. and W. R. Fischer 1973. *Geoderma* **10** (3), 237–47.

Schwertmann, U. and R. M. Taylor 1973. In *Pseudogley and gley. Trans Comms V and VI ISSS*, E. Schlichting and U. Schwertmann (eds), 45–54. Weinheim: Chemie.

Thomasson, A. J. and P. Bullock 1975. *Soil Sci.* **119** (5), 339–48.

Vaneman, P. L., M. P. Vepraskas and J. Bouma 1976. *Geoderma* **15** (2), 103–18.

Wang, M. K. 1972. *J. Agric. For.*, Taiwan **21**, 43–52.

Zaidelman, F. R. 1974. *Podzolisation and gleyisation.* Moscow: Nauka.

Zonn, S. V. 1973. In *Pseudogley and gley. Trans Comms V and VI ISSS*, E. Schlichting and U. Schwertmann (eds), 221–7. Weinheim: Chemie.

# Chapter 12

# Sesquioxide-rich soils

## I General characters

Soils formed in hot climates (subtropical, tropical, equatorial) have a certain number of common characteristics. Weathering of primary minerals is more powerful than in a temperate climate and it occurs to a greater depth. Conversely, organic matter remains in the near-surface zone and, with a few exceptions, is subject to rapid biodegradation. *The zone of weathering is outside the influence of acid organic compounds produced at the surface*. In these circumstances weathering is geochemical, i.e. a neutral or slightly acid hydrolysis, which causes a higher concentration of free oxides than is characteristic of temperate soils. Essentially this involves iron oxides, but also aluminium oxides in the case of ferrallitic soils.

In terms of environment, these soils are typical of regions that are sufficiently humid to allow the development of woody plants, either hygrophilic forest (ferrallitic soils of humid climates), or xerophilic forest (fersiallitic soils of mediterranean evergreen forests) or, finally, bush or mixed savanna and thorn bush (ferruginous tropical soils). When the climate becomes even dryer, more or less xerophilic steppe takes the place of the forest and the soils are the transitional **isohumic, sub-arid brown soils** (tropical) and **reddish chestnut soils** (mediterranean) which have been discussed previously (see Ch. 8).

### 1 The nature of weathering

The geochemical weathering typical of hot climates was discussed in Chapter 1 and compared to weathering in a temperate climate. It should be recalled that a neutral or very slightly acid hydrolysis is involved, that is at most only very slightly influenced by surface organic matter, which is different from the situation in temperate or cold climates where the weathering by acidolysis and complexolysis is initially more rapid but remains less complete than the neutral hydrolysis of hot climates. This has an indirect effect on pedogenesis as a whole, for in fact the speed of development is very variable and, compared to temperate climate pedogenesis, it is more closely related to site conditions, such as the parent material and topography. Consequently, the variety of soils in a given region is often greater than in a temperate climate and the concept of climatically controlled development is generally more difficult to discern (see Ch. 4).

**Origin of clays: inheritance, transformation, neoformation.** As previously stated, the weathering of primary minerals is more complete than in a temperate climate, so that in a hot climate, on equivalent materials, the clays and sesquioxides of the *weathering complex* are more important. In addition, *the transformed or inherited clays that are dominant in temperate soils tend to decrease in importance and be replaced by neoformed clays (particularly kaolinite in the most humid climates)*. The proportion of neoformed clays compared to the transformed clays increases regularly as the mean temperature increases. Neoformation, while still being of moderate importance in the subtropical regions (fersiallitic soils), becomes the dominant process in tropical climates (ferruginous soils) and almost the only process in equatorial climates (ferrallitic soils) (see Table 1.1).

**Special development of iron oxides.** The colour of soils in hot climates is generally much more intense than that in temperate climates, being either a strong ochreous or red colour, which is dependent on both the type of iron oxide and its amount.

In terms of the amount, on equivalent materials there are always more iron oxides in a hot climate than in a temperate climate, for which there are two reasons: first, the amount of weathering is greater and, secondly, the freed iron oxides (and also aluminium oxides in ferrallitic soils) remain *in situ*, or almost so, within the profile, while in contrast in a temperate climate, acid hydrolysis causes a certain loss of $Fe^{3+}$ and $Al^{3+}$ ions in drainage waters (Lelong & Souchier 1972). In general, the ratio of free iron : total iron of a weathering horizon on granite never exceeds 50% in a temperate acid brown soil, but reaches 60–70% in a fersiallitic red soil and 100% in a ferrallitic soil.

Another cause of the intense colouration of soils of hot climates is the kind of iron mineral involved. According to Schwertmann *et al.* (1974), this colour is dependent on the fact that the processes of crystallisation of the free iron oxides are not prevented by the presence of organic matter, as they are in temperate soils. In the (B) horizon of temperate brown soils, the coexistence of amorphous forms of iron and a certain amount of cryptocrystalline goethite bound to the clays gives the soil a dull brown colour.

In the (B) horizon of the soils of a hot climate, the processes of crystallisation of iron oxides tend to be general and they result in the formation either of **goethite**, which gives the soil an ochreous colour, or **haematite**, which gives a red colour. Goethite results from a *gradual crystallisation* that occurs in an almost constantly humid environment, while haematite *crystallises rapidly* in an environment subject to phases of marked desiccation. More detailed explanations of these different physicochemical processes will be given when the processes of fersiallitisation (rubification) are discussed.

Note that, as far as very old ferrallitic soils are concerned, it seems that a part of the goethite is very slowly transformed into haematite by dehydration. Thus there would seem to be two quite distinct ways of haematite formation, one that is rapid and the other slow.

## 2   *The three phases of weathering in hot climates*
Three basic types of weathering, in which organic matter is not involved to any significant extent, are recognised in hot climates and each of them is

generally characteristic of different climatic zones. However, as all of them can occur side by side in the same climatic zone, as they do in the humid Tropics where site factors determine the relative duration or local speed of pedogenesis (see Ch. 4), they can be considered also as being *phases in the same weathering process.*

*These three phases are characterised by an increasing degree of weathering of primary minerals, an increasing loss of combined silica, and finally an increasingly marked dominance of neoformation of clays, formed from entities previously in solution.*

**Phase 1: fersiallitisation.** There is a dominance of 2 : 1 clays rich in silica, partially inherited and partially of neoformation (or of a very special kind of transformation). Considerable amounts of free iron oxides are formed that are generally more or less **rubified**. The absorbent complex is saturated or almost saturated by the movement towards the surface of calcium in the dry season. An **argillic** (Bt) horizon occurs as a result of fine clay pervection, often complicated by an *impoverishment* in clay of the surface horizons.

**Phase 2: ferrugination.** Weathering is stronger, but certain primary minerals still persist (orthoclase, muscovite). De-silication is more marked and there are more neoformed 1 : 1 clays (kaolinite) than 2 : 1 transformed clays, but free gibbsite does not generally occur (except in certain transitional soils). Iron oxides may or may not be rubified (red or ochreous colour). Base saturation is very variable, depending on the humidity of the climate and the importance of the dry season. The processes of pervection, preferentially affecting the 2 : 1 clays, are still active, even though to a lesser extent than in the fersiallitic soils.

**Phase 3: ferrallitisation.** There is complete weathering of primary minerals (except for quartz) and clays are all neoformed, consisting solely of kaolinite. Free gibbsite occurs frequently, although its presence is not absolutely essential. Clay pervection decreases as clay is increasingly resistant to dispersion by water and no true argillic horizon is formed. However, more or less marked *lateral impoverishment* can occur at the surface.

Note that these three processes are closely interrelated with one another, the fersiallitic soils representing the first stage which is characterised particularly by a marked formation of free oxides that are subject to a particular kind of development (rubification). Neoformation is not as yet very great and the clays, still fairly rich in silica, are subject to fairly considerable pervection. In the succeeding stages, loss of silica increases, neoformed kaolinites increase in proportion and then become the only clays. As they are resistant to movement by water, pervection tends to decrease or disappear:

fersiallitisation ⟶ ferrugination ⟶ ferrallitisation

## 3   The environmental control of the three phases
While climate remains the fundamental factor of pedogenesis in hot climates, its speed of development as related to site conditions is particularly impor-

tant. It is also essential to take account of the time factor which here plays a much more important part than in temperate climates.

**Climate.** In hot humid climates, phase 3 is generally reached, particularly on those materials that are the oldest or on which very rapid development can occur. Phase 2 is the final phase of pedogenesis in climates that are either less hot (humid sub-Tropics) or characterised by a marked dry season (dry Tropics). Finally, phase 1 is typical of subtropical or mediterranean climates, with a dry season. It is only under exceptional circumstances (particular kind of parent material or a humid climate) that a transition to phase 2 occurs.

**Parent material.** This involves both the mineralogical composition (crystalline parent rocks) and the age of the deposit (Quaternary sediments).

Account must be taken of the amount of weatherable minerals, rich in iron and bases, together with the general climate and also with local drainage conditions. In terms of pedogenesis, a high percentage of these minerals can act as a brake (in a dry climate) or as an accelerator (in humid climates on well drained sites). In the first case, the bases ($Ca^{2+}$ and $Mg^{2+}$) freed at the same time as the iron favour the immobilisation of the silica within the 2 : 1 clays, so that only fersiallitisation occurs. In contrast, in the second case, the low silica reserve is rapidly exhausted by transport to depth together with the bases, and phases 2 and 3 are very quickly reached. The abundance of silica (even as quartz) favours kaolinisation and prevents the formation of free gibbsite.

The age of the parent material is also of great importance, particularly if Quaternary deposits are involved. As ferrallitisation is a slow process it can only be completed on sufficiently old materials, even in the equatorial humid zone. Several examples of this will be given in the detailed study of the ferruginous soils.

**Topography.** Topography has important and varied effects. Where slopes are steep enough, rejuvenation is caused by erosion so that in the humid tropical zone, soils on slopes are often less developed than those on old peneplained surfaces. However, indirect effects are more important in which relief favours lateral transport of soluble materials resulting from weathering. Depending on the kind of parent material, this can be either silica and bases or alternatively iron oxides that are transported laterally. On basic materials poor in reducing and complexing organic matter, silica and bases tend to accumulate at the base of slopes in impeded environments where they cause a very special kind of pedogenesis (**vertisol formation**, see Ch. 8). In an acid environment, water tables containing organic matter circulate laterally and mobilise and transport the iron oxides which move down slope and, by their precipitation and crystallisation, give rise to a marked accumulation of concretionary ironstone (**ferruginous cuirasse**, Figs 12.3 & 4).

## 4 Classification

It would appear to be absolutely essential to differentiate clearly between the soils resulting from hot climate pedogenesis and those resulting from temperate climate pedogenesis. The way in which the processes of weathering, in the first case, are independent of the effects of organic matter and also the way that clays are formed and iron oxides develop would appear to justify this point of view. However, as previously stated in Chapter 5, neither the USA nor FAO classification has considered it necessary to separate these two types of weathering at the level of the orders. In the Soil Taxonomy there are inceptisols and alfisols in practically all climates in the same way that the cambisols and luvisols of the FAO classification can also be both temperate and tropical. This often causes climatic correlations between the various classifications to be very difficult.

This comment being made, it would seem to be logical to base the classification of soils of hot climates on the three basic processes previously described, each of them characterising a class, despite the occurrence of numerous intermediate types and the possibility, where the time factor is involved, of more than one class occurring under identical environmental conditions.

In these circumstances the definition given of ferrallitic soils corresponds more or less with that of the **oxisol** order (USA) or the **ferralsols** (FAO). Lack of pervection and low exchange capacity of the clays (less than 16 mEq/100 g clay) are essential characteristics of this order.

The **ultisols** (USA) or the **acrisols** (FAO) are also soils that are confined to regions with a hot climate, with some exceptions, and they have a kind of pedogenesis intermediate between the extremes and thus correspond to a large extent to the previously described phase 2. They are strongly weathered, very unsaturated and, differing from the oxisols, have an argillic horizon. Thus there are reasons to group them with the tropical ferruginous soils of the French classification, except that this would exclude from the ultisols all those tropical ferruginous soils of dry climates that are only slightly desaturated, and place them in the alfisols. It would be more logical, as proposed by Sys (1967, 1978), to use not the base saturation but the exchange capacity of the clays to separate this class, on the one hand from the brunified and fersiallitic soils and on the other from the ferrallitic soils (oxisols). *The exchange capacity, according to Sys, is intermediate, i.e. 16–25 mEq/100 g clay* (measured at pH 7). This criterion will be adopted to define the ferruginous class of soils and to distinguish them on the one hand from the fersiallitic soils (CEC of clays >25 mEq/100 g) and on the other from the ferrallitic clay (CEC clays <16 mEq/100 g).

## A   FERSIALLITIC SOILS

Most fersiallitic soils can be considered as being characteristic, not of the Mediterranean region in the strict sense, but rather of a mediterranean type of climate, that is to say one with a marked contrast between a relatively cold wet season and a hot and dry summer. According to Emberger (1939), fersiallitic soils are characteristic only of the more humid mediterranean climates, i.e. where the climax is clearly a forest (generally holly oak or cork), with the previously studied **reddish chestnut soils** in dryer lower rainfall areas which have a sparse forest, thorn bush or steppe.

In fact, climates favourable to fersiallitisation have a fairly large range of both temperature (mean temperatures 13–20°C) and precipitation (500 mm to more than 1 m), with the only common character being the occurrence of a hot dry season, and are often referred to as subtropical to differentiate them from the hotter tropical climates.

However, note particularly that in humid subtropical climates with a reduced dry season, there are more developed soils (acid ferruginous soils or American ultisols). Conversely, fersiallitic soils occur in tropical climatic zones (even of the humid type) where special circumstances (basic parent material rejuvenated by erosion, or deposits of recent age) have prevented normal climatic development. This is the case for some of the tropical eutrophic brown soils and fersiallitic red soils which, in the author's opinion, should be placed in this class.

In these climatic conditions, the occurrence of the wet season is responsible for the liberation of the iron by weathering or by decarbonation, if the parent material contains carbonates. In fact, as will be seen, decarbonation is a necessary preliminary to all developments of the fersiallitic type; the dry season plays an important part by maintaining in the profile by capillary rise most elements freed by weathering, including the more soluble silica and bases. *In addition, it causes the special type of development of the iron oxides bound to the clays which is responsible for the characteristic red colour; this is the process of rubification.*

Fersiallitisation affects a great range of materials, from the most calcareous to the most acid. But this process occurs at a very variable speed, depending on the type of parent material, which is of great importance with regard to the geographic distribution of red soils, and also the possibilities of the superimposition of subsequent kinds of pedogenesis in cases where the process is particularly slow. However, extremely quartzose materials without weatherable minerals and poor in iron are resistant to fersiallitisation, no matter what the climate.

Apart from the fairly well developed red colour of the (B) horizon, or rather a Bt horizon enriched in clay (for pervection almost always occurs), fersiallitic profiles are differentiated from temperate lessived brown soils by certain definite characters:

(a) The absence of marked acidification results not only from an effective biogeochemical cycling of bases and a very favourable humification, which assures a reserve of bases (*eutrophic mull*), but also from a movement towards the surface of bases, particularly $Ca^{2+}$ and $Mg^{2+}$, by capillarity at the beginning of the dry season, which causes the absorbent complex to be resaturated. *In these circumstances, on equivalent parent materials, fersiallitic soils are always less acid than temperate brown or lessived brown soils.*

(b) The amount of clay and particularly of iron is higher (maximum value of the iron : clay ratio, Bornand 1978). As for the clay, to the processes of inheritance and transformation characteristic of a temperate climate is added a significant neoformation of 2 : 1 clays, even in well drained conditions.

(c) As in the case of the temperate brunified soils, the profiles are decarbonated in the A and B horizons, but many cases differ from temperate examples in that the transported calcium bicarbonate is reprecipitated to form a calcareous layer (a more or less chalky Cca horizon), but which is not absolutely essential and it is always thinner and at greater depth than that in the reddish chestnut soils (see Ch. 8).

# I The morphological and geochemical characters of fersiallitic profiles (Fig. 12.1)

Three basic types of fersiallitic soil will be considered, corresponding to the three subclasses that will be differentiated at the end of this section. In fact, these three subclasses correspond to three phases of development: the least developed is still very similar to a temperate brunified soil or to certain calcimagnesian soils; the most developed already has certain characteristics of ferruginous soils, in particular their acidity. To begin with, the most characteristic and widespread of these soils, a **modal red soil** that has not been markedly acidified, will be considered. When it has not been modified by erosion or reworking of the surface, it is characteristic of a wide range of rock

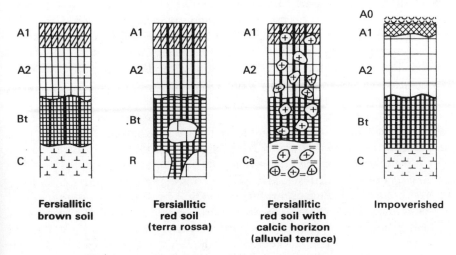

| Fersiallitic brown soil | Fersiallitic red soil (terra rossa) | Fersiallitic red soil with calcic horizon (alluvial terrace) | Impoverished |

**Figure 12.1** Fersiallitic profiles (see general key, p. ix).

outcrops in the humid and subhumid Mediterranean zone, such as indurated limestones (terra rossa), alluvial terraces, colluvium of basic crystalline rocks or of iron-rich schists, calcareous sandstones, etc.

## 1   Fersiallitic red soils (Duchaufour 1978: XVI$_5$ & XVI$_6$)

Under climatically determined vegetation such as a forest of holly oak with a very dense substorey of bushes, the surface mull is eutrophic, fairly thick with a good structure. With cultivation, the upper horizons are homogenised and changed into an Ap horizon with a considerable decrease in organic matter content and partial structural degradation.

**Morphology.** As an example, a profile will be considered that is still under forest vegetation, and hence one in which the original humus-rich horizon is preserved.

(a) A1:  crumb mull, partially brunified, 5YR Munsell colour; porous, very impoverished in clay.

(b) A2:  (or A/B): less humus rich, less dark, often less red than the B horizon (5YR colour); nutty structure and a little richer in clay.

(c) Bt:  clay-rich horizon, bright red (2.5YR and often 10R); strongly developed polyhedra or angular prisms with larger shining surfaces both of *argillans*, which are evidence of pervection, and *slickensides*, particularly if the soil is rich in clays (occurrence of montmorillonites); in certain cases the argillans are fragile and can be difficult to identify (Reynders 1972).

(d) Cca:  a chalky calcareous deposit, frequent on parent materials that contain calcareous material or simply those that free many $Ca^{2+}$ ions in weathering; this horizon does not occur under more acid conditions; at the bottom of deep fissures in indurated limestone where pervection has filled up the pores in the material and it is constantly humid, this horizon takes the form of rounded concretions, covered by an *ochreous clay*, rich in goethite and montmorillonite.

**Geochemical characters.** While the profile is normally decarbonated, it remains rich in bases, particularly $Ca^{2+}$ and $Mg^{2+}$. Base saturation remains about 100% (70–100%) except on very acid materials (Dachary 1975). On very calcareous materials there is often a *supersaturation* of the exchange capacity, which is very specific to these soils (Bruin 1970, Bottner 1972). As yet, this supersaturation, which results from a capillary rise of the bicarbonates in the dry season, is poorly explained.

Weathering is stronger than in temperate soils, the ratio free iron : total iron generally exceeding 60% except in certain young profiles with very rapid rubification (where it happens suddenly before weathering is complete). To the micaceous clays, inherited and very slightly transformed, are added clays of neoformation (or of aggradation) also of 2 : 1 type, except on certain basic materials, poor in silica, that give rise to considerable amounts of kaolinite (basaltic colluvium; Duchaufour 1978: XVI$_5$). As a whole, the amount of clay is greater than in a temperate brown soil on equivalent parent materials.

These clays have a high exchange capacity, averaging 50 mEq/100 g clay and often more. It can be markedly lower than this on basic rocks or when the soil is very developed, but remains greater − 25 mEq/100 g clay (Bruin 1970).

*Pervection* of a mechanical type always occurs in undisturbed profiles: the index of clay movement is about 1 : 2. In certain soils, according to Bottner (1972), transport of iron is less important than that of clay. At the surface, there is a general *impoverishment* by lateral movement, which, however, is less marked than in the acid fersiallitic type.

**Variations in the character of the red soil profile according to the parent material** (Fig. 12.1). The profile that has been described occurs on many parent materials; apart from characters that are indicative of pedogenesis, secondary characters inherited from the parent material can also occur. The amount of clay is very variable, and attains its maximum on terra rossa formed on hard limestones (often 80% if the terra rossa is not contaminated by wind-blow silt which is very common; Yaalon and Ganor 1973, McLeod 1980). On the other hand, on parent materials poorer in calcareous materials or which occur at the top of slopes, calcic horizons are not formed.

Terra rossa is formed by the decarbonation of hard limestones by the process of *surface dissolution* described by Lamouroux (1971); in each humid period, a skin of silicate impurities is detached from the corroded surface of the rock; these impurities are dominantly clays and iron oxides, which rubify more or less quickly. The process is generally slow, but in certain climatic conditions of high humidity it can be rapid. If, in addition, the decarbonated layer is well protected by the forest vegetation against erosion, the profile preserves completely the characters that have been described, and the A1, A/B and Bt horizons are well differentiated. However, because of the forest environment, which maintains a cool and humid pedoclimate, a marked brunification of the upper horizons can occur as a result of a greater incorporation of organic matter, causing a partial regression of the rubification: the soil is then a **brunified fersiallitic red soil**.

When a terra rossa is formed in less favourable climatic conditions (dryer climate), its development is slower and it is then an *old soil*. In these circumstances the original forested profile is rarely preserved, human activity having very frequently caused a degradation of the climax forest. The soil has then been subject to various kinds of reworking and transformation: erosion, truncation reducing the profile to discontinuous pockets, colluviation and secondary recarbonation leading to the formation of calcareous brown soils on rubified materials (dolines).

Finally, certain terrae rossae are very old. These are true palaeosols going back to the early Quaternary or the Tertiary and are distinguished from recent terra rossa by their desaturation and great abundance of kaolinite resulting from a gradual loss of silica by the originally formed 2 : 1 clays (Spaargaren 1979, McLeod 1980). The profile thus formed is related to the acid fersiallitic soils, or even the ferruginous soils: most often, it acts as a *parent material* and has been subject to a varied recent pedogenesis, according to the new environmental conditions occurring at a particular site (*polycyclic soil*, see Ch. 4).

## 2 Fersiallitic brown soils

These soils are incompletely rubified intergrades, either as a result of climate (lowlands or mountains on the border of the Mediterranean region where the climate is colder or the dry season less pronounced, such as in the case of the Danubian profile of Duchaufour 1978: $XVI_1$), or owing to the parent material being of such a recent age that decarbonation is still incomplete, which is the case of the Würmian terraces of the Midi of France (Duchaufour 1978: $XVI_3$).

These soils have the same general characters as the fersiallitic red soils. However, if pervection has occurred it remains limited (Bornand 1978); in addition, owing to the more humid pedoclimate (Danubian soil) or the presence of a small amount of limestone persisting in the Bt, rubification is not complete and the colour is never redder than 5YR on the Munsell scale.

## 3 Acid (or desaturated) fersiallitic soils (Duchaufour 1978: $XVI_4$)

These soils are not only acidified by the removal of a part of the exchangeable bases, but have been subject to much more pervection and particularly a greater amount of impoverishment in fine-grained materials than the modal red soil type. Some result from prolonged development and are thus very old soils, as is the case for the early Quaternary terraces of the Midi of France or of the Guadalquivir (Bornand 1978, Torrent *et al*. 1980). Others are formed on parent materials with lower base reserves and a dominantly quartzose skeleton, which favours the three processes of acidification, pervection and impoverishment (Bottner 1972). They can be considered as being transitional to the desaturated ferruginous soils; the Soil Taxonomy considers them to be ultisols (xerult, or an ultic xeralf at a minimum). They clearly differ from the modal type in a certain number of quite definite morphological and geochemical characters:

(a) The complex is partially desaturated, the amount of base saturation being about 50%.

(b) Deeper weathering (the profile is sometimes 2 m deep) and the index of weathering (free iron : total iron) is higher. However, the ratio iron : clay is lower owing to the beginning of the kaolinisation of the feldspars (Bornand 1978).

(c) Horizons markedly differentiated both in colour and texture, which is an indication *not only of pervection but also of strong lateral impoverishment*; the pebbly or sandy A2 has a pale colour (greyish or pinkish) and has lost not only most of its clay but often its silt. It overlies a bright red Bt horizon (sometimes yellowish ochre) across a very sharp boundary.

(d) The clays are often partially de-silicated; even kaolinite appears in the weathering complex and the exchange capacity is considerably lowered, although it is always above 25 mEq/100 g clay.

(e) In certain cases, the base of the Bt horizon, choked up by the deposition of large amounts of pervected clay, becomes impermeable and takes on hydromorphic characters (Bornand 1978). Thus, localised reduction occurs, causing a partial de-rubification: the general colour becomes ochreous (formation of goethite) with the occurrence of decolourised patches.

## 4  Fersiallitic soil properties with regard to agronomy and forestry

Brown and red fersiallitic soils that have resisted degradation as a result of cultivation or of deforestation by grazing and fire, have properties that are favourable to plant growth. The crumb structure of the surface horizons (brunified, owing to the humus) provides good aeration and aids the penetration of rain to depth, which maintains water reserves that plants use in the dry season. From a chemical point of view, their high content of bases, the effectiveness of the biogeochemical cycle of all the nutritive elements and their reserves in the A1 horizon are favourable factors.

But these soils are fragile and when they are badly treated, either when being used agriculturally or when under forest, they degrade very rapidly.

**In agricultural areas.** Intensive cultivation of the soil decreases the amount of humus and makes the structure unstable. Violent rainstorms increase the speed of impoverishment in fine-grained materials of the surface, leaving behind only coarser materials, sands or rounded pebbles (on alluvial terraces). The clays and silts carry with them some of the bases and soil organic matter, which is thus considerably impoverished.

**In forested areas.** Intensive grazing and deliberate burning, which has gone on in certain regions for thousands of years, has caused the original holly oak forests (or cork oak forests) to disappear and to be replaced by a xerophilic bush formation: *garrigue* on limestone, *maquis* on siliceous rocks. The

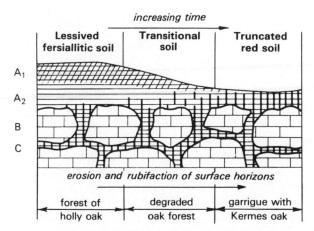

**Figure 12.2**  Degradation of a fersiallitic soil (terrạ rossa) (see general key, p. ix).

brunified surface horizon disappears by mineralisation of the organic matter and by complete dehydration of the amorphous iron oxides originally protected by the organic matter. Generally, as a result of erosion there is a complete truncation of the profile which becomes a degraded red soil, often consisting of only the Bt horizon (see Fig. 12.2).

In both cases, the reserves of water at depth are formed with greater difficulty, which increases the disastrous effects of dry periods on the vegetation.

## II   The dynamics of fersiallitisation

The constancy of the characters of fersiallitic red soils on a range of parent materials is a good illustration of the fact that development is controlled to a large extent by bioclimatic factors, but as for all soils of hot climates, the speed of the process is closely dependent on conditions of parent material and site. As stated in Chapter 4, fersiallitic red soils can be either present day soils, or old soils, or palaeosols with polycyclic development; in addition, certain parent materials are resistant to all processes of rubification.

The most fundamental question that can be asked is whether rubification and fersiallitisation can be considered as characteristic processes of the mediterranean climate. The question has been asked many times and the answers that have been given have caused a great deal of argument. From recent research, it would seem to be confirmed that mediterranean-type climates are particularly favourable to this type of pedogenesis, for it is in the Mediterranean region that profiles of red soils were studied initially and it is in this zone that the most characteristic examples have been described. But fersiallitic soils are not absolutely restricted to the Mediterranean region: they occur elsewhere in the world, particularly where a climate of the same type occurs, that is to say, warm temperate or subtropical, with an alternation of wet and dry seasons, the latter having a high *PET*, even if, as in the case of Mexico (Boulaine 1966), the wet season coincides with the summer and not the winter, as in the Mediterranean region.

In addition, the occurrence of fersiallitic soils in regions where the mean temperature is lower (Danubian sub-Mediterranean region for example) or higher (tropical climate) than that of the mediterranean climate *sensu stricto* is to be noted, but such soils occur only at favourable sites and on suitable parent materials.

In these circumstances it would seem that a study of the environmental conditions should precede that of the physicochemical processes, which will enable a better explanation of the latter to be given. As the interrelationship of parent materials and site conditions with the general climate is so close, such a study need not consider them separately and it can be said that *within any one climatic zone the very variable rate of rubification is dependent on the local environmental conditions*. Having done this, it will then be possible to assess reasonably accurately, in terms of the *time* factor, the difficult problem of whether a fersiallitic red soil is contemporary or is an old soil or palaeosol, that has caused such controversy in the recent past and which has still not been completely resolved.

Before doing this, it should be remembered that when active carbonate is present, it needs to be removed before rubification can occur, for, as will be seen, this factor plays an essential part in the solution of this problem.

## 1  Environment of rubification

In the Mediterranean region most, if not almost all, red soils have been considered by many authors as palaeosols; therefore, to specify the optimal conditions of rubification, it would seem to be necessary to refer to local detailed studies which give evidence *in specific cases of recent rubification, i.e. subsequent to the Würmian period* (the short cycle, see Ch. 4). Three examples will be discussed: the first on fluvioglacial deposits of the southern Jura, the second on hard limestones of the mountains of Lebanon, and finally the third on schists in Greece and Portugal.

**Rubification in the southern Jura** (Bresson 1976, Gratier & Pochon 1976). This process has been described by these authors in the southern Jura of France and Switzerland, on Würmian periglacial deposits, and they have thus developed in less than 10 000 years. However, this is a region where the climate is not really mediterranean (mean temperature of about 10°C; vegetation, pubescent oak with box). But this is partially compensated for by the local climate (warm site, at an elevation of 600 m), which has a high degree of insolation in a dry atmosphere; the seasons are markedly different, the rainfall very high (more than 1000 mm and sometimes 1500 mm), the summers often being hot and dry. In addition, the parent material is very porous owing to the abundant content of calcareous and siliceous pebbles. In these circumstances decarbonation occurs with extraordinary rapidity, by corrosion and then the elimination of all calcareous pebbles over a depth of 60–80 cm, accompanied by a marked surface acidification (pH often 4.5), and the formation of a calcareous horizon at the base of the profile by precipitation of the transported carbonate. Rubification immediately follows the decarbonation, but it is still moderate (colour 5YR, exceptionally 2.5YR) and is *preferentially concentrated in more porous dissolution pockets*. The clays and their ferric covering rubify in the surface horizons, probably before the strong acidification occurs, then they are pervected and form a Bt horizon with red ferriargillans.

**Hard limestones of the mountains of Lebanon** (Lamouroux 1971). Here the parent material is less favourable, for it is a hard limestone which weathers by very gradual surface dissolution and only gives in each rainy season a thin skin of silicate residues. The production of the clay-rich material, of the terra rossa type, susceptible to rubification, is thus necessarily slow on this type of material, so that in general, terrae rossae are considered as old and fossil materials. But Lamouroux has shown that where climatic conditions are much more active with regard to limestones than those of the southern Jura, the formation of terrae rossae soils can occur in less than 10 000 years. High rainfall (more than 1000 mm, often 1200–1500 mm) and also a very high mean summer temperature, increased in the mountains at the surface of the soil by intense radiation, are necessary. The author was able to show that the dryness of summer had a twofold effect: not only does it cause rubification itself, but also it increases the rate of surface weathering of the limestones and thus the formation of decarbonated clay material. Hubschmann (1967) – for the mountains of Morocco – and Guerra (1972) – for those of Spain – have confirmed the work of Lamouroux.

**Schistose parent rocks:** Portugal (Dachary 1975), Greece (Duchaufour 1969). Schists, particularly the acid ones poor in calcium and iron, are considered unfavourable to rubification. In the French Mediterranean region they are not generally rubified, except locally on some very porous colluvium (Servant 1970). The acidity and weak porosity (because of the high clay content) are sometimes considered responsible for the slowness (or the absence) of fersiallitic pedogenesis on this parent material. In fact, recent fersiallitic red soils on schists are not uncommon, but their geographic distribution is related to very special mineralogical and climatic conditions: *large amounts of weatherable minerals rich in iron* (**chlorite schists** of Thrace in Greece; **biotite schists** of the Beja region of Portugal), and a very definite Mediterranean climate with high average temperatures (in Portugal a mean of 16°C and 24°C in August), *but where heavy precipitation is not necessary, doubtless because the question of prior decarbonation has no longer to be faced and rainfall often decreases to only 500–600 mm*. According to Guerra (1972), the acidity owing to the lack of calcium in the parent material here does not appear to be a limiting factor to rubification, if pedoclimatic conditions are favourable and if the amount of iron freed by weathering is sufficient.

**Summary of the environmental conditions necessary for rubification.** As has also been stated by Hubschmann (1975), *the very porous but silicate-rich calcareous parent materials like moraines, fluvioglacial deposits and alluvial terraces are most favourable to rapid rubification: but, as the prior condition to this rubification is complete decarbonation of the parent material, it cannot develop in climates that are too dry, where rubification is slowed down or prevented.* This is the situation on the Würmian terraces of the south of France (Costieres), where the rainfall is low (600 mm); the very slow dissolution of the carbonates that occurs in these conditions means the persistence (differing from the fluviocalcareous pebbles of the Swiss Jura) of a small quantity of calcareous pebbles, which prevent the absolute decarbonation of the fine earth so that, as yet, rubification is only in its initial stage (Duchaufour 1978: XVI$_3$).

On hard limestones it is the same lack of precipitation that slows down the surficial weathering of the rock (on outcrops of Urgonian age, for example) so that the red soils on terra rossa develop slowly, not as a result of slowness of rubification but owing to slowness with which the silicate material is freed, which is fundamental to the formation of the soil. For this reason, the terrae rossae of the French Mediterranean region are old or even polycyclic soils, the same as those that are characteristic of the dry areas of Spain (Duchaufour 1978: $XVI_2$ is a completely preserved forest profile of a **brunified terra rossa**).

The calcareous sandstones of the mollasse, which are very porous, rich in quartz sand and poor in carbonates, can rubify rapidly, even under this type of relatively dry climate (Duchaufour 1978: $XVI_6$), for decarbonation of this rock requires smaller amounts of water than material containing hard limestone pebbles, which form a slow weathering *reserve*.

Among the silicate-rich parent materials, only the most porous and richest in calcium and in weatherable minerals rubify as easily as parent materials containing limestone; this is the case of colluvium formed from chlorite schists and basic volcanic rocks, but here, as the prior elimination of carbonate is no longer necessary, rubification can occur rapidly even at a rainfall as low as 500–600 mm. This is the case of the profile on basaltic colluvium (Duchaufour 1978: $XVI_5$), in which the calcareous horizon comes not from the dissolution of carbonates but from the partial weathering of silicate minerals, which themselves are no obstacle to rubification.

*Parent materials freeing a great deal of clays, or those which are very acid or very poor in iron, are resistant to rubification.* Thus it is for different reasons that marls (Guerra 1972), acid quartzose granites and non-calcareous sandstones do not rubify. The first are not permeable enough, decarbonation is very slow and erosion removes the surface before it can be rubified. Acid granites and sandstones free too little iron and clay, while in addition fersiallitic development is opposed by the acidity (Taylor & Graley 1967).

*Conclusion.* Calcareous parent materials are most favourable to fersiallitic development, but the necessity of prior decarbonation postpones the initiation of the process: *the speed of fersiallitisation is increased in more humid climates and on more porous parent materials* (short cycle recent red soils). If these conditions do not occur, fersiallitic pedogenesis is very slow (old or polycyclic soils).

Schistose rocks rich in iron, or basic crystalline rocks, are also favourable to rubification, in as far as they produce well aerated weathered material which increases the seasonal contrasts of wetting and drying. For this type of parent material the process is rapid, and it can occur in a less humid climate than for calcareous rocks. In contrast, clay-rich sediments of all kinds which decrease the pedoclimatic contrasts owing to their poor aeration and are prone to erosion, together with the acid quartzose rocks, poor in iron, are resistant to the process of rubification. All these conclusions have been completely confirmed by the experimental studies of rubification by Williams and Yaalon (1977).

## 2   *Physicochemical process of rubification*

Although most authors who have thought about this problem are agreed that rubification is a dehydration (or rather a dehydroxylation) of the iron oxides bound to the clays, resulting from a more or less sudden drying out of the environment, serious disagreements still persist in the literature as to whether the iron oxides are crystalline or amorphous. Ségalen (1969) thinks that the red, skin-like coverings around the clays are mostly amorphous. Lamouroux (1971) differs somewhat and suggests that the amorphous iron is mixed with a considerable amount of haematite. Finally, other authors (Fischer & Schwertmann 1974, Juo *et al*. 1974, Schwertmann *et al*. 1974) are of the opinion that the dehydration of the iron oxides is accompanied by their crystallisation and transformation into haematite.

According to Schwertmann *et al*. (1974), the main difference between brunification and rubification (in well drained conditions) lies in the effect of organic matter, which plays an important part in brunification, but in contrast, little or no part in rubification. When the organometal complexes free their iron by slow biodegradation, it forms brown or ochreous goethite by very gradual crystallisation, a part of the iron remaining in the amorphous state. This occurs in the organic-rich, brunified surface of forest profiles, where only the B horizon is red. If, in contrast, the iron is freed by rapid weathering, without the formation of organomineral complexes, it appears first of all in a brownish-red cryptocrystalline form, called **ferrihydrite** by these authors, which develops by simultaneous internal dehydration and crystallisation to very red *haematite*.

This idea can be compared to that of Lamouroux, who speaks of the formation of *pre-haematite*, a cryptocrystalline precursor of true haematite. It can be said that these intermediate cryptocrystalline forms, pre-haematite and ferrihydrite, which are as yet poorly defined, are in fact characteristic of the first phase of still incomplete rubification, and give the soil a brownish-red colour. In contrast, *true rubification occurs when there has been a complete crystallisation of the dehydrated iron in the form of haematite*. It is now recognised and admitted that the iron oxides arranged as skins around the clay particles can very well be crystalline.

This rapid crystallisation of dehydrated iron oxides in aerated and well drained conditions can be compared with the slower crystallisation of goethite, where the iron oxides remain hydrated in an almost constantly humid environment not subject to periodic phases of desiccation. These soils have a more vivid ochreous-brown colour than temperate brown soils, for here the goethite is not mixed with amorphous forms bound to organic matter (Schwertmann & Taylor 1973). Lamouroux showed that **hydrated, fersiallitic brown soils** of impeded environments, adjacent to rubified soils of well drained and aerated sites, are of this type, as are the ochreous pockets of terra rossa in fissures filled by pervected material, at the base of karst profiles below rubified Bt horizons. The carbonate that occurs in this impeded environment causes a considerable rise in the pH and catalyses the formation of goethite.

## 3    The characters of fersiallitic weathering: development of clays

As previously stated, one of the characters of fersiallitic weathering is the retention of basic cations ($Ca^{2+}$, $Mg^{2+}$) within the profile, because of the constant return to the surface by the biogeochemical cycle and of capillary rise in the dry season. The environment is really acidified only in extreme cases of high rainfall and low *PET*.

One of the indirect effects of this base retention is the retention also of silica freed by weathering (Lamouroux 1971), for the more the sudden alternations of wetting and drying favour the processes of aggradation or very rapid neoformation of clays, the more quickly silica is integrated within the clay mineral lattice – a cause which is similarly responsible, as was seen, for the crystallisation of iron oxides. To the micaceous clays (illites) inherited from the parent material and very slightly transformed (sometimes vermiculised; Paquet 1969), are added the montmorillonites of neoformation or those resulting from *aggradation by ion substitution* and re-integration of silica into the gaps in the tetrahedral layer of some vermiculites; this process, described by Lamouroux et al. (1973) and Makumbi (1972), is favoured by the great abundance of $Mg^{2+}$ and $Ca^{2+}$ ions in the environment. It should be recalled that this process occurs to a lesser extent in the temperate eutrophic brown soils on diorite (see Ch. 9).

Neoformation of kaolinite appears to be excluded in most fersiallitic soils, still rich in calcium. In contrast, it can happen, as previously stated, by slow loss of silica from the 2 : 1 clays, when the profile ages and acidifies (acid fersiallitic soils: Spaargaren 1979, McLeod 1980). In addition, parent materials poor in silica, such as basalts, provide an important exception that requires notice; from the initial stage of weathering in neutral conditions and still base-saturated, kaolinite is formed, for *in well drained conditions, weathering frees an insufficient quantity of silica to allow only 2 : 1 clays to be formed; in these circumstances a considerable amount of kaolinite appears in the profile* (Dan & Singer 1973, Lamouroux et al. 1973, Beckmann et al. 1974, Duchaufour 1978: XVI_5, Paton 1978).

It can be concluded that fersiallitic weathering is characterised by a fairly moderate neoformation of 2 : 1 clay, while the freeing of iron is at a maximum. The ratio iron : clay reaches its maximum in this class (Bornand 1978).

Note that, as stated previously, the neoformation (or aggradation) of montmorillonite which is a constant feature of most red soils but not in large amounts, is greatly increased under conditions of impeded drainage, owing to the normally much greater concentration of silica and bases which tends to occur in these conditions. Not only does rubification not occur, but a part of the iron is integrated into the clays that are thus neoformed; the soil then takes on *vertic* characters (vertic fersiallitic brown soils).

## 4    The processes of transport in fersiallitic soils

Pervection of clay and iron is a characteristic and constant process of those fersiallitic soils that are sufficiently developed. The formation of a suspension of fine clays with their covering of iron oxides, owing to the very intense

Mediterranean rainfall, occurs even when the environment is saturated by calcium and thus an essentially mechanical pervection is favoured. The presence of a Bt horizon with prisms or polyhedra with coverings of red ferriargillans is characteristic of fersiallitic soils. The first rains after the dry season seem to play a determinative part, for they move clays into the *contraction cracks* which have not as yet closed up in the Bt horizon.

But clay balance determinations show that only a part of the clays pervected from the A horizon occur in the Bt; the rest are lost by lateral movement. This process of impoverishment is characteristic of all hot climates with markedly contrasted seasons, for with the closing up of the contraction cracks in the wet season, the B horizon cannot cope with all of the most intense rainfall: the surface horizons are waterlogged for a short time and some of this water, with the suspended clay, moves laterally by surface flow (Servat 1966, Roose 1970).

Bottner (1972) showed that clay pervection is of two kinds, depending on the degree of development (acidification) of the fersiallitic soils.

(a) *Modal fersiallitic* red soils are, it will be recalled, decarbonated but still base saturated (sometimes supersaturated) in the upper horizons. In these circumstances, pervection remains limited to the finest iron-rich clays; some of the iron remains locked up in the small concretions or **pseudosilts**. Kubiena (1953) has stated, as a result of microfabric work, that the clay-rich plasma gels are more dispersible than the fine-grained aggregates cemented by the crystalline iron. The index of movement of the clay remains moderate (about 1 : 2) while that of iron in these circumstances is even less. *These fersiallitic red soils are, in fact, moderately lessived.*

(b) *Acid fersiallitic soils* are older or formed on parent materials that are rich in unweatherable siliceous skeletal materials and are, as previously stated, much more strongly pervected and even impoverished at the surface; the very stable aggregates of the pseudosilt type cemented by iron are partially destroyed by the acidity, which causes the clays contained in them to be freed. The index of movement increases and becomes comparable for both the iron and the clay (often 1 : 7). In addition, abundance of skeletal silica (coarse sands and pebbles), makes the impoverishment (of silts and clays) all the greater where it makes the lateral circulation of rainwater easier, which in these climates is very intense. Very frequently, the Bt horizon becomes impermeable owing to the pores being filled up by pervected clays, so that a secondary hydromorphic development of the pseudogley type is added to the fersiallitic type of development (Duchaufour 1978: $XVI_4$).

## III   Main types of fersiallitic soils: classification

The three subclasses are related to the degrees of development recognised at the beginning of this chapter (see Fig. 12.1):

(a) *fersiallitic brown soils* with incomplete rubification: soils that are transitional to temperate brown soils;

(b) *fersiallitic red soils*, modal type, absorbent complex practically base saturated and moderately pervected;

(c) *acid fersiallitic soils*, very pervected and impoverished: many can be considered as intergrades to the ferruginous class.

Within each subclass, groups are differentiated in terms of secondary characters related to various special features of the environment: (i) hot pedoclimate (tropical regions); (ii) more or less impeded conditions preventing rubification; (iii) undoubted hydromorphic conditions (iron segregation); (iv) surface developments modifying the A horizons: brunification under the moister conditions of forests or, in contrast, reworking and truncation by erosion, resulting from degradation of the forest vegetation by man.

## 1 Fersiallitic brown soils

Two major groups are differentiated related to two different climatic zones, the tropical and the subtropical (or with mediterranean-type climate).

**Tropical brown eutrophic soils.** These are in fact young soils at the start of their development. *They should be carefully differentiated from the vertic eutrophic brown soils which belong to the vertisols (see Ch. 8) and which are, in contrast, very developed but restricted to special sites on basic rocks that are poorly drained.* These initial eutrophic brown soils which are now being considered have been clearly distinguished from the vertic eutrophic brown soils by Kaloga and Thomann (1971) in Senegal. In addition, Makumbi (1972), Herbillon *et al.* (1973) and Makumbi and Jackson (1977) consider the tropical eutrophic brown soils, on the schists of Zaire, as an initial stage of the tropical red soils (see Fig. 12.3, phase I). These soils, less developed and consequently not as deep as the red or ferruginous soils, occur on outcrops of basic rocks and they have several characters that are similar to those of the temperate brunified soils, to such an extent that the French classification of 1967 put them in the same class. They are still influenced by the organic matter and are thus not rubified; the dominant clays, particularly those that are inherited, are of the 2 : 1 type; while those resulting from neoformation or aggradation are mainly montmorillonites, as the environment is still rich in $Ca^{2+}$ and $Mg^{2+}$ and the absorbent complex is practically base saturated (it should be recalled that montmorillonites are also characteristic of the initial phase of weathering of basic rocks in a hot climate). However, the freeing of considerable amounts of iron, which here undergoes a more rapid development than can occur in a temperate climate, would seem to make it possible to consider these soils as having been subject to a very slight degree of fersiallitisation.

**Subtropical or mediterranean brown soils.** They have a brownish red colour because an environmental factor retards the rubification which, here, is incomplete.

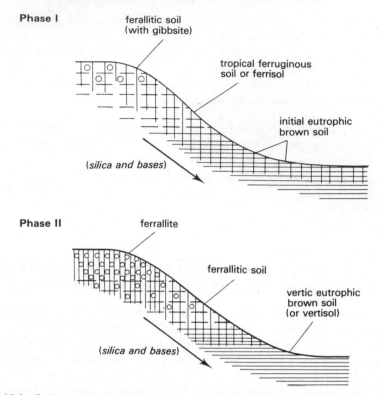

**Figure 12.3** Soil catena on basic volcanic rocks in a humid tropical climate (after the work of Bocquier 1973, Makumbi & Herbillon 1973, Levêque 1975) (see general key, p. ix).

*Climatic factor:* fersiallitic soils of sub-mediterranean zones, for example the Danubian fersiallitic brown soil (Duchaufour 1978: $XVI_1$) as well as some fersiallitic soils of mountain regions.

*Site factors* preventing the drying out of the profile (impeded drainage): the hydrated brown soils differentiated by Lamouroux (1971 and 1972); some can be referred to as being vertic.

*Site and climatic factors together:* on calcareous parent material if the climate is too dry, the decarbonation is slow and incomplete, such as on the Würmian terraces of the Midi of France (Bornand 1973, Duchaufour 1978: $XVI_3$): **fersiallitic brown soil with calcareous crust on a terrace.**

## 2   *Modal fersiallitic red soils*

Here again, it will be necessary to differentiate a group in hot climates (tropical) and one in the sub-Tropics, which are differentiated from one another particularly in terms of the clays, the first being generally richer in kaolinites than the second, on identical parent materials.

**Tropical fersiallitic red soils.** They have been described and differentiated by numerous authors, particularly by Martin *et al.* (1966) and Makumbi (1972) in Africa. Common on basic rocks, they represent a stage of evolution inter-mediate between the eutrophic brown soils and the ferruginous soils, differing from the first in that rubification has occurred and the 2 : 1 clays have started to degrade to kaolinite by loss of silica (Makumbi 1972, Herbillon *et al.* 1973). However, kaolinite, is not, as in the more developed ferruginous soils, the dominant element of the weathering complex:

initial eutrophic brown soils ⟶ tropical red soils ⟶ tropical ferruginous soils

**Subtropical or mediterranean fersiallitic red soils.** The subdivisions of this great group can be made either on the basis of the secondary development of the upper part of the profile, resulting from the nature of the soil–vegetation equilibria, or the kind of the parent material which gives the soil certain second-order characteristics, such as the presence or absence of a calcic horizon.

*Secondary development of the upper part of the profile.* The most typical example is to be seen on the terra rossa characteristic of outcrops of hard limestone.

*Secondary brunification:* owing to abundant organic matter coming from for-ests which prevents rubification in the A1 and A/B horizons and results in the formation of a **brunified fersiallitic red soil** (Duchaufour 1978: $XVI_2$).
*Reworking and truncation:* **Truncated red soils** (see Fig. 12.2) occur under degraded forests (garrigues) in lower Provence, where severe soil erosion has resulted from a lack of effective protection by the forest and its litter. The soil occurs as discontinuous pockets of red clay between limestone blocks that have been stripped bare; in these pockets the upper horizons have been reworked and frequently removed by erosion, the old Bt horizon is thus exposed; a very thin, humus-rich horizon is reformed under the influence of xerophilic bushes which occur within the karst pockets.
*Mechanical reworking colluviation and secondary recarbonation* (Bottner 1972) is similar to the preceding case, but here there are additions, particu-larly of calcareous materials of all sizes and of all kinds, resulting from the breakdown of the bedrock which causes a moderate degree of secondary recarbonation of the previously rubified profile. These recarbonated soils occur at the bottom of slopes, particularly in dolines, or even in some karst pockets where the surrounding limestone weathers to form fragments of all sizes. This recarbonation, particularly if there is a considerable amount of it and it is of some age, causes the original red colour to decrease and become somewhat browner. Soils of this kind must be classed with the calcimag-nesian soils (**calcareous brown soils on rubified parent materials**).

*Development dependent on parent materials.* **Fersiallitic red soils with calcic horizon** (Duchaufour 1978: $XVI_5$). The calcic horizons form at the base of the calcareous materials (or even of the colluvium of volcanic rocks or schists very rich in calcium) by decarbonation (or decalcification) of the upper part of the profile, and appear as calcareous crusts necessarily situated beneath the rubified argillic horizon.

### 3 Acid fersiallitic soils (impoverished, strongly pervected)

Here there is no need to differentiate a *tropical* group, because such acid tropical soils necessarily belong to the tropical ferruginous soils. In fact, even in the mediterranean zone, the soils of this subclass can already be considered as ferruginous intergrades. No matter what the parent materials, even limestone, the acidification is generally associated with a certain amount of kaolinite formed by (i) degradation of 2 : 1 clays and (ii) beginnings of the kaolinisation of feldspars (Bornand 1978).

Several of these soils belong to the ultisols of the Soil Taxonomy (xerult), or to the acrisols of the FAO classification. The exchange capacity thus falls considerably below 50 mEq/100 g clay (Bruin 1970); however, they always remain above 25 mEq (Bornand 1978).

There are two subgroups:

(a) *The modal subgroup:* non-hydromorphic, on permeable materials or on slopes; the Bt horizon remains porous despite the large accumulation of clay owing to pervection.

(b) *Hydromorphic subgroup* (with pseudogley): result from the choking up of pores in the Bt horizon causing a local de-rubification as a result of hydromorphism, which forms whitish or ochreous patches next to the zone that remains red. An example is provided by soils formed on the oldest terraces in Provence and of the Rhone valley (Vigneron & Rutten 1967, Bornand 1973 & 1978) or of the Guadalquivir (Torrent et al. 1980).

## B   FERRUGINOUS SOILS

Ferruginous soils have a form of development intermediate between that of fersiallitisation and ferrallitisation. General weathering of primary minerals is stronger than that of fersiallitic soils, but it is not as complete as that of ferrallitisation. The degree of elimination of soluble silica by drainage to depth lies between that of the other two classes: *in these circumstances most clays are produced by neoformation and are kaolinite*. A certain amount of 2 : 1 clay still persists, but this is exactly the fraction that is most easily dispersible by water (see Ch. 3, Sec. II). Therefore, these ferruginous soils are generally lessived, even though to a somewhat lesser degree than the fersiallitic soils, so that in certain of them the argillic Bt horizon is not too well developed. Apart from exceptions provided by the transitional forms, de-silication never goes so far as to form free gibbsite.

In addition to being intermediate between the other two classes of soils of hot climates in terms of characters and properties, the ferruginous soils are also intermediate in terms of *climate*. In fact they occur in two climatic zones, one more humid than the mediterranean zone (*humid subtropical* without a marked dry season), the other hotter than the mediterranean zone (*tropical with dry season*, more or less intense, and thus less humid than the *equatorial* zone of the oxisols). In fact, this is true only in a broad sense and it is no more than an approximation, *for the slowness of the processes of ferrugination and ferrallitisation is such that the degree of development is strongly influenced by the age and the kind of parent material*. In addition, for the oldest of soils there are often factors of rejuvenation which disturb the climatic equilibria. If account is taken of important Quaternary climatic variations of hot regions, particularly in Africa, it can be readily seen why the soil map of tropical regions is a complete puzzle that is difficult to interpret.

It can be concluded from this discussion that in tropical regions the climax vegetation is not as characteristic of the degree of pedogenesis as in temperate climates, all the more so as the original vegetation has often been modified and disturbed by man. In general, however, it can be said that ferruginous soils are characterised in the subtropical humid zone by evergreen forest of the *laurel type*, and in the tropical zone by a xerophilic bush formation (acacias, thorn bush) or open, semi-woody, formations which are not forests in the strict sense and are generally referred to as **savanna**.

In fact, it is possible to distinguish clearly two stages in the development of ferruginous soils, the first giving rise to soils that are still relatively close to fersiallitic soils, the second, in contrast, having clearly ferrallitic characters: these are the **ferrisols**. These two stages are naturally the basis of the differentiation of the two subclasses that are distinguished:

tropical fersiallitic red soils $\longrightarrow$ tropical ferruginous
soils $\longrightarrow$ ferrisols $\longrightarrow$ ferrallitic soils

# I  Profiles of ferruginous soils: morphological and geochemical characters

The ferruginous soil, *sensu stricto*, will be described first and then the ferrisol.

## 1  Ferruginous soil

**Profile morphology** (Duchaufour 1978: XV$_6$ & XVII$_1$). As far as the general organisation of the profile is concerned, the morphological characters of ferruginous soils are relatively constant (see Fig. 12.5). They can be summarised as follows. The weathered profile is deeper than that of fersiallitic soils (it reaches and often exceeds 2 m). There are three main horizons: an A1, 15–20 cm thick, dark brown, humus rich with a crumb structure; a pervected A2 horizon (and often impoverished) down to a depth of 60–80 cm of a browny-yellow colour with little structure and, finally, a clay-enriched Bt horizon, with a polyhedral structure, in which the argillans are often not too well developed. The colour of this horizon is always strong: red, brown or ochreous, with a variable degree of rubification depending on the parent material. The considerable amount of kaolinite neoformed from feldspars causes the iron : clay ratio to decrease compared to the fersiallitic soil, which weakens the degree of rubification, even in aerated conditions.

The A1 horizon becomes darker and increases in thickness as the dry season becomes more intense; if the variation in texture and presence of argillans is taken as evidence of pervection, it is much less apparent than in temperate lessived soils and fersiallitic soils: the accumulation is more diffuse, more gradual and the argillans less clearly defined.

Very often a marked impoverishment occurs in the A1 and A2 horizons which can cause local planosol formation (see Duchaufour 1978: $XV_6$, which is placed among the planosolic soils for this reason). The Bt horizon such as that which occurs below 0.90 m has the character of a eutrophic ferruginous soil that is clearly more rubified than profile $XVII_1$ (Duchaufour 1978), probably because of the more marked dry season and the higher amount of bases.

**Geochemical characters.** Mineralogical studies show that for the clays kaolinite is dominant, but that also a certain amount of 2 : 1 micaceous clays and some resistant minerals (orthoclase) persist; there is no free gibbsite. Exchange capacity is between 15 and 25 mEq/100 g clay (Sys 1967) which is a good reflection of the mineralogical composition of the weathering complex; this exchange capacity increases in moving from the A to the Bt as a result of the preferential pervection of 2 : 1 clays, which are most dispersible in water, so that kaolinite is more abundant at the top of the profile as a result of its relative accumulation. But if the profile has been impoverished as a result of much more intense selective erosional processes, this clay is also eliminated from the upper horizons together with a large part of the silt.

When the profile is simply *pervected*, clay movement generally remains moderate and the index rarely exceeds 1 : 1.5; however, it can be 1 : 2 on very porous materials and in more humid climates. Another character that distinguishes ferruginous from ferrallitic soils is the silt : clay ratio, being greater than 0.20 for the first group (Jamagne 1963).

Surface humification varies to a large extent as a function of the general climate and of the conditions of the parent material and the local vegetation: generally, *surface accumulation of organic matter and the degree of maturation (formation of very stable humic compounds) increase as the dry season becomes more intense* (Duchaufour & Dommergues 1963, Perraud 1971). As will be seen, this can have important consequences on the base saturation of the absorbent complex, which can vary considerably, according to the type of ferruginous soil, from less than 10% (pH less than 5) to 75% (pH of about 6) in the Bt. Complete saturation of the absorbent complex is rare compared to fersiallitic soils where this is generally the case. These large variations of base saturation allow the differentiation of acid ferruginous soils (or desaturated) and eutrophic ferruginous soils (or saturated). The USA and FAO classifications have made this a fundamental criterion of the classification of ferruginous soils, which will be discussed later.

## 2    Ferrisols (Duchaufour 1978: XVIII$_1$, XVIII$_2$ and XVIII$_3$; Jamagne 1963)

This soil is more deeply weathered than the tropical ferruginous soils (often more than 3 m) and has a number of ferrallitic characters, both morphologically and analytically; but the weathering of primary minerals is not complete and in many cases a Bt horizon again occurs, enriched in clays by pervection, which is very like the ferruginous soils (Fig. 12.5).

In fact, ferrisols only occur in the humid tropical climatic zones and on parent materials that are older than, or where development occurs more quickly than, in the case of ferruginous soils. In very humid climatic zones with ferrallitisation dominant, they tend to occur in the mountains, for the lowering of the mean temperature slows down the ferrallitisation; in addition, the occurrence of steep enough slopes rejuvenates some profiles and prevents their complete development. There are also very humus-rich types characteristic of high-altitude forests with tree ferns (Duchaufour 1978: XVIII$_2$).

**Morphology.** As in the ferrallitic soils, the lower part of the profile, sometimes more than 3 m thick, is a zone of weathering often varicoloured in very irregular ochreous, red or white patches (Duchaufour 1978: XVIII$_3$), which has been called **saprolite** or mottled zone which sometimes, in a temporarily hydromorphic environment, has the appearance of **plinthite**, which will be discussed later (Duchaufour 1978: XVIII$_1$).

With regard to the upper horizons, two cases can be distinguished: either it is still possible to detect by morphological examination and analysis a Bt horizon enriched in clay as in the ferruginous soils (and thus of the argillic type) or, alternatively, no movement of clay can be detected between the A and B horizons (apart from the sharp decrease in the surface owing to impoverishment), with the B horizon only more intensely coloured, and often redder, than the ochreous yellow A horizon. Thus there are two types of ferrisol depending on whether a Bt horizon is present or not. Again it needs to be emphasised that when the Bt horizon does occur, it is less well developed than in a ferruginous soil: argillans are restricted and not too easily seen. The decrease in the amount of clay in going from the Bt towards the A and C horizon is very gradual. It is on this basis alone that the *nitosol* class has been differentiated in the FAO classification.

As previously stated, ferrisols of high altitude are very humus rich in the surface and are often very acid: the black A1 horizon can be 40 cm thick, and is sometimes covered by an A0 horizon of 10 cm (a true mor). Sometimes, a light-coloured zone appears in the A1 as a result of impoverishment (start of planosol formation).

**Geochemical characters.** The geochemistry and mineralogy of a ferrisol can be outlined as follows: while the A horizon and upper part of the B horizon have characters that are slightly ferrallitic (weathering almost complete, absence of clay other than kaolinite, sometimes even the presence of a small amount of gibbsite), the bottom of the profile (that is to say the lower part of the B and the mottled zone of weathering) has the character of ferruginous soils: absence of gibbsite, presence of 2 : 1 clays (particularly in the argillans if there is a Bt), weathering of primary minerals still incomplete (Daniels & Gamble 1978).

However, the two basic criteria used in most classifications (particularly the USA and FAO) to define the ferrallitic class of soils (oxisols or ferralsols), by means of the exchange capacity of clays and silt : clay ratio from grain size analysis, would not allow ferrisols to be considered as true ferrallitic soils, for their exchange capacity is between 16 and 25 mEq/100 g clay and the silt : clay ratio greater than 0.20 (Jamagne 1963); this clearly places the ferrisols outside of the ferrallitic class and nearer in fact to the tropical ferruginous soils.

Humus-rich ferrisols of high altitude have frequently, in the lower part of the humus-rich horizons, a marked maximum of gibbsite, even though the remainder of the profile contains little or none of it: this is **secondary gibbsite**, resulting from the degradation of kaolinites at the top of the profile under the influence of acid organic matter. This process, studied by Lelong (1969), is also seen in some lowland ferrallitic soils in a very humid climate; this will be discussed in the ferrallitic soil section.

## II   Development of ferruginous soils and ferrisols

Before considering the processes, it is necessary to examine the environmental conditions that control the development of ferruginous soils, so that account can be taken of the fact that here the time factor is fundamental. In climatic terms, it should be recalled that an increase in rainfall and mean temperature increases the speed of weathering as well as the elimination by drainage of freed silica and bases, with the dry seasons acting as a brake on these processes. *Ferrallitic soils, the most weathered and the poorest in silica, are only able to occur in regions with high temperature and very humid climate* (equatorial or tropical humid climate with a very short dry season). But because of the slowness of pedogenesis in a hot climate, which was emphasised repeatedly in Part I (Ch. 4), a ferrallitisation can be completed only on the oldest of parent materials or under site conditions where the processes are accelerated: *otherwise, the phase of ferrugination is not exceeded even where the climate is suitable for ferrallitisation.* An evaluation of this problem requires a discussion first of all of the age of parent materials and then of the influence of site, and finally of the part played by biological factors.

*1   The part played by the age of the parent material*
This factor is particularly involved when recent sedimentary deposits (Quaternary) are concerned or older deposits that have been mechanically or biologically reworked: three examples will be considered.

(a)  The first is a very good example from the humid zone of Madagascar (Bourgeat & Ratsimbasafy 1975). Ferrallitic soils *sensu stricto* occur only on the oldest alluvial terraces which are about 100 000 years or more old. The soils on the middle terraces, where pedogenesis has been going on for 25 000 to 50 000 years, according to the authors, are **rejuvenated ferrallitic soils** which correspond exactly to the definition of ferrisols.

(b) In Duchaufour (1978), another example was given coming from the Llanos lowland at the foot of the eastern cordillera of Colombia. Immediately at the foot of the cordillera on alluvium that is still young, the soil is a tropical ferruginous soil ($XVII_1$) while 200–300 km further east under a similar humid tropical climate and on material of the same kind but older, there is a ferrallitic soil ($XVII_3$).

(c) Finally, Daniels and Gamble (1978) have studied the age of ultisols on the coastal plains of the south-east of the USA. Only the oldest (ferrisols) have gibbsite in the top of their profiles.

## 2   Site factors

The part played by parent material composition and that of relief (slopes) will be discussed.

**The nature of parent material, its permeability.** Whereas in dry, tropical climates basic materials rich in iron develop less quickly than acid materials poor in iron (the case of tropical fersiallitic red soils), it is the opposite when the climate is humid: *de-silicification occurs very rapidly on basic parent materials, provided that the permeability is very high*; the part played by parent material has been emphasised by Lelong (1969). Eswaran and de Coninck (1971) described soils at three levels of weathering, fersiallitic, ferruginous or ferrallitic, on the same basaltic parent material. The fundamental factor of their development, according to them, is the importance and the speed of percolation through the profile. When they are high, they increase the speed at which bases and silica can be eliminated.

**Relief.** The relief can act in a variable fashion, sometimes even in ways that are opposed to one another. A gentle slope increases the speed of lateral drainage and favours impoverishment, thus increasing the loss of silica. Conversely, steep slopes can cause a complete rejuvenation of the whole profile, by the erosion and reworking that it causes. The examples given by Makumbi and Herbillon (1973), Verheye (1974), Levêque (1975) and Lersch et al. (1977) are illustrative of this. In many cases the soils on crystalline parent materials reach the ferrallitic soil stage only on summit plateaux, while the soils on the slopes never get past the tropical ferruginous stage (or perhaps ferrisol). It is the same where soils have been reworked at the surface either by lateral transport (Paton 1978) or by the effect of termites (see Figs 12.3 & 4).

In addition, pedogenesis in impeded conditions at the bottom of slopes differs more or less from that of well drained sites up slope. On basic rocks poor in quartz, the contrast is absolute, the neoformation of swelling clays being favoured by the addition of silica and bases from upslope (vertisol formation, see Ch. 8). In contrast, on acid rock the difference between down

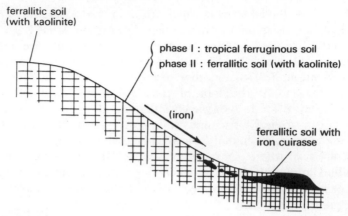

ferrallitic soil
(with kaolinite)

phase I : tropical ferruginous soil
phase II : ferrallitic soil (with kaolinite)

(iron)

ferrallitic soil with
iron cuirasse

**Figure 12.4** Soil catena on acid crystalline rocks in a humid tropical climate (after the work of Lelong 1969, Fölster et al. 1971, Bocquier 1973) (see general key, p. ix).

slope and up slope is less marked, with ferrallitisation tending to be general, but with cuirasse formation frequent on lower slopes (Fölster et al. 1971, Fig. 12.4).

### 3 Biological factors

The mechanical effect of termites is frequently emphasised. With regard to biological humification, they play an essential part in the biogeochemical cycle of bases and, indirectly, in determining the base saturation. The termites rework the upper part of the soils and often this is responsible for a rejuvenation. According to Bachelier (1977) and Levêque (1975), termites bring to the surface considerable amounts of fine earth from deeper down, while coarser material remains at depth. An accumulation of stones, more or less aligned (**stone lines**) often indicates the limit between the reworked upper part and the *in-situ* deeper part of the profile.

Humification plays an important part in soil development. In mountains or in some regions with a very humid climate, the formation of a thick moder (or even mor) humus causes a biological degradation of clays, with the formation of secondary gibbsite which is related to an undoubted podzolisation (humic ferrisols of high altitude). But this is exceptional, for in most cases the tropical humus is a mull of a special type, described by Perraud (1971). In a humid climate decomposition is rapid, humification weak, and the total organic matter is not very great, with slightly polymerised fulvic acid being the dominant fraction, and its mineralisation is rapid. As the dry season becomes more intense, a part of the organic matter polymerises and stabilises and becomes more resistant to microbial biodegradation (maturation, see Ch. 2). Grey humic acids and developed humin form, which retain some of the bases, particularly $Ca^{2+}$ and $Mg^{2+}$, contributed by the biogeochemical cycle (Duchaufour & Dommergues 1963). *Thus one of the indirect consequences of the occurrence of a dry season is a greater storage of bases in the humus-rich horizons, which raises the pH and percentage base saturation* (see Ch. 3): *this phenomenon acts against the acidification which is normally associated with the change from fersiallitisation to ferrallitisation* (Perraud 1971).

But it must not be forgotten that here pedogenesis is a slow process, while the equilibria resulting from the biogeochemical and humification cycles are established much more rapidly. In these circumstances, it is not surprising that there is no direct relationship between this cycle and thus the pH and percentage base saturation, on the one hand, and the amount of weathering and development on the other. This is all the more so when account is taken of the multiple climatic variations (particularly with regard to precipitation) which have occurred during the Quaternary to disturb the soil–vegetation equilibria, and thus to modify the biogeochemical cycle of bases in one way or the other. The lack of agreement between the amount of weathering (and thus of profile development) and the percentage base saturation has been emphasised by several authors (de Coninck *et al.* 1977, Moormann & van Wambecke 1978).

## 4 Physicochemical processes: ferruginous weathering

After what was said in the introduction, ferrugination must be considered as a stage in the general development of soils in hot climates:

fersiallitisation  ⟶  ferrugination  ⟶  ferrallitisation

**Weathering.** The weathering that characterises tropical ferruginous soils and ferrisols can be considered as an incomplete ferrallitisation; **kaolinisation**, which is the dominant process, can occur in two ways: (i) gradual degradation of more vulnerable 2 : 1 minerals (montmorillonites, illites and interstratified clays) which lose part of their silica and form kaolinite; (ii) the beginning of kaolinisation of feldspars, which is the major differentiating phenomenon between ferrugination and fersiallitisation. The only things persisting along with the kaolinite (and sometimes a little gibbsite in the top of the ferrisols) are the more resistant primary minerals, such as quartz, orthoclase and certain micas and also sericite clays. They only disappear (except for quartz) during ferrallitisation.

**Pervection.** Weathering is associated with a moderate pervection of clays which in a tropical (or subtropical) climate goes through three phases, of which the first is characterised by ferruginous development while the other two also affect the ferrallitic soils.

*First phase:* preferential migration of 2 : 1 clays, which are much more mobile than kaolinites (see Ch. 3), to form argillans in the Bt, where they raise the exchange capacity compared to the A horizon.
*Second phase:* large-scale kaolinisation (often associated with strong acidification), causing a gradual decrease in aeration, a partial temporary reduction of the iron, haematite changes to goethite: *the top of the profile changes from a brownish red to an ochreous yellow.* At the same time the finest

kaolinite is partially destabilised, which can then be transported. This process has been well described by Alwis and Pluth (1976), Chauvel (1977), Bornand (1978) and Eswaran and Sys (1979).

*Third phase:* it can begin before the end of the second phase and involves a gradual degradation of the argillans deposited in the Bt in the first phase. The 2 : 1 clays are changed to kaolinite which is well seen in microfabric studies (Bennema *et al.* 1970, Boulangé *et al.* 1975, Brook & van Schuylenborgh 1975).

Roose (1980) has investigated, by detailed lysimetry, the mechanism of kaolinite pervection in ferruginous and ferrallitic soils. The kaolinite is dispersed as a result of organic matter forming a covering around the individual clay particles and, in a humid sub-equatorial climate, most are transported to considerable depth to form a diffuse and poorly differentiated accumulation, which is undetectable by analysis and does not give rise to true argillans, while some (10%) is lost by drainage to depth. However, in a tropical climate with a dry season, the accumulation of clay, although it is qualitatively more limited, occurs nearer the surface and is more concentrated, so that a better Bt horizon is developed.

## III   Classification of ferruginous soils

### 1   General considerations

This group of soils is particularly difficult to classify and the difficulties do not seem to have been resolved in a satisfactory manner by the various classifications currently in use.

To begin with the three basic criteria that are used in the USA and FAO classifications will be restated:

(a) *Presence or absence of an argillic Bt horizon:* this separates the ultisols or acrisols (acid ferruginous soils with a Bt) from oxisols or ferralsols (ferrallitic soils without a Bt).

(b) *Exchange capacity of clays* (measured with ammonium acetate at pH 7); less than 16 mEq/100 g for oxisols, greater than 16 mEq/100 g for inceptisols, alfisols and most of the ultisols.

(c) *Percentage base saturation of the absorbent complex of the Bt horizon:* less than a certain value (35% USA, 50% FAO) for the most developed lessived ferruginous soils (ultisols); greater than this value for less well developed lessived ferruginous soils (alfisols).

The inconvenience of splitting tropical soils between four different classes despite their undoubted similarity of development, which has been commented on in this book, needs to be emphasised first of all. This is all the more so in that two of these classes (alfisols and cambisols) also contain soils that have been subject to the totally different temperate pedogenesis. In addition, the use of the three basic criteria given above is responsible for several difficulties that will now be summarised.

**The criterion of clay pervection.** This criterion is often difficult to apply. Smith *et al.* (1975) and Isbell (1977) emphasised the uncertainties that arise in its application. Argillans are often weathered, not too easy to see, and difficult to distinguish from simple surfaces of friction. In addition, there are cases where argillic horizons only contain kaolinite which has an exchange capacity of less than 16 mEq/100 g clay, which is in fact one of the oxisol criteria; however, the soil taxonomy places such a soil in the ultisols. Thus, in this particular case, there is a contradiction between the two criteria, which has been emphasised by several authors (Juo *et al.* 1974, Gallez *et al.* 1975).

**The criterion of percentage base saturation.** By itself, this would seem to be insufficient to establish a valid distinction between the ultisols and alfisols. The reasons for this have been given in the preceding section. The percentage base saturation is dependent on the biogeochemical cycles and thus on the present-day soil–vegetation equilibrium, very often disturbed by man. This equilibrium can be totally unrelated to the old pedogenesis and thus to the weathering type. It is acknowledged in the Soil Taxonomy text that in ultisols the biogeochemical cycle can frequently raise the base saturation of the surface horizons, which should therefore not be taken into account. But a certain transport of the bases can occur and raise abnormally the base saturation of the Bt horizon, which also requires care in evaluation.

On this point, Moormann and van Wambecke (1978) showed that most of the tropical alfisols (or luvisols) are, in fact, nearer in terms of their weathering complex to the ultisols than to the temperate alfisols.

**The criterion of exchange capacity of clays.** This criterion would appear to be more valid than the preceding one, although the capacity is not always strictly related to the mineralogical nature of the clays, for in the oxisols it needs to be emphasised particularly that kaolinite of silt size occurs, which has a certain exchange capacity, so that when added to that of the fine fraction, the total exchange capacity can exceed the prescribed 16 mEq/100 g.

However, despite this reservation this criterion would appear to be more reliable than base saturation and, in agreement with Sys (1967) and Buurman (1980), *it can be used as a basis for the differentiation of the ferruginous class of soils where the exchange capacity of the clays ranges from 16 to 25 mEq/100 g (measured at pH 7).*

In conclusion, only an exact mineralogical study together with a quantitative determination of the types of clay can serve as a reliable basis for the classification of tropical soils. The presence or absence of an argillic Bt horizon and the exchange capacity of clays give very important additional information. With regard to percentage base saturation, it is proposed that, as for temperate soils, it will only be used at the group level to distinguish *eutrophic* or *oligotrophic* groups.

## 2   *Proposed classification* (Fig. 12.5)

**Subclass of ferruginous soils** (*sensu stricto*). Presence of an argillic Bt horizon.

(a) *Eutrophic ferruginous soils* (slightly desaturated): percentage base saturation greater than 50% (Duchaufour 1978: $XV_6$).
(b) *Oligotrophic ferruginous soils* (desaturated): percentage base saturation less than 50% (Duchaufour 1978: $XVII_1$).
(c) *Hydromorphic ferruginous soils* (with iron segregation): ferruginous soils with plinthite, with pseudogley, indurated.

**Subclass of ferrisols.** Presence of Bt horizon not essential.

(a) *Ferrisol with Bt* (diffuse accumulation of clay).
(b) *Ferrisol with weathered (B) horizon* (without diffuse accumulation of clay).
(c) *Hydromorphic ferrisol* (with segregation of iron): subgroups the same as for the ferruginous soils.
(d) *Humic ferrisol of high altitude:* two subgroups according to whether secondary gibbsite does or does not occur at the base of the A1.

There will be an additional discussion of the hydromorphic groups.

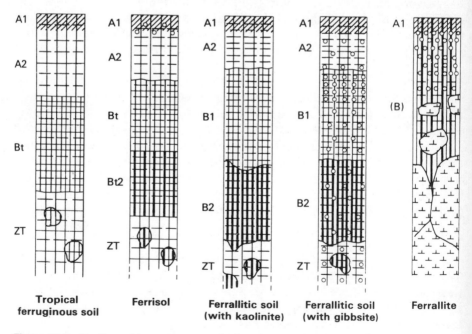

**Figure 12.5**   Profiles of ferruginous soils, ferrisols and ferrallitic soils (see general key, p. ix).

## 3 Hydromorphic ferruginous soils or ferrisols

As in a temperate climate, iron is mobilised in hydromorphic conditions as a result of redox processes: but here the processes are more complex than in a temperate climate. In temperate hydromorphic soils it is rare for organic matter not to be involved, particularly in the formation of pseudogleys. In contrast, in tropical regions, particularly when hydromorphism is at depth, organic matter is never involved even though it can still influence redox processes in horizons nearer the surface. Apart from this, the types of hydromorphism are the same as in a temperate climate: temporary or permanent, very mobile or slightly mobile water tables etc. and depending on the different types of hydromorphism, iron segregation can assume very different forms.

In summary, three main types can be differentiated:

(a) *Plinthite* (Wood & Perkins 1976): a very clay-rich and iron-rich horizon, with red and white patches which can indurate when it is intensely dried. Plinthite appears to form in horizons not subject to the effect of organic matter and subject to marked seasonal alternations of wetting and drying; it contains haematite particularly (Duchaufour 1978: $XVIII_1$).

(b) *Pseudogley* (Brabant 1973): formation of rusty patches and concretions as in a temperate climate. This time it is a question of goethite produced by an *oxidation and gradual crystallisation of iron*, which is mobile over short distances. As in a temperate climate, the environment is acid and soluble organic matter is involved, particularly when the pseudogley is fairly near the surface.

(c) *Cuirasse formation* by water tables on slopes where lateral circulation can occur. Iron is transported in the ferrous state over great distances and it accumulates in the ferric form at low points in the landscape. Organic matter is frequently involved, which slows down the recrystallisation of the ferric oxides and favours goethite formation initially (Fölster *et al*. 1971, Schwertmann *et al*. 1974). When the cuirasse ages, it changes partially to haematite (Schwertmann 1974). As in temperate climates (see Ch. 11, the white gleys), the upslope ferruginous soils are, in contrast, impoverished in iron and almost entirely decolourised (Brabant 1973).

## C  FERRALLITIC SOILS

## 1  General characters

The ferrallitic class of soils represents the final phase of development and weathering of soils in a hot and humid climate with the profiles being up to several metres thick. Almost all of the primary minerals, except for quartz, have been subject to a total hydrolysis (in neutral or very slightly acid conditions, without organic matter being involved), liberating their essential constituents: oxides of iron and aluminium, silica and bases (Leneuf 1959, Pedro

1964, Robert 1970). Profiles retain most of the iron and aluminium while, in contrast, an important fraction of the silica and almost all of the bases are removed in the soluble state, from the profile which acidifies rapidly above the level of the weathering horizon *sensu stricto*. Because of the acidity, the affinity of alumina for silica is reduced to a minimum so that *the neoformed clays are kaolinites* poor in silica which are formed by a recombination of alumina with silica that has not been transported. If the latter is insufficient, a certain amount of free alumina as gibbsite (primary gibbsite) forms in the profile.

As a whole, the primary minerals, including the micaceous minerals, have effectively disappeared, the mineralogical composition of a ferrallitic profile is reduced to four entities: quartz, kaolinite, gibbsite and ferric oxides (haematite or goethite).

This definition restricts the class of ferrallitic soils to those in which weathering is practically complete and it excludes, consequently, transitional types, i.e. the ferrallitised soils which still contain a certain amount of primary minerals. The *ferrisols* which correspond to this category have been, as stated previously, classed with the ferruginous soils, which is not accepted by all authors. This restricted definition of ferrallitic soils sensibly corresponds to that of oxisols (Soil Taxonomy) and ferralsols (FAO), which allows the basic criteria used to define oxisols and ferralsols to be used: *exchange capacity of clays less than 16 mEq/100 g*. With regard to the argillic horizon, it is generally absent and, if it occurs, it has very special characters which are different from the temperate argillic horizon (Roose 1980).

Ferrallitic soils are always very old as complete weathering of silica-rich, very resistant minerals (orthoclase and muscovite) takes a very long time, often several tens of thousands of years (on this subject, see Ch. 4). In fact, this time varies considerably depending on the kind of parent material: *it is much shorter for well drained basic volcanic materials than for poorly drained acid crystalline materials*. One of the consequences of the long time involved in the development of ferrallitic soils is to be seen in their distribution, which in certain cases appears to be aberrant. The only type of climate suited to ferrallitisation is in fact the most *active*, that is to say the hottest and most humid, and it corresponds to the evergreen rainforest (equatorial and humid tropical with a very short dry season). However, ferrallitisation extends outside of this quite considerably into drier climatic zones, characteristic of the semi-rainforest or xerophilic forest, or even the savannas (Isbell & Field 1977). Undoubtedly, this is the result of climatic variations occurring in the Quaternary, in which certain former humid phases have allowed ferrallitisation of old materials. Conversely, even in the presently most favourable climatic zones for ferrallitisation, soils that are still young or formed on reworked materials are as yet only in the ferruginous weathering stage.

From this introduction, two special characters of ferrallitic profiles should be noted: the first is the contrast between pedogenesis at the top and the bottom of the profile; the second concerns the parent material and how it affects both the speed and the kind of pedogenesis.

*1 Contrast between pedogenesis at the top and bottom of the profile*
Weathering, as previously stated, is a neutral hydrolysis (sometimes slightly alkaline), but the bases freed are transported rapidly from the profile: *this results in a gradient of decreasing pH from the bottom to the top of the profile.* The pH of the middle part, called saprolite or plastic regolith, often several metres thick, stabilises at about 5: this is the zone of maximum neoformation of kaolinites. With regard to the upper part of the profile, it is influenced by the organic matter coming from the litter and acidifying it yet more, the pH (water) very often reaches 4 or less.

*In these circumstances, surface pedogenesis is controlled by the effect of soluble organic compounds coming from the litter and is related to temperate pedogenesis characteristic of the most acid conditions. This is totally different from the process of neutral or slightly alkaline weathering which is general in the deep weathering zone, at the base of the profile.*

Ferrallitisation at depth causes the elimination of silica and bases, the hydroxides of iron and aluminium (insoluble at pH 7) being, in contrast, immobilised *in situ*. In contrast, the surface acidification is often associated with a process of hydrolysis or of acid complexolysis, which mobilises the oxides of iron and aluminium and causes their redistribution; it also causes a *destabilisation of kaolinites*, which can lead to their mobilisation, and for some of them the clay lattice is degraded, freeing silica and alumina; it is a process comparable to temperate podzolisation but which has been much increased by the very high temperatures. The alumina thus freed crystallises after moving a limited distance and forms **secondary gibbsite**, the origin of which is very different to that which results from the initial weathering of primary minerals (primary gibbsite): this process has already been described, it should be re-called, in the case of the humic ferrisols at high altitude.

In these circumstances the complete profile of a ferrallitic soil can be split into three main zones from the bottom to the top:

(a) *Deep zone of weathering:* mealy rock or regolith, as yet not very clay rich, with pH of about 7 (*geochemical* weathering).
(b) *The middle zone*, called, depending on the authors, plastic regolith, mottled zone or saprolite; often several metres thick; large-scale neo-formation of kaolinite.
(c) *The upper horizons*, forming a profile of the A(B)C or ABC type, the B horizon coming from the redistribution of clays and iron and aluminium complexes mobilised by organic matter: an acid pedogenesis of the acidolysis and complexolysis type dominant (biochemical weathering).

*2 Influence of parent material on weathering*
Pedogenesis on basic materials poor in combined silica and without quartz is not only more rapid but it is also different from the pedogenesis of acid materials rich in quartz: granitic regoliths or sedimentary materials made up of a mixture of quartz and clay (see Figs 12.3 and 12.4).

*Basic volcanic rocks* poor in combined silica and quartz: the considerable loss of silica subsequent to weathering leads to a deficit such that it is almost always insufficient to saturate the whole of the free alumina in the neoformation of kaolinite; generally, there remains a considerable amount of free alumina which occurs as primary gibbsite. The soils are ferrallites *sensu stricto*, when there is not a great deal of neoformed kaolinite, or at most a ferrallitic soil with gibbsite when the amount of kaolinite is greater.

*Granitic regolith or sedimentary rock containing quartz:* the de-silicification is more gradual and incomplete for two fundamental reasons: (i) the general reserve of combined silica is greater and it occurs as very resistant primary minerals that weather only slowly; (ii) the quartz also forms a far from negligible reserve; without doubt it is resistant to weathering, but a part (15–20%, Lelong 1969) is dissolved at a very slow rate. The silica thus dissolved has the possibility of combining with the free alumina from the weathering of primary minerals; in these conditions, gibbsite cannot persist in the free state. In the presence of quartz there is enough silica in solution for all the alumina to occur as kaolinite (Leneuf 1959, Fritz & Tardy 1976). The ferrallitic soils formed on acid materials are thus generally **ferrallitic soils with kaolinite**, without primary gibbsite; but owing to their low reserve of bases and their acidity, these materials can undergo a process of surface degradation of clays more frequently than basic materials. Although primary gibbsite is rare or absent in this type of profile, it is not so for secondary gibbsite which occurs frequently in the upper part of the profile (horizons A and B). This secondary gibbsite can be formed in two ways: (i) acid hydrolysis of kaolinite at the surface under the effect of active, soluble organic compounds; (ii) degradation of old argillans able to go as far as freeing gibbsite.

## II   The morphology and geochemistry of ferrallitic profiles

Two types of profile will be considered: (i) the commonest, that is to say ferrallitic soil with kaolinite, formed on outcrops of acid rocks (granite or gneiss) with subdued relief on plateaux; (ii) a less common profile of the *ferralite* type, formed on basic rock, poor in silica (for example, dolerite or gabbro), and at a very well drained site, often sloping. Finally, a short description of the indurated types (with cuirasse) will be given.

*1   Ferrallitic soils with kaolinite* (Duchaufour 1978: XVII$_3$, Fig. 12.5) As already stated, this profile consists of two parts subject to two different types of pedogenesis: (i) the relatively well drained upper part, with acid pedogenesis involving surface organic matter, causing redistribution of iron and surface depletion in clay; (ii) the lower part of ferrallitisation *sensu stricto* where the semi-impeded, often waterlogged environment allows the almost complete saturation by silica of the alumina freed by weathering and the neoformation of kaolinite.

**Upper part of the profile: A and B horizons.** It consists of three horizons – A1, A2 and B – or sometimes of two only, for often the A1 and A2 horizons are not differentiated. The A horizons have an acid mull that is not very thick, despite the considerable addition of plant debris; litter decomposition is very rapid, but humification is weak, owing to the lack of a marked dry season; a great amount of strongly complexing, soluble organic compounds are formed, that after insolubilisation remain as fulvic acids. These horizons are highly depleted in clay and iron. The underlying (B) horizon is generally enriched in iron, as is seen in the study of balances, and it is strongly coloured. *In contrast, the polyhedral structure is not too well developed and argillans of illuviation are not apparent*; while sometimes there are shining surfaces that are the result of the degradation of old argillans or are the result of friction (**stress cutans**), *thus the B horizon does not have an argillic character*.

The colour of the (B) horizon is characteristic of the pedoclimate of the profile and it can be either ochreous, red, or made up of red and white patches (plinthite). The ochreous colour results from the formation of goethite in constantly moist but aerated conditions and affected by organic matter; it is associated with the process of kaolinite destabilisation (Fölster *et al.* 1971, Schwertmann *et al.* 1974). The red colour is caused by haematite, generally as a result of intense cycles of wetting and drying, with no involvement of organic matter; in addition, it tends to be dominant in the oldest soils. In most cases the upper part of the (B) horizon is dominated by goethite (ochreous colour) and the lower part by haematite (red colour), a phenomenon already seen in the ferrisols.

Finally, *plinthite* occurs in hydromorphic conditions by localised segregation of iron; this is a clay with red and white patches, either as irregular spots or as a polygonal network, that remains soft when moist but can indurate irreversibly when dried (van Wambecke 1973). According to Wood and Perkins (1976), it is the haematite that is dominant in the oldest and most developed plinthite, while goethite is still abundant in younger types.

**Lower part of the profile: mottled zone and weathering zone.** The deeper part of the profile itself consists of two horizons:

(a) *The mottled zone* (or saprolite or again plastic regolith: Lelong 1969): thick, sometimes up to several metres, with large irregular white, ochreous or red patches, and a massive structure: *this is the zone of kaolinite neoformation where there is a practically constant capillary hydromorphism*. If there is a plinthite in the B horizon, it extends into the mottled zone, but as iron segregation is less complete, the general colour is lighter and the contrast between the mottles less; acidity is moderate and about pH 5. (For an example of the deep mottled zone see Duchaufour 1978: XVIII₃.)

(b) *The weathering zone* at the bottom of the profile, in contrast, has a pH of about 7. Weathering is often irregular, with fragments of mealy rocks coexisting with zones already very weathered. The environment is more porous and this allows the removal of materials both vertically and laterally by water (transport of silica and bases).

## 2    *Ferrallites* (Duchaufour 1978: XVII$_6$, Fig. 12.5)

These soils, formed on basic rocks on slopes, are considerably less deep than the preceding soils: the mottled zone is lacking and is replaced by a transitional horizon which is not too thick and is formed of fragments of weathered rock and a **plasma** containing amorphous materials still rich in silica which is removed by drainage (Tercinier 1969). The structure of the (B) horizon consists of strongly developed aggregates (or even concretions if the aggregates are partially indurated). It thus differs from the ferrallitic soils. There is little neoformation of kaolinite because of the lack of silica, *so that kaolinite is of minor importance compared to gibbsite in the weathering complex*. Gibbsite and iron oxides (cryptocrystalline goethite and haematite) are in reasonable equilibrium in most materials; however, on certain parent materials with a special composition one of the two oxides, aluminium or iron, becomes much more important than the other (see Sec. III). It should be noted that on certain basic rock with very rapid weathering in a humid climate at high altitude, the abundance of organic matter which decomposes slowly prevents the crystallisation of sesquioxides and maintains a particular abundance of amorphous materials. The soil then has a more or less pronounced andic character (Troy 1979). With regard to the varieties that occur at high altitude, the weathering cannot be as complete and the soil still has some ferrisol characters (Duchaufour 1978: XVII$_6$).

## 3    *Variations in the characters of ferrallitic soils and ferrallites according to the environment*

**Ferrallitic soils with gibbsite** (Duchaufour 1978: XVII$_4$, Fig. 12.5). Between the ferrallitic soils with kaolinite and the ferrallites *sensu stricto*, there is an intermediate type, the *ferrallitic soil with gibbsite*, in which gibbsite is present but always in minor amounts compared to the kaolinite. The presence of gibbsite can be the result of a variety of causes: it could be primary gibbsite, when it is distributed in all horizons, or secondary, when it is restricted to the upper part of the profile (A and B). In many cases the two types coexist.

The presence of primary gibbsite is related either to a low amount of silica in the parent material or to very intense drainage conditions (Ségalen 1973). With regard to secondary gibbsite, it occurs most frequently in the most acid and most humid climatic conditions. The process of formation will be discussed in the following section.

**Percentage base saturation and reworked profiles.** The weathering of primary minerals being total, the initial reserve of bases is virtually completely exhausted in the ferrallitic soils. The only bases remaining are those accumulated in the surface as a result of the biogeochemical cycle and occurring either in the form of complexes with organic matter or as exchangeable cations. But the biogeochemical cycle is related to the present-day climate and vegetation and is independent of the degree of weathering, which reflects a slow process, which is thus much older. Therefore, it is for essentially practical reasons that ORSTOM (Aubert & Ségalen 1966) adopted the percentage base saturation to split ferrallitic soils into three subclasses and they differentiated: (i) very saturated; (ii) moderately saturated; and (iii) slightly saturated. Perraud (1971) showed that these subclasses were arranged in the Ivory Coast in latitudinal zones in which the percentage base saturation increases regularly from south to north as the rainfall decreases and the dry season lengthens. Studying the humification shows that this process plays a fundamental part in building up a reserve of bases derived from the litter (see Ch. 3): *the more intense the dry season, the more the organic matter of the surface increases in quantity and degree of maturation and in the polycondensation of the humic compounds*. Thus there occurs as a function of increasing climatic dryness, a close correlation between the kind and quantity of organic matter and the percentage base saturation of the absorbent complex.

In addition, ferrallitic soil profiles diverge very frequently (particularly in Africa) from the ideal profile described above. As these profiles are very old, the surface has been frequently disturbed by erosion or deposition resulting from a possible variety of causes: the bringing to the surface of fine earth by termites; lateral movement of the surface layer (Paton 1978, Troy 1979); erosion or disturbance of the surface horizons by the action of man (bush fires). The importance of this surface disturbance has been recognised by ORSTOM pedologists, who for practical purposes have used it as a basis for the differentiation of so-called *groups*. In addition to the normal *humic* and *impoverished* groups, *disturbed* groups, in which the disturbed surface is generally separated from the underlying *in-situ* material by a stone line, and *pseudo-developed groups*, that have been truncated by erosion and rejuvenated at the surface by colluviation or alluviation, have been recognised. Finally, there is a group of *indurated* ferrallitic soils, with a more or less indurated cuirasse, that is particularly important and will be discussed in greater detail.

**Indurated ferrallitic soils: with a cuirasse.** The indurated horizons, usually called **carapaces** (moderate induration, the concretionary mass can still be broken in the hands) or **cuirasses** (indurated horizon that can only be broken with a hammer), are common in ferrallitic profiles. They generally form at a certain depth, at the level of the B horizon. But frequently, as the profile ages, they are denuded by erosion and outcrop. This is the case of the old cuirasses of erosion, often very thick, which cover great areas of the Guinean zone of Africa which is still humid but where the dry season exceeds five months. Throughout this zone, the primary forest is less stable than in more humid regions and, because of man's activities, has degraded into a **secondary savanna**.

It will be seen that the process of cuirasse formation is the result of a considerable accumulation of free sesquioxides which crystallise and indurate as a result of high temperatures: *in fact, it is the iron oxides that play the dominant part in cuirasse formation*. Gibbsite is absent or subordinate in most cuirasses of *absolute accumulation*, that is to say, enriched as a result of the addition by water of iron oxides which act as a cement. Only cuirasses of *relative accumulation*, called bauxitic, contain a considerable amount of gibbsite and are related to the ferrallites, but they are not so common.

The ferruginous cement of cuirasses is either goethite or haematite, with goethite being dominant in those of recent age (Fölster *et al*. 1971) and haematite, in contrast being dominant in very old cuirasses (Schwertmann 1974). Depending on the way in which cementation has occurred, the structures have a characteristic appearance: **pisolitic cuirasses** result from the welding together of concretions that form in isolation in the profile in semi-hydromorphic conditions, becoming cemented when the whole of the horizon dries out; **scoriaceous** or **vesicular cuirasses** (Maignien 1964, Duchaufour 1978: XVIII$_5$) that result from accumulation of iron in the old network of fissures of a polyhedral or prismatic (B) horizon. This iron forms the indurated framework of the cuirasse and isolates the softer material which is then partially removed by circulating water; the **petroplinthites** that result from the induration of plinthites often have such a vesicular structure, for only the areas enriched in iron are capable of being indurated, as compared to the decoloured patches poor in iron that remain softer.

## 4   Agronomic properties of ferrallitic soils and their utilisation

In general, the fertility of soils of intertropical regions varies inversely with the degree of profile development. It is high for soils that have undergone little or no ferrallitisation, such as the eutrophic brown soils and the fersiallitic red soils; it decreases in tropical ferruginous soils and becomes very low both in physical and chemical terms for ferrallitic soils and even more so for ferrallites, which is easily explained in terms of the accumulation of unfavourable properties. The reserve of bases becomes lower and lower and then nil, the structure is degraded, the exchange capacity is gradually lowered, and finally, the clays become more and more inert, losing their ability to form clay–humus aggregates with favourable properties. At the limit, the ferrallites have a mineral fraction where the exchange capacity is sensibly nil (less than 1 mEq/100 g), and consequently incapable of retaining exchangeable bases, $Ca^{2+}$, $Mg^{2+}$, $K^+$, contained in the litter and freed during its decomposition. *The importance of the biogeochemical cycle and of the organic matter becomes considerable, for it is the only source of assimilable elements such as nitrogen, phosphorus, exchangeable bases, minor elements etc*. Thus this is the only natural source of fertility in these soils. These nutritive elements remain assimilable while they are retained in the humus-rich horizons; however, on being freed by mineralisation they are rapidly removed from the profile and lost or, alternatively, insolubilised in an unassimilable form (retrogradation of phosphates, irreversibly precipitated by iron oxides).

Only the equatorial forest grows in a satisfactory manner on ferrallitic soils, as a result of the effectiveness of the biogeochemical cycle which is a closed cycle and, owing to its well developed roots that penetrate to the deeper horizons, including the weathering zone which is richer in mineral reserves, it can concentrate in the surface within the humus a part of the mineral reserves transported from this depth; this has not been overlooked by the natives who understand that the only horizon that can be used for primitive agriculture is the humus-rich surface horizon. This is the reason why shifting cultivation of long duration (associated with clearing by bushfires) has been for a long time the only method of using these soils agriculturally. Native cultivation uses the nutrient reserves (N-P-bases) stored in the forest humus over several years, then the soil becomes sterile and it is abandoned to the forest.

This practice is responsible for the degradation of ferrallitic soils that is so well seen in the Guinean zone of Black Africa. In overpopulated areas the time allowed for the forest to reform the humus, during the intervals between cultivation of the same area, has a tendency to shorten more and more and be reduced to only a few years. In these circumstances, the forest is not reformed and is reduced to a bushy vegetation which gradually becomes more and more open, as grasses and woody xerophilic vegetation take the place of hygrophilic species, and thus a *secondary savanna* is formed. But this savanna is distinctly less effective than the primary forest, firstly in maintaining the biogeochemical cycle on a permanent basis (rooting is too superficial) and secondly in providing sufficient protection against erosion (litter is reduced, shade is weak or nil). This causes a partial disappearance of the humus-rich horizons and a new mobilisation of iron, associated with an induration of the mineral horizons – processes discussed with regard to cuirasse formation.

Modern methods of cultivation use ferrallitic soils both as a mechanical support and as a water reservoir during the dry season. The nutritive elements are added as amendments and fertilisers; the crops – coffee, cocoa, pineapples, bananas, etc. – have the advantage of protecting the soil and favouring the formation of humus by providing abundant plant debris (Godefroy 1974).

## III  Physicochemical process of ferrallitisation

Avoiding intermediate types, this discussion of process will deal with two extreme situations: (i) parent material with large reserves of silica and imperfect drainage, conditions that cause large-scale neoformation of kaolinite (ferrallitic soil with kaolinite: development very slow); (ii) parent material with low reserves of silica and good drainage (slopes) causing the formation of gibbsite (ferrallites: relatively rapid development; Lelong & Souchier 1979). In both cases transitional types occur, i.e. ferrallitic soils with gibbsite, but it will be a question mainly of secondary gibbsite in the first case and primary gibbsite in the second.

### 1  Ferrallitic soils: imperfectly drained acid parent materials
The method of *geochemical balances* is applied particularly to ferrallitic soils, for the simplicity of their composition allows a mineralogical reconstruction to be made easily and in sufficiently precise terms from information given by a general geochemical analysis. It is sufficient to differentiate clearly between free silica (quartz) and combined silica, which in ferrallitic soils necessarily

occurs as kaolinite. Lelong (1969) determined such balances on Guyanese soils and his main conclusions will be given (see also Fig. 1.2). However, it must be admitted that there are difficulties in this approach. Parent rocks are very often heterogeneous and surface disturbance is frequent. It is necessary to use homogeneous profiles and to confirm this homogeneity by an investigation of the grain-size distribution of sand-size quartz. On the other hand, because of variations in density and volume which occur as a result of weathering, it is necessary to determine these balances in terms of a standard mineral, which is quartz, but this is still relative, for in hot climates the solution of fine-grained quartz is by no means negligible. Nevertheless, Lelong has managed to determine a sufficiently exact balance, in calculating the loss of quartz in terms of a standard volume, which amounts to some 15–20% from top to bottom of the profile, and making the necessary correction.

There are three aspects of the results obtained: (i) the environmental and pedological conditions in the successive *zones* of the profile; (ii) general results of geochemical balances; and (iii) interpretation of these balances in terms of the overall development process.

**Environment and pedoclimate of the different zones** (Fig. 12.6). It should be remembered that three zones with different physical and chemical properties were differentiated in the profile description: the deep zone, the zone of weathering ('D' on Figs 12.6 & 7); the middle zone, the mottled zone, or, as it is called by Lelong, plastic regolith (which is 'C' in the figures); and finally, the surface zone, subdivided into A and B horizons.

The deep zone (zone D) has a high lateral porosity and permeability; the pH is near neutrality, owing to the abundance of cations freed by weathering; and water circulates easily both laterally and vertically as a result of the jointing of the granite. In the middle zone (zone C), the plastic regolith is

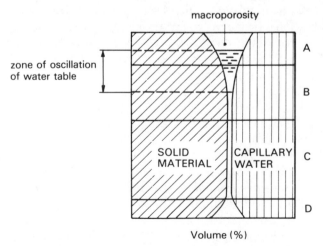

**Figure 12.6** Diagram of the porosity of a ferrallitic clay.

several metres thick and is the focus of large-scale kaolinite neoformation; the non-capillary porosity becomes very low (5–6%; Lelong 1969). It is highly acid as a result of the removal of bases; hydromorphism owing to *capillary adsorption* is almost permanent and there is little or no circulation of water as a result of the low permeability. The surface zone is divided into two horizons, one that is depleted in all entities (A), the other enriched (B). Compared to the preceding zone in terms of environmental conditions, this zone has two important differences: (i) it is influenced by surface organic matter which is generally very acid; the highly humid climate favours the formation of abundant water-soluble compounds which remain as fulvic acids and only polymerise slowly; these compounds can form complexes with certain cations to a considerable extent; (ii) this zone is more aerated than the preceding zone and from the bottom of the B horizon hydromorphism is not permanent: porosity increases towards the surface, the water of the water table becomes more mobile and can circulate laterally in periods of high humidity. In dry periods, this mobile water table disappears or falls in level, which explains why the position of the B horizon is determined by the limit of rainfall penetration.

In fact, variations in these pedoclimatic conditions occur, particularly with regard to the A and B horizon; the importance of the redistribution of iron and the general colour of the horizon resulting from this has been emphasised in the profile description. Generally, the upper part of the B horizon, enriched in goethite by movement of iron from the surface horizon, is ochreous; the lower part is red owing to the formation *in situ* of haematite. But goethite can occur throughout the profile, which is then ochreous coloured; in contrast, when the pedoclimate is dry and organic matter present in only small amounts, it is the red colour of haematite that is dominant throughout the profile. Plinthite, with local segregation of iron, is characteristic of very intense temporary hydromorphic conditions.

**Geochemical balances** (Fig. 12.7). The basic geochemical balances determined by Lelong for eight homogeneous clay profiles in Guyana are summarised in Fig. 12.7; the value of a 100 for each of the oxides is equivalent to the amount present in the parent material.

The deep zone of weathering has lost a great deal of combined silica (60% or more of the initial parent rock content) and almost all of the bases. In the overlying zone (plastic regolith), the composition remains constant over a great thickness (several metres): the amount of combined silica and alumina corresponds almost exactly with the formula for kaolinite, the silica : alumina molecular ratio having been reduced to about 2 by the previous loss of silica. The decrease in the amount of iron compared to the zone of departure is to be noted. With regard to the upper part of the profile, as previously stated it is divided into two main horizons: A (pervected or rather impoverished, Ségalen 1965) and B, which always has a very evident accumulation of iron which explains the very intense colour of this horizon. But when silica and alumina are considered, the profiles fall into two groups: some have no accumulation in the B, while the others show, in the shape of the graphs, considerable accumulation at this level. Not only is the accumulation of iron very great, but alumina is slightly in excess of silica, if the formula for kaolinite is taken as a

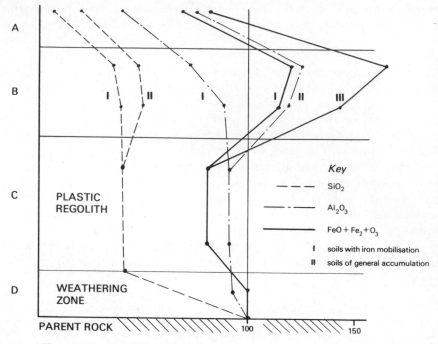

**Figure 12.7** Geochemical balance of ferrallitic soil (after Lelong 1969).

basis of comparison (generalised accumulation). The surface A horizon is markedly depleted in all entities – combined silica, alumina and iron – and in the most developed profiles quartz becomes overwhelmingly dominant at the surface.

**Interpretation of balances:** phases of development. It is now possible to reconstruct the phases of development of the ferrallitic clays by comparing, as was done by Lelong, the results of the geochemical balances with the mineralogical investigations of each zone.

*At the base of the profile.* Weathering occurs under neutral conditions (locally slightly alkaline) with large-scale losses of silica and bases, removed by lateral transport, or infiltration into the joints in conditions that are still permeable. The *iron and most of the alumina remain in situ*: this is ferrallitic weathering *sensu stricto*. In the plastic regolith zone, which is more acid and normally waterlogged, conditions favour the neoformation of kaolinite by fixation of what remains of the silica (in the soluble state) by the alumina (insoluble). With regard to the iron, the formation of ochreous or red patches is not necessarily the result of hydromorphic segregation, but weathering *in situ* of material richer in iron (Fölster *et al.* 1971). However, the whitish zones correspond to those where reduction and solution of iron are more marked, which is the source of the iron deficit to be seen in the balance determinations.

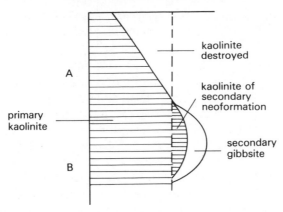

**Figure 12.8**  Details of aluminium balance, type II (upper part of the profile).

*The surface horizons*. These are the centre of the very different biochemical weathering, owing to the influence of local environmental conditions, which closely resemble those which occur in temperate soils in very acid conditions. The first point, which has already been emphasised, is that iron is always mobilised. The marked decrease in the amount of kaolinite seen at the top of the profile is generally not balanced by an equivalent accumulation in the B; it may be completely lacking or, if present, it is always less than the surface loss. In addition, this accumulation, which is difficult to recognise morphologically (as it occurs as poorly developed argillans), is in the form of poorly crystallised clay and gibbsite (Lelong 1969, see Fig. 12.8).

*Profiles of type I, without accumulation in the B other than that of iron*, have the most widespread distribution in moderately humid climates and in environments poor in organic matter. Pedogenesis is easy to interpret; only iron is mobilised in appreciable amounts by surface chcluviation and a part of it accumulates in the B. The loss of kaolinite at the surface with no detectable accumulation in the B is explicable, either by a marked lateral impoverishment or by a very deep and very diffuse deposit of the transported clays (Roose 1980).

*Profiles of type II (those with generalised accumulation)*, characteristic of the most humid climates, favour the formation of a reasonably thick moder surface humus, or even mor. The phases 2 and 3 of destabilisation of fine kaolinites and the weathering of argillans (formed previously), which have been described in the section on ferrugination, here reach their full development. But there is added to it a process of surface podzolisation which is specific to the most acid and most developed ferrallitic soils. This involves the hydrolysis by complexolysis of some of the kaolinites, owing to the effect of fulvic acids and the accumulation in the B of the freed entities ($SiO_2$ and $Al_2O_3$), as secondary newly formed kaolinite and gibbsite (Hughes 1980, Fig. 12.8).

**Note** that the balance *for each horizon* of Lelong and Souchier (1972) confirms very clearly this overall interpretation; gibbsite of primary weathering is almost completely absent from both the plastic regolith and the weathering zone (2–3%); while, in contrast, secondary gibbsite amounts to 16% in the A and 29% in the B for a Guyana profile on granite.

## 2   Ferrallites: basic, well drained parent materials

On basic materials poor in silica, the rate of pedogenesis is more rapid. The contrast that occurs between well drained slopes and those at the bottom of slopes with impeded drainage is spectacular, much more so than in catenas on acid rocks; the reader is referred to Chapter 8 for a discussion of impeded conditions, here the soil of well drained slopes will be discussed.

The general interpretation of the process is easy: most of the silica and bases are removed by vertical or lateral drainage; there is little or no kaolinite neoformation owing to a lack of silica. The excess alumina crystallises as gibbsite, which this time is *primary*, i.e. it results from the direct weathering of rock minerals. Finally, only two basic entities remain in the oxic horizon, gibbsite and goethite (or haematite), often also with titania which is equally as stable in these conditions. The amorphous materials that are more abundant at the base of the profile, and often still contain silica, can be considered as the initial phase in the formation of the weathering complex which develops by crystallisation and loss of silica (Troy 1979).

The *pilot* role played by iron (always very abundant) in the formation of these soils has been emphasised by Lelong (1968): under the influence of seasonal microclimatic contrasts, more intense than in the preceding case as a result of the better internal drainage, this element has been subject to a very rapid development in this soil. In particular, it is immobilised very quickly by crystallisation; this insolubilisation of the iron carries with it a great part of the alumina, the extraction of the latter necessitating the prior solution of the former (Duchaufour & Souchier 1966, Schwertmann *et al.* 1978, who showed the isomorphous substitution of Al in the iron oxides). This has two consequences: (i) the creation of an aggregate structure that maintains a high soil porosity (Greenland *et al.* 1968); (ii) a slowing down of kaolinite neoformation caused by the too rapid crystallisation of the hydroxides. In these circumstances, it is not surprising that a large part of the silica is removed by lateral drainage in the soluble state.

In extreme situations such as at the top of slopes, where drainage is excessive and where only weak protection is afforded by forest vegetation, there is little or no neoformation of kaolinite (fully developed ferrallite; Tercinier 1969). If the drainage is less good, at mid-slope for example, or if the forest vegetation decreases the pedoclimatic contrasts, the crystallisation of hydroxides is then slowed down and neoformation of kaolinite can occur. It is favoured by a supplementary addition of silica carried by the lateral flow of the water table from higher levels (ferrallitic soils with gibbsite: Delvigne 1965, see Fig. 12.3).

Comparing the soils formed on basic rocks with the ferrallitic soils formed on acid rocks, it can be said that clay neoformation is slowed down; however, this clay is stable and is not degraded, for it is bound within the aggregates of which the free hydroxides form the cement and which protect the clay against the activity of the soluble acid organic compounds.

## 3 Process of cuirasse formation

As previously stated, cuirasse formation is the result of the accumulation of sesquioxides (mainly iron) within one or several ferruginous or ferrallitic soil horizons. There are two ways in which this can occur: by *relative* accumulation, in which the other constituents are removed, or by *absolute* accumulation, where sesquioxides are added after transport in solution (D'Hoore 1955).

Cuirasse formation by relative accumulation is in fact only a special case of ferrallite formation on basic rocks in well drained conditions (Duchaufour 1978: XVIII$_6$); in most cases, however, ferrallite formation causes the formation of an aggregate structure that is not favourable to cuirasse formation *en masse*. This process only occurs in some particular, and fairly exceptional, circumstances (Bonifas 1958, Blot *et al*. 1976). This type of cuirasse is often very rich in gibbsite (bauxite cuirasse). On rocks poor in alumina, iron oxides are dominant, associated, however, with various trace elements (titania, oxides of chromium, manganese, nickel, etc. – Duchaufour 1978: XVIII$_6$).

Cuirasse formation by absolute accumulation, i.e. by the addition of iron oxides mobilised in acid conditions, is the most common: *the iron is mobilised by the effect of acid organic matter in the A horizon and is transported laterally for more or less large distances* (Maignien 1958, Lelong 1969, Fölster *et al*. 1971). It is concentrated and precipitated as goethite, where conditions of aeration are better (at the top of the water table, zone of emergence, etc.). This mobility of iron is very favourable to cuirasse formation and explains why, paradoxically, the process of cuirasse formation is more frequent on acid rocks poor in iron than on iron-rich basic rocks (see Fig. 12.4).

Although this kind of induration can be considered as a general process, it is, in fact, subject to considerable variations in which the time factor plays an important part. Rapidly formed cuirasses generally found at restricted sites need to be differentiated from the very slow-forming plateaux cuirasses that occur over large areas.

**Rapidly formed water-table cuirasses.** These occur at particular sites such as at breaks of slope, basin margins, etc. The iron carried by lateral circulation of the water tables on slopes accumulates in lower topographic positions where it crystallises as goethite or haematite. In many cases, iron accumulation occurs in conditions that are still too hydromorphic (even within the water table) for induration to be immediate. The alternation of wetting and drying only allows a very localised segregation of iron, with the formation of red patches of haematite. This is plinthite (Duchaufour 1978: XVIII$_1$) which is still soft. It is only when the level of the water table is lowered that this plinthite irreversibly indurates to form a **petroplinthite** horizon (van Wambecke 1973, Duchaufour 1978: XVIII$_4$).

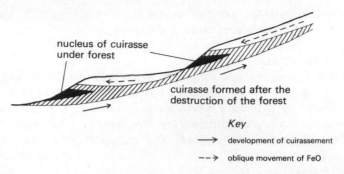

nucleus of cuirasse
under forest

cuirasse formed after the
destruction of the forest

*Key*

⟶      development of cuirassement

⟶      oblique movement of FeO

**Figure 12.9**   The processes of degradation of forest soils on slopes (after Maignien 1958).

**Slowly formed plateaux cuirasses.** These are very thick and occupy immense peneplained surfaces. They could also be called **erosion cuirasses** because the upper, less indurated horizons have been gradually removed. These cuirasses are concentrated in zones with marked climatic contrasts (Guinean zone) in tropical Africa, today occupied by secondary savanna, which was referred to previously. The gradual destruction of the forest by man, by shifting cultivation and bush fires has certainly increased the speed of cuirasse formation. In this sense it can be called a **cuirasse of degradation**. In all cases the formation of these cuirasses has been very slow and has taken several tens of thousands of years (Ségalen 1965).

Maignien (1958) showed that the process of induration for this type of cuirasse was slow and gradual: in the forest phase, incompletely indurated nuclei are formed at low points in the landscape or where changes of slope occur. Clearing and destruction of the forest favours the mobilisation of more iron, causing the lateral extension and thickening of the originally formed indurated nuclei (Fig. 12.9). The cuirasse becomes more and more widespread and can be several metres thick. It is associated with a change in the iron oxides: goethite is gradually replaced by haematite which becomes dominant (Schwertmann 1974). Thus there are two possible origins of haematite in tropical soils: (i) rapid crystallisation of amorphous iron oxides in the absence of organic matter; (ii) gradual formation by slow dehydration of goethite.

The opposite kind of development is also possible: the haematite of old cuirasses can be reduced and dissolved as an iron complex, then transported laterally and recrystallised lower in the landscape as ochreous goethite, which often cements colluvial blocks of old cuirasse with red haematite (Schwertmann 1974).

## IV   Classification and main types of ferrallitic soils

The ferrallitic soils that correspond to the oxisols (Soil Taxonomy) or ferralsols (FAO) can be subdivided into three subclasses.

(a) *Ferrallitic soils, sensu stricto*, with kaolinite dominant in the weathering complex.
(b) *Ferrallites* with sesquioxides (of aluminium or iron or both) dominant in the weathering complex.
(c) *Ferrallitic soils with hydromorphic segregation of iron*, including in particular the indurated ferrallitic soils, resulting from an *absolute accumulation*.

## 1 Ferrallitic soils

They correspond to the profiles described previously and two groups can be differentiated.

(a) *Ferrallitic soils with kaolinite* (gibbsite absent or in very small amounts).
(b) *Ferrallitic soils with gibbsite:* kaolinite remains dominant but gibbsite makes up a significant part of the weathering complex throughout the profile (its distribution is different according to whether the gibbsite is primary or secondary).

Subgroups can be differentiated into classes as function of (i) colour (red or ochreous); (ii) origin of the gibbsite (primary or secondary); (iii) the superficial processes of disturbance that have affected the profile: pervection, podzolisation, impoverishment, reworking, pseudo-development, induration (facies in which there is little or no hydromorphism, with relative accumulation); and (iv) percentage base saturation.

## 2 Ferrallites

The modal type described previously is one of relative accumulation in which there are about equal amounts of both iron and aluminium oxides. But in certain cases one of the two oxides can be considerably more important than the other, while the other can be almost completely absent from the profile; this is so in the cases of **ferrites** (or ferritic soils) and **allites** (or allitic soils).

(a) *Ferrites* (Latham 1975, Trescases 1975). These soils are formed on ultrabasic rocks, poor in aluminium (peridotite), by removal of silica and magnesium; practically the only thing left is iron as goethite (humid climate, abundant organic matter); iron development then follows the general rule.

   Certain ferrites are subject to cuirasse formation, but this is a fairly exceptional case (Duchaufour 1978: $XVIII_6$).
(b) *Allites*. These soils are in fact hydromorphic ferrallites formed in well drained conditions with a permeable parent material in a very humid climate. The iron in these conditions is mobilised by reduction and removed from the profile, so that the only thing remaining *in situ* is gibbsite, the profile being uniformly white.

Certain allitic soil profiles have been described in Colombia on outcrops of very old volcanic ashes: the phase of andosol formation preceded that of ferrallitisation, then a ferrallite is formed by complete removal of the silica. Then, in its turn, the iron is mobilised as a result of hydromorphism (plinthite, formed by the iron accumulated in zones lower in the landscape, attests to this process). Finally, gibbsite alone remains *in situ* (Faive & Luna, personal communication; Zebrowski 1975).

## 3   Ferrallitic soils with hydromorphic segregation of iron

In the modal ferrallitic soils that have been described, the zone of plastic regolith (or mottled zone) is generally almost permanently waterlogged, and because of this is often referred to as being hydromorphic. In fact, it is a case of capillary water that moves very little and only mobilises iron very slightly (Fölster *et al*. 1971). It is different from water tables with lateral circulation which cause considerable iron mobilisation and are responsible for cuirasse formation by absolute accumulation.

**Hydromorphic ferrallitic soils** formed in very humid climates and in well drained conditions (Duchaufour 1978: $XVII_5$) have almost entirely lost their free iron; it no longer occurs in the completely decolourised profile that is made up of kaolinite with a little quartz and gibbsite.

**Ferrallitic soils with plinthite** are frequent down slope from the preceding soil; the mobilised iron has been transported over fairly short distances and has formed plinthite on the badly drained lower slopes.

**Indurated ferrallitic soils** (or with cuirasses) result from the transport of iron over greater distances within acid water tables; in the zones of emergence of these water tables, various types of cuirasse are formed (that have been described) either by cementation of concretions or by precipitation within pre-existing fissures where the iron crystallises in forming a solid framework.

## References

Alwis, K. A. de and D. J. Pluth 1976. *Soil Sci. Soc. Am. J.* **40** (6), 912–20 and 920–28.

Aubert, G. and P. Ségalen 1966. *Cah. ORSTOM, Sér. Péd.* **IV** (4), 97–112.

Bachelier, G. 1977. *Science du Sol* **1**, 3–11.

Beckmann, G. G., C. H. Thompson and G. D. Hubble 1974. *J. Soil Sci.* **25** (3), 265–81.

Bennema, J., A. Jongerius and R. C. Lemos 1970. *Geoderma* **4** (3), 333–55.

Blot, A., J. C. Leprun and J. C. Pion 1976. *Bull. Soc. Géol. Fr.* **XVIII** (1), 51–4.

Bocquier, G. 1973. *Genèse et évolution de deux toposéquences de sols tropicaux du Tchad. Interprétation biogéodynamique*. State doct. thesis. Univ. Strasbourg. Mémoire ORSTOM, no. 72.

Bonifas, M. 1958. *Contribution à l'étude géochimique de l'altération latéritique*. State doct. thesis. Univ. Strasbourg.

Bornand, M. 1973. *Ann. Scient. Univ. Besancon, Géologie*, 3$^e$ sér., fasc. **2**, 15–18.

Bornand, M. 1978. *Altération des materiaux fluvio-glaciaires, génèse et evolution des sols sur terrasses quaternaires dans la moyenne vallee du Rhône*. State doct. thesis. Univ. Montpellier and ENSA Montpellier.

Bottner, P. 1972. *Evolution des sols en milieu carbonaté*. State doct. thesis. Univ. Strasbourg. Mémoires no. 37, Sciences géologiques.

Boulaine, J. 1966. *C.R. Conf. Sols. Médit*., Madrid 281–4.

Boulangé, B., H. Paquet and C. Bocquier 1975. *C.R. Acad. Sci. Paris* **280D**, 2183–6.

Bourgeat, G. and C. Ratsimbasafy 1975. *Bull. Soc. Géol. Fr.* (7) **XVII** (4), 554–61.

Brabant, M. 1973. In *Pseudogley and gley. Trans Comms V and VI ISSS*, E. Schlichting and U. Schwertmann (eds), 371–7. Weinheim: Chemie.

Bresson, L. M. 1976. *Science du Sol* **1**, 3–22.

Brook, R. H. and J. van Schuylenborgh 1975. *Geoderma* **14** (1), 3–13.

Bruin, J. H. S. 1970. *A correlation study of red and yellow soils in areas with a Mediterranean climate*. FAO: World Soil Resources Report 39.

Buurman, P. 1980. *Red soils in Indonesia*. PUDOC. Wageningen: Centre for agricultural publishing and documentation.

Chauvel, A. 1977. *Recherches sur la transformation des sols ferrallitiques dans la zone tropicale à saisons contrastées*. Document ORSTOM, no. 62 Bondy.

Coninck, F. de, G. Stoops and R. K. Chatterjee 1977. *Proc. CLAMATROPS*, Kuala Lumpur, 41–50.

Dachary, M.-C. 1975. *Science du Sol* **4**, 231–48.

Dan, J. and A. Singer 1973. *Geoderma* **9** (3), 165–93.

Daniels, R. B. and E. E. Gamble 1978. *Geoderma* **21** (1), 41–65.

Delvigne, J. 1965. *Pédogénèse en milieu tropical, la formation des minéraux secondaires en milieu ferralitique*. Thesis. Fac. Sci. Paris.

Duchaufour, Ph. 1969. *Rapport de mission en Grèce*. Institut de recherches, forestières d'Athènes (project UNSF/FAO/FRE 20/230).

Duchaufour, Ph. 1978. *Ecological atlas of soils of the world*. New York: Masson.

Duchaufour, Ph. and Y. Dommergues 1963. *Sols Africains* **VIII** (1), 5–16.

Duchaufour, Ph. and B. Souchier 1966. *Science du Sol* **1**, 17–27.

Emberger, L. 1939. *Mem. Soc. Sci. Nat. Maroc., Inst. Rubel, Zurich* **14**, 40–157.

Eswaran, H. and C. de Coninck 1971. *Pédologie*, Ghent **XXI** (2), 181–210.

Eswaran, H. and C. Sys 1979. *Pédologie*, Ghent **XXIX** (2), 175–90.

Fischer, W. R. and U. Schwertmann 1974. *Clays Clay Mineral.* **23**, 33–7.

Fölster, H., M. Moshrefi and A. G. Owenuga 1971. *Pédologie*, Ghent **XXI** (1), 95–124.

Fritz, B. and Y. Tardy 1976. *Bull. Soc. Géol. Fr.* (7) **XVII** (1), 7–12.

Gallez, A., A. S. R. Juo, A. J. Herbillon and F. R. Moormann 1975. *Soil Sci. Soc. Am. Proc.* **39** (3), 577–85.

Godefroy, J. 1974. *Evolution de la matière organique du sol sous culture du bananier et de l'ananas. Relation avec la structure et la capacité d'echange cationique*. Engng doct. thesis. Univ. Nancy.

Gratier, M. and P. Pochon 1976. *Les sols rubéfiés du pied du Jura*. Soc. Suisse de Pédologie, session of 12 March 1976.

Greenland, D. J., J. M. Oades and T. W. Sherwin 1968. *J. Soil Sci.* **19** (1), 123–34.

Guerra, A. 1972. *Los solos rojos en España*. Inst. Edap. Biol. Veg.

Herbillon, A. J., M. N. Makumbi and R. Frankart 1973. *Cah. ORSTOM, Sér. Péd.* **XI** (1), 15–18.

Hoore, J. d' 1955. *African Soils* **3** (1), 66–81.

Hubschmann, J. 1967. *Sols, pédogénèses et climats quaternaires dans la plaine des Triffa (Maroc)*. Engng doct. thesis. Univ. Toulouse.

Hubschmann, J. 1975. *Morphogénèse et pédogénèse quaternaires dans le piémont des Pyrénées garonnaises et ariégeoises*. State doct. thesis. Toulouse-le-Mirail.

Hughes, J. C. 1980. *Geoderma* **24** (4), 317–25.

Isbell, R. F. 1977. *Proc. CLAMATROPS*, Kuala Lumpur (*Soils and Fertilizers* (1979) **42** (11), 701).

Isbell, R. F. and J. B. F. Field 1977. *Geoderma* **18** (3), 155–75.

Jamagne, M. 1963. Contribution à l'étude des sols du Congo oriental, *Pédologie*, Ghent **XII** (2), 271–414.

Juo, A. S. R., F. R. Moormann and H. O. Makuador 1974. *Geoderma* **11** (3), 167–79.

Kaloga, B. and C. Thomann 1971. *Cah. ORSTOM, Sér. Péd.* **IX** (4), 461–505.

Kubiena, W. 1953. *The soils of Europe*. London: Thomas Murby.

Lamouroux, M. 1971. *Etude de sols formés sur roches carbonatées. Pédogénèses fersiallitique*. State doct. thesis. Univ. Strasbourg. Mém. ORSTOM, no. 56.

Lamouroux, M. 1972. *Cah. ORSTOM, Sér. Péd.* **X** (3), 243–51.

Lamouroux, M., H. Paquet and G. Millot 1973. *Pédologie*, Ghent **XXIII** (1), 53–71.

Latham, M. 1975. *Cah. ORSTOM, Sér. Péd.* **XIII** (2), 159–72.

Lelong, F. 1968. *Science du Sol* **2**, 93–104.

Lelong, F. 1969. *Nature et genèse des produits d'altération de roches crystallines sous climat tropical humide*. State doct. thesis, Fac. Sci. Nancy; Mém. science de la terre, Nancy, no. 14.

Lelong, F. and B. Souchier 1972. *C.R. Acad. Sci. Paris* **274D**, 1896–1900.

Lelong, F. and B. Souchier 1979. In Alteration des roches crystallines en milieu superficiel: INRA Seminar. *Science du Sol* no. 2 and 3, 267–79.

Leneuf, N. 1959. *L'altération des granites calco-alkalins et des granodiorites en Côte-d'Ivoire forestière*. Mém. ORSTOM.

Lersch, I. F., S. W. Buol and R. B. Daniels 1977. *Soil Sci. Soc. Am. J.* **41** (1), 104–9 and 109–15.

Levêque, A. 1975. *Pédogénèse sur socle granito-gneissique du Togo. Différentiation des sols et remaniements superficiels*. State doct. thesis. Univ. Strasbourg.

McLeod, D. A. 1980. *J. Soil Sci.* **31** (1), 125–36.

Maignien, R. 1958. *Contribution à l'étude du cuirassement des sols en Guinée française*. State doct. thesis. Fac. Sci. Strasbourg.

Maignien, R. 1964. *Compte rendu de recherches sur les latérites*. Paris: Mém. Unesco.

Makumbi, L. 1972. *Contribution à la pédogénèse tropicale. Etudes des sols développés sur chloritoschistes de Gangila (Zaïre)*. Doct. thesis. Univ. Louvain.

Makumbi, L. and A. J. Herbillon 1973. *Pédologie*, Ghent **XXIII** (1), 5–26.

Makumbi, L. and M. L. Jackson 1977. *Geoderma* **19** (3), 181–97.

Martin, D., G. Sieffermann and M. Valérie 1966. *Cah. ORSTOM Sér. Péd.* **4**, 3–26.

Millot, G. 1964. *Géologie des argiles*. Paris: Masson.

Moormann, F. R. and A. van Wambecke 1978. *11th Congr. ISSS Edmonton*, Plenary papers **2**, 272–91.

Paquet, H. 1969. *Evolution géochimique des minéraux argileux dans les altérations et les sols de climats tropicaux et méditerranéens à saisons contrastées*. State doct. thesis. Fac. Sci. Strasbourg.

Paton, T. R. 1978. *The formation of soil material*. London: George Allen & Unwin.

Pedro, G. 1964. *Contribution à l'étude expérimentale de l'altération géochimique des roches crystallines*. State doct. thesis. Fac. Sci. Paris.

Perraud, A. 1971. *La matière organique des sols forestiers de la Côte d'Ivoire*. State doct. thesis. Univ. Nancy.

Reynders, J. J. 1972. *Geoderma* **8**, 267–79.

Robert, M. 1970. *Etude expérimentale de la désagrégation du granite et de l'évolution des micas*. State doct. thesis. Fac. Sci. Paris.

Roose, E. J. 1970. *Cah. ORSTOM, Sér. Péd.* **VIII** (4), 469–92.

Roose, E. J. 1980. *Dynamique actuelle de sols ferrallitiques et ferrugineux tropicaux d'Afrique occidentale*. State doct. thesis. Univ. Orléans.

Schwertmann, U. 1974. *Mitt. Deutsch. Bodenk. Ges.* **20**, 87–9.

Schwertmann, U. and R. M. Taylor 1973. In *Pseudogley and gley. Trans Comms V and VI ISSS*, E. Schlichting and U. Schwartmann (eds), 45–54. Weinheim: Chemie.

Schwertmann, U., W. R. Fischer and R. M. Taylor 1974. *10th Congr. ISSS*, Moscow **VI** (1), 237–47.

Schwertmann, U., R. W. Fitzpatrick and R. M. Taylor 1978. *11th Congr. ISSS*, Edmonton **1**, 177.

Ségalen, P. 1965. *Cah. ORSTOM Sér. Péd.* **III** (3), 179–94.

Ségalen, P. 1969. *Cah. ORSTOM Sér. Péd.* **VII** (2), 225–37.

Ségalen, P. 1973. *L'aluminium dans les sols*. Mém. ORSTOM, Doc. Techn. no. 22, Paris.

Servant, J. 1970. *Notice carte pédologique France 1/100,000*, sheet Argeles–Perpignan. INRA.

Servat, E. 1966. *C.R. Conf. Sols Médit.*, Madrid, 407–11.

Smith, G., C. Sys and A. Van Wambecke 1975. *Pédologie*, Ghent **XXV** (1), 5–24.

Spaargaren, O. C. 1979. *Weathering and soil formation in a limestone area near Pastena (Italy)*. Doct. thesis. Univ. Amsterdam, publ. no. 30.

Sys, C. 1967. *Pédologie*, Ghent **XVII** (3), 284–325.

Sys, C. 1978. *Pédologie*, Ghent **XXVII** (3), 307–35.

Taylor, R. M. and A. M. Graley 1967. *J. Soil Sci.* **18** (2), 341–53.

Tercinier, G. 1969. *Cah. ORSTOM, Sér. Péd.* **VII** (4), 583–94.

Torrent, J., U. Schwertmann and D. G. Schulze 1980. *Geoderma* **23** (3), 191–208.

Trescases, J. J. 1975. *L'évolution géochimique supergène des roches ultrabasiques en zone tropicale*. Mém. ORSTOM, no. 78. State doct. thesis. Univ. Strasbourg, 1973.

Troy, J. P. 1979. *Pédogénèse sur roches charnockitiques en region tropicale humide de montagne dans le sud de l'Inde*. State doct. thesis. Univ. Nancy.

Verheye, W. 1974. *Pédologie*, Ghent **XXIV** (3), 266–82.

Vigneron, J. and P. Rutten 1967. *Chronologie des paleosols de Bas-Languedoc*. Mém. CNA Bas–Rhône–Languedoc.

Wambecke, A. van 1973. In *Pseudogley and gley. Trans Comms V and VI ISSS*, E. Schlichting and U. Schwertmann (eds), 357–62. Weinheim: Chemie.

Williams, C. and D. H. Yaalon 1977. *Geoderma* **17** (3), 181–91.

Wood, B. W. and H. F. Perkins 1976. *Soil Sci. Soc. Am. Proc.* **40** (1), 143–6.

Yaalon, D. H. and E. Ganor 1973. *Soil Sci.* **116** (3), 146–55.

Zebrowski, C. 1975. *Cah. ORSTOM, Sér. Péd.* **XIII** (1), 49–59.

*Chapter 13*

# Salsodic soils

## I  General characters

When the sodium ion is sufficiently abundant, it gives the soil special properties which, since early times, in western European classifications has been responsible for the creation of a special class for all soils affected by the presence of this ion. But it is important to emphasise that there are two types of sodium ion with very different properties: the *saline* type, generally sodium chloride (also sodium sulphate) which does not give rise to any alkalinity, and the *exchangeable* type associated with the absorbent complex which, in contrast, causes the soil solution to be alkaline by enriching it in sodium carbonate or bicarbonate which raised the pH considerably. The occurrence of these two types of sodium ion justifies the use of the term **salsodic**, suggested by Servant (1975), for this whole class (previously referred to as **sodic soils**).

The exchangeable sodium itself can come from two sources: (i) sodium chloride (and sodium sulphate) from a saline water table, gradually saturating the absorbent complex by exchange with the alkaline earth cations $Ca^{2+}$ and $Mg^{2+}$; (ii) direct saturation of the absorbent complex by weathering of rocks containing sodium minerals.

In a humid climate, no matter what its origin or type, the sodium ion, owing to the great solubility of all sodium salts, is rapidly removed from the profile by drainage waters (see Ch. 3): *the sodium ion can only persist in a profile in a dry climate, when the high potential evapotranspiration prevents all climatically controlled drainage*. In terms of environment, two conditions are necessary for salsodic soil formation: (i) a particular type of climate, for salsodic soils can only occur in steppe, semi-desert or tropical dry climates; (ii) a particular type of site, for a source of sodium is absolutely necessary (presence of salts or parent material rich in sodium). These environmental conditions, of the *intrazonal* type, have been discussed in Chapter 4.

There is an important exception to this where salsodic soils occur in a humid climate but in the coastal zone, when the source of sodium is a shallow saline water table of marine origin. This is the situation of the polders in a temperate climate and the mangrove soils of the humid Tropics. Even if the water table is strongly diluted by rain in wet periods, the sodium reserve of the water table is such that it cannot be eliminated by the climatic conditions and it continues to control soil development in the dry period.

As stated above, the properties of salsodic soils are very different depend-

ing on whether sodium is present as neutral salts (NaCl, $Na_2SO_4$) or it saturates more or less completely the absorbent complex in the exchangeable
form. This has resulted in two subclasses being differentiated in most classifications: (i) **saline soils** where the soil solution is rich in neutral sodium salts in
which the pH never rises above 8.5; and (ii) **alkali soils** (or *with alkali*) where
exchangeable $Na^+$ is dominant and in which the pH, at least of certain horizons and at certain seasons, exceeds 8.5 (a process called **alkalisation**).

However, there is a major difficulty in an *intergrade* group of soils in which
the sodium ion occurs both as a neutral salt and as an exchangeable cation. It
is a saline soil with a sodic complex (Servant 1975) or sodic solontchak
(Szabolcs 1974). This group poses a difficult classificatory problem, for some
soils are more like saline soils and others nearer to alkali soils according to the
criteria selected. In fact, it would seem that this rather hazy boundary area
between saline and alkaline soils should be divided into two intergrade
groups: those that are clearly saline, generally weakly alkaline (pH less than
8.5), and where the profile shows little differentiation, should be placed in the
saline subclass; those with low salinity and generally high exchangeable
sodium (pH above 8.5) and a structural (B) horizon, can be considered as
belonging to the alkaline subclass.

In addition, several saline soils of coastal areas (polder and mangrove) have
the special property of being strongly reducing, at least in their initial stages,
and having a particular type of sulphur cycle with a sulphide–sulphate
equilibrium depending on the Eh conditions. This particular group, distinguished by Brummer *et al*. (1971) and Durand (1973), can be considered as
belonging to a subclass of saline soils. This leads to the differentiation of
subclasses in line with the criteria suggested by Szabolcs (1974) and the FAO
(1974) in the following manner.

**Subclass of saline soils.** Conductivity greater than 4 mmho/cm in the surface
horizons (25 cm), 15 mmhos in the lower horizons (from 25 to 75 or 125 cm,
depending on the texture). AC profile (sometimes the beginning of a structural (B) horizon), pH generally remaining below 8.5.

(a) *Saline soils with calcic complex* (calcic solontchak): ratio $Na^+$ : exchange
    capacity always less than 15%.
(b) *Saline soils with sodic complex* (sodic solontchak): ratio $Na^+$ : exchange
    capacity greater than 15%.
(c) *Saline soils with sulphate reduction* (polder and mangrove).

**Subclass of alkali soils** (with alkali). Weakly saline (sometimes nil); ratio
$Na^+$ : exchange capacity always greater than 15%; A(B)C or ABC profiles;
markedly alkaline conditions with a pH greater than 8.5. Depending on the
degree of profile differentiation, alkali soils with a structural (B) horizon are
distinguished from the more developed solonetz and soloth profiles with a
textural B horizon.

## II   Study of salsodic profiles: environment, morphology, geochemistry

No matter what the classification, four basic groups are differentiated: saline soils with calcic complex, saline soils with sodic complex, saline soils with sulphate reduction, alkali soils with a more or less differentiated profile (Fig. 13.1).

### 1   Saline soils with calcic complex (calcic solontchaks) (Porta 1975, Duchaufour 1978: $XIX_1$)

These soils occur in steppe or subdesertic regions in which the saline water table contains, in addition to the sodium salts, an important amount of calcium salts (often gypsum). In these circumstances, the absorbent complex is preferentially saturated in alkaline earth cations ($Ca^{2+}$, often also $Mg^{2+}$). The percentage of sodium in the exchange complex is always less than 15%, the pH never being very high in the presence of fresh water (but there is an exception when there is a great deal of exchangeable magnesium).

The profile is thus saturated in bivalent $Ca^{2+}$ and $Mg^{2+}$, which maintain a flocculated structure in the humic horizons, and it remains only slightly differentiated and of the AC type. Frequently, salts accumulate at the surface during the dry season and form white efflorescences and sometimes even a saline crust, which on crystallising can partially destroy the structure (*powdery* structure of Servant 1974).

### 2   Saline soils with sodic complex (sodic solontchaks) (Duchaufour 1978: $XIX_2$)

Again, these soils have a saline water table but this time the $Na^+$ ion is markedly dominant over the alkaline earths, $Ca^{2+}$ and $Mg^{2+}$. Differing from the preceding case, the ionic ratio of the water table is near that of sea water.

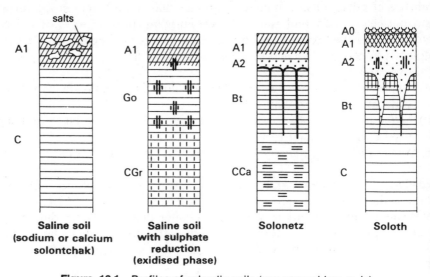

**Figure 13.1**   Profiles of salsodic soils (see general key, p. ix).

This generally occurs in lagoonal areas fairly near to the coast, in which the salt content is near to that of the sea water from which it is derived (Gaucher 1967).

*The profile has the twofold character of strong salinity and a partially sodic complex (sodium saturation greater than 15% and able to reach 30%).*

However, because of the great salinity at all seasons, the influence of the partial saturation of the complex in sodium is only slight. The profile remains slightly differentiated (no true (B) horizon), the crumb structure of the A1 horizon is preserved and the pH never exceeds about 8.5. In many cases the sodium salts ($NaCl$, $Na_2SO_4$) rise with the water table to the surface and accumulate, as in the preceding case, and this gives the A1 a powdery structure (Servant 1974). This AC-type profile is a good example of a **modal saline soil** *sensu stricto*, but it represents an unstable equilibrium which develops very quickly into an alkali soil when the influence of rainwater increases, with the natural or artificial lowering of the saline water table (Duchaufour 1978: $XIX_2$). Compared to an undrained soil, the upper and middle parts of such a profile are partially desalinised and this has caused the alkalinity to begin to increase (slight rise of pH) and a (B) or even a true Bt horizon, with a well developed prismatic structure, to appear. According to Szabolcs (1974), such a profile would belong to the alkali subclass.

## 3   Saline soils with sulphate reduction (Duchaufour 1978: $XIX_5$)

This is a very special group of soils formed from marine muds on coasts and estuaries. The initial material (*schlick* of German authors) is a mixture of clay loam and organic matter coming from marine organisms, having very marked powers of reduction. The Eh falls markedly below zero at certain periods, which at neutrality allows the reduction of sulphates to sulphides and their accumulation as black iron sulphide (2% to 5% pyrites; van Breemen 1973, van Breemen & Harmsen 1975). The free iron oxides are reduced, which gives the profile a general bluish-grey colour comparable to that of a gley (Gr horizon). These characters occur both in the saline polders of temperate coasts (Brummer *et al.* 1971) and the mangroves of the tropical coastal lowlands (Vieillefon 1974).

If the saline water table is lowered, the sulphides oxidise and hydrolyse to produce sulphuric acid and ferric hydrate, which forms rusty patches; the profile is greatly acidified by the $H_2SO_4$, particularly if the material is only slightly buffered (lack of carbonates) and the pH can fall to 2 or 3 in some extreme cases.

The polders of the Baltic coast described by Brummer *et al.* (1971) are classified as follows:

(a) *Schlick:* initial material very reducing; FeS present forming black patches on a blue–grey background (horizon Gr).
(b) *Saline polder:* aerated surface causing the partial oxidation of the sulphide and the formation of rusty patches (Go horizon).
(c) *Calcic polder* (embanked): under the influence of $H_2SO_4$ formed by the oxidation of sulphide, the $Na^+$ is eliminated, partial decarbonation and saturation of the complex by the

$Ca^{2+}$ ion occurs. The saline water table is greatly lowered so that the soil cannot be classified as a salsodic soil: generally it is a **brunified intergrade**.

(d) *'Knick' polder and peaty polder:* marked decalcification of the complex, acidification and compaction; symptoms of secondary hydromorphism.

## 4   Alkali soils (Fig. 13.1)

The amount of salt is lower than in saline soils *sensu stricto* (in some types, the saline water table is completely absent), and consequently the alkalinity is more marked: the pH is greater than 8.5, the ratio of $Na^+$ : exchange capacity is high, often reaching 50% and sometimes more. In this slightly or non-saline type of soil, the pH reflects very closely the percentage sodium saturation of the complex (pH 9 for 30%, pH 10 for 50% and more; El Nahal & Whittig 1973).

The alkaline conditions created by the $Na^+$ ion (freeing of $Na_2CO_3$) initiate certain geochemical processes such as the dissolution of part of the organic matter and degradation and dispersion of the sodic clays, which will be examined in the following section. These are responsible for the development of certain special structures which are characteristic of the different alkali soil phases.

**Non-lessived alkali soils.** (Initial phase: **solontchak–solonetz**; Szabolcs 1974). In wet periods the material has a *massive* structure, and in dry periods the marked contraction of the sodic clays causes well developed prisms to form (horizon (B) similar to that of Duchaufour 1978: $XIX_2$). As a result of evaporation, black efflorescences of organic matter and sodium carbonate form at the surface.

**Lessived alkali soils. (Solonetz**: Duchaufour 1978: $XIX_3$). The pervection of sodic soils occurs as a result of rainwater, so that the profile becomes differentiated; an A2 horizon, partially decolourised, silty and poorly structured, is formed in the upper part of the profile. In this horizon the pH is considerably lowered as a result of the partial removal of the $Na^+$ ions (pH 7 to 8). It immediately overlies the **natric horizon**, with very characteristic rounded columns, the tops of which are often covered in a pulverescent white layer while their sides often have an amorphous covering formed of a mixture of partially degraded and non-optically oriented clays, silicate gels and sodium humates. This horizon is strongly alkaline, the sodium ion representing more than 15% of the exchangeable complex (Fig. 13.2).

**Degraded alkali soils. (Soloth** or **solod**: Duchaufour 1978: $XIX_4$). This is the ultimate stage of development of a solonetz, which leads to a marked acidification of the upper horizons; the base of the Bt horizon can still have its original alkaline state. It is not unusual for the pH to be about 5 at the surface, 7 in the upper Bt and 9 at its base. Some profiles (Duchaufour 1978: $XIX_4$) are more developed and in these the $Na^+$ has almost completely disappeared while, in contrast, the less mobile $Mg^{2+}$ ion is still abundant. This relates such a profile to a **solodic planosol**, the ultimate profile of alkali soil development

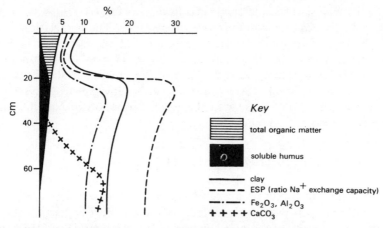

**Figure 13.2** Physicochemical properties of a solonetz profile (after Szabolcs 1974).

in hydromorphic conditions in the presence of fresh water (see Duchaufour 1978: $XV_5$ and also Ch. 11).

Soloths still have an ABtC-type profile, but overall they are characterised by a process of degradation, either of the structure in general or more specifically of the clays under the successive effects of alkalinity and acidity; the whitish A2 horizon takes on a lamellar structure, sometimes interspersed with rusty patches. The formation of pulverescent bleaches at the A2/Bt boundary develops as almost glossic structures between the columns, which are less well developed than in the solonetz, as this horizon has a more massive structure.

## III   The dynamics of ionic equilibria

The two basic physicochemical processes that are characteristic of sodic soils – alkalisation and sulphate reduction – will now be discussed.

### 1   Process of alkalisation

**The part played by exchangeable sodium.** The alkalisation of the profile is the result of the presence of a relatively high concentration of alkaline salts such as sodium carbonate (often together with a considerable amount of bicarbonate) in the soil solution. How does this $Na_2CO_3$ form and whence is it derived? Several theories have been advanced, particularly those that depend on simple reactions such as that between NaCl and active carbonate:

$$CaCO_3 + 2NaCl \longrightarrow Na_2CO_3 + CaCl_2$$

However, it has now been demonstrated that this reaction does not occur in soils (Kovda 1965, Pankowa *et al.* 1973) and that a high concentration of sodium carbonate in soil solution is generally a secondary process resulting from the hydrolysis of sodic clays: *alkalisation is closely related to a partial saturation by the Na⁺ ion of the absorbent complex*. The correlation between the pH and the percentage sodium saturation of the complex, in low-salt concentrations (which has been discussed above), is proof of the fundamental part played by the exchangeable $Na^+$ ion in the process of alkalisation.

In the presence of fresh water (rain), the sodic clays are hydrolysed, which frees the $Na^+$ and $OH^-$ ions, with the rapid formation of $Na_2CO_3$ and the environment becomes alkaline:

$$Na^+_{clay} + H_2O \overset{(1)}{\underset{(2)}{\rightleftharpoons}} H^+_{clay} + Na^+ + OH^-$$

It must be emphasised that in the presence of saline solutions (rich in NaCl or $Na_2SO_4$) this hydrolysis is prevented, for the equilibrium has a tendency to be displaced in direction (2) and alkalisation remains moderate. In addition, the excess $Na^+$ ion has a flocculating effect on the clays, which maintain an aggregated structure. This is the situation of saline soils with a sodic complex; but under the influence of a desalinisation, owing to either a dilution or a lowering of the saline water table, the sodic clays are able to be hydrolysed again, which causes alkalisation and a dispersion of the aggregates. Saline soils with a sodic complex are thus changed into alkali soils (Fig. 13.3).

As already stated in the introduction, most people agree that the percentage saturation of the complex in sodium of 15% is critical. Below this, alkalisation remains low; above it, the pH exceeds 8.5 in the presence of non-saline water. It increases as the ratio of the $Na^+$ : exchange capacity increases.

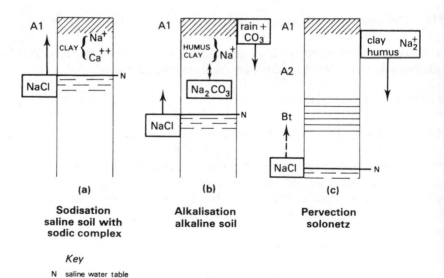

**Figure 13.3** Development of saline soils by lowering of the saline water table (indirect alkalisation).

The mechanisms whereby the exchange complex becomes gradually saturated with sodium (**sodisation** according to Servant 1975) need to be discussed, for, as is well known, it is the bivalent alkaline earths, $Ca^{2+}$ and $Mg^{2+}$, because of their great absorption energy, that are dominant in the exchange complex of most saturated soils that have been discussed up to this point.

**Mechanisms whereby the exchange complex is saturated in sodium (sodisation,** Servant 1975). The exchange of bivalent ions, particularly $Ca^{2+}$, of the absorbent complex for $Na^+$ ions *is only possible when the concentration of the $Na^+$ ion becomes very much greater than that of the $Ca^{2+}$ ion in the soil solution.* According to Durand (1954), if the concentration of $Na^+$ ions is the same as that of the bivalent ions in the soil solution, the saturation of $Na^+$ in the complex never exceeds 10%, i.e. there is neither sodisation nor alkalisation. Other authors (Kelley 1962, Kotin 1962) have given similar data, for they have determined that the concentration of sodium in solution must be greater than 70% of the total cations for its influence to be seen, and this corresponds sensibly to the threshold value defined previously: $Na^+$ : exchange capacity of 15%.

The Riverside group of the US Department of Agriculture (1969) use a more exact parameter than the simple ratio of concentrations to define the composition of soil solutions or saline water tables. This involves the **sodium adsorption ratio** (*SAR*):

$$Na \bigg/ \sqrt{\frac{Ca + Mg}{2}}$$

This ratio is closely related to the *ESP* (**exchangeable sodium percentage**). Several authors (e.g. Belkhodja 1972, Servant 1975) have used it with profit. In practice, the graph indicating the relation between *SAR* and *ESP* is a straight line through the origin and very nearly bisecting it. It can be said with a reasonable approximation that the *SAR* and *ESP* values are very near to one another:

(a) *SAR* 10: *ESP* around 12;
(b) *SAR* 20: *ESP* around 22;
(c) *SAR* 30: *ESP* slightly above 30.

But as will be seen in the following section, the $Na^+$ ion of the absorbent complex can come from two sources: either by *direct alkalisation* by weathering of rocks containing sodium minerals (in a dry climate), or by *indirect alkalisation* by the exchange of $Ca^{2+}$ ions with the $Na^+$ ions of neutral salts (NaCl or $Na_2SO_4$) present in a saline water table; *in both cases it is necessary, so that sodisation of the complex can occur, that the sodium concentration of the soil solutions is much greater than that of calcium, at least at certain periods of the year.*

**Consequences of alkalisation: dispersion and degradation of the sodic clays.** The alkalisation of the profile, the origin of which has been explained, is associated with, as will be remembered, particularly unfavourable effects on the physicochemical properties of the soil. The humus and the sodic clays disperse so that the aggregate structure is destroyed and replaced by a non-aerated massive structure. The clays are sodium montmorillonites and are subject to considerable variations in volume, depending on whether it is the dry or wet season. If the soil dries out, contraction is important and a prismatic structure is developed in which the individual prisms are separated from one another by vertical cracks. In addition, in very alkaline conditions, the crystallinity of the clay decreases and a part becomes amorphous, which increases to an even greater extent the unfavourable, characteristically columnar, structural development (Gerei *et al*. 1966, Paquet *et al*. 1966, Klages 1969, Robert 1970).

Frankart and Herbillon (1971) made a detailed study of the degradation of sodium montmorillonites in alkaline conditions and showed that the poorly crystalline zeolite **analcime** is formed as an intermediate phase before it becomes completely amorphous.

In humid periods, if certain local environmental conditions occur (which will be discussed in the following section), the simultaneous movement of amorphous organic compounds (sodium humates) and still-existing dispersed clays becomes possible. The deposition of all of these entities thus transported makes up the characteristic blackish coverings of the columns of the **natric horizon** of the solonetz. The rounded tops of the columns and their covering by a pulverescent siliceous residue is evidence of this alkaline degradation of clays.

The oldest soils (soloths) are formed when the source of sodium is exhausted; then the transported $Na^+$ ions are gradually replaced, first of all at the surface, then at greater depths by $H^+$ and $Al^{3+}$ ions which cause an acidification of the upper part of the profile; the structure is completely destroyed and a seasonal surface hydromorphism develops: *the degradation of clays by alkaline hydrolysis is followed by a new degradation by acid hydrolysis, this time occurring in more or less hydromorphic conditions*. Rusty patches are distributed throughout the A2, while analyses detect the formation of free alumina and amorphous silica. It is a kind of hydromorphic pseudopodzolisation (Tursina 1966). Some **solodic planosols** (see Ch. 11) have a similar development.

## 2   Process of sulphate reduction

**Initial phase of reduction.** In coastal soils (polder and mangrove) the process of salinisation is not the only one. The conditions, at least initially, are characterised by the proximity of an extremely reducing saline water table. It is not uncommon for the Eh to fall to $-100$ or $-200$ mV, which not only causes the reduction of free iron to the ferrous state but also, and more importantly, the reduction of sulphates; the black patches of iron sulphide scattered throughout the profile are the reason why certain authors have called this a **black gley**. This mechanism has been described by many authors: Callame and Dupuis

(1972) for the polders of western France, Engler and Patrick (1973) in the USA, and McCleod (1973) in Wales.

At this stage, most polders of marine origin are saline soils and their alkalisation is only moderate because of their high salt concentration. However, there are exceptions; certain polders (inland seas) are richer in sodium sulphate than chloride (Kovda 1965, Cheverry 1974). The reduction of $Na_2SO_4$ gives unstable sodium sulphides which decompose in the presence of $CO_2$ to give $Na_2CO_3$ and $H_2S$ and the pH rises. This alkalisation rapidly ceases when oxidation occurs.

**Phase of oxidation.** The process of sulphide oxidation occurs either naturally, in mangroves, by lowering of the saline water table (**tannes**; Vieillefon 1974, van Breeman & Harmsen 1975) or as a result of embankment of the polder soils (Brummer *et al.* 1971, Pons & Van der Molen 1973).

The physicochemical processes are fundamentally the same in both cases: pyrite oxidises to jarosite (ferric sulphate) which rapidly hydrolyses to give iron hydrates (appearance of rusty patches in the upper part of the profile) and sulphuric acid, which tends to acidify more or less strongly the absorbent complex. *In these conditions it is evident that alkalisation is completely prevented*. In addition, the acidification of the profile by $H_2SO_4$ occurs at a very variable rate and intensity, depending on whether or not the environment is well buffered by the presence of carbonate in the sediments.

*Buffered environment.* Sulphuric acid first of all rapidly eliminates the $Na^+$ ion from the absorbent complex, which prevents all alkalisation (Cheverry 1974); then it very gradually dissolves the active carbonate. The environment is only slightly acidified, or only very gradually: the ferric hydrates associate with the clays and there is a tendency to brunification. However, for the oldest polders, when all the carbonates are eliminated from the profile, the loss of structure and the pervection of the clays favour hydromorphism and a gradual acidification of the profile (knick or peaty polder).

*Non-buffered environment* (non-calcareous). The environment acidifies very strongly and very quickly; the pH reaches 2 or 3: there is then a partial hydrolysis of the clay with the formation of free aluminium (van Breeman & Harmsen 1975).

## IV   Development of salsodic soils: role of environmental factors

In this section the very special sulphate reduction processes will not be considered again, but the environmental conditions of salinisation, sodisation and alkalisation of the absorbent complex will be discussed. According to the source of the sodium, there are three possibilities:

(a) Saline water tables that contain in addition to the sodium salts a considerable amount of calcium and magnesium: this is *salinisation* without alkalisation.

(b) Saline water table with a marked dominance of the sodium ions, which in certain conditions causes a gradual alkalisation (*indirect alkalisation*).

(c) Freeing of sodium ions by weathering of sodium-containing minerals: rapid alkalisation and simultaneous formation of a natric horizon (*direct alkalisation*).

## 1 Saline water table containing calcium and magnesium salts (calcic solontchak)

Around lagoons in subdesertic regions, the water tables are rich in calcium, magnesium and sodium. As the $Ca^{2+}$ and $Mg^{2+}$ are preferentially absorbed by the complex, the percentage saturation by the $Na^+$ ion never exceeds 15%. The soil has the characteristics of a saline soil, but it is not subject to alkalisation. In these circumstances, the lowering or desalinisation of the water table does not cause a great modification of soil properties. However, in many cases a part of the calcium in solution precipitates as gypsum (soils with gypsum; Dutil 1971, Porta Casanallas 1975).

## 2 Saline water table with sodium salts dominant (indirect alkalisation)

This involves water tables with compositions similar to that of sea water. They are often inherited from more or less old lagoons that have been gradually isolated from the sea as a result of its retreat. Several authors have described sodic soils derived from such water tables in which the concentration of sodium is about 5 to 6 times higher than that of calcium (coastal zone of Roussillon: Servant 1975; the lower Volga with a water table inherited from the Caspian: Slavnyi & Kauricheva 1967). In these conditions, the adsorbed sodium can amount to about 30% of the exchange capacity and this amount is even greater if the lagoonal water table is richer in sodium.

However, alkalisation is not immediate, for as said previously, the presence of sodium salts in high concentrations ($NaCl$ and $Na_2SO_4$) in the soil solution maintains the clays in a flocculated state and prevents the process of alkalisation, which can only occur when the *desalinisation of the profile* occurs, at least seasonally. In the presence of fresh water the processes of hydrolysis and dispersion and pervection of sodic clays are initiated (Fig. 13.3):

saline soil with sodic complex ⟶ alkaline soil ⟶ solonetz

The solonetz can eventually develop into a soloth by surface acidification, the $Na^+$ ion tending to be gradually eliminated from the profile (**secondary solonetz**; Ivanova & Bolshakov 1972).

Permanent or temporary desalinisation of the profile can be the result of various causes.

(a) *General lowering of the saline water table:* this is the case of the **steppe solonetz** of the south-east of the USSR (Kovda 1965).

(b) *Important seasonal variations in the salinity of the soil solutions:* in the dry season the complex is sodium saturated; in the wet season the sodium clays are dispersed and pervected – **humic solonetz with saline water table** (Szabolcs 1969).

(c) *Solution and lateral transport of a saline deposit coming from an old lagoon:* alkali soils develop downslope. As the addition of salt decreases, the sodium clays are pervected and development can go as far as surface acidification (Danubian terraces: Obrejanu *et al*. 1967; Tunisian salsodic soils: Belkhodja 1972, Gallali 1979). This last author has shown that lateral transfer affects not only sodium but also soluble organic compounds.

## 3   Weathering of minerals containing sodium (direct alkalisation)

Here, the source of sodium is no longer a neutral sodium salt but a sodium-containing mineral (feldspar or feldspathoid) which directly frees the $Na^+$ ion by weathering. Numerous examples of alkaline soils having this origin have been given by various authors: some came from loess with sodium feldspars (Nebraska, USA: Lewis & Drew 1973) or the albitic granites of Mongolia (Gusenkov 1966) and others, finally, from the weathering of volcanic rocks (phonolites, albitophyres).

*In this type of development, alkalisation is immediate and it is obvious that the saline phase does not occur.* It is the same for the pervection of sodium clays in wet periods: the natric horizon is formed at the outset (**primary solonetz**: Ivanova & Balshakov 1972). In contrast, the development to the soloth stage and the replacement of $Na^+$ ions by $H^+$ and $Al^{3+}$ are more gradual. They only occur at a very late stage, when increasingly intense weathering of the parent material gradually exhausts the source of sodium.

This type of development requires very particular climatic conditions. Given the extreme mobility of sodium ions which, as seen in Chapter 3, are rapidly removed from the profile in a humid climate, it requires a dry climate with a low *PET*, preventing all loss of free ions by deep drainage.

It remains to explain how the $Na^+$ can reach a relative concentration in the soil solution sufficient to be able to exchange against the $Ca^{2+}$ ions and gradually saturate the complex; this can result either from a twofold vertical movement of soil solutions or from a lateral movement (soil catena).

**Vertical movements of saline solutions.** The explanation proposed by Kovda and Samoilova (1969) depends on the differential solubility of $Ca^{2+}$ and $Mg^{2+}$ ions and $Na^+$ ions, when the concentration increases, i.e. in the dry season; the least soluble ions are precipitated – $Ca^{2+}$ first of all and then $Mg^{2+}$. In these circumstances the complex absorbs preferentially the ions that remain soluble for the longest time: $Na^+$ particularly, and to a lesser extent $Mg^{2+}$ (which also plays a part in the alkalisation of the profile, although to a lesser degree than the $Na^+$).

Seasonal alternation plays an important part in these circumstances, the dynamics of the ions being very different in the wet and the dry seasons.

*Wet season.* Weathering of primary minerals and freeing of $Na^+$, $Ca^{2+}$ and $Mg^{2+}$ ions, are all moved downwards, but are not removed completely

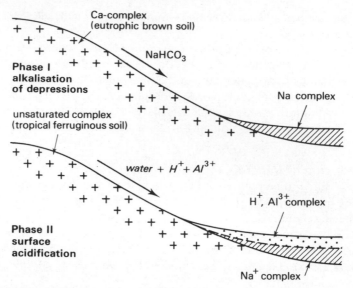

**Figure 13.4** Phases in the development of a soil catena on rocks with sodium minerals.

because of a lack or complete absence of drainage. In addition, the clays and the humic compounds which have previously been subject to sodisation are dispersed and pervected (formation of natric horizon).

*Dry season.* Movement towards the surface by capillarity of soil solutions owing to evaporation. The concentration of the solutions increases and the precipitation of calcium salts occurs, which allows the saturation by sodium of a part of the humus and clays, which can be transported in the following wet period.

**Development by lateral movement: soil catenas** (Paquet *et al.* 1966, Bocquier 1973, Kovda 1973). The same differential solubility of $Mg^{2+}$ and $Na^+$ ions explains their lateral movement over variable distances downslope from inselbergs of sodium-rich crystalline rocks in a tropical climate. Calcium precipitates in the middle part of the catena and only sodium reaches the bottom, where it has the possibility of alkalising the complex (alkali soils and solonetz, Fig. 13.4). As the weathering progresses, the source of the $Na^+$ ion diminishes and thus the initial solonetz is gradually solodised by surface acidification.

## V Summary: classification of salsodic soils

As stated in the introduction, two subclasses are differentiated, one having a strong salinity and little or no alkalisation, the other on the contrary having a weak salinity and a more or less marked alkalisation.

## 1   Subclass of saline soils (solontchaks)

AC or AG profiles, with strong salinity (see introduction for conductivity criteria), and weak alkalinity.

(a) *Saline soils*, sensu stricto (without sulphate reduction): *AC profile*.
   (i) *Saline soil with calcic complex:* calcic solontchak: no sodisation or alkalisation.
   (ii) *Saline soil with sodic complex:* sodic solontchak: sodisation and alkalisation moderate.
(b) *Saline soils with sulphate reduction: AG profile*.
   (i) *Reduced saline soil* (with iron sulphide: AGr)
   (ii) *Oxidised saline soil* (with rusty patches and sulphates: AGoGr): eutrophic or calcic subgroup (eutrophic polders and mangroves); very acid subgroup (very acid polders and mangroves).

## 2   Subclass of alkali soils

A(B)C or ABC profile; little or no salinity; marked sodisation of the complex and more or less strong seasonal alkalisation (pH greater than 8.5) at least in certain horizons.

(a) *Alkaline soils*, sensu stricto, *with structural (B):* A(B)C profile still only slightly differentiated.
   (i) *Saline alkaline soil*.
   (ii) *Alkaline soil with little or no salinity*.
(b) *Lessived alkaline soil* profile ABtC: *solonetz*.
   (i) *Primary solonetz*.
   (ii) *Secondary solonetz:*
      steppe solonetz;
      humic solonetz with a saline water table.
(c) *Degraded alkaline soils:* profile ABtC: soloth (degradation of structures in the A2 and Bt by acidification).

## VI   Properties and utilisation of salsodic soils

As is well known, salsodic soils are very unfavourable to plant growth, the alkali soils to an even greater extent than the saline soils with weak alkalisation. While the alkali soils are often completely lacking in natural vegetation, saline soils are capable of carrying a halophytic vegetation, the salsolaceae, well adapted to these conditions; certain cultivated plants are resistant to a moderate degree of salinity, while there is practically none that can stand up to strong alkalinity, as shown by Thorup (1967) for the tomato.

In the saline soils, sodium salts (NaCl, $Na_2SO_4$) cause an increase in the osmotic pressure of soil solutions, which prevents the absorption of water and causes physiological drought; in addition the excess of $Na^+$ ions has an antagonistic effect with regard to the other ions ($Ca^{2+}$ and $Mg^{2+}$) that are present in small amounts in the absorption complex. Because of this, the sodium solontchaks are much more infertile than the calcic solontchaks.

For saline soils with sulphate reduction (polders), the very low Eh and insufficiency of oxygen add their deleterious effects to that of the salinity. As long as a saline water table at little depth persists, which is periodically supplied with salts by spring tides, they remain practically uncultivatable.

## 1   Utilisation of saline soils (sensu stricto)

The problem of the use of saline soil is easier to solve for calcic solontchaks than for sodic solontchaks. In the first case, a dilution of the water table by irrigation or its lowering by drainage is sufficient to eliminate the bad effects of the salinity without any unfavourable properties developing. It is not the same for sodium solontchaks, which are much more difficult to improve. Thus the artificial lowering of the saline water table is often accompanied by, as has been said, an increase in the alkalinity and a degradation of the structure: the remedy can be worse than the illness.

Generally, the method used is to irrigate with fresh water (or one of very low salinity) together with, if possible, drainage to prevent the rise of salts in the neighbouring non-irrigated areas. But such drainage is often difficult or even impossible in coastal lowlands, where ways of removing the water are lacking.

Brackish water can be used for irrigation in so far as it contains sufficient $Ca^{2+}$ ions (as well as $Na^+$) in solution. In these circumstances, the $Ca^{2+}$ ion replaces the $Na^+$ ion of the exchange complex. If the water used does not contain calcium, it is prudent to add gypsum ($CaSO_4$), which will prevent all alkalisation of the profile.

In the Carmargue, certain crops have been raised on saline soils with a sodic complex (rice, grasslands with *Agropyrum*) in elevated zones along the alluvial levée banks of the Rhone distributaries, i.e. where the saline water table is at a great enough depth. In lower basin areas with surface saline water table, all cultivation is impossible. In the cultivated areas, irrigation has created a water table of fresh water which overlies a saline water table. Alkalisation occurs but at depth, which allows rice roots to develop in the weakly alkaline surface horizons.

## 2   Utilisation of saline soils with sulphate reduction (polders)

*Embankment* removes polders from the effects of sea water. After embankment only rain water, rich in dissolved oxygen, is involved and corrects both the excess of salinity and the insufficiency of oxygen. It should be remembered that moderate salinity is not unfavourable to raising sheep (saline meadows). The major problem of embankment, as already stated, is the production of $H_2SO_4$ capable of causing strong acidification: fortunately in most cases, marine muds are rich in carbonates (shell debris), which mitigates this effect. However, it is necessary to monitor the development of the absorbent complex after embankment and, in case of acidification, to add large amounts of lime as an amendment to prevent structural degradation, secondary hydromorphism, or even peat development.

# References

Belkhodja, K. 1972. *Origine, évolution et caractères de la salinité dans les sols de la plaine de Kairouan (Tinisie)*. Thesis, Univ. Toulouse; *Bull. Division des Sols de Tunisie*.

Bocquier, G. 1973. *Genèse et évolution de deux toposéquences de sols tropicaus du Tchad. Interprétation biogéodynamique*. State doct. thesis. Univ. Strasbourg; Mémorie ORSTOM, no. 72.

Breemen, N. van 1973. *Soil Sci. Soc. Am. Proc.* **37** (5), 694–7.

Breemen, N. van and K. Harmsen 1975. *Soil Sci. Soc. Am. Proc.* **39** (6), 1140–53.

Brummer, G., H. S. Grunwald and D. Schroeder 1971. *Z. Pflanzener. Bodenk.* **129** (2), 92–108.

Callame, B. and J. Dupuis 1972. *Science du Sol* **2**, 33–60.

Cheverry, C. 1974. *Contribution à l'étude pédologique des polders du lac Tchad*. State doct. thesis. Univ. Strasbourg.

Duchaufour, Ph. 1978. *Ecological atlas of soils of the world*. New York: Masson.

Durand, J. H. 1954. *Publ. Serv. Etudes Scient. Péd.*, Clairbois-Birmandreis **2**.

Durand, J. H. 1973. *Agron. Trop.* **28** (6–7), 640–64.

Dutil, P. 1971. *Contribution à l'étude des sols et des paléosols du Sahara*. State doct. thesis. Univ. Strasbourg.

El Nahal, M. A. and L. D. Whittig 1973. *Soil Sci. Soc. Am. Proc.* **37** (6), 956–8.

Engler, R. M. and W. H. Patrick 1973. *Soil Sci. Soc. Am. Proc.* **37** (5), 685–8.

FAO 1974. *Soil map of the world* **I**, Legend.

Frankart, R. P. and A. J. Herbillon 1971. *Ann. Musée R. Afr.*, Tervuren.

Gallali, T. 1979. *Transport sels-matière organique en zones arides mediterraneennes*. State doct. thesis. Inst. Nal. polyt. Lorraine.

Gaucher, G. 1967. *Pédologie*, Ghent **XVII** (2), 153–64.

Gerei, L., K. Darab and M. Remenyi 1966. *Agrokem. Taljt.* **15**, 469–80.

Gusenkov, Y. P. 1966. *Pochvovedeniye* **7**, 50–61. (*Soviet Soil Sci.* **7**.)

Ivanova, Y. E. and A. V. Bolshakov 1972. *Pochvovedeniye* **4**, 88–104. (*Soviet Soil Sci.* **4** (2), 156–71.)

Kelley, W. P. 1962. *Soil Sci.* **94** (1), 1–5.

Klages, M. G. 1969. *Soil Sci. Soc. Am. Proc.* **33** (4), 543–6.

Kotin, N. I. 1962. *Pochvovedeniye* **7**, 67–76.

Kovda, V. A. 1965. *Symposium on Sodic Sols*, Budapest **14**, 15–48.

Kovda, V. A. 1973. *The principles of pedology* (2 vols).

Kovda, V. A. and E. M. Samoilova 1969. *International Symposium on the Reclamation of Sodic Soils*, Yerevan, 32–45.

Lewis, D. T. and J. V. Drew 1973. *Soil Sci. Soc. Am. Proc.* **37** (4), 600–6.

McCleod, D. A. 1973. In *Pseudogley and gley. Trans Comms V and VI, ISSS*, E. Schlichting and U. Schwertmann (eds), 647–56. Weinheim: Chemie.

Obrejanu, G., G. Sandu, I. Aksenova and R. Rudzik 1967. *Pochvovedeniye* **4**, 55–65. (*Soviet Soil Sci.* **4**, 474–81.)

Pankowa, Y. N., V. P. Ignatova and T. I. Abaturova 1973. *Pochvovedeniye* **5**, 15–25. (*Soviet Soil Sci.* **5** (3), 269–78.)

Paquet, H., G. Bocquier and G. Millot 1966. *Bull. Serve. Carte Géol Als. Lorr.* **19** (3–4), 295.

Pons, L. J. and W. H. Van der Molen 1973. *Soil Sci.* **116** (3), 228–35.

Porta Casanellas, J. 1975. *Redistribuciones ionicas en suelos salinos. Influencia sobre la vegetacion halofila y las posibilidades de recuperacion de los suelos con horizonte gypsico*. Thesis. Univ. Polytech. Madrid.

Robert, M. 1970. *Etude expérimental de la désagregation du granite et de l'évolution des micas*. State doct. thesis. Fac. Sci. Paris.

Servant, J. 1974. *C.R. Acad. Sci. Paris* **278D**, 589–91.

Servant, J. 1975. *Contribution a l'étude pédologique des sols halomorphes. L'example des sols salés du Sud et du Sud Ouest de la France*. State doct. thesis. Univ. Montpellier.

Slavnyi, Y. A. and Z. N. Kauricheva 1967. *Pochvovedeniye* **5**, 121–30. (*Soviet Soil Sci.* **5**, 685–92.)

Szabolcs, I. 1969. *Agrokemia es Talastan* **18** (Suppl.), 37–68.
Szabolcs, I. 1974. *Salt-affected soils in Europe*. The Hague: Martinus Nijhoff.
Thorup, J. T. 1967. *Diss. Abstr.* **28B** (1), 20–21.
Tursina, T. V. 1966. *Pochvovedeniye* **5**, 7–17. (*Soviet Soil Sci.* **5**, 491–504.)
US Department of Agriculture 1969. *Saline and alkali soils*. Washington, DC: USDA.
Vieillefon, J. 1974. *Contribution à l'étude de la pédogénèse dans le domaine fluvio-marin en climat tropical d'Afrique de l'Ouest*. State doct. thesis. Univ. Paris VI.

# Index

Bold page numbers refer to text sections. Italic numbers (e.g. *5.3*) refer to text figures. References to table and plates are given as, e.g., 'Table 5.3' and 'Plate 16c'.